Rice Improvement

Jauhar Ali • Shabir Hussain Wani
Editors

Rice Improvement

Physiological, Molecular Breeding
and Genetic Perspectives

 Springer

Editors
Jauhar Ali
Hybrid Rice Breeder, Senior Scientist II,
Leader, Hybrid Rice Breeding Cluster,
Head, Hybrid Rice Development
Consortium (HRDC)
International Rice Research Institute (IRRI)
Metro Manila, Philippines

Shabir Hussain Wani
Assistant Professor/Scientist, Mountain
Research Centre for Field Crops
Sher-e-Kashmir University of Agricultural
Sciences and Technology
Kashmir, Jammu and Kashmir, India

Dedicated to the 60th Anniversary of the International Rice Research Institute (IRRI) and to the global rice research community.

Foreword

Rice is one of the most significant cereal crops globally, intertwined with food and human culture. Ninety percent of the rice produced and consumed in Asia is linked to poverty. Rice is a model crop for geneticists, physiologists, and biotechnologists. The recent advances in their areas got a boost from the sequenced 3000 rice genomes that are placed in the public domain for exploitation and will provide greater depth and a more complete picture of the genetic information. A deeper understanding of rice physiology, molecular breeding, and genetics could pave the way for more sustainable varietal products for the benefit of humanity, particularly for those living in the developing world. On this subject, the editors of this book have attempted to highlight rice research advances in the fields of physiology, molecular breeding, and genetics, with a focus on increasing productivity, improving biotic and abiotic stress tolerance, and improving the nutritional quality of rice. This book offers a balanced set of chapters after the authors in the opening chapter give an overview of the advances in genetics and breeding of rice. It is widely understood that increasing plant biomass and its efficient translocation to the sink hold the key to increasing grain yield. Another chapter targets the strategies for engineering photosynthesis for enhanced plant biomass production. It is vital to use the green traits concerning multiple abiotic and biotic stresses, including water- and nutrient-use efficiencies. Breeding of climate-resilient rice varieties could effectively provide insurance to rice farmers to combat against climatic turbulence. Hybrid rice technology is becoming the most viable option to meet global food security concerns as it assures a 20–25% yield advantage over the best inbred varieties. It faces two major challenges for its wide-scale adoption. First, heterosis per se needs to be attractive for farmers by assuring them a stable yield advantage of >25% over inbred varieties besides addressing market requirements. Second, the higher hybrid rice seed reproducibility (>3 t/ha) and decreased production costs should be attractive to the seed industry. In this regard, two-line hybrid rice breeding is reclaiming attention to make rice hybrids more heterotic and more efficient for hybrid seed production. This two-line hybrid technology is likely to reach farmers in a big way at an affordable cost. In recent times, irrigation water scarcity for agriculture, particularly for rice, requires the development of water-use-efficient and drought-tolerant rice cultivars. Understanding drought physiology and developing drought-tolerant varieties are the need of the hour. Recently, the COVID-19 pandemic showed a massive shortage of labor for transplanted rice that has sparked interest in adopting direct-seeded rice (DSR) in many parts of India and other rice-producing countries. However, systematic breeding for DSR is still in its infancy. The development of DSR varieties with the appropriate set of traits such as anaerobic germination, herbicide tolerance, weed competitiveness, effective germination under deeper seed placement, and uniform seedling establishment can help improve and popularize this technology. The use of DSR varieties and associated genetics and management technologies is going to improve rice cultivation with less water without losing grain yield. Also, there will be a considerable decrease in greenhouse gas emissions and a decrease in global warming to some extent. Increasing global temperatures are going to cause havoc to agriculture in particular, and breeding heat-tolerant varieties is going to be a challenging task while sustaining current yield rates. Global climatic aberrations are causing more areas to experience cold spells, particularly in temperate regions, where higher rice productivity is attained. Breeding for rice varieties with tolerance of low-temperature stress is another important objective to be fulfilled. Many parts of the world are facing toxic elements such as arsenic entering into the food system. In many places in Eastern India and Bangladesh, rice is being cultivated in arsenic-polluted environments. It is important for breeders to develop suitable rice varieties that would restrict arsenic from entering into rice grain and straw, thereby making this rice safer for human consumption and straw for cattle.

Likewise, breeding for tolerance against insect pests and diseases is an area of much concern, especially under the changing climatic scenario that is triggering the evolution of new pathogen strains and leaving rice more susceptible than ever. Molecular approaches to breed varieties with insect pest and disease resistance are essential and provide an opportunity for using broad-spectrum resistance genes. However, with the recent advances in genome editing tools, it will be relatively easier to incorporate tolerance by understanding the molecular basis of this tolerance. To maximize the genetic gains, researchers are attempting to speed up breeding in many crops. Recent technological innovations promise to accelerate the growth and life cycle of the rice crop considerably to allow four generations per year. Interestingly, another powerful alternative technology is doubled haploids (DH), which could fix segregating lines in less than 6 months. Scaling up of the DH approach is one of the most important approaches and using new genetic technologies around this could be a game changer for future breeding programs.

Rice is often linked to poverty in many parts of the world, where it constitutes the primary source of calories. Hidden hunger arises when essential nutrients such as zinc and iron are missing from a staple diet. Therefore, an increased effort is required for the biofortification of rice varieties with adequate concentrations of zinc and iron by mainstreaming rice breeding itself.

I would like to congratulate the editors of this book for bringing out a valuable collection of chapters concerning the most important aspects of rice research. This book will serve as a vital reference tool of benefit to rice scientists, students, policy-makers, and other researchers in academia and industry.

Gurdev S. Khush FRS
Member US National Academy of Sciences
Adjunct Professor Emeritus
University of California,
Davis, CA 95616 USA

Former Head, Plant Breeding Genetics and Biotechnology, IRRI
Los Baños, Philippines

Preface

The global human population is rapidly increasing. This is placing enormous pressure on all available natural resources needed to feed 9.7 billion people by 2050, which poses a severe threat of hunger and chaos to all of humanity in the coming decades. The food security situation will become worse with global climatic aberrations. Rice, among the staple cereal crops, could be much affected, especially in Asia, where more than 90% of the crop is produced and consumed. Rice production and productivity need to be augmented even though we have limited land, water, and chemical inputs. The challenge is to develop climate-resilient rice varieties that will provide insurance for the farmers growing them, and in a sustainable manner. Most biotic and abiotic stresses are complex in nature. Because of the quantitative mode of inheritance and tolerance traits being governed by genes with minor effects, this limits their accelerated and precise improvement using conventional plant breeding methods. With the recent developments in whole-genome sequencing, molecular marker technology, and high-throughput phenotyping methods, it is now possible to develop breeding products in less time and in a more efficient manner. New breeding strategies involving genomic selection and accelerated breeding approaches are being adopted. Concerted efforts are being made at various leading rice research institutes worldwide to tackle the abovementioned drawbacks in rice production, productivity, nutritional quality, and resilience to various biotic and abiotic stresses. Through this volume on *Rice Improvement: Physiological, Molecular Breeding and Genetic Perspectives*, an effort is made to put all the state-of-the-art technological accomplishments in rice physiology, molecular breeding, and genetics in one basket. We have included chapters from leading authors from various international institutes recognized for their rice improvement work. The chapters cover advances in rice genetics and breeding, and deep insights into rice physiology to increase assimilates and efficient partitioning of the photosynthates. Detailed chapters cover Green Super Rice breeding technology, advances in two-line hybrid rice technology, and breeding of direct-seeded rice. Water is soon going to be the scarcest resource; therefore, another chapter covers growing rice with less water. Breeding for climate resilience is a universal goal among rice scientists globally; therefore, other chapters explain the development of rice with climate resilience against heat, drought, cold, and heavy metal stress, written by subject matter experts. Molecular breeding for biotic stress tolerance such as disease and insect pest resistance is broadly discussed. Also, speeding up the fixation of segregating materials through doubled haploids is well covered to maximize genetic gains. Last but not least, the objective of breeding rice for high grain iron and zinc is dealt with comprehensively in two chapters discussing recent updates on micronutrient biofortification of iron and zinc in rice to decrease malnutrition in women and children. This book will serve as a comprehensive reference material for rice researchers, teachers, and graduate students involved in rice improvement through physiological, molecular, and genomic approaches for yield improvement and tolerance aiming for climate resilience and increased nutritional quality. We express our sincere thanks and gratefulness to our esteemed authors. Without their determined efforts, this book project would not have been possible. We are also grateful to the Bill & Melinda Gates Foundation for providing funding for this project, from which this book will be an open access publication for the benefit of the scientific community and students. We would like to kindly thank Ms. Kristine Alexis Arellano for providing secretarial support for this book. We also acknowledge all the reviewers who helped to improve the chapters. We would like to thank Bill Hardy for meticulously editing all the chapters. We appreciate Springer Nature and its editorial staff for timely completing the production process for this book.

Metro Manila, Philippines   Jauhar Ali
Kashmir, Jammu and Kashmir, India   Shabir Hussain Wani

Contents

Abstract

Rice remains the staple food source for a majority of the global population and especially in Asia where 90 percent of rice is grown and consumed. Human population is rapidly increasing and by 2050, it is expected to reach 9.7 billion; therefore the demand for increased rice production needs to be met from ever reducing resources like land, water, and chemical inputs. In addition to the escalating demand, the changing climate scenario has increased several production constraints multifold that includes abiotic and biotic stresses. Rice researchers worldwide have been continuously making research efforts to provide technological solutions to counter the above challenges. Among them, the 'green super rice' breeding strategy has been successful for leading the development and release of multiple abiotic and biotic stress tolerant rice varieties for both favorable and unfavorable environments. Recent advances in plant molecular biology and biotechnologies have led to the identification and use of thousands of genes involved in biotic and abiotic stress tolerances over the last decade, which have opened up new vistas for increased rice production. Many of these are the regulatory genes regulating stress responses (e.g., transcription factors and protein kinases) and functional genes that guard and maintain the cell (e.g., enzymes for generating protective metabolites and proteins). These genes are primarily used to augment the stress tolerance pathways in rice. In addition, numerous quantitative trait loci (QTLs) associated with elevated stress tolerance have been cloned, resulting in the detection of considerably imperative genes for biotic and abiotic stress tolerance. Also, the molecular understanding of the genetic basis of traits such as N and P use is allowing rice researchers to engineer nutrient use efficient rice varieties, which would result in higher yields with lower inputs. Further, knowledge of the biosynthesis of micronutrients in rice permits genetic engineering of metabolic pathways to enhance the availability of micronutrients. Advances in genome sequencing tools have led to the improvement in rice molecular markers, their number was significantly increased, their physical order was deciphered, and closeness to annotated genes was valuable to forecast gene-trait associations. Rice genome sequencing efforts over time will be rapidly scaled from the currently available 3000 to 10,000 genomes with the advancement of low cost third-generation sequencing techniques. This book on rice emphasizes on the quarters of rice science that are predominantly applicable to crack the foremost limitations on rice production. It would bring out the advances in rice research in the fields of physiology, molecular breeding, and genetics, especially with a focus on increasing productivity, improving biotic and abiotic stress tolerance, and nutritional quality of rice.

Advances in Genetics and Breeding of Rice: An Overview

E. A. Siddiq and Lakshminarayana R. Vemireddy

Abstract Rice (*Oryza sativa* L.) is life for more than half of the human population on Earth. In the history of rice breeding, two major yield breakthroughs or leaps occurred, which phenomenally revolutionized rice breeding: the Green Revolution in the 1960s and hybrid technology in the 1970s. However, the fruits of these technologies have not spread globally to all rice-growing areas, especially African countries, for diverse reasons. It is estimated that at least 50% more rice yield is needed to feed the anticipated nine billion people by 2050. This clearly warrants another breakthrough in rice. It is apparent that the currently used conventional and molecular marker-assisted methods need to be updated with multi-pronged approaches involving innovative cutting-edge technologies for achieving the next breakthrough in rice. Here, we attempt to discuss the exciting avenues for the next advances in rice breeding by exploiting cutting-edge technologies.

Keywords Rice · Green revolution · Hybrid rice · Multi-pronged approaches · Gene editing

1 Introduction

Rice is the source for more than 20% of the total calorie intake for more than half of the world population. More than 90% of it is produced and consumed in Asia. Chronically food-deficit Asia became self-sufficient in this crop by the early 1980s following the introduction and extensive adoption of high-yielding varieties with dwarf plant type starting in the mid-1960s. To sustain this self-sufficiency, it is estimated that the global rice requirement by 2050 will be 70% more than what is produced now (Fig. 1). Meeting such a huge demand projection sustainably in the face

E. A. Siddiq
Institute of Biotechnology, Professor Jayashankar Telangana State Agricultural University, Hyderabad, India

L. R. Vemireddy (✉)
Department of Genetics and Plant Breeding, SV Agricultural College, Acharya NG Ranga Agricultural University (ANGRAU), Tirupati, Andhra Pradesh, India

© The Author(s) 2021
J. Ali, S. H. Wani (eds.), *Rice Improvement*,
https://doi.org/10.1007/978-3-030-66530-2_1

1

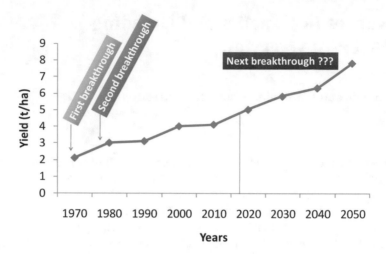

Fig. 1 Rice yield trends and demand projections toward 2050

of the shrinking favorable growth of the 1970s and 1980s, especially for natural resources such as arable land, irrigation water, and genetic resources, is the most challenging task ahead. This situation warrants the discovery of novel gene sources and innovative breeding-selection strategies to develop varieties that would enable the world to meet this challenge.

Systematic breeding for the improvement of Asian rice (*Oryza sativa* L.), although begun more than a century ago, has been witnessing rapid advances for the past 60 years, with landmark achievements in both applied and mission-oriented basic research. In keeping with the objective of this publication, "Molecular and physiological breeding strategies toward sustained self-sufficiency in rice," this introductory chapter offers an overview of the significant achievements made during this period.

2 First Breakthrough: The Green Revolution

Raising the ceiling for genetic yield had been the major breeding objective until the 1950s, when Chinese breeders succeeded with the first-ever dwarf variety, Guang-Chang-Ai, using the spontaneous dwarf mutant Ai-zi-zhan (Huang 2001), followed by Taichung (Native)-1 in Taiwan using yet another spontaneous dwarf mutant, Dee-Gee-Woo-Gen. Impressed with its yield performance and period-bound maturity, the International Rice Research Institute (IRRI), using the same dwarfing gene source in a cross with tropical *japonica* variety Peta, developed the miracle yielder, IR8, by the mid-1960s. The extensive adoption of this variety and its derivatives heralded Asia's Green Revolution. Dwarf stature, inherited as a simple recessive trait, together with a set of favorable physiological traits such as increased leaf area index (LA1), photo-insensitivity, higher harvest index, and higher fertilizer responsiveness, enabled rice breeders worldwide to develop hundreds of "IR8 plant type" varieties combining the desired maturity range and grain quality. Thus, the DGWG dwarfing gene (*sd1*) provides short stature in more than 90% of the high-yielding dwarf varieties being planted globally in the past 50 years. The dwarf varieties developed for the relatively risk-free irrigated ecosystem are not adapted to rainfed upland and lowland ecosystems, which account for more than 45% of global rice area (Mackill et al. 1996). Efforts to raise the genetic yield of temperate *japonica* and African rice varieties through the same plant type strategy employing the DGWG dwarfing gene, however, did not succeed except for limited success achieved through variety Tongil in South Korea. Understanding that germplasm of *indica* origin would not be of help to achieve the plant type goal, the United States and Japan used dwarfing gene sources of spontaneous and induced origin identified in the respective germplasm. Designated as *sd2* and *sd3*, and found to be non-allelic to *sd1* of DGWG, they have been extensively employed in breeding for higher genetic yield in American and Japanese varieties.

3 Second Breakthrough: Hybrid Rice Technology

Ever since the first yield breakthrough achieved through dwarf plant type varieties, keeping in view the need to meet huge future demand projections, especially under limited scope for horizontal growth, breeders have been looking for strategies that would enable them to make a second yield breakthrough. The search took place amid reservations that the chances of finding one such strategy would be difficult as the physiological limit for genetic yield in terms of sink-source equilibrium had already been reached through dwarf varieties. However, this notion was soon proved wrong when Chinese breeders succeeded in the commercial exploitation of hybrid vigor in self-pollinated rice in the late 1970s. Of the more than 20 different cytoplasmic male sterility (CMS) sources, researchers discovered only Wild Abortive (WA), suiting *indica* rice, and Boro Tai (BT) in *japonica* rice, which are widely used for commercial hybrid seed production (Li and Yuan 2000; Fuji and Toriyama 2009). More than 90% of the hybrids cultivated in China are based on WA cytosterility (Sattari et al. 2007). The yield advantage of about 15% over the best high-yielding dwarf varieties marked the second major yield breakthrough. The adoption of hybrid technology exceeding 18 million ha in a short period of 10–12 years enabled China to add 20 million tons annually to its rice production. Sadly, this proven technology could not be replicated sufficiently in countries outside China. Among the reasons for the slow adoption of the technology, the still unsatisfactory yield advantage, inconsistent yield performance, less acceptable grain quality, and non-suitability of many varieties for the long wet season in countries such as India and Bangladesh are important. Given the recent successes achieved in parental line improvement, some of these deficiencies could be rectified in future hybrids and thereby the pace of adoption of the technology is expected to increase in the coming years.

With the less attractive yield advantage being one of the major limitations against hybrid rice, various breeding strategies have been attempted across countries to raise yield vigor. Among them, the shift from excessively depending on intra-subspecific combinations *(indica/indica or japonica/japonica)* to inter-subspecific *(indica/tropical japonica or indica/japonica)* combinations has been rewarding. The persistent sterility characteristic in *indica/japonica* hybrids has been overcome following the discovery and use of sterility neutralizing wide compatibility gene (WCG) loci. Using an early discovered WCG such as S^n5 obtained from traditional varieties such as Dular, Keta Nanga, and others, many *indica-japonica* hybrids (Liangyou Pei9, Xieyou 9308, etc.) with yield surpassing that of intra-subspecific hybrids have been developed. Success achieved in overcoming the sterility problem in inter-subspecific hybrids through the use of WCG loci prompted rice geneticists to search for more such genes, leading to the discovery of as many as 50 loci for hybrid fertility. Of these, some were identified in inter-subspecific crosses of *O. sativa* while others were found in crosses between *O. sativa* and other species of the genus *Oryza* (Ouyang et al. 2009). Among the loci causing female sterility in inter-subspecific hybrids, *S5* is a major locus (Song et al. 2005). The locus with three alleles—*indica* allele *S5*-i, *japonica* allele *S5*-j, and neutral allele *S5*-n—has been mapped on chromosome 6 (Yanagihara et al. 1995). The tightly linked flanking markers of the *S5* gene, RM253 and RM276, have been found quite valuable in developing appropriate parents for producing sterility-free inter-subspecific hybrids (Singh et al. 2006; Siddiq and Singh 2005). Pyramiding of *S5*-n and *f5*-n genes was also demonstrated to cumulatively improve percentage seed-setting in *indica-japonica* hybrids (Mi et al. 2016). Priyadarshi et al. (2017) successfully introgressed a major gene for wide compatibility (*S5^n*) into the maintainer line IR58025B through marker-assisted breeding.

Yet another development toward strengthening hybrid rice technology has been hybrid seed production by environment-sensitive genic male sterility (EGMS) using a two-line approach as an alternative to conventional cytoplasmic male sterility using three-line breeding. This was possible following the discovery of photoperiod-sensitive male sterility (PGMS) gene sources such as NK58s, PMS1, PMS3, and TMS5, and their non-sensitive gene sources such as Annong IS, Norin12, SA2, and F61 (Siddiq and Ali 1999; Ali 1993; Ali et al. 1995). Dispensing with the need for a male sterility maintainer line, the EGMS system enables the use of a large number of varieties as male parents and thereby increases the probability of identifying more heterotic hybrids. Finding the two-line breeding strategy more efficient and economical in the past three decades, many EGMS system–based two-line hybrids with higher yield, improved grain quality, and resistance to major biotic stresses have been released for commercial planting in China. Now, approximately three million ha are planted to two-line hybrids in China. It is a good sign that interest is growing for the two-line approach in countries such as Vietnam and in multinational seed companies such as RiceTec Inc. Many of the TGMS lines, including SA2 and F61 identified in India, have been linked with robust microsatellite markers enabling the rapid development of inter-subspecific hybrids (Reddy et al. 2000; Hussain et al. 2011).

4 Next Breakthrough: Strategies

After the grand success of the first and second breakthroughs, in the form of semi-dwarf varieties and hybrid technology, the yield levels of rice have almost reached stagnation. However, the basic understanding of trait inheritance has been enhanced tremendously with the advances in cutting-edge molecular technologies. The next breakthrough requires a multipronged approach involving diverse disciplines and methods (Fig. 2). The following are some of the concepts or pilot studies that have potential to achieve the next breakthrough in rice yield improvement.

4.1 Enrichment of the Rice Gene Pool

The most important prerequisite for progressive crop improvement is the availability of genetic variability. Most crop plants are endowed with rich variability and rice is no exception. Wild/weedy species, landraces, modern cultivars, induced mutants, etc. constitute the major source of variability. Despite such rich diversity, more than 80% of current rice cultivars owe their making to a few parental lines. Strong sexual barriers make it difficult to introgress genes of interest from distant relatives and the possibility of introducing undesirable traits into cultivars via linkage drag dissuaded breeders all along from resorting to the strategy of wide hybridization. As a result, 85–90% of the variability remains unused in landraces and wild/weedy relatives. Given the rising need for additional/novel variability to meet the unfolding challenges, strategies to bring out the hidden genes in distant rice gene pools and induce variation are inevitable. Whereas association mapping, sequence-based mapping, etc. have been found rewarding in bringing out still undiscovered variability lying in the natural gene pool, induced (CRISPR) and inserted (activation tagging) mutagens could help generate novel variability (Wei et al. 2013).

4.2 Discovery and Stacking of Yield Genes Hidden in Wild/ Weedy Species

Wild-weedy gene pools are rich reservoirs of gene sources for breeders looking to progressively improve crop plants and rice is no exception. Aside from finding and using several Mendelian genes largely governing resistance to biotic stresses, the search for genes/QTLs governing polygenically controlled yield and its major components began on the assumption that many of these genes/QTLs might not have been captured in modern varieties during the course of evolution of rice and they could still remain in the wild gene pool (Xiao et al. 1996). If they could be identified by QTL mapping and the harmonious ones stacked in the current high-yielding varieties by marker-assisted breeding, genetic yield could be further increased. This

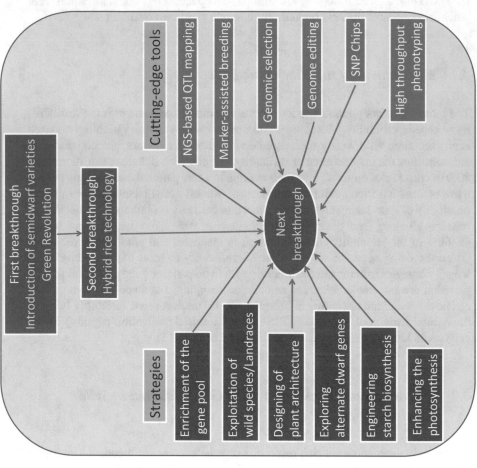

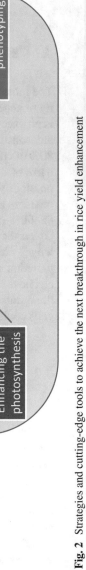

Fig. 2 Strategies and cutting-edge tools to achieve the next breakthrough in rice yield enhancement

followed the successful mapping and validation of two yield QTLs, *yld 1.1* and *yld 2.1*, in *O. rufipogan* by the marker-associated QTL approach by Cornell University, USA (Xiao et al. 1996). Worldwide interest was aroused among rice geneticists to search for more and more yield genes/QTLs in wild/weedy relatives and primitive cultivars, resulting in the identification of many promising yield QTLs in *O. rufipo-gan, O. nivara*, and landraces (Marri et al. 2005; Kaladhar et al. 2008; Swamy and Sarla 2008; Sudhakar et al. 2012; Swamy et al. 2011, 2012, 2014). This pioneering work culminated in the development of higher-yielding varieties such as Jefferson using *yld2.1* from *O. rufipogan* (Imai et al. 2013), DRR Dhan 40 involving yield genes/QTLs derived from *O. nivara* (Haritha et al. 2017), and the high-yielding salinity-tolerant IET21943 based on a yield QTL from *O. rufipogon* (Ganeshan et al. 2016) (Table 1).

4.3 Designing of Plant Architecture or Ideotype Breeding

On the strength of findings from simulation modeling, IRRI physiologists believed that the potential yield of 10 t/ha achieved through dwarf varieties could be further increased by 25% (Dingkuhn et al. 1991) through enhancement of biomass/unit area without altering the harvest index ($\approx$45%). This prompted IRRI breeders to conceptualize and tailor a morpho-physiologically more efficient new plant type (NPT) suited to high-density planting. Characterized by less profuse tillering habit, long and upright top leaves, heavy panicles, and robust and active root system, the

Table 1 List of wild species/landraces used for trait enhancement in rice

Donor species	Recipient species	Trait transferred	Reference
Oryza rufipogon	*Oryza sativa*	Yield	Xiao et al. (1996)
O. glumaepetula	*O. sativa*	Yield	Brondani et al. (2002)
O. grandiglumis	*O. sativa*	Yield	Ahn et al. (2003)
O. rufipogon	*O. sativa*	Yield	Liang et al. (2004)
Landrace (FR13A)	*O. sativa (Swarna)*	Submergence tolerance	Xu et al. (2006)
O. nivara (IRGC81848)	*O. sativa* (Swarna)	Yield and contributing traits	Swamy and Sarla (2008)
O. nivara (KDML 105)	*O. sativa* (Rathuheenathi)	BPH resistance (*Bph3*)	Jairin et al. (2010)
Landrace (Basmati 370)	*O. sativa* (Manawthukha)	Fragrance and amylase content	Yi et al. (2009)
O. rufipogon (IRGC 104814)	*O. sativa* (Koshihikari)	Blast resistance	Hirabayashi and Sato (2010)
Landrace (FL 478)	*O. sativa* (AS 996)	Salt tolerance	Luu et al. (2012)
O. rufipogon (Coll-4)	*O. sativa* (B 29-6)	Blast resistance (*Pi9*)	Ram and Majumder (2007)
Landrace (Tetep)	*O. sativa* (PRR78)	Blast resistance (*Pi54*)	Singh et al. (2012)
O. meridionalis	*O. sativa*	Blast resistance (*Pi-cd*)	Fujino et al. (2019)

NPT lines developed at IRRI have been reported to yield 11–12 t/ha vis-à-vis 13–14 t/ha reportedly achieved by Chinese breeders in *indica/tropical japonica* hybrids in the further improved NPT background (Yuan and Fu 1995).

4.4   Designing of Shoot and Panicle Architecture

Evidence suggests that, by altering shoot and panicle architecture, genetic yield can be substantially increased. Shoot architecture includes traits that affect plant height and leaf length, width, and thickness, whereas panicle architecture involves traits that affect panicle number, panicle length, number of grains per panicle, and grain weight. With precise information available on the mapping positions of QTLs and access to many cloned genes for the key traits, it is now possible to design the architecture of the rice plant by pyramiding appropriate QTLs/genes. The genes that control grain number (*Gn1a, Ghd7, DEP1,* and *WFP*), grain weight (*GS3* and *GW2*), grain filling (*GIF1*), grain size (*GS3* and *GW5*), and panicle number (*DEP1* and *WFP*) are fortunately located on different chromosomes, which would enable combining favorable genes/QTLs easily into elite varietal backgrounds.

Earlier, many researchers showed that pyramiding of multiple yield-related genes enhanced yield significantly. Ashikari et al. (2005) were successful in increasing grain number by 45% and decreasing plant height by 20% by combining the grain number QTL (*Gn1a*) and the semi-dwarfing gene (*sd1*) by a pyramiding strategy. Later, Ando et al. (2008) reported that a pyramided near-isogenic line (NIL) containing two QTLs (*qSBN1* for secondary branch number on chromosome 1 and *qPBN6* for primary branch number on chromosome 6) developed by introgressing a QTL from Habataki (*indica*) into Sasanishiki (*japonica*) produced more spikelets than the independent NIL harboring the QTLs *qSBN1* and *qPBN6*. Further, this pyramided line (*qSBN1* + *qPBN6*) showed 4–12% higher yield than the recurrent parent Sasanishiki because of greater translocation of carbohydrates from stem to panicle (Ohsumi et al. 2011). Wang et al. (2012a, b) also demonstrated that the pyramid line (*qHD8* + *GS3*) had higher yield potential, longer grains, and more suitable heading date than the recipient parent, Zhenshan97.

In addition to gene pyramiding with the aid of marker technology, some successful attempts have been made to increase yield through genetic engineering of plant architecture traits. One such effort employed light-regulated overexpression of the *Arabidopsis phytochrome A* gene by Cornell University (Garg et al. 2006). Phytochromes are a family of red/far red light-absorbing photoreceptors, which control plant development and plant metabolic activities. The group demonstrated that, by splicing the *Arabidopsis PHY A* gene into rice by employing light-regulated tissue-specific *rbc* promoter, plant stature could be altered by further decreasing the height of the already semi-dwarf variety and simultaneously increasing the number of productive tillers, resulting in significantly higher yield than for the control variety. In another study, Wang et al. (2015) reported that transgenic rice plants expressing the *Arabidopsis* phloem-specific sucrose transporter (AtSUC2), which loads sucrose into the phloem under control of phloem protein2 promoter (pPP2), showed 16% higher grain yield than the wild type in field trials. Park et al. (2017) demonstrated overexpression of the gene *OsGS* to improve redox homeostasis by enhancing the glutathione pool, which resulted in greater tolerance of environmental stresses in addition to higher grain yield and total biomass.

4.5   Modification of Root Architecture

Genetic improvement of the root system is important for developing tailor-made varieties tolerant of abiotic stresses. A deeper, thicker, and more branched root system with high root to shoot ratio is usually preferred for plants to withstand drought stress. Although drought tolerance has been extensively investigated for the past few decades, an in-depth study to understand its genetics and breeding behavior has hardly been attempted because of its complex nature and the tedious work involved in phenotyping of the root system. However, it is known that rice germplasm is rich in variability for root traits and that the root system and related traits are governed by many genes with small effects, often regarded as QTLs. Now, more than 600 QTLs have been identified for various root-related traits (www.gramene.org). Of the 162 functionally characterized root-related genes, most are annotated as being related to transport and transcriptional or hormonal regulation. The vast majority (98%) of these genes have been identified through reverse-genetic approaches and, of these, only three (*PSTOL1*, *DRO1*, and *Bet1*) identified based on natural allelic variation affecting phenotype. Despite such a large number of QTLs available, only two related to nutrient uptake, *PSTOL1* at the *Pup1* locus (Chin et al. 2010) and the root length QTL *qRL6.1* that governs nitrate uptake from deeper soil layers (Obara et al. 2010), have been used.

The recent development of several non-invasive 2D and 3D root imaging systems has enhanced our ability to accurately observe and quantify architectural traits in complex whole-root systems. Coupled with the powerful marker-based genotyping and sequencing platforms currently available, root phenotyping technologies lend themselves to large-scale genome-wide association studies, and can speed up the identification and characterization of the genes and pathways involved in root system development.

4.6   Green Super Rice for Sustainable Performance

As a massive breeding effort placing emphasis on developing ecologically and economically sustainable varieties in high-yield backgrounds for resource-poor rice farmers in Asia and Africa, the Green Super Rice (GSR) Project was launched jointly by the Bill & Melinda Gates Foundation, Chinese Academy of Agricultural Sciences, and IRRI. The breeding strategy consists of developing advanced backcross populations involving selected popular high-yielding varieties (40–50) chosen from major rice-growing countries as recurrent parents and around 500 varieties with traits of unique adaptive value as donors. Advanced backcross generations (BC$_2$F$_2$) are then screened under targeted stress conditions for transgressive segregants exceeding the respective parents and local checks in their trait performance. This is followed by pyramiding of complex trait-specific non-allelic QTLs of promise derived from different donor sources into country-specific popular varieties such as IR64, BR11, BG300, and Huang-Hua-Zhan. The recovery of promising lines in large numbers is attributed to the harmonious complementation of genetic networks for complex traits, which otherwise are incomplete in both parents (Ali et al. 2012). While the breeding emphasis of the GSR strategy is for developing eco-friendly/farmer-friendly varieties/hybrids, the massive exercise is bound to produce as well super-yielding varieties/hybrids.

4.7 Physiological Breeding Approaches

Exploring Alternative Sources of Dwarfing Genes Given the experience with maize (corn) in the United States, that genetic uniformity for even one gene would make any crop plant vulnerable to a sudden outbreak of any pest, rice breeders have been apprehensive of such an eventuality due to the widespread and excessive use of the dwarfing gene *sd1* in rice breeding, thus warranting diversification of the dwarfing gene. In rice, as many as 192 dwarfing genes are known. They include both dominant/semi-dominant and recessive inheritance: *D53, Ssi1, Sdd(t), Dx, TID1, LB4D, Slr-f, D-h, d13, Sdt97*, etc. (http://shigen.nig.ac.jp/rice/oryzabase). Dwarf mutants characterized at the molecular level have been found for their short stature as a result of defective signal transduction molecules such as heterotrimeric-G protein (Ueguchi-Tanaka et al. 2000), homeobox-like OSH15 (Sato et al. 1999), brassinosteroids (Yamamuro et al. 2000), and various GA biosynthesis genes (Sasaki et al. 2002; Itoh et al. 2004).

The *sd1* gene-based modern semi-dwarf varieties, aside from being short in stature, are associated with some pleiotropic effect on other traits that reportedly include decreased spikelet number and grain weight (Murai et al. 2002), decreased root length (Lafitte et al. 2007), and poor response to applied nutrients (Zhang et al. 2013). Yet other research underway is looking for a novel allele of *sd1* that would be devoid of such negative effects on yield components and thereby help develop a dwarf plant type variety with still higher potential yield. Differing from the traditional tall-statured native varieties, wherein the GA pathway remains intact vis-à-vis the modern semi-dwarf varieties, wherein the GA pathway has been suppressed (GA-repressed), an allele of *sd1* or some other height-reducing genetic mechanism that would affect the GA pathway (GA-independent) might help to find a new yield threshold. A novel and valuable dwarfing gene, *asd-1* (*alternate semi-dwarf gene*), was identified on chromosome 1 employing a QTL-seq approach (Gopalakrishna et al. 2017).

Engineering of Starch Biosynthesis The first yield breakthrough achieved in semi-dwarf rice took place through partitioning of photosynthate in favor of grains while the second occurred through an NPT variety that enabled an increase in biomass by manipulating crop geometry but not by enhanced photosynthesis. Recent findings reveal the possibilities of redirecting biosynthetic pathways through recombinant DNA technology enabling plants to produce more or altered quality of natural products such as starch, protein, and lipids. It is now possible to manipulate source-sink equilibrium either by overexpression of endogenous or heterologous enzymes or by down-regulation of endogenous enzymes by using gene silencing techniques.

Manipulation of ADP glucose pyrophosphorylase (ADPGPPase), the key rate-limiting allostearic enzyme in the biosynthetic pathway of starch, for instance, is regarded as a potential strategy to improve yield and starch quality in cereal crops. This enzyme catalyzes glucosyl phosphate into ADP glucose, which is the precursor of starch. Down-regulation of the gene encoding the enzyme by an antisense RNA approach resulting in a drastic decrease in its activity as well as starch accumulation was the first report to establish the crucial role of the enzyme in starch biosynthesis (Lin et al. 1988). The observed increase in starch yield due to antisense inhibition of resident ADPGPPase proved the key regulatory function of the enzyme. Following this report, many researchers explored the possibility of raising yield in other crops by manipulating the enzyme. Believing that natural variability in sink components would be a reflection of the varied behavior of the enzyme, an effort was made to study the extent of variability in the nature of the enzyme in rice. The findings revealed enzyme activity (total and specific), its response to effectors (activator 3PGA and inhibitor Pi), gene expression (transcription), and starch synthesis (total and rate of accumulation) to vary with the genotype and stage of development of the endosperm. Overall evaluation suggests that the highest yielding varieties and highly heterotic hybrids are the most promising in nearly all respects of the enzyme behavior (Devi et al. 2010). More intensive further study of wild/weedy species and ecotypes from different rice ecosystems might result in valuable sources for efficient ADPGPPase for exploitation by even conventional recombination breeding.

Enhancement of Photosynthesis Enhancement of photosynthetic efficiency could be one of the potential means for raising the yield ceiling in rice. In C_3 rice, CO_2 is assimilated into a 3-carbon compound by the enzyme ribulose-1,5-bisphosphate carboxylase/oxygenase (Rubisco). This enzyme also catalyzes oxidation of RuBP in a wasteful process known as photorespiration, which results in a loss of as much as 25% of the previously fixed carbon. At temperatures above 30 °C, which are typical of tropical rice-growing areas, the rate of oxygenation increases substantially and this considerably decreases the photosynthetic efficiency of C_3 plants by up to 40% (Ehleringer and Monson 1993). On the contrary, C_4 plants have very much decreased rates of photorespiration and thus are adapted to thrive in hot and dry environments; this offers valuable insights for yield improvement strategies. Rice with a C_4 photosynthesis mechanism would have increased photosynthetic efficiency while using scarce resources such as land, water, and fertilizer (specifically nitrogen) more effectively (Hibberd et al. 2008). As it would perform well under high temperature and require less water and nitrogen, C_4 rice would benefit varied rice ecosystems, including marginal lands. Engineering the photosynthetic pathway of C_3 rice into a C_4 plant is quite a challenging and time-consuming task. The main challenge of converting the photosynthetic pathway of a C_3 plant into that of a C_4 plant lies in decreasing photorespiration and modifying the leaf canopy (anatomy). IRRI, through an ambitious collaborative project (International C_4 Rice Consortium) involving advanced countries/laboratories, is engaged in converting C_3 rice into a C_4 plant by introducing appropriate genes from maize and other C_4 plant species. The C_4 pathway genes such as CA (carbonic anhydrase), PEPC (phosphoenolpyruvate carboxylase), PPDK (PEP carboxykinase), NADP-ME (NADP-dependent malic enzyme), and NADP-MDH (NAD-dependent malate dehydrogenase) cloned from maize are being engineered into rice. Also, the transporters that were overexpressed in the C_4 metabolic pathway such as 2-oxoglutarate/malate transporter (OMT1), dicarboxylate transporter1 (DiT1), dicarboxylate transporter2 (DiT2), PEP/phosphate transporter (PPT1), mesophyll envelope protein (MEP), and triose-phosphate phosphate translocator (TPT) and that were identified through proteomics of maize bundle sheath and mesophyll cells (Friso et al. 2010) were transformed into rice. Models show that increased water and nitrogen use efficiencies from this engineering effort could result in yield increases of 30% to 50% (Karki et al. 2013). Wang et al. (2017a, b), in their theoretical analysis of biochemical and anatomical factors, demonstrated that integrating a C_4 metabolic pathway into rice leaves with a C_3 metabolism and mesophyll structure may lead to increased photosynthesis under current ambient CO_2 concentration. Also, they concluded that the partitioning of energy between C_3 and C_4 photosynthesis and the partitioning of Rubisco between mesophyll and bundle sheath cells would be decisive factors controlling photosynthetic efficiency in an engineered C_3–C_4 leaf.

In yet another attempt to enhance photosynthesis and thereby genetic yield, Ambavaram et al. (2014) identified a master regulator, HYR (HIGHER YIELD RICE), a transcription factor associated with photosynthetic carbon metabolism (PCM). It directly activates the photosynthetic pathway genes and other downstream genes involved in PCM and yield stability under drought and high-temperature environmental stress conditions. Haritha et al. (2017) reported wild introgressions from *O. rufipogon* to increase the photosynthetic efficiency of KMR3 rice lines.

4.8   Defending Against Biophysical Stresses

Hardly any crop plant is challenged by diverse biotic and abiotic stresses as rice is. The crop is vulnerable to more than one dozen pathogens and as many insect pests, many of which exist in virulent/viruliferous races/biotypes. As for abiotic stresses, diverse stressful water regimes in rainfed lowlands, moisture-deficit rainfed uplands, coastal saline and inland sodic soils, temperature extremes, etc. constitute the major stresses, covering 45% of the world's rice area. Climate change is yet another threat to agriculture in general and rice in particular.

Breeding for Resistance to Biotic Stresses Since the introduction of high-yielding semi-dwarf varieties, diseases and insect pests have been increasingly causing severe yield losses year after year. Among pathogens, blast, bacterial leaf blight, sheath blight, and Rice Tungro Virus (RTV), and insect pests yellow stem borer, brown planthopper, leaf folder, and gall midge are the most devastating. Management of these pests has been largely through resistance breeding, taking advantage of race-/biotype-specific resistance genes identified in the gene pool and the adoption of rational gene deployment strategies. Frequent breakdown of resistance in multi-racial/multi-biotype pests has made pest management all the more difficult and challenging.

Breeding for Resistance against Diseases Rice blast caused by the fungus *Magnaporthe oryzae* is the most widespread and devastating disease of rice. Existing in as many as 30 races, it has been managed for many decades mainly by resistance breeding using more than 100 resistance genes identified in the rice gene pool. Nevertheless, the disease is still a challenge because of the frequent breakdown of resistance warranting the need for varieties with broad-spectrum resistance. Pyramiding of genes matching region-specific races was chosen as the strategy to manage the problem and molecular marker technology has been found handy in this effort. Following the pioneering attempts to introgress resistance genes (*Pi1, Pi-5, Piz,* and *Pita*) by marker-assisted backcross breeding into varieties such as Co39 (Hittalmani et al. 2000), IR50 (Narayanan et al. 2002), and Zhenshan 97A (Liu et al. 2003), this strategy is being adopted globally to make varieties stable against the disease.

Bacterial leaf blight (BLB), caused by *Xanthomonas oryzae* pv. *oryzae* (*Xoo*), is the most serious disease worldwide and it causes yield losses up to 70%. Based on analyses of phenotypic responses to *Xoo* races and molecular mapping results, 41 genes (29 dominant and 12 recessive) conferring resistance to the disease have been registered in the Oryzabase database (www.shigen.nig.ac.jp/rice/oryzabase/gene/list). Additionally, some R genes/alleles have been generated by mutation breeding. They include nine isolated genes (*Xa1, xa5, xa13, Xa21, Xa23, xa25, Xa26/Xa3, Xa27,* and *xa41*) and nine fine-mapped genes (*Xa2, Xa4, Xa7, Xa22, Xa30, Xa33, Xa38, Xa39,* and *Xa40*) (www.shigen.nig.ac.jp/rice/oryzabase/gene/list). In addition to such Mendelian genes, QTLs for resistance have been reported. Many pathotype-specific resistance genes linked to molecular markers have been successfully used for selective improvement of popular high-yielding quality rice varieties such as BPT5204 (Samba Mahsuri) introgressed with *Xa21, xa13,* and *xa5* (Sundaram et al. 2008) and Pusa Basmati-1 introgressed with *Xa21* and *xa13* (Joseph et al. 2004). The strategy of marker-assisted introgression of dominant genes (*Xa4, Xa7,* and *Xa21*) against BLB has been extended as well to the parental lines of popular hybrids (Borines et al. 2000; Zhang et al. 2006). Recently, Hajira et al. (2016) developed a single-tube, functional marker-based multiplex PCR assay for simultaneous detection of the major BLB resistance genes *Xa21, xa13,* and *xa5* in rice. Pyramiding diverse resistance genes against any disease is considered a promising strategy for ensuring broad-spectrum resistance and slowing down the breakdown of resistance. As for the choice of gene sources, many of those accessed from wild/weedy germplasm appear to be of multi-racial/multi-pathotype resistance. For instance, *Xa21*, which provides resistance to as many as 30 pathotypes and shows synergism with country-/region-specific critical resistance genes, is from *O. longistaminata*, an African A-genome species, while *Xa30* and *Xa31*, offering resistance to many pathotypes, are from *O. nivara*, and *Xa34* is from *O. rufipogon*. Better understanding of the molecular mechanisms underlying bacterial pathogenesis that has revealed the role of several factors facilitating infection and progression of the disease might pave the way in the near future to the development of an effective and environmentally safe strategy to manage BLB.

Rice Tungro Virus (RTV), the most dreaded disease induced by mixed infection of Rice Tungro Bacilliform Virus (RTBV) and Rice Tungro Spherical Virus (RTSV), is transmitted by green leafhopper. Screening of a large number of rice germplasm accessions reveals many traditional varieties to be resistant to RTSV and only a few to RTBV (Shim et al. 2015). RTSV resistance is a recessive trait controlled by a translation initiation factor4 gamma (eIF4G) located on chromosome 7 (Lee et al. 2010). A few years ago, an RTSV resistance gene was transferred to *japonica* rice by marker-assisted selection (Shim et al. 2015). Initially, the disease was managed by host-plant resistance breeding using a few accessions as donors. Recent studies show that the RNA interference (RNAi) technique would be more effective and could be used to develop virus-resistant transgenic rice. Le et al. (2015) generated transgenics capable of producing small interfering RNA specific against RTSV sequences. In order to develop transgene-based resistance against RTBV, Valarmathi et al. (2016) used the ORF IV gene by RNA-interference in rice variety Pusa Basmati-1, and the transgene was subsequently introgressed into ASD 16, a variety popular in southern India, by marker-assisted breeding.

Against sheath blight (ShB), which was once a minor disease but is now a major one, no source of resistance has as yet been found. However, as many as 50 QTLs with moderate resistance have been mapped. Several studies have reported candidate genes for resistance such as chitinase, glucanase, glutathione S-transferase, and kinase protein to be within the mapped QTL region (Yadav et al. 2015). Although most of the sheath blight resistance QTLs identified so far are of only limited effect on ShB, reports showing an expected level of resistance were not uncommon. For instance, Zuo et al. (2007) reported introgression of the QTL *qSB-11LE* to decrease grain loss by 10.71% in the background of variety Lemont under severe disease conditions in field trials. Pinson et al. (2005) predicted *qSB-9^{TQ}* and *qSB-3^{TQ}* to decrease crop loss due to the disease by 15% when introduced into the same variety. Wang et al. (2012a, b) found pyramiding of diverse sheath blight resistance QTLs such as *qSB9-2* and *qSB12-1* to increase resistance. Interestingly, Zuo et al. (2014) reported that pyramiding of the QTLs *qSB-9^{TQ}* and *TAC1^{TQ}*, governing stem borer resistance and tiller angle (TA), respectively, improved resistance to ShB. Thus, in the absence of R genes, pyramiding of ShB moderate resistance QTLs and of other QTLs governing unrelated traits/stresses could be worth attempting as an alternative strategy to manage the disease, as reported by Zuo et al. (2014). In addition to such host resistance-based strategies, a transgenic approach through inhibition of chitin metabolism in fungi such as *Rhizoctonia solani*, by expression of rice *chitinase*, could be a strategy for controlling the disease. Karmakar et al. (2017) demonstrated that transgenics overexpressed with constructs pyramided with two genes, *OsCHI11* (*chitinase* gene) and *AtNPR1 (Arabidopsis NPR1)*, were superior to a single-gene cassette in enhancing sheath blight tolerance.

Breeding for Resistance against Insect Pests Rice hosts more than one dozen insect pests, of which stem borer, brown planthopper, white backed planthopper, leaf folder, and gall midge cause serious yield losses. The host-plant resistance available in abundance against all the major insect pests and their biotypes with the exception of stem borer and leaf folder has enabled breeders to manage them so far by sequentially releasing resistant varieties matching newly emerging biotypes. Nevertheless, the emergence of newer and increasingly viruliferous biotypes requiring matching resistance gene sources, occurrence of more than one pest/biotype in any given region, and still no way to find resistance gene sources against yellow stem borer and leaf folder have made insect pest management a challenging task. Given the limitations of the conventional breeding-selection approaches, more rational gene deployment such as molecular marker-aided resistance gene pyramiding and introgression of novel alien genes by recombinant DNA technology are now employed/contemplated for effective management of insect pest problems. Many of the insect pest-/biotype-specific resistance genes have been mapped and linked to closely placed markers. For engineering resistance, many different insecticidal proteins and molecules known to be highly selective in their action against a given pest, causing no harm to non-target organisms, are being experimented with. Among the widely used genes encoding insecticidal proteins/molecules against rice insect pests, endotoxin crystal proteins of *Bacillus thuringiensis*, digestive enzyme-specific protease inhibitors, plant lectins, α amylase inhibitors, insect chitinases, and insecticidal viruses are important. Of these, *Bt* toxin genes (*cry IA, cry IB, cry IC*, etc.) and protease inhibitors (*cowpea serine P1*) against stem borer and lectin protein gene (*gna, asa lectin*) against hoppers have been reported to be effective. Advances made in managing various insect pests are presented hereunder.

Yellow stem borer (*Scirpophaga incertulas*) (Walker) is the most important insect pest globally in rice-producing areas. As there are no strong host-resistance gene sources against it, genetic engineering approaches have been attempted and found promising. Datta et al. (1996) and Nayak et al. (1997) were the first to report transformation of rice with the *Bt* gene against yellow stem borer. Since then, several workers have successfully engineered rice with different *Bt* genes (*cry IA(b), cry IA(c)*, and *cry IAb*) alone as well as in fusion forms against the pest. Liu et al. (2016) pyramided two foreign genes, *cry1Ac* driven by rice *Actin I* promoter and *lysine-rich protein* (*LRP*) driven by endosperm-specific *GLUTELIN1* (*GT1*) promoter, into elite *indica* cultivar 9311. In the pyramided line, *cry1Ac* has been found to efficiently express in leaves and stems against striped stem borer (*Chilo suppressalis* Walker) under laboratory conditions and against rice leaf folder (*Cnaphalocrociṣ medinalis* Guenee) under field conditions. Despite such success stories elsewhere, including in China, Spain, Pakistan, etc., *Bt* rice is yet to be deregulated for commercial planting.

Brown planthopper (BPH), *Nilaparvata lugens* Stål, has been a threat to rice production in Asia since the advent of high-yielding varieties. Experience with its management shows host-plant resistance to be the most efficient and sustainable strategy. As of now, 31 BPH resistance genes have been identified in cultivars and wild species and all except *bph5* and *bph8* have been mapped to various chromosomes of rice. To date, 13 BPH resistance genes (*Bph14, Bph3, Bph15, Bph26/2, bph29, Bph18, Bph9/1/7/10/21,* and *Bph32*) have been identified and characterized via a map-based cloning approach (Jing et al. 2017). Many of the genes have been introduced alone or in combination into modern rice varieties/parental lines of hybrids by marker-assisted selection. Wang et al. (2017a, b) pyramided *Bph6* and *Bph9* into elite restorer line 93-11, while Fan et al. (2017) developed three broad-spectrum BPH-resistant restorer lines by pyramiding big-panicle gene *Gn8.1*, BPH resistance genes *Bph6* and *Bph9*, and fertility restorer genes *Rf3, Rf4, Rf5,* and *Rf6* through molecular marker-assisted breeding.

Besides the gene pyramiding strategy to develop broad-spectrum hopper-resistant varieties, genetic engineering approaches have been attempted. For instance, Nagadhara et al. (2004) successfully engineered Chaitanya variety with *gna lectin* protein and another variety with onion/garlic *lectin* (*asa lectin*) (Saha and Majumder 2006). To develop durable resistance against BPH, green leafhopper, and white-backed planthopper, *ASACI* and *GNA* protein genes have been pyramided by crossing single-gene-based transgenic lines. The lines developed so far have been found to surpass in their level of resistance in all three hoppers vis-à-vis single-gene-based transgenics (Rao et al. 1998).

Asian rice gall midge (GM) (*Orseolia oryzae* Wood-Mason) is a serious pest in rice-growing countries, especially in China, India, and Sri Lanka. To date, 11 GM resistance genes (*Gm1, Gm2, gm3, Gm4, Gm5, Gm6, Gm7, Gm8, Gm9, Gm10,* and *Gm11(t)*) have been identified and characterized. As for their pyramiding to realize broad-spectrum resistance against more than one biotype, Nair et al. (2011) demonstrated that stacking of *Gm1* with any one of the genes in group II, which exclude *Gm4* and *Gm7*, would confer resistance to five biotypes (*GMB1, GMB2, GMB3, GMB5,* and *GMB6*). To cover all the biotypes, at least three genes, preferably *Gm1, Gm2,* and *Gm4*, would be required.

Rice leaffolder (RLF) (*Cnaphalocrocis medinalis* Guenee) is another major insect pest. Developing RLF-resistant lines in rice through conventional breeding has been a challenge due to the non-availability of a host-plant resistance gene. Alternatively, genetic engineering has been attempted by introducing heterologous insecticidal genes. Manikandan et al. (2016) succeeded in developing transgenic rice resistant to the pest with codon optimized synthetic *cry2AX1* gene fused with a rice chloroplast transit peptide sequence. In another report, Chakraborty et al. (2016) demonstrated that transgenic rice expressing the *cry2AX1* gene conferred resistance to multiple lepidopteran pests, including RLF. Pradhan et al. (2016) also reported transgenic rice expressing vegetative insecticidal protein (Vip) of *Bacillus thuringiensis* to show broad insecticidal properties.

Breeding for Tolerance of Abiotic Stresses Among abiotic stresses that severely depress productivity, drought, submergence, and salinity are important. The plant's ability to withstand such stresses results from the cumulative effects of a network of physiological and biochemical functions. Negative effects of abiotic stresses include broadly stress-imposed homeostasis imbalance, disruption of growth and metabolic activities, and generation of cell-damaging reactive oxygen species (ROS). At the molecular level, plants under abiotic stresses adapt to the conditions by triggering a cascade of events that start with stress perception and end with the expression of a battery of genes of adaptive response. Adaptation at the molecular level is through restoration of homeostasis (ion and osmotic gradient), control of damage, and detoxification of ROS. Knowledge of the genetics governing these stresses is a pre-requisite for finding solutions to the problem. Unlike resistance to biotic stresses that largely follows a Mendelian mode of inheritance, tolerance of abiotic stresses is quite complex and polygenically controlled. Added to the inadequate knowledge of genetics, a lack of reliable and reproducible screening/selection techniques has made breeding all the more challenging, despite having sources of resistance in the rice gene pool. As a result, more than 60% of the global rice area, especially in the rainfed ecosystem does not have as yet high-yielding varieties ideally suited to this harsh environment. For the past 15 years, advances in plant molecular biology have provided breeders with a variety of genomic tools and resources capable of over-coming the technological constraints that have been impeding progress in finding genetic solutions to such stresses.

Of the estimated huge future rice demand, most production has to come from rainfed lowland and upland rice ecosystems, where drought is the major yield con-straint. Unlike the irrigated ecosystem, no truly high-yielding varieties are adapted to drought stress. In drought-prone parts of the world, farmers have no option but to grow traditional low-yielding but well-adapted varieties. All efforts to develop high-yielding drought-tolerant varieties by using traditional drought-tolerant varieties as donor sources proved a futile exercise. This is largely on account of the genetically complex nature of drought tolerance and dependence on phenotype-based selection. Major indices of tolerance QTLs now mapped and linked to robust markers have made genotype-based selection and pyramiding of varieties with tolerance QTLs easy and efficient.

Several QTLs relating to different parameters of drought tolerance such as osmotic adjustment, cell membrane stability, relative water content, root character-istics, and stress recovery as well as yield per se under stress have been mapped and linked to robust markers. Among the traits that govern drought tolerance, the direct measure of grain yield under drought rather than its component traits is promising. As of now, a large number of QTLs governing grain yield under drought (DTY) have been identified, such as $qDTY12.1, qDTY3.1, qDTY6.1, qDTY2.2,$ and $qDTY9.1.$ The various combinations of these DTY QTLs resulted in an average yield advan-tage of 300–500 kg/ha under stress conditions. For instance, the major-effect QTL $qDTY_{12.1}$ has been introgressed into Vandana and it has a yield advantage of about 500 kg/ha over its donor parent under reproductive-stage stress (Kumar et al. 2014).

Similar success has been achieved by introgressing the QTL *qDTYs* into drought-susceptible mega-variety IR64 (Swamy and Kumar 2013). Two QTLs (*qDTY3.2* and *qDTY12.1*) with large effects for grain yield under drought have been trans-ferred into Sabitri, a popular variety of Nepal, through marker-assisted breeding (Dixit et al. 2017). Many more QTLs have been identified for yield under drought conditions and they are being transferred to the background of popular high-yielding varieties such as Swarna, IR64, Vandana, Sabitri, TDK1, Anjali, Samba Mahsuri, MRQ74, MR219, Jinmibyeo, Gayabyeo, Hanarumbyeo, and Sangnambatbyeo by marker-assisted breeding at IRRI and other centers. Besides these DTY QTLs, Uga et al. (2013) reported that the QTL *Deeper Rooting 1 (DRO1)* increased root growth angle in rice, leading to higher yield under drought conditions.

Given the availability of many novel genes conferring high tolerance of the stress along with efficient transformation-regeneration protocols in place, genetic engi-neering for drought tolerance has been found to be a distinct possibility. Candidate genes used with some success in rice are trehalose (*tps*) cloned from *Arabidopsis*, trehalose-6-phosphate synthase, and trehalose-6-phosphate phosphatase (*tps1*) from yeast, pyrroline carboxylate synthase (*P5cb*) from *Vigna aconitifolia,* chloroplast glutamine synthetase (*GS2*) from rice, and choline oxidase (glycine-betaine synthe-sis) (*cod A*) from *Arthrobacter globiformis*. Among the regulatory genes, DRE binding protein (*dreb1α*), calcium-dependent protein kinase *(OsCDPK7)*, and those identified with cellular-level tolerance such as helicases (*PDH45*) and APZ/ERF family DREB transcription factors are important (Saijo et al. 2000; Sahoo et al. 2012; Rashid et al. 2012).

Although many sources of tolerance of salt stress are available in native germ-plasm such as SR26B, Nona Bokhra, and Pokkali, efforts to combine the desired level of tolerance in high-yield backgrounds by conventional breeding yielded no tangible results. Convinced of the potential of molecular marker-assisted breeding and gene transfer technology, possibilities have been explored to find genetic solu-tions to the problem. As no single QTL/gene could provide the desired level of tol-erance, pyramiding of genes for tolerance-related morpho-physiological features (Na and Cl exclusion, K uptake, N/K ratio, and tissue tolerance) and biochemical pathways (glyoxalase, abscisic acid, proline, glycine-betaine, and polyamine) has been employed. More than 80 salt-tolerance QTLs with large effects have been mapped so far. Of these, *Saltol* mapped in Pokkali is a major one and it has already been transferred to high-yielding varieties such as BR11, BRRI dhan28, IR64, and AS996 (Babu et al. 2017).

The genetic engineering strategy is based on genes/enzymes involved in the pathways of (a) osmosis homeostatic balance, wherein ion and osmatic gradient are restored; (b) detoxification by scavenging of ROS; and (c) stress damage control restoring growth and metabolic activities. Transgenics with relevant salt-tolerance genes obtained from various sources have been reported to adapt well to and survive in saline soils. Many important genes conferring salt tolerance belong to the salt overly sensitive (SOS) pathway. Of the several candidate and regulatory genes over-expressed in rice, calcium-dependent protein kinase (*Os CDPK*), a transcription factor, choline oxidase (*cod A*) involved in glycine-betaine synthesis, chloroplast glutamine synthetase (GS2), pyrroline carboxylate synthetase, etc. are some of the successfully employed sources. Expression of two genes of the glyoxalase pathway, *glyI* and *glyII,* preferably together, has been shown to confer tolerance of salinity (Singla-Pareek et al. 2006). Overexpression of several stress-induced genes, including helicases, has been shown to provide tolerance of the stress in rice. Sahoo et al. (2012) have demonstrated pea DNA helicase 45 to promote salinity tolerance in IR64 with higher yield.

Flash floods causing submergence are the major yield-depressing factor in rainfed shallow lowland rice. Depending on the growth stage and period under water, crop damage could vary from 10 to 100%. Unlike traditional tall varieties such as FR13A and FR43B, which are known to adapt to submergence, short-statured high-yielding varieties suffer the most, even when the period of submergence is not even a few days. Hence, farmers in the rainfed lowland ecosystem prefer traditional varieties, not minding their low productivity. Despite years of effort to breed submergence tolerance into high-yield backgrounds using FR13A-like donor sources, no progress could be made. This was largely due to the lack of precise knowledge of the genetics of submergence tolerance per se and of the morpho-physiological parameters that govern it and the non-reliability of phenotype-based selection under stress conditions. The major QTL identified in flood-tolerant landrace FR13A and designated as *Sub1,* which explains 70% of the phenotypic variance, was mapped onto chromosome 9. A joint international effort involving NRRI and NDUAT in India, BRRI in Bangladesh, University of California-Davis in the United States, and IRRI began for the characterization and use of the QTL in marker-assisted breeding for developing submergence-tolerant varieties. Dissection of the QTL revealed it to be a cluster of three genes encoding ethylene responsive factors designated as *Sub1A, Sub1B,* and *Sub1C.* Of these, *Sub1A* was identified as the key gene conferring tolerance of submergence. Two alleles of *Sub1A, Sub1A-1* (tolerance specific) and *Sub1A-2* (intolerance specific) were identified with a single nucleotide polymorphism changing proline to serine (Septiningsih et al. 2009). Swarna selectively introgressed with *Sub1A* was formally released for general cultivation as Swarna-Sub1A in India (Odisha, Uttar Pradesh, etc.), Bangladesh, the Philippines, and Indonesia. The *Sub1A* gene has also been deployed into Thai fragrant rice Khao Dawk Mali 105 and rice restorer line Wanhui 6725 (Luo et al. 2016). This was probably the first example of a biotech product ever developed for abiotic stress conditions by marker-assisted breeding in rice.

Enrichment of Nutritive Quality Enrichment of nutritive quality is as important as yield enhancement on account of people in Asia depending on rice for a sizable part of their nutritional requirement. Because about one-half of the global population is suffering from one or more nutrient deficiency–related health disorders and more than three million children die each year because of malnutrition, there is a need and urgency to pay due attention to the nutritive quality in rice.

Among the nutritional limitations of rice, low protein content (PC) is the foremost. Nevertheless, the need to raise the protein content received no serious attention from breeders after experiencing early failures. Following the identification of high-protein donor sources, despite the complex inheritance and the earlier reported negative relationship between yield and protein content, breeders have started believing that high protein content could be combined with high yield. By using ARC10075 as a donor, CR Dhan310 (IET24780) with PC of 11% and rich in threonine and lysine content has been developed and released for general cultivation in India (Mahender et al. 2016). Although many efforts have been made for developing marker-assisted breeding for high PC, success has so far been eluded. Xu et al. (2017) were successful in developing transgenic rice with enhanced high-quality protein content by expressing the *AmA1* gene from *Amaranthus* sp.

Vitamin deficiency–related health disorders are widespread in third world countries. Vitamin A deficiency causing vision impairment is especially rampant. In the absence of exploitable natural variability, in all food grains except maize, alternative strategies have to be found. Professor Ingo Potrykus of the Federal Institute of Plant Sciences, Switzerland, jointly with Professor Peter Beyer of the University of Freiburg, Germany, succeeded in raising β carotene content by an ingenious genetic engineering strategy. The strategy lay in restoring three critical genes that are missing in the isoprenoid pathway of carotene synthesis by accessing them from daffodil (phytoene synthase (*psy*), phytoene desaturase (*pds*), and lycopene cyclase (*lyc*)), and the bacterium *Erwinia uredovora* (carotene desaturase, *crt1*) (Ye et al. 2000). So, the transformed rice popularly known as *Golden rice*, although a great scientific achievement, accumulates far less (1.6 mg/g) beta carotene. Understanding that it was due to the use of relatively less efficient daffodil gene *psy* encoding phytoene synthase, the multinational biotech company Syngenta succeeded in increasing the content of β carotene by several fold (37 mg/g) by using the *psy* gene cloned from maize (Paine et al. 2005). Sadly, the Golden rice developed close to 20 years ago has yet to be deregulated for commercial planting because of regulatory hurdles.

As for mineral nutrients, iron and zinc are the most crucial as their deficiency is the main cause of malnutrition. Modern high-yielding varieties are low in these two mineral nutrients. On average, Fe content in polished rice is 2 mg/kg vis-à-vis the recommended dietary intake for humans of 10–15 mg/kg. In the case of Zn, males within the age bracket of 15–74 years require on average 12–15 mg/day vis-à-vis 68 mg/day required for females (Mahender et al. 2016). Conventional breeding for developing Fe-enriched rice has not progressed to the desired extent due to limited variability for Fe content in polished rice. Evaluation of more than 20,000 rice accessions from Asia, Latin America, and the Caribbean for Fe and Zn content revealed a maximum of only 8 mg/kg in polished grains. This is because most of the Fe and Zn are concentrated in the aleurone and sub-aleurone layers of rice kernels, and is lost upon polishing. Taking advantage of Zn-rich donor sources, under the HarvestPlus project, the Bangladesh Rice Research Institute in collaboration with IRRI developed and released two Zn-enriched (19 and 24 mg/kg) varieties. The Indian Institute of Rice Research, Hyderabad, has as well succeeded in developing three rice varieties (DRR Dhan 45, DRR Dhan 48, and DRR Dhan 49) with high Zn content: 22.3, 20.91, and 26.13 mg/kg, respectively (Rao et al. 2020). Many low to moderate Zn QTLs have now been mapped in the germplasm. In addition, genome-wide association mapping and QTL mapping enabled the identification of several loci associated with grain Fe and Zn content (Norton et al. 2014; Swamy et al. 2018; Calayugan et al. 2020).

Transgenic approaches to enhance Fe content in rice grains were first explored more than a decade ago. Since then, attempts have been made to increase Fe content in rice endosperm by overexpressing the genes involved in Fe uptake from soil and those involved in translocation from aerial parts to grains. Among these studies, a concomitant increase in Fe and Zn content in rice grains was obtained by overexpression or activation of *NAS* (*nicotianamine synthase*) genes, either *in solo* or in combination with other transporters or Fe storage genes. Constitutive expression of *OsNAS2* has been reported to result in increased Fe content (19 mg/kg) and Zn concentration (76 mg/kg) in polished rice grains (Johnson et al. 2011).

4.9 Selective Modification of Traits by Gene Editing

Aside from the marker-assisted introgression of genes of interest discussed in the foregoing, an alternative genetic engineering approach for selectively transforming crops with targeted genes has opened up yet another molecular strategy known as genome editing. Genome editing tools such as TALENs (Transcription Activator-Like Effector Nucleases) and CRISPR/Cas9 (Clustered Regularly Interspaced Short Palindromic Repeats) enable selective interchange of genes of interest. Considered as a non-transgenic method, products thereby are devoid of any foreign material in their final product DNA. Employing these techniques, targeted gene editing for important traits is on the rise. TALENs, for instance, have been used for enhancing seed storability and herbicide resistance in rice by precisely editing the *LOX3* gene (Ma et al. 2015) and acetolactate synthase gene (*OsALS*) (Li and Liu 2016), respectively. Likewise, CRISPR/Cas9 has been used to manipulate traits such as herbicide resistance (Xu et al. 2014), yield components (Li et al. 2016), blast resistance (Wang and Wang 2016), TGMS line development (Zhou et al. 2016), stomatal development (Yin and Biswal 2017), and modifying amylose content (Sun et al. 2017; Perez et al. 2019) (Table 2).

5 Conclusions

For predominantly rice-consuming Asia to emerge and remain a food-/nutrition-secure continent, sustained self-sufficiency is crucial. Meeting future demand projections sustainably is the challenging task in the face of shrinking natural resources—arable land area, irrigation water, genetic variability—and the inevitable adverse effects of climate change. For the next breakthrough to meet future food demand projections, it seems imperative to exploit advanced cutting-edge tools, which enables development of high yielding, nutrient rich and input-use-efficient designer rice varieties.

Table 2 Targeted editing of important genes employing genome editing techniques

Trait	Targeted gene	Gene editing method	Reference
Herbicide resistance (resistance to bentazon and sulfonylurea herbicides)	BEL	CRISPR-Cas9 system	Xu et al. (2014)
Improvement of seed storability	LOX3	TALENs	Ma et al. (2015)
Regulators of grain number, panicle architecture, grain size, and plant architecture	Gn1a, DEP1, GS3, and IPA1	CRISPR/Cas9 system	Li et al. (2016)
Herbicide resistance	Acetolactate synthase gene (OsALS)	TALENs	Li and Liu (2016)
Blast resistance	OsERF922	CRISPR/Cas9	Wang and Wang (2016)
TGMS line development	TMS5	CRISPR/Cas9	Zhou et al. (2016)
Stomatal development	EPFL9	CRISPR/Cas9 and CRISPR/Cpf1	Yin et al. (2017)
High amylose content	SBEI and SBEII	CRISPR/Cas9	Sun et al. (2017)
Low amylase content	GBSS	CRISPR/Cas9	Perez et al. (2019)

References

Ahn SN, Kwon SJ, Suh JP, Kang KH, Kim HJ, Hwang HG, Moon HP (2003) Introgression for agronomic traits from *O grandiglumis* into rice, *O sativa*. In: Mew TW, Brar DS, Peng S, Dawe D, Hardy B (eds) Rice science: innovations and impact for livelihood. International Rice Research Institute (IRRI), Los Baños, pp 265–274

Ali J (1993) Studies on temperature sensitive genic male sterility and chemical induced sterility towards development of two-line hybrids in rice (*Oryza sativa* L.). PhD thesis, Indian Agricultural Research Institute, New Delhi

Ali J, Siddiq EA, Zaman FU, Abraham MJ, Ahmed IM (1995) Identification and characterization of temperature-sensitive genic male sterile sources in rice (*Oryza sativa* L). Indian J Genet 55:243–259

Ali J, Xu JL, Gao Y, Fontanilla M, Li Z (2012) Enhanced productivity across different rice ecologies through green super Rice (GSR) breeding strategy. International Dialogue on Designer Rice for Future, ICRISAT, Hyderabad, 9–10 July 2012

Ambavaram MMR, Basu S, Krishnan A, Ramegowda V, Batlang U, Rahman L, Baisakh N, Pereira A (2014) Coordinated regulation of photosynthesis in rice increases yield and tolerance to environmental stress. Nat Commun 5:5302. https://doi.org/10.1038/ncomms6302

Ando T, Yamamoto T, Shimizu T et al (2008) Genetic dissection and pyramiding of quantitative traits for panicle architecture by using chromosomal segment substitution lines in rice. Theor Appl Genet 116(6):881–890. https://doi.org/10.1007/s00122-008-0722-6

Ashikari M, Sakakibara H, Lin S, Yamamoto T, Takashi T, Nishimura A, Angeles ER, Qian Q, Kitano H, Matsuoka M (2005) Cytokinin oxidase regulates rice grain production. Science 309(5735):741–745. https://doi.org/10.1126/science.1113373

Babu NN, Krishnan SG, Vinod KK, Krishnamurthy SL, Singh VK, Singh MP, Singh R, Ellur RK, Rai V, Bollinedi H, Bhowmick PK, Yadav AK, Nagarajan M, Singh NK, Prabhu KV, Singh AK (2017) Marker aided incorporation of *Saltol*, a major QTL associated with seedling stage salt tolerance, into *Oryza sativa* 'Pusa Basmati 1121'. Front Plant Sci 8:41. https://doi.org/10.3389/fpls.2017.00041

Borines LM, Veracruz CM, Redoña ED, Hernandez JF, Natural MP, Raymundo AD, Leung H (2000) Marker aided pyramiding of bacterial blight resistance genes in maintainer lines for hybrid rice production. International Rice Research Institute Conference Abstract 4:162

Brondani C, Rangel PHN, Brondani RPV, Ferreira ME (2002) QTL mapping and introgression of yield-related traits from *Oryza glumaepatula* to cultivated rice (*Oryza sativa* L.) using microsatellite makers. Theor Appl Genet 104:1192–1203

Calayugan MIC, Formantes AK, Amparado A et al (2020) Genetic analysis of agronomic traits and grain iron and zinc concentrations in a doubled haploid population of rice (*Oryza sativa* L.). Sci Rep 10:2283. https://doi.org/10.1038/s41598-020-59184-z

Chakraborty M, Reddy PS, Mustafa G, Rajesh G, Narasu VM, Udayasuriyan V, Rana D (2016) Transgenic rice expressing the cry2AX1 gene confers resistance to multiple lepidopteran pests. Transgenic Res 25(5):665–678. https://doi.org/10.1007/s11248-016-9954-4. Epub 2016 Mar 26

Chin HJ, Lu X, Haefele SM, Gamuyao R, Ismail AM, Wissuwa M, Heuer S (2010) Development and application of gene-based markers for the major rice QTL *Phosphate uptake 1* (*Pup1*). Theor Appl Genet 120:1073–1086

Datta K, Torrizo L, Oliva N, Alam MF, Wu C, Abrigo E, Vasquez A, Tu J, Quimia C, Alejar M, Nicola Z, Khush GS, Datta SK (1996) Production of transgenic rice by protoplast, biolistic and *Agrobacterium* systems. In: Proceedings of the fifth international symposium on rice molecular biology. Yi Sein Publishing Co., Taipei, pp 159–167

Devi T, Anjana N, Sarla B, Siddiq EA, Sirdeshmukh R (2010) Activity and expression of adenosine diphosphate glucose pyrophosphorylase in developing rice grains: varietal differences and implications on grain filling. Plant Sci 178:123–129

Dingkuhn M, Penning de Vries FWT, De Datta SK, van Laar HH (1991) Concepts for a new plant type for direct seeded flooded tropical rice. Selected papers from the International Rice Research conference, Seoul, Korea, 27–31 August 1990

Dixit S, Yadaw RB, Mishra KK, Kumar A (2017) Marker-assisted breeding to develop the drought-tolerant version of Sabitri, a popular variety from Nepal. Euphytica 213:184

Ehleringer JR, Monson RK (1993) Evolutionary and ecological aspects of photosynthetic pathway variation. Annu Rev Ecol Syst 24:411–439

Fan F, Li N, Chen Y, Liu X, Sun H, Wang J, He G, Zhu Y, Li S (2017) Development of elite BPH-resistant wide-spectrum restorer lines for three and two line hybrid rice. Front Plant Sci 8:986. https://doi.org/10.3389/fpls.2017.00986

Friso G, Majeran W, Huang MS, Sun Q, van Wijk KJ (2010) Reconstruction of metabolic pathways, protein expression, and homeostasis machineries across maize bundle sheath and mesophyll chloroplasts: large-scale quantitative proteomics using the first maize genome assembly. Plant Physiol 152:1219–1250

Fuji S, Toriyama K (2009) Suppressed expression of *Retrograde-Regulated Male Sterility* restores pollen fertility in cytoplasmic male sterile rice plants. Proc Natl Acad Sci U S A 106:9513–9518. https://doi.org/10.1073/pnas.0901860106

Fujino K, Hirayama Y, Obara M, Ikegaya T (2019) Introgression of the chromosomal region with the *Pi-cd* locus from *Oryza meridionalis* into *O. sativa* L. during rice domestication. Theor Appl Genet 132:1981–1990

Ganeshan P, Jain A, Parmar B, Rao AR, Sreenu K, Mishra P, Mesapogu S, Subrahmanyam D, Ram T, Sarla N, Rai V (2016) Identification of salt tolerant rice lines among interspecific BILs developed by crossing *Oryza sativa* × *O. rufipogon* and *O. sativa* × *O. nivara*. Aust J Crop Sci 10(2):220–228

Garg AK, Sawers RJ, Wang H et al (2006) Light-regulated overexpression of an *Arabidopsis phytochrome A* gene in rice alters plant architecture and increases grain yield. Planta 223(4):627–636. https://doi.org/10.1007/s00425-005-0101-3

Gopalakrishna K, Vemireddy LR, Srividhya A, Nagireddy R, Jena SS, Gandikota M, Patil S, Veeraghattapu R, Deborah DAK, Reddy GE, Shake M, Dasari A, Ramanarao PV, Durgarani CV, Neeraja CN, Siddiq EA, Sheshumadhav M (2017) QTL-Seq-based genetic analysis identifies a major genomic region governing dwarfness in rice (*Oryza sativa* L.). Plant Cell Rep 37(4):677–687. https://doi.org/10.1007/s00299-018-2260-2

Hajira SK, Sundaram RM, Laha GS, Yugander A, Balachandran SM, Viraktamath BC, Sujatha K, Balachiranjeevi CH, Pranathi K, Anila M, Bhaskar S, Abhilash V, Mahadevaswamy HK, Kousik M, Dilip Kumar T, Harika G, Rekha G (2016) A single-tube, functional marker-based multiplex PCR assay for simultaneous detection of major bacterial blight resistance genes *Xa21*, *xa13* and *xa5* in rice. Rice Sci 23(3):144–151

Haritha G, Vishnukiran T, Yugandhar P, Sarla N, Subrahmanyam D (2017) Introgressions from *Oryza rufipogon* increase photosynthetic efficiency of KMR3 rice lines. Rice Sci 24(2):85–96

Hibberd JM, Sheehy JE, Langdale JA (2008) Using C4 photosynthesis to increase the yield of rice: rationale and feasibility. Curr Opin Plant Biol 11(2):228–231. https://doi.org/10.1016/j.pbi.2007.11.002

Hirabayashi H, Sato H (2010) Development of introgression lines derived from *Oryza rufipogon* and *O. glumaepatula* in the genetic background of japonica cultivated rice (*O. sativa* L.) and evaluation of resistance to rice blast. Breed Sci 60:604–612

Hittalmani S, Parco A, Mew TV, Zeigler RS, Huang N (2000) Fine mapping and DNA marker-assisted pyramiding of the three major genes for blast resistance in rice. Theor Appl Genet 100:1121–1128

Huang Y (2001) Rice ideotype breeding of Guangdong Academy of Agricultural Sciences in retrospect. Guangdong Agric Sci 3:2–6. (in Chinese)

Hussain AJ, Ali J, Siddiq EA, Gupta VS, Reddy UK, Ranjekar PK (2011) Mapping of *tms8* gene for temperature-sensitive genic male sterility (TGMS) in rice (*Oryza sativa* L.). Plant Breed 131(1):42–47

Imai I, Kimball JA, Conway B et al (2013) Validation of yield-enhancing quantitative trait loci from a low-yielding wild ancestor of rice. Mol Breed 32:101–120

Itoh H, Tatsumi T, Sakamoto T, Otomo K, Toyomasu T, Kitano H, Ashikari M, Ichihara S, Matsuoka M (2004) A rice semidwarf gene, *Tan-Ginbozu* (*D35*), encodes the gibberellin biosynthesis enzyme, *ent*-kaurene oxidase. Plant Mol Biol 54(4):533–547. https://doi.org/10.1023/B:PLAN.0000038261.21060.47

Jairin J, Sansen K, Wongboon W, Kothcharerk J (2010) Detection of a brown planthopper resistance gene *bph4* at the same chromosomal position of *Bph3* using two different genetic backgrounds of rice. Breed Sci 60:71–75. https://doi.org/10.1270/jsbbs.60.71

Jing S, Zhao Y, Du B, Chen R, Zhu L, He G (2017) Genomics of interaction between the brown planthopper and rice. Curr Opin Insect Sci 19:82–87

Johnson AAT, Kyriacou B, Callahan DL, Carruthers L, Stangoulis J, Lombi E et al (2011) Constitutive overexpression of the *OsNAS* gene family reveals single-gene strategies for effective iron- and zinc-biofortification of rice endosperm. PLoS One 6(9):e24476. https://doi.org/10.1371/journal.pone.0024476

Joseph M, Gopalakrishnan S, Sharma RK, Singh VP, Singh AK, Singh NK, Mohapatra T (2004) Combining bacterial blight resistance and basmati quality characteristics by phenotypic and molecular marker-assisted selection in rice. Mol Breed 13(4):377–387

Kaladhar K, Swamy BPM, Babu AP, Reddy CS, Sarla N (2008) Mapping quantitative trait loci for yield traits in BC$_2$F$_2$ population derived from Swarna x O. *nivara* cross. Rice Genet Newsl 24:34–36. https://doi.org/10.1093/jhered/esr145

Karki S, Rizal G, Quick WP (2013) Improvement of photosynthesis in rice (*Oryza sativa* L.) by inserting the C4 pathway. Rice 6:28. https://doi.org/10.1186/1939-8433-6-28

Karmakar S, Molla KA, Das K, Sarkar SN, Datta SK, Datta K (2017) Dual gene expression cassette is superior than single gene cassette for enhancing sheath blight tolerance in transgenic rice. Sci Rep 7:7900. https://doi.org/10.1038/s41598-017-08180-x

Kumar A, Dixit S, Ram T, Yadaw RB, Mishra KK, Mandal NP (2014) Breeding high-yielding drought-tolerant rice: genetic variations and conventional and molecular approaches. J Exp Bot 65(21):6265–6278. https://doi.org/10.1093/jxb/eru363

Lafitte HR, Guan YS, Shi Y, Li ZK (2007) Whole plant responses, key processes, and adaptation to drought stress: the case of rice. J Exp Bot 58:169–175

Le DT, Chu HD, Sasaya T (2015) Creation of transgenic rice plants producing small interfering RNA of Rice tungro spherical virus. GM Crops Food 6(1):47–53

Lee J-H, Muhsin M, Atienza GA, Kwak D-Y, Kim S-M, De Leon TB, Angeles ER, Coloquio E, Kondoh H, Satoh K et al (2010) Single nucleotide polymorphisms in a gene for translation initiation factor (eIF4G) of rice (*Oryza sativa*) associated with resistance to Rice tungro spherical virus. Mol Plant-Microbe Interact 23:29–38

Li T, Liu B (2016) TALEN-mediated homologous recombination produces site-directed DNA base change and herbicide-resistant rice. J Genet Genomics 43:297–305

Li JM, Yuan LP (2000) Hybrid rice: genetics, breeding, and seed production. Plant Breed Rev 17:15–158

Li M, Li X, Zhou Z et al (2016) Reassessment of the four yield-related genes *Gn1a*, *DEP1*, *GS3*, and *IPA1* in rice using a CRISPR/Cas9 system. Front Plant Sci 7:377. https://doi.org/10.3389/fpls.2016.00377

Liang F, Deng Q, Wang Y et al (2004) Molecular marker-assisted selection for yield-enhancing genes in the progeny of "9311×*O. rufipogon*" using SSR. Euphytica 139:159–165

Lin TP, Caspar T, Somerville CR, Preiss J (1988) Isolation and characterization of a starchless mutant of *Arabidopsis thaliana* (L.) Heynh lacking ADP glucose pyrophosphorylase activity. Plant Physiol 86:1131–1135

Liu SP, Li X, Wang CY, Li XH, He YQ (2003) Improvement of resistance to rice blast in Zhenshan 97 by molecular marker-aided selection. Acta Bot Sin 45:1346–1350

Liu X, Zhang C, Li X, Tu J (2016) Pyramiding and evaluation of both a foreign *Bacillus thuringiensis* and a *Lysine-rich protein* gene in the elite *indica* rice 9311. Breed Sci 66(4):591–598

Luo Y, Ma T, Zhang A, Ong KH, Li Z, Yang J, Yin Z (2016) Marker-assisted breeding of the rice restorer line Wanhui 6725 for disease resistance, submergence tolerance and aromatic fragrance. Rice (N Y) 9:66. https://doi.org/10.1186/s12284-016-0139-9

Luu T, Huyen N, Cuc LM, Ismail AM, Ham LH (2012) Introgression of the salinity tolerance QTL *Saltol* into AS996, the elite rice variety of Vietnam. Am J Plant Sci 3:981–987

Ma L, Zhu F, Li Z et al (2015) TALEN-based mutagenesis of lipoxygenase LOX3 enhances the storage tolerance of rice (*Oryza sativa*) seeds. PLoS One 10(12):e0143877. https://doi.org/10.1371/journal.pone.0143877

Mackill DJ, Coffman WR, Garrity DP (1996) Rainfed lowland rice improvement. International Rice Research Institute, Los Baños

Mahender A, Anandan A, Pradhan SK et al (2016) Rice grain nutritional traits and their enhancement using relevant genes and QTLs through advanced approaches. Springerplus 5:2086. https://doi.org/10.1186/s40064-016-3744-6

Manikandan R, Balakrishnan N, Sudhakar D, Udayasuriyan V (2016) Transgenic rice plants expressing synthetic *cry2AX1* gene exhibit resistance to rice leaffolder (*Cnaphalocrosis medinalis*). 3 Biotech 6(1):10

Marri PR, Sarla N, Vemireddy LR, Siddiq EA (2005) Identification of yield enhancing quantitative trait loci (QTLs) from *Oryza rufipogon* of Indian origin. BMC Genet 6:33

Mi J, Li G, Huang J, Yu H, Zhou F, Zhang Q, Ouyang Y, Mou T (2016) Stacking S5-n and f5-n to overcome sterility in indica-japonica hybrid rice. Theor Appl Genet 129(3):563–575. https://doi.org/10.1007/s00122-015-2648-0. Epub 2015 Dec 24

Murai M, Takamure I, Sato S, Tokutome T, Sato Y (2002) Effects of the dwarfing gene originating from 'Dee-geo-woo-gen' on yield and its related traits in rice. Breed Sci 52:95–100

Nagadhara D, Ramesh S, Pasalu IC et al (2004) Transgenic rice plants expressing the snowdrop lectin gene (*gna*) exhibit high-level resistance to the whitebacked planthopper (*Sogatella furcifera*). Theor Appl Genet 109(7):1399–1405. https://doi.org/10.1007/s00122-004-1750-5

Nair S, Bentur JS, Sama VSAK (2011) Mapping gall midge resistance genes: towards durable resistance through gene pyramiding. In: Muralidharan K, Siddiq EA (eds) Genomics and crop improvement: relevance and reservations. Institute of Biotechnology, ANGR Agricultural University, Hyderabad, pp 256–264

Narayanan NN, Baisakh N, Vera Cruz N, Gnananmanickam SS, Datta K, Datta SK (2002) Molecular breeding for the development of blast and bacterial blight resistance in rice cv. IR50. Crop Sci 42:2072–2079. https://doi.org/10.2135/cropsci2002.2072

Nayak P, Basu D, Das S et al (1997) Transgenic elite indica rice plants expressing CryIAc delta-endotoxin of Bacillus thuringiensis are resistant against yellow stem borer (Scirpophaga incertulas). Proc Natl Acad Sci U S A 94(6):2111–2116. https://doi.org/10.1073/pnas.94.6.2111

Norton GJ, Douglas A, Lahner B, Yakubova E, Guerinot ML, Pinson SRM et al (2014) Genome wide association mapping of grain arsenic, copper, molybdenum and zinc in rice (Oryza sativa L.) grown at four international field sites. PLoS One 9:e89685. https://doi.org/10.1371/journal.pone.0089685

Obara M, Tamura W, Ebitani T, Yano M, Sato T, Yamaya T (2010) Fine-mapping of qRL6.1, a major QTL for root length of rice seedlings grown under a wide range of NH4+ concentrations in hydroponic conditions. Theor Appl Genet 121:535–547

Ohsumi A, Takai T, Ida M, Yamamoto T, Arai-Sanoh Y, Yano M, Ando T, Kondo M (2011) Evaluation of yield performance in rice near-isogenic lines with increased spikelet number. Field Crops Res 120:68–75. https://doi.org/10.1016/j.fcr.2010.08.013

Ouyang Y, Chen J, Ding J, Zhang Q (2009) Advances in the understanding of inter-subspecific hybrid sterility and wide-compatibility in rice. Chin Sci Bull 54:2332–2341

Paine JA, Shipton CA, Chaggar S et al (2005) Improving the nutritional value of golden rice through increased provitamin A content. Nat Biotechnol 23:482–487

Park S-I, Kim Y-S, Kim J-J, Mok J-E, Kim Y-H, Park H-M, Kim I-S, Yoon H-S (2017) Improved stress tolerance and productivity in transgenic rice plants constitutively expressing the Oryza sativa glutathione synthetase OsGS under paddy field conditions. J Plant Physiol 215:39–47

Pérez L, Soto E, Farré G, Juanos J, Villorbina G, Bassie L, Medina V, Serrato AJ, Sahrawy M, Rojas JA, Romagosa I, Muñoz P, Zhu C, Christou P (2019) CRISPR/Cas9 mutations in the rice Waxy/GBSSI gene induce allele-specific and zygosity-dependent feedback effects on endosperm starch biosynthesis. Plant Cell Rep 38(3):417–433. https://doi.org/10.1007/s00299-019-02388-z. Epub 2019 Feb 4. PMID: 30715580.

Pinson SRM, Capdevielle FM, Oard JH (2005) Confirming QTLs and finding additional loci conditioning sheath blight resistance in rice using recombinant inbred lines. Crop Sci 45:503–510. https://doi.org/10.2135/cropsci2005.0503

Pradhan S, Chakraborty A, Sikdar N, Chakraborty S, Bhattacharyya J, Mitra J, Manna A, Dutta Gupta S, Sen SK (2016) Marker-free transgenic rice expressing the vegetative insecticidal protein (Vip) of Bacillus thuringiensis shows broad insecticidal properties. Planta 244(4):789–804. https://doi.org/10.1007/s00425-016-2535-1. Epub 2016 May 10

Priyadarshi R, Arremsetty HPS, Singh AK, Khandekar D, Ulaganathan K, Shenoy V (2017) Molecular stacking of wide compatibility gene, S5ⁿ and elongated uppermost internode (eui) gene into IR58025B, an elite maintainer line of rice. J Plant Biochem Biotechnol 26(4):425–435

Ram T, Majumder ND (2007) Introgression of broad-spectrum blast resistance gene(s) into cultivated rice (Oryza sativa ssp. indica) from wild rice O. rufipogon. Curr Sci 92(2):225–230

Rao KV, Rathore KS, Hodges TK, Fu X, Stoger E, Sudhakar D, Williams S, Christou P, Bharathi M, Brown DP, Powell KS, Spence J, Gatehouse AM, Gatehouse JA (1998) Expression of snowdrop lectin (GNA) in transgenic rice plants confers resistance to rice brown planthopper. Plant J 15(4):469–477

Rao SD, Neeraja CN, Madhu Babu P, Nirmala B, Suman K, Rao LVS, Surekha K, Raghu P, Longvah T, Surendra P, Kumar R, Babu VR, Voleti SR (2020) Zinc biofortified rice varieties: challenges, possibilities, and progress in India. Front Nutr 7:26. https://doi.org/10.3389/fnut.2020.00026

Rashid M, Guangyuan H, Guangxiao Y, Hussain J, Xu Y (2012) AP2/ERF transcription factor in rice: genome-wide canvas and syntenic relationships between monocots and eudicots. Evol Bioinformatics Online 8:321–355. https://doi.org/10.4137/EBO.S9369

Reddy O, Siddiq E, Sarma N et al (2000) Genetic analysis of temperature-sensitive male sterility in rice. Theor Appl Genet 100:794–801. https://doi.org/10.1007/s001220051354

Saha P, Majumder P (2006) Transgenic rice expressing *Allium sativum* leaf lectin with enhanced resistance against sap-sucking insect pests. Planta 223:1329–1343

Sahoo RK, Gill SS, Tuteja N (2012) Pea DNA helicase 45 promotes salinity stress tolerance in IR64 rice with improved yield. Plant Signal Behav 7(8):1042–1046

Saijo Y, Hata S, Kyozuka J, Shimamoto K, Izui K (2000) Overexpression of a single Ca^{2+}-dependent protein kinase confers both cold and salt/drought tolerance on rice plants. Plant J 23:319–327

Sasaki A, Ashikari M, Ueguchi-Tanaka M, Itoh H, Nishimura A, Datta S, Ishiyama K, Saito T, Kobayashi M, Khush GS, Kitano H, Matsuoka M (2002) Green revolution: a mutant gibberellin-synthesis gene in rice. Nature 416(6882):701–702. https://doi.org/10.1038/416701a

Sato Y, Sentoku N, Miura Y, Hirochika H, Kitanoc H, Matsuoka M (1999) Loss-of-function mutations in the rice homeobox gene *OSH15* affect the architecture of internodes resulting in dwarf plants. EMBO J 18(4):992–1002. https://doi.org/10.1093/emboj/18.4.992

Sattari M, Kathiresan A, Gregorio G, Hernandez JE, Nas TM, Virmani SS (2007) Development and use of a two-gene marker-aided selection system for fertility restorer genes in rice. Euphytica 153:35–42

Septiningsih EM, Pamplona AM, Sanchez DL, Neeraja CN, Vergara GV, Heuer S, Ismail AM, Mackill DJ (2009) Development of submergence-tolerant rice cultivars: the Sub1 locus and beyond. Ann Bot 103(2):151–160. https://doi.org/10.1093/aob/mcn206

Shim J, Torollo G, Angeles-Shim RB, Cabunagan RC, Choi I-R, Yeo U-S, Ha W-G (2015) Rice tungro spherical virus resistance into photoperiod-insensitive japonica rice by marker-assisted selection. Breed Sci 65:345–351. https://doi.org/10.1270/jsbbs.65.345

Siddiq EA, Ali J (1999) Innovative male sterility systems for exploitation of hybrid vigour in crop plants: 1. Environment sensitive genetic male sterility system. PINSA B65(6):331–350

Siddiq EA, Singh S (2005) Wide-compatibility system for yield enhancement of tropical rice through intersubspecific hybridization. Adv Agron 87:157–209

Singh SP, Sundaram RM, Biradar SK, Ahmed MI, Viraktamath BC, Siddiq EA (2006) Identification of simple sequence repeat markers for utilizing wide-compatibility genes in inter-subspecific hybrids in rice (*Oryza sativa* L.). Theor Appl Genet 113:509–517. https://doi.org/10.1007/s00122-006-0316-0

Singh VK, Singh A, Singh SP, Ellur RK, Choudhary V, Sarkel S, Singh D, Krishnan SG, Nagarajan M, Vinod KK, Singh UD, Rathore R, Prashanthi SK, Agrawal PK, Bhatt JC, Mohapatra T, Prabhu KV, Singh AK (2012) Incorporation of blast resistance into 'PRR78', an elite Basmati rice restorer line, through marker assisted backcross breeding. Field Crops Res 128:8–16

Singla-Pareek SL, Yadav SK, Pareek A, Reddy MK, Sopory SK (2006) Transgenic tobacco over-expressing glyoxalase pathway enzymes grow and set viable seeds in zinc-spiked soils. Plant Physiol 140(2):613–623. https://doi.org/10.1104/pp.105.073734

Song X, Qiu S, Xu C, Li X, Zhang Q (2005) Genetic dissection of embryo sac fertility, pollen fertility, and their contributions to spikelet fertility of intersubspecific hybrids in rice. Theor Appl Genet 110:205–211

Sudhakar T, Panigrahy M, Lakshmanaik M, Babu AP, Reddy CS, Anuradha K, Swamy BPM, Sarla N (2012) Variation and correlation of phenotypic traits contributing to high yield in KMR3: *Oryza rufipogon* introgression lines. Int J Plant Breed Genet 6:69–82

Sun YW, Jiao GA, Liu ZP, Zhang X, Li JY, Guo XP, Du WM, Du JL, Francis F, Zhao YD, Xia LQ (2017) Generation of high-amylose rice through CRISPR/Cas9-mediated targeted mutagenesis of starch branching enzymes. Front Plant Sci 8:298. https://doi.org/10.3389/fpls.2017.00298

Sundaram RM, Vishnupriya MR, Biradar SK, Laha GS, Reddy AG, Rani NS, Sarma NP, Sonti RV (2008) Marker assisted introgression of bacterial blight resistance in Samba Mahsuri, an elite indica rice variety. Euphytica 160(3):411–422

Swamy BPM, Kumar A (2013) Genomics-based precision breeding approaches to improve drought tolerance in rice. Biotechnol Adv 31(8):1308–1318. https://doi.org/10.1016/j.biotechadv.2013.05.004

Swamy BPM, Sarla N (2008) Yield-enhancing quantitative trait loci (QTLs) from wild species. Biotechnol Adv 26(1):106–120

Swamy BPM, Kaladhar K, Ramesha MS, Viraktamath BC, Sarla N (2011) Molecular mapping of QTLs for yield and yield-related traits in *Oryza sativa* cv. Swarna × *O. nivara* (IRGC81848) backcross population. Rice Sci 18(3):178–186

Swamy BPM, Kaladhar K, Shobharani N, Prasad GSV, Viraktamath BC, Reddy GA, Sarla N (2012) QTL analysis for grain quality traits in 2 BC_2F_2 populations derived from crosses between *Oryza sativa* cv. Swarna and 2 accessions of *O. nivara*. J Hered 103(3):442–452

Swamy BPM, Kaladhar K, Reddy GA, Viraktamath BC, Sarla N (2014) Mapping and introgression of QTL for yield and related traits in two backcross populations derived from *Oryza sativa* cv. Swarna and two accessions of *O. nivara*. J Genet 93(3):643–654

Swamy BPM, Descalsota GIL, Nha CT, Amparado A, Inabangan-Asilo MA, Manito C, Tesoro F, Reinke R (2018) Identification of genomic regions associated with agronomic and biofortification traits in DH populations of rice. PLoS One 13(8):e0201756

Ueguchi-Tanaka M, Fujisawa Y, Kobayashi M, Ashikari M, Iwaski Y, Kitano H, Matsuoka M (2000) Rice dwarf mutant *d1*, which is defective in the alpha subunit of the heterotrimeric G protein, affects gibberellin signal transduction. Proc Natl Acad Sci U S A 97(21):11638–11643. https://doi.org/10.1073/pnas.97.21.11638

Uga Y, Sugimoto K, Ogawa S et al (2013) Control of root system architecture by *DEEPER ROOTING 1* increases rice yield under drought conditions. Nat Genet 45:1097–1102

Valarmathi P, Kumar G, Robin S, Manonmani S, Dasgupta I, Rabindran R (2016) Evaluation of virus resistance and agronomic performance of rice cultivar ASD 16 after transfer of transgene against Rice tungro bacilliform virus by backcross breeding. Virus Genes 52(4):521–529. https://doi.org/10.1007/s11262-016-1318-x. Epub 2016 Mar 16

Wang F, Wang C (2016) Enhanced rice blast resistance by CRISPR/Cas9-targeted mutagenesis of the ERF transcription factor gene OsERF922. PLoS One 11(4):1–18

Wang Y, Pinson SRM, Fjellstrom RG, Tabien RE (2012a) Phenotypic gain from introgression of two QTL, *qSB9-2* and *qSB12-1* for rice sheath blight resistance. Mol Breed 30:293–303

Wang P, Xing Y, Li Z et al (2012b) Improving rice yield and quality by QTL pyramiding. Mol Breed 29:903–913. https://doi.org/10.1007/s11032-011-9679-2

Wang L, Lu Q, Wen X, Lu C (2015) Enhanced sucrose loading improves rice yield by increasing grain size. Plant Physiol 169:2848–2862

Wang S, Tholen D, Zhu X-G (2017a) C4 photosynthesis in C3 rice: a theoretical analysis of biochemical and anatomical factors. Plant Cell Environ 40:80–94

Wang Y, Jiang W, Liu H, Zeng Y, Du B, Zhu L, He G, Chen R (2017b) Marker assisted pyramiding of *Bph6* and *Bph9* into elite restorer line 93-11 and development of functional marker for *Bph9*. Rice (N Y) 10(1):51. https://doi.org/10.1186/s12284-017-0194-x

Wei F-J, Droc G, Guiderdoni E, Hsing YC (2013) International consortium of rice mutagenesis: resources and beyond. Rice (N Y) 6(1):39

Xiao J, Grandillo S, Ahn SN, McCouch SR, Tanksley SD, Li J, Yuan L (1996) Genes from wild rice improve yield. Nature 384:223–224

Xu K, Xia X, Fukao T, Canlas P, Maghirang-Rodriguez R, Heuer S et al (2006) *Sub1A* is an ethylene response factor-like gene that confers submergence tolerance to rice. Nature 442:705–708

Xu R, Li H, Qin R, Wang L, Li L, Wei P et al (2014) Gene targeting using the *Agrobacterium tumefaciens* mediated CRISPR-Cas system in rice. Rice 7:5. https://doi.org/10.1186/s12284-014-0005-6

Xu R, Qin R, Li H, Li D, Li L, Wei P et al (2017) Generation of targeted mutant rice using a CRISPR-Cpf1 system. Plant Biotechnol J 15:713–717. https://doi.org/10.1111/pbi.12669

Yadav S, Anuradha G, Kumar RR, Vemireddy LR, Sudhakar R, Donempudi K, Venkata D, Jabeen F, Narasimhan YK, Marathi B, Siddiq EA (2015) Identification of QTLs and possible candidate genes conferring sheath blight resistance in rice (*Oryza sativa* L.). Springerplus 4:175

Yamamuro C, Ihara Y, Wu X, Noguchi T, Fujioka S, Takatsuto S, Ashikari M, Kitano H, Matsuoka M (2000) Loss of function of a rice brassinosteroid insensitive1 homolog prevents internode elongation and bending of the lamina joint. Plant Cell 12(9):1591–1606

Yanagihara S et al (1995) Molecular analysis of the inheritance of the *S-5* locus, conferring wide compatibility in *indica/japonica* hybrids of rice (*Oryza sativa* L.). Theor Appl Genet 90:182–188

Ye X, Al-Babili S, Klöti A, Zhang J, Lucca P, Beyer P, Potrykus I (2000) Engineering the provitamin A (β-carotene) biosynthetic pathway into (carotenoid free) rice endosperm. Science 287:303–305

Yi M, Nwe KT, Vanavichit A, Chaiarree W, Toojinda T (2009) Marker assisted backcross breeding to improve cooking quality traits in Myanmar rice cultivar Manawthukha. Field Crops Res 113:178–186

Yin X, Biswal AK, Dionora J, Perdigon KM, Balahadia CP, Mazumdar S, Chater C, Lin HC, Coe RA, Kretzschmar T, Gray JE, Quick PW, Bandyopadhyay A. (2017) CRISPR-Cas9 and CRISPR-Cpf1 mediated targeting of a stomatal developmental gene EPFL9 in rice. Plant Cell Rep. 36(5):745–757. https://doi.org/10.1007/s00299-017-2118-z.

Yuan LP, Fu XQ (1995) Technology of hybrid rice production. FAO, Rome

Zhang J, Li X, Jiang G, Xu Y, He Y (2006) Pyramiding of *Xa7* and *Xa21* for the improvement of disease resistance to bacterial blight in hybrid rice. Plant Breed 125:600–605. https://doi.org/10.1111/j.1439-0523.2006.01281.x

Zhang F, Jiang Y-Z, Yu S-B, Ali J, Paterson AH, Khush GS, Xu J-L, Gao Y-M, Fu B-Y, Lafitte R, Li Z-K (2013) Three genetic systems controlling growth, development and productivity of rice (*Oryza sativa* L.): a reevaluation of the 'Green Revolution'. Theor Appl Genet 126:1011–1024

Zhou H, He M, Li J, Chen L, Huang Z, Zheng S, Zhu L, Ni E, Jiang D, Zhao B, Zhuang C (2016) Development of commercial thermo-sensitive genic male sterile rice accelerates hybrid rice breeding using the CRISPR/Cas9-mediated *TMS5* editing system. Sci Rep 6:37395. https://doi.org/10.1038/srep37395

Zuo SM, Yin YJ, Zhang L, Zhang YF, Chen ZX, Pan XB (2007) Breeding value and further mapping of a QTL *qSB-11* conferring the rice sheath blight resistance. Chin J Rice Sci 21(2):136–142

Zuo M, Zhu YJ, Yin YJ, Wang H, Zhang YF, Chen ZX, Gu SL, Pan XB (2014) Comparison and confirmation of quantitative trait loci conferring partial resistance to rice sheath blight on chromosome 9. Plant Dis 98(7):957–964

Strategies for Engineering Photosynthesis for Enhanced Plant Biomass Production

Wataru Yamori

Abstract Crop productivity would have to increase by 60–110% compared with the 2005 level by 2050 to meet both the food and energy demands of the growing population. Although more than 90% of crop biomass is derived from photosynthetic products, photosynthetic improvements have not yet been addressed by breeding. Thus, it has been considered that enhancing photosynthetic capacity is considered a promising approach for increasing crop yield. Now, we need to identify the specific targets that would improve leaf photosynthesis to realize a new Green Revolution. This chapter summarizes the various genetic engineering approaches that can be used to enhance photosynthetic capacity and crop productivity. The targets considered for the possible candidates include Rubisco, Rubisco activase, enzymes of the Calvin–Benson cycle, and CO_2 transport, as well as photosynthetic electron transport. Finally, it describes the importance of considering ways to improve photosynthesis not under the stable environmental conditions already examined in many studies with the aim of improving photosynthetic capacity, but under natural conditions in which various environmental factors, and especially irradiation, continually fluctuate.

Keywords Calvin–Benson cycle · CO_2 assimilation · CO_2 transport · Electron transport · Photosynthesis · Rubisco

1 Introduction

Crop productivity would have to increase by 60–110% compared with the 2005 level by 2050 to meet both the food and energy demands of the growing population (Tilman et al. 2011; Alexandratos and Bruinsma 2012). At the same time, the CO_2 concentration in the atmosphere is increasing and is predicted to reach 550 μmol/mol by 2050 (IPCC 2013; Ballantyne et al. 2012), which will lead to an increase in air temperature. Thus, it is considered that approaches designed to improve plant

W. Yamori (✉)

Institute for Sustainable Agro-Ecosystem Services, Graduate School of Agricultural and Life Sciences, The University of Tokyo, Tokyo, Japan

e-mail: yamori@g.ecc.u-tokyo.ac.jp

© The Author(s) 2021

J. Ali, S. H. Wani (eds.), *Rice Improvement*,

https://doi.org/10.1007/978-3-030-66530-2_2

biomass and crop yield should take account of global climate change and the predicted future environmental conditions.

It has been reported that in most cases leaf photosynthetic rate does not correlate positively with grain yield (Richards 2000). Some critical reviews suggest that improving photosynthesis would not be a useful strategy for enhancing crop productivity (Gu et al. 2014; Sinclair et al. 2004). However, a meta-analysis of several studies on elevated CO_2 experiments in various crops has indicated that any strategy for increasing photosynthesis can enhance crop yield (Ainsworth et al. 2008). Similarly, it has been proposed that altering photosynthetic electron transport rates by manipulating the cytochrome b_6/f complex can improve both the photosynthetic capacity and crop yield of transgenic plants (Yamori et al. 2016a; Fig. 1). Enhancing photosynthetic capacity in plants is now considered a promising approach for increasing crop yield and decreasing the atmospheric concentration of CO_2, which is the primary component of greenhouse gases.

This chapter summarizes the various genetic engineering approaches that can be used to enhance photosynthetic capacity and plant production. The targets considered for the possible candidates include Rubisco, Rubisco activase, enzymes of the Calvin–Benson cycle, and CO_2 transport, as well as photosynthetic electron

Fig. 1 Relationship between CO_2 assimilation rate at a CO_2 concentration of 390 μmol/mol, total plant dry weight at the final stage, and grain yield in rice. Wild type: open triangles; transgenic plants that contain variable amounts of Rieske FeS protein in the cytochrome b_6/f complex from 10 to 100% of wild-type levels: filled circles. The regression lines are shown

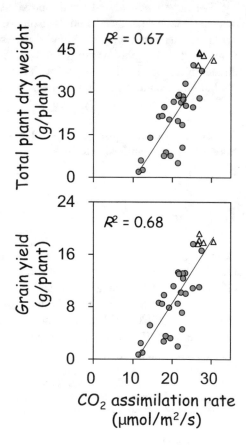

transport. Finally, it describes the importance of considering ways to improve photosynthesis not under the stable environmental conditions already examined in many studies with the aim of improving photosynthetic capacity, but under natural conditions in which various environmental factors, and especially irradiation, continually fluctuate.

2 Improving Rubisco Performance

2.1 Rubisco Kinetics

Rubisco (ribulose-1,5-bisphosphate carboxylase/oxygenase) is an enzyme involved in the first step of CO_2 fixation in photosynthesis (Fig. 2). Rubisco has a low catalytic efficiency and can only fix approximately two to four CO_2 molecules per second per active site in higher C_3 plants. Thus, 20–30% of the nitrogen in the leaves of C_3 plants is invested in Rubisco to compensate for its low activity (Spreitzer and Salvucci 2002). There is a strong positive correlation between leaf Rubisco content and photosynthetic rate (Evans 1989; Makino et al. 1997; Wright et al. 2004), indicating that Rubisco would be rate-limiting as regards photosynthesis at the current CO_2 concentration. Rubisco can fix CO_2 in photosynthesis and O_2 in photorespiration (Fig. 2). Photosynthetic CO_2 fixation produces two molecules of phosphoglycerate (PGA) for every carbon fixed, while photorespiration produces one PGA and one phosphoglycolate (PGO). PGO must be recycled to PGA, with a loss of CO_2 and NH_3 via a photorespiratory pathway. Although the released CO_2 may be re-fixed by the chloroplasts and the NH_3 re-assimilated in the leaves (Morris et al. 1988; Busch et al. 2013), photorespiration is considered to be a wasteful reaction. Thus, it may be possible to improve photosynthetic efficiency by modifying Rubisco in plants to increase catalytic activity and/or decrease oxygenation rate.

In plants, Rubisco usually consists of two types of protein subunit: a chloroplast-encoded large subunit, which contains the active site, and nuclear-encoded small subunits. The introduction of Rubisco variants with high specificity values such as that from C_4 plants and cyanobacteria into plants could improve the photosynthetic efficiency of crop plants. Previously, transgenic tobacco plants expressing *Flaveria bidentis* (C_4) and *F. pringlei* (C_3) Rubisco large subunit chimeras revealed that the substitution of methionine-309 with isoleucine is responsible for increases in the carboxylation rate of Rubisco (Whitney et al. 2011). However, the CO_2 assimilation rate and plant growth were lower in transgenic plants than in wild-type plants since transformants decreased the Rubisco content of the former compared with the latter. Lin et al. (2014) successfully produced transgenic tobacco plants with functional Rubisco by replacing the Rubisco with the large and small subunit genes found in cyanobacterium. The transgenic plants increased the CO_2 assimilation rate per Rubisco content, but they grew more slowly than wild-type plants. Thus, although mutated forms of Rubisco protein have been achieved in tobacco plants, the

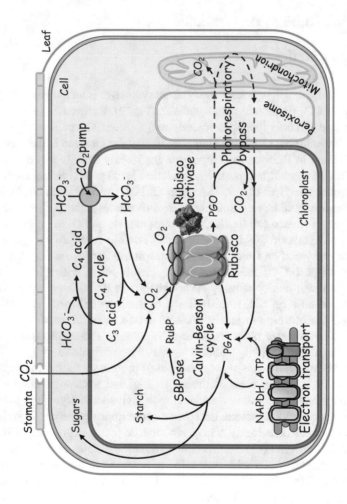

Fig. 2 Strategies for engineering photosynthesis for enhanced plant biomass production. The targets considered for the possible candidates include Rubisco catalysis, Rubisco activase, photorespiratory bypass, stomatal opening, introduction of C_4 cycle, introduction of CO_2 pump from cyanobacteria, Calvin–Benson cycle, and electron transport

site-directed mutagenesis of Rubisco has as yet been largely unsuccessful (Furbank et al. 2015). If the replacement of the Rubisco variants of C_3-type Rubisco (i.e., a low catalytic turnover rate for Rubisco, *kcat*, and a low Michaelis–Menten constant, *Km*; a high *Km* for CO_2 indicates low CO_2 affinity) with C_4-type or cyanobacteria-type Rubisco variants (i.e., high *kcat* and high *Km*) is successful, the transgenic C_3 plants could enhance their photosynthetic efficiency and plant growth toward the high-CO_2 world of the near future.

Although the evidence from transplastomic studies of Rubisco indicates that the catalytic variability resides within its large subunit, the importance of its small sub-units to Rubisco catalysis has also attracted attention. Recent success has demon-strated that the introduction of a C_4-Rubisco small subunit (*rbcS*) gene from sorghum into rice successfully produced chimeric Rubisco with a greater *kcat* in transgenic rice (Ishikawa et al. 2011). This breakthrough could provide future ways to engineer Rubisco in various important crops such as wheat and rice.

2.2   Photorespiration Bypass

Rubisco is a dual-function enzyme that fixes CO_2 or O_2, and these functions are known as photosynthesis and photorespiration, respectively. While photosynthesis results in a net fixation of CO_2, the photorespiratory pathway requires ATP and releases previously fixed CO_2 (Fig. 3). The photorespiration rate is affected by the concentration of CO_2 in the chloroplast (C_c) relative to the O_2 concentration, and increases with increasing temperature. At current atmospheric CO_2 concentrations and a temperature of 30 °C, the rate of photorespiratory CO_2 release from the mito-chondria is approximately 25% of the CO_2 assimilation rate (Sage et al. 2012). Thus, lowering photorespiratory flux could alleviate the decrease in photosynthetic efficiency in C_3 plants. However, manipulations aimed at blocking the photorespira-tory pathway had detrimental effects on plant growth (Kozaki and Takeba 1996; Walker et al. 2016). Nonetheless, advances have been made for engineering plants that can make better use of the CO_2 released from photorespiration via photorespira-tory bypasses (Carvalho et al. 2012; Kebeish et al. 2007; Maier et al. 2012; Peterhansel et al. 2013).

To date, three different strategies have been designed to bypass photorespiration in C_3 plants (Fig. 3). The first pathway was engineered using *Escherichia coli* encoded genes from the glycerate pathway that convert glycolate to glycerate and release CO_2 within the chloroplast (Kebeish et al. 2007; Peterhansel et al. 2013). Transgenic plants engineered with this pathway decreased photorespiration and enhanced photosynthesis, resulting in improved plant growth (Kebeish et al. 2007). With the second approach, transgenic plants engineered with a glycolate catabolic cycle designed to oxidize glycolate to CO_2 in chloroplasts (Fig. 3) displayed higher photosynthetic rates and greater plant growth (Maier et al. 2012; Peterhansel et al. 2013). These observations show that shifting glycolate metabolism from the photo-respiratory pathway via peroxisome and mitochondria to the chloroplast is

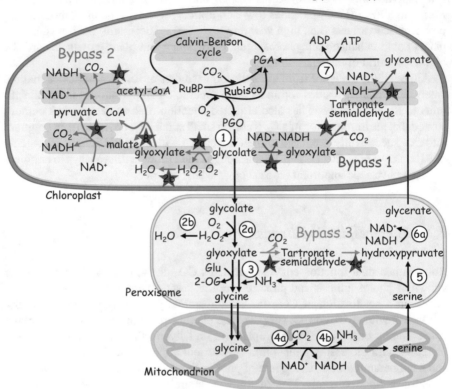

Fig. 3 Schematic diagram of photorespiration in plants (black), with three bypasses to minimize photorespiratory expenses engineered in plants (red, blue, green). Enzymatic reactions or metabolite transport steps are indicated by arrows. (1) phosphogycolate phosphatase, (2a) glycolate oxidase, (2b) catalase, (3) glyoxylate/glutamate aminotransferase, (4a) glycine decarboxylase, (4b) serine hydroxymethyl transferase, (4c) glyoxylate carboligase, which catalyzes the decarboxylation of glyoxylate and ligation to a second molecule of glyoxylate to form tartronate semialdehyde, (4d) hydroxypyruvate isomerase, (5) serine/glyoxylate aminotransferase, (6a) hydroxypyruvate reductase in photorespiration, (6b) tartronic semialdehyde reductase, (7) glycerate kinase, (8) malate synthase, (9) NADP-malic enzyme, (10) pyruvate dehydrogenase. *Rubisco* ribulose-1,5-bisphosphate carboxylase/oxygenase; *RuBP* ribulose-1,5-bisphosphate; *PGA* phosphoglycerate; *PGO* phosphoglycolate

beneficial for plants and can enhance photosynthesis. The third bypass was created by short-circuiting the original C_2 cycle to avoid NH_3^+ release and to prevent energy loss in its refixation (Carvalho et al. 2012). The glyoxalase in peroxisomes can be converted to hydroxypyruvate by introducing glyoxylate carboligase and hydroxypyruvate isomerase from *E. coli* into the plant peroxisomes, and feeding them back to the C_2 cycle (Fig. 3) (Carvalho et al. 2012; Peterhansel et al. 2013). However, in transgenic plants, the photorespiratory cycle has not yet been completely bypassed and the short-circuiting led to damage of the photosynthetic apparatus and thus deleterious phenotypes (Carvalho et al. 2012).

Facilitating photorespiratory flux through the overexpression of subunits of glycine decarboxylase (GDC), which produces CO_2 by the photorespiratory process, could be another approach for improving photorespiration (Timm et al. 2016). GDC comprises four proteins, three enzymes (P-protein, T-protein, and L-protein), and a small lipoylated protein known as H-protein, which has no catalytic activity and interacts with the other proteins. The overexpression of either GDC-H protein or GDC-L protein in *Arabidopsis thaliana* resulted in increases in CO_2 assimilation rate and plant biomass (Timm et al. 2012, 2015, 2016). Additionally, the overexpression of GDC-H contributed to greater plant growth in tobacco (*Nicotiana tabacum*) in both a controlled environment and under field conditions (Lopez-Calcagno et al. 2018). Although the underlying mechanism responsible for these effects has not been fully elucidated, it has been proposed that the Calvin–Benson cycle is stimulated by the increase in GDC activity, resulting in a decrease in the steady-state levels of photorespiratory metabolites.

3   Improving Thermotolerance of Rubisco Activase

The Rubisco catalytic sites must be activated to fix CO_2 (Fig. 2). This requires the carbamylation of a lysine residue at the Rubisco catalytic site, allowing the binding of Mg^{2+} and ribulose-1,5-bisphosphate (RuBP). Rubisco activase facilitates carbamylation and the maintenance of Rubisco activity by removing inhibitors such as tight-binding sugar phosphates from the Rubisco catalytic sites in an ATP-dependent manner (Spreitzer and Salvucci 2002; Portis Jr 2003; Parry et al. 2008).

In many plant species, the Rubisco activation state decreases at high temperatures (Crafts-Brandner and Salvucci 2000; Salvucci and Crafts-Brandner 2004a; Yamori et al. 2006b, 2014; Yamori and von Caemmerer 2009). Rubisco deactivation at high temperature could have occurred because Rubisco activase is insufficiently active to keep pace with the faster rates of Rubisco inactivation at high temperature due to its thermolability (Salvucci and Crafts-Brandner 2004b). A decrease in Rubisco activase content resulted in decreases in photosynthetic rate at high temperature when using mutants/transgenic plants in *Arabidopsis* (Salvucci et al. 2006), rice (Yamori et al. 2012), and tobacco (Yamori and von Caemmerer 2009). Also, the overexpression of Rubisco activase from maize into rice stimulated the Rubisco activation state and photosynthetic rate at high temperature (Yamori et al. 2012). Moreover, transgenic *Arabidopsis* expressing thermotolerant Rubisco activase isoforms generated by either gene shuffling technology (Kurek et al. 2007) or chimeric Rubisco activase constructs (Kumar et al. 2009) improved photosynthesis, biomass production, and seed yield. In addition, the introduction of Rubisco activase from cotton into a cool-season species such as *Camelina* resulted in improvement in the thermotolerance of photosynthesis (Carmo-Silva and Salvucci 2012). This is also supported by a recent report stating that genes encoding thermostable Rubisco activase from a wild relative (*Oryza australiensis*) were overexpressed in domesticated rice (*O. sativa*), leading to an improvement in plant growth and seed yield in rice under heat stress (Scafaro et al. 2018). Taken together, Rubisco activase activity would constitute a major limiting factor for photosynthesis under high temperature and engineering Rubisco activase would be an efficient way to improve crop yield under high temperatures. The structure of Rubisco activase has already been determined, providing insight into its interactions with Rubisco (Stotz et al. 2011) and its counterpart CbbX in red algae (Mueller-Cajar et al. 2011). This structural information coupled with the knowledge of regulation in Rubisco activase will help to improve its thermostability and catalytic properties.

4 Increasing CO_2 Concentration Around Rubisco

Photosynthesis in C_3 plants is limited by the large drawdown in CO_2 concentrations from the atmosphere to the Rubisco catalytic sites in chloroplasts. The CO_2 diffusion conductance responsible for this drawdown is attributed to the stomatal pores and the paths across the mesophyll from the cell surface to the Rubisco catalytic sites in chloroplasts (Evans et al. 2009). Increasing CO_2 concentration in chloroplasts and thereby minimizing photorespiration is therefore a promising target in terms of increasing photosynthetic rate in crops. CO_2 diffusion to the chloroplast can be influenced by modifying conductance through the stomata (stomatal conductance) to the intercellular air space, either by increasing stomatal density (Tanaka et al. 2013) or by preventing stomatal closure (Kusumi et al. 2012; Yamori et al. 2020). Both approaches would result in increases in photosynthetic rate at the cost of higher transpiration rates and lower water-use efficiency.

An alternative approach addresses the other major diffusion conductance route for CO_2 from the intercellular air space into the mesophyll cell chloroplasts (mesophyll conductance). In contrast to modifying stomatal conductance, increasing mesophyll conductance does not negatively affect water-use efficiency. The resistance of the cell wall (25–50%) and chloroplast (24–76%) accounts for most of the total resistance (Evans et al. 2009), meaning that CO_2 diffusion can potentially be improved by modifying plants so that they have smaller mesophyll cells (i.e., a higher surface area of the chloroplasts is exposed to intercellular air spaces, Sc) with thinner cell walls (Terashima et al. 2011). The second important component of mesophyll conductance involves CO_2 diffusion through the plasma and chloroplast membranes (Evans et al. 2009), and several approaches are being developed to increase CO_2 concentration in chloroplasts in C_3 plants by increasing membrane permeability for CO_2. Aquaporins that are permeable to CO_2 are proteins that assist CO_2 diffusion through the membranes by providing pores through which CO_2 can be channeled (Kaldenhoff 2012). It has been shown that disruption to the aquaporin *AtPIP1;2* gene limits CO_2 transport across the membrane (Heckwolf et al. 2011), while the overexpression of different aquaporin genes results in increased g_m (Hanba et al. 2004; Flexas et al. 2006). Furthermore, it has been shown that the expression of an aquaporin in *A. thaliana* stimulates CO_2 flux through a mesophyll membrane (Uehlein et al. 2012).

Once CO_2 is transferred to the cytosol, it is partially converted into HCO_3^- to facilitate its diffusion into the chloroplast, and the HCO_3^- is then dehydrated back to CO_2 by carbonic anhydrase to maintain a high CO_2 flux through the chloroplast membrane. Thus, carbonic anhydrase plays a role in facilitating the diffusion of CO_2 in the chloroplast stroma by interconverting between CO_2 and HCO_3^- (Evans et al. 2009). It has been suggested that the amount of carbonic anhydrase found in plants somewhat limits conductance in the stroma of C_3 crops, and thus there would be a possibility to improve this aspect by molecular engineering (Tholen and Zhu 2011).

A substantial increase in the CO_2 concentration around Rubisco to enhance photosynthesis and water-use efficiency has been expected as the result of the installation of a carbon concentrating mechanism (CCM) in C_3 plants (Fig. 4). Cyanobacteria have evolved a CCM in which Rubisco is encapsulated in a cellular compartment known as a carboxysome (Price et al. 2011). In carboxysomes, CO_2 concentration is enriched by up to 1000-fold, thus significantly decreasing the photorespiration rate.

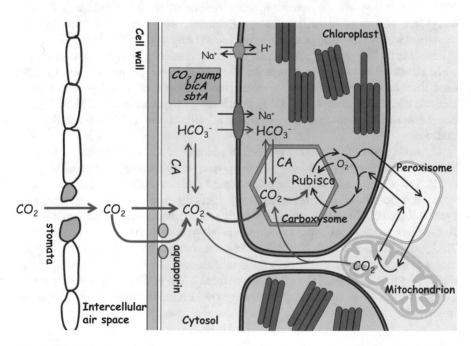

Fig. 4 Schematic diagram of mechanisms for concentrating CO_2 around Rubisco. The diagram shows CO_2 transfers from the outside to the intercellular air space through the stomatal pore and the CO_2 diffuses through the cell wall and plasma membrane into the cytosol. Aquaporins assist the CO_2 diffusion into the cytosol of the mesophyll cell through the membranes by providing pores through which CO_2 can be channeled. Introducing a cyanobacterial HCO_3^- transporter (e.g., *bicA* and *sbtA*) into the chloroplast envelope could improve CO_2 transport. The introduction of a Rubisco- and carbonic-anhydrase-containing compartment, such as the carboxysome, could further increase the CO_2 concentration around Rubisco, resulting in minimization of the photorespiration rate. (The figure is adapted from Price et al. (2011) and Yamori et al. (2016b))

To incorporate CCM from cyanobacteria into C_3 plants, the following distinct features need to be addressed: (1) CO_2 and HCO_3^- transport mechanism and (2) functional carboxysome assembly. Incorporating cyanobacterial HCO_3^- transporters into the chloroplast envelope of C_3 plants would provide a parallel route for inorganic carbon to enter the chloroplast, in addition to the diffusion of dissolved CO_2 (Price et al. 2011, 2013). To date, five different inorganic carbon transport mechanisms have been identified in cyanobacteria (Price et al. 2011, 2013). A previous study showed that overexpressing the *ictB* gene, an HCO_3^- transporter in cyanobacteria, in *A. thaliana* and *N. tabacum* plants contributed more to increases in photosynthesis and water-use efficiency than in the wild type (Lieman-Hurwitz et al. 2003). Furthermore, the overexpression of the *ictB* gene in soybeans led to increases in mesophyll conductance, photosynthesis, and plant productivity in both ambient and elevated CO_2 environments under both greenhouse and field conditions (Hay et al. 2017). It is now considered that a fully functional CCM in C_3 plants would require the introduction of HCO_3^- transporters, adjustments in the expression of chloroplast carbonic anhydrase to allow HCO_3^- accumulation, and the establishment of a Rubisco- and carbonic-anhydrase-containing compartment, such as a carboxysome (Price et al. 2011, 2013). Recently, well-assembled carboxysome structures were successfully expressed in plants (Long et al. 2018). Incorporation of cyanobacterial Rubisco large and small subunit genes along with genes for carboxysome structural proteins could improve Rubisco catalytic properties, but decrease total Rubisco content, resulting in lower photosynthetic rates and growth than in tobacco wild-type (Long et al. 2018). Since the incorporation of CCM into crops has been expected to improve crop yields, efforts toward transplantation are under way.

C_4 plants evolved CCM in two types of photosynthetic cells, where CO_2 is initially fixed in the mesophyll cells by the enzyme phosphoenolpyruvate carboxylase (PEPC) to produce a C_4 acid. The organic acid diffuses to the bundle-sheath cells, where it is decarboxylated, resulting in significantly increased CO_2 concentrations around Rubisco. Currently, considerable efforts are under way to incorporate the features of the complex C_4 pathway into C_3 crops such as rice (Covshoff and Hibberd 2012; von Caemmerer et al. 2012). Challenges associated with this approach include morphological adjustments, such as the establishment of a Kranz(-like) anatomy, as well as the introduction of C_4 biochemistry into C_3 leaves. The benefits of the introduction of the C_4 photosynthetic pathway would include higher yield as well as improved nitrogen-use efficiency and water-use efficiency.

5 Enhancing Activity of Calvin–Benson-Cycle Enzymes

The Calvin–Benson cycle uses ATP and NADPH from photosynthetic electron transport to fix CO_2 in carbon skeletons that are mainly used for sucrose and starch production (Fig. 2). The Calvin–Benson cycle also supplies intermediates to many other pathways in the chloroplast, including the shikimate pathway for the biosynthesis of amino acids, lignin, isoprenoid, and precursors for nucleotide metabolism and cell wall synthesis. This cycle comprises 11 different enzymes, catalyzes 13 reactions, and is initiated by Rubisco (Raines 2003). Four of the 11 enzymes are regulated by thioredoxins: glyceraldehyde 3-phosphate dehydrogenase (GAPDH), fructose 1,6-bisphosphatase (FBPase), sedoheptulose-1,7-bisphospha-tase (SBPase), and phosphoribulokinase (PRK). Two of the 11 enzymes catalyze reversible reactions: aldolase and transketolase.

Previous studies have demonstrated that moderate reductions in Calvin–Benson-cycle enzymes such as SBPase and fructose 1,6-bisphosphate aldolase (FBPA) induce significant decreases in photosynthetic rate and plant growth, indicating that these enzymes would limit photosynthesis (Ding et al. 2016; Haake et al. 1998, 1999; Harrison et al. 1998, 2001; Lawson et al. 2006; Ölcer et al. 2001; Raines 2003; Raines and Paul 2006; Raines et al. 1999; Hatano-Iwasaki and Ogawa 2012). Furthermore, the disruption of the chloroplastic FBPase was also shown to nega-tively affect photosynthetic rate (Kossmann et al. 1994; Rojas-González et al. 2015; Sahrawy et al. 2004). These results strongly suggest that photosynthetic CO_2 fixa-tion could be improved by increasing the activity of individual Calvin–Benson-cycle enzymes. Evidence supporting this hypothesis was provided by transgenic tobacco plants overexpressing SBPase (Lefebvre et al. 2005; Tamoi et al. 2006), FBPase (Tamoi et al. 2006), the cyanobacterial bifunctional SBPase/FBPase (Miyagawa et al. 2001), or FBPA (Uematsu et al. 2012). These single manipulations resulted in increases in photosynthetic rate and plant growth. Recently, SBPase has been receiving a lot of attention, and its role in determining carbon flux in the Calvin–Benson cycle under natural environmental conditions has been revealed. Transgenic tobacco plants overexpressing SBPase from *A. thaliana* exhibited an enhanced photosynthetic rate and biomass production when grown under free-air CO_2 enrichment (FACE) conditions at a CO_2 concentration of 585 μmol/mol (Rosenthal et al. 2011). Moreover, the expression of cyanobacterial bifunctional FBPase/SBPase increases photosynthetic rate in soybeans grown under field condi-tions and prevents yield losses under high-CO_2 and high-temperature conditions (Köhler et al. 2016). In addition, transgenic lines with increased SBPase exhibited improvement of leaf photosynthesis, total biomass, and seed yield in wheat under greenhouse conditions (Driever et al. 2017). Taken together, the manipulation of SBPase could increase photosynthetic capacity and could be an efficient way to improve photosynthetic rate and crop yield, especially in a future high-CO_2 world.

6 Enhancing Electron Transport Rate in Thylakoid Membranes

ATP and NADPH generated during photosynthetic electron transport in thylakoid membranes are used to power photosynthetic carbon reduction. In a future high-CO_2 world, CO_2 assimilation rate would be limited by the RuBP regeneration rate in the Calvin–Benson cycle (Farquhar et al. 1980), which in turn will be limited by chloroplast electron transport capacity (Yamori et al. 2011). The cytochrome b_6/f complex has a unique role in chloroplast electron transport (Fig. 5) as it can act in both linear electron transport (production of ATP and NADPH) and cyclic electron transport (ATP generation only). There is a strong linear relationship between chloroplast electron transport rate and cytochrome b_6/f complex content at any leaf temperature (Yamori et al. 2011). Thus, this could be a suitable target for genetic manipulation to improve photosynthesis and thus plant yield.

Previous experiments with antisense lines have shown that even a moderate decrease in the amounts of chloroplastic ferredoxin NADP(H) oxidoreductase (FNR), which catalyzes the terminal reaction of the photosynthetic electron transport chain by transferring electrons from reduced ferredoxin to NADP$^+$, has a negative impact on photosynthetic rate under both low and high light conditions (Hajirezaei et al. 2002). However, the overexpression of FNR (Rodriguez et al.

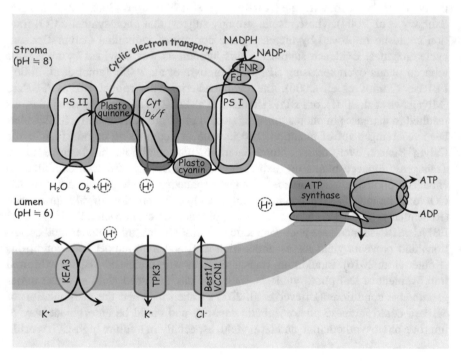

Fig. 5 Schematic diagram of electron transport in thylakoid membranes. Electron transport, driven by the excitation of photosystem I (PS I) and photosystem II (PS II), results in the reduction of NADP$^+$ to NADPH and the accumulation of protons in the thylakoid lumen. The resulting proton motive force (*pmf*), which constitutes ΔpH across the thylakoid membrane as well as membrane potential ($\Delta\psi$), is used to produce ATP through ATP synthase. Linear electron transport generates both ATP and NADPH, whereas cyclic electron transport produces ATP without producing NADPH. Several ion channels, such as the thylakoid K$^+$ channel TPK3, K$^+$ efflux antiporters KEA, and Cl$^-$ channel Best1/VCCN1, would adjust $\Delta\psi$ and ΔpH and function to fine-tune *pmf* and thus electron transport via pH-dependent NPQ. PS II: photosystem II, Cyt b_6f, cytochrome b_6/f complex, PS I: photosystem I, Fd: ferredoxin. FNR: ferredoxin-NADP$^+$ reductase

2007) or ferredoxin (Yamamoto et al. 2006) did not increase photosynthesis or plant growth in tobacco, irrespective of growth light conditions. Electron transfer between the cytochrome b_6/f complex and photosystem I is mediated by plastocyanin in higher plants, whereas, in many algae, it is mediated by cytochrome $c6$. Variations in plastocyanin levels have been reported to coincide with variations in photosynthetic electron transport activity (Burkey 1994; Burkey et al. 1996; Schöttler et al. 2004), leading to the conclusion that plastocyanin pool size could limit photosynthetic electron transport. It has been reported that the introduction of a parallel electron carrier between the cytochrome b_6/f complex and photosystem I through the expression of an algal cytochrome $c6$ gene in A. *thaliana* improved electron transport rate, leading to improved plant growth (Chida et al. 2007). An analysis of knockout plants for two homologous plastocyanin isoforms (*PETE1* and *PETE2*) *in* A. *thaliana* showed that plastocyanin content can be significantly decreased with no apparent changes in photosynthetic rate, suggesting that the concentration of plastocyanin does not limit photosynthetic electron transport rate (Pesaresi et al. 2009). However, the overexpression of either *PETE1* or *PETE2* results in an increase in biomass production (Pesaresi et al. 2009). Thus, there is still a discrepancy between the experimental knockout data and the overexpression lines.

It was also shown in antisense studies that decreasing Rieske FeS protein content resulted in a decrease in cytochrome b_6/f complex level, leading to a decrease in photosynthetic electron transport, plant biomass, and seed yield in tobacco and rice (Price et al. 1998; Yamori et al. 2016a). These findings identified the cytochrome b_6/f complex as a limiting step in electron transport and would suggest that the overexpression of Rieske FeS protein could be a suitable target for increasing photosynthesis and yield. This has been proven by recent work showing that the overexpression of Rieske FeS protein had a substantial and significant impact on electron transport, plant biomass, and seed yield in *Arabidopsis* plants (Simkin et al. 2017).

Other reports have documented an enhancement in plant biomass realized by the genetic manipulation of photosynthetic electron transport. In plant cells, NADP is mainly located in the chloroplast, where $NADP^+$ functions as the final electron acceptor of the photosynthetic electron transport chain (Wigge et al. 1993). NAD kinase regulates the NAD(H)/NADP(H) balance through its catalysis of NAD phosphorylation in the presence of ATP (Kawai and Murata 2008). In A. *thaliana*, one of the NADK isoforms localized in the chloroplast (NADK2; Chai et al. 2005) catalyzes a key step in the regulation of NAD/NADP ratio (Kawai and Murata 2008). The overexpression of chloroplastic *NADK2* from *Arabidopsis* plants into rice succeeded in enhancing electron transport and CO_2 assimilation rates (Takahara et al. 2010).

In situations in which the electron transport rate is limited by the amount of available light that can be absorbed by the plant, increased light harvesting might enhance photosynthetic rate and plant productivity. Land plants use chlorophyll *a* and *b*, which absorb light at wavelengths of 400–700 nm. Chlorophyll *d*, which is used by *Acaryochloris* (Miyashita et al. 1996), and chlorophyll *f*, which was discovered in the cyanobacterial communities of stromatolites (Chen et al. 2010), have red-shifted absorption spectra that enable their host organisms to perform oxygenic photosynthesis at the much longer wavelengths of 700–750 nm, which are inaccessible to other organisms. Introducing these chlorophylls into higher plants to supplement or replace the existing chlorophylls could potentially increase the amount of usable photon flux by up to 19% (Chen and Blankenship 2011). The up-regulation of *Arabidopsis* chlorophyllide *a* oxygenase (CAO), involved in chlorophyll *b* biosynthesis, in tobacco has been shown to increase electron transport rate, CO_2 assimilation, and plant biomass (Biswal et al. 2012). In addition, plants with a mutation in TAP38, an enzyme involved in the dephosphorylation of the light harvesting complex of photosystem II, exhibited an increased photosynthetic electron flow, leading to improved plant growth under low-light conditions (Pribil et al. 2010). In the same manner, facilitation of the chloroplast accumulation response, which shows that chloroplasts accumulate along periclinal cell walls at low light, led to improved leaf photosynthesis and plant biomass production in *A. thaliana* (Gotoh et al. 2018). Since the photosynthetic electron transport chain provides energy and reducing equivalents for the reduction of fixed CO_2 to carbohydrates in the Calvin–Benson cycle as well as for nitrogen assimilation and other processes, the genetic manipulation of photosynthetic electron transport could be a candidate for improving the entire photosynthetic system, and thus plant yield.

7 Improving Photosynthetic Performance Under Fluctuating Light in Natural Environments

Research into finding ways to increase crop yield has focused on improving steady-state photosynthesis. However, leaves in natural plant canopies experience a highly variable light environment over the course of a day because of changes in cloud cover and overshadowing canopy cover (Fig. 6; Yamori 2016). By contrast, transgenic plants have not yet been used to clarify the limiting step of non-steady-state photosynthesis, and thus few studies address the improvement of non-steady-state photosynthesis. When light intensity is increased suddenly after a prolonged period of low light or darkness, photosynthetic rate increases gradually over several minutes and approaches a steady state (Pearcy 1990; Yamori 2016). This phenomenon has been termed "photosynthetic induction," and it is typically divided into three limiting phases: (1) electron transport systems; (2) activation of Calvin-cycle enzymes, especially Rubisco; and (3) CO_2 diffusion into the chloroplast (Fig. 6). The first of these three phases is often completed within 1–2 min of increases in irradiance, the second requires 5–10 min, and the third could take 10–30 min to reach a steady-state (Pearcy 1990). The slow induction results in a time lag between the changes in irradiance and those in the photosynthetic rate. This delay may cause damage to the photosynthetic apparatus and eventually decrease plant productivity if excess energy accumulates during repeated fluctuations in light intensity (Murchie and Niyogi 2011; Tikkanen et al. 2012; Yamori 2016; Yamori et al. 2016c). Daily photosynthetic rates under fluctuating light conditions can be up to 20–35% lower

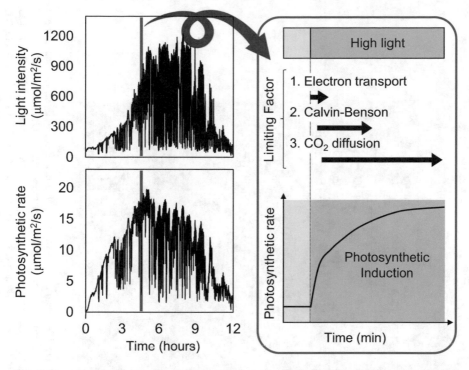

Fig. 6 Representative responses of photosynthetic rate under field light conditions. The natural light fluctuations were mimicked in portable photosynthesis systems (LI-6400XT), and the photosynthetic rates at a CO_2 concentration of 400 μmol/mol were measured under these light conditions. Three arrows in the right figure indicate major biochemical processes limiting photosynthetic induction with different time courses

than the optimal photosynthetic rates under constant light (Naumburg and Ellsworth 2002; Taylor and Long 2017). Therefore, characterization of the mechanisms that regulate photosynthetic responses to fluctuating light intensities may lead to improved photosynthetic induction and crop yield under natural conditions (Tanaka et al. 2019). The following section summarizes the various genetic engineering approaches that can be used to enhance photosynthesis under fluctuating light conditions (Fig. 7).

7.1 Electron Transport

Photosynthetic electron transport systems consist of linear and cyclic electron transport around photosystem I (Fig. 7). Linear electron transport generates both ATP and NADPH for a Calvin–Benson cycle, photorespiration, and other metabolisms. On the other hand, cyclic electron transport produces ATP without producing NADPH to balance the ATP/NADPH production ratio and is now considered to be

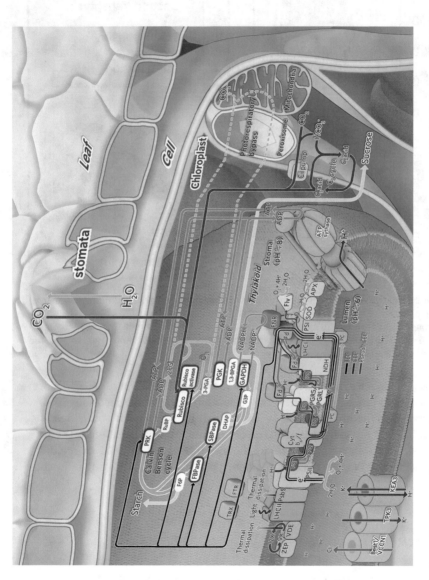

Fig. 7 Strategies for engineering steady-state and non-steady-state photosynthesis. The possible targets would be (1) Rubisco catalysis; (2) Rubisco activase; (3) photorespiratory bypass; (4) stomatal opening; (5) introduction of C_4 cycle; (6) introduction of CO_2 pump from cyanobacteria; (7) light-induced activation of enzymes in the Calvin–Benson cycle by ferredoxin–thioredoxin (TRX) system; (8) electron transport including cyclic electron flow via PGR5-PGRL1 and NDH complex, flavodiiron protein (Flv), and various ion channels/transporters across chloroplast envelopes and thylakoid membranes (i.e., Best1/VCCN1, TPK3, KEA3); and (9) thermal dissipation of excess absorbed light via the protonation of PsbS protein and the activation of a xanthophyll cycle

essential in providing protection from photodamage via the thermal dissipation of excess absorbed light (NPQ, non-photochemical quenching) (Yamori and Shikanai 2016). There are two cyclic electron flows around photosystem I: the main pathway depends on PGR5/PGRL1 proteins and the minor pathway depends on a chloroplast NADH dehydrogenase-like (NDH) complex. It has been shown in rice that PGR5/PGRL1-dependent cyclic electron transport is a key regulator of rapid photosynthetic responses to high light intensity under fluctuating light, and that both PGR5/PGRL1-dependent and NDH-dependent cyclic electron transport have physiological roles for photoprotection in sustaining photosynthesis and plant growth in rice under repeated light fluctuations (Yamori et al. 2016c). In cyanobacteria, pseudocyclic electron transport by flavodiiron protein (*Flv*) mediates the photoreduction of O_2 to H_2O and is essential for photosystem-I photoprotection in fluctuating light (Allahverdiyeva et al. 2013). Recent work indicated that the introduction of the *Flv* gene from moss (*Physcomitrella patens*) into *Arabidopsis* and rice led to the enhancement of cyclic electron transport, resulting in successful improvement of the resistance of photosynthetic machinery under fluctuating light conditions (Fig. 7; Yamamoto et al. 2016; Wada et al. 2018).

NPQ can be activated and relaxed within minutes and is a highly regulatory process involving multiple factors, such as the protonation of PsbS protein and the activation of a xanthophyll cycle that converts the pigment violaxanthin (V) to antheraxanthin (A) and zeaxanthin (Z) (for a review, see Yamori and Shikanai 2016). In tobacco, the simultaneous overexpression of PsbS, violaxanthin de-epoxidase, and zeaxanthin epoxidase increases the rate of NPQ relaxation, which subsequently increases growth under fluctuating light in field conditions (Fig. 7; Kromdijk et al. 2016). Thus, plant productivity and crop yield appear to be highly dependent on NPQ under fluctuating light conditions in nature.

It has been reported that ion channels/transporters across chloroplast envelopes and thylakoid membranes play fundamental roles in the regulation of photosynthetic electron transport (Figs. 5 and 7; Finazzi et al. 2015). Photosynthetic electron transport is coupled with proton translocation across the thylakoid membrane, resulting in the formation of transmembrane H^+ concentration (ΔpH) and electrical potential ($\Delta \Psi$) gradients. Although both ΔpH and $\Delta \psi$ contribute to ATP synthesis as a proton motive force (*pmf*), only the ΔpH component can activate the PsbS- and xanthophyll-cycle-dependent NPQ while down-regulating electron transport during the plastoquinol oxidation step at the cytochrome b_6/f complex (photosynthetic control, Kramer et al. 2003; Yamori and Shikanai 2016). Recent evidence suggests that several ion channels, such as the thylakoid K^+ channel TPK3, K^+ efflux antiporter KEA3, and Cl^- channel Best1/VCCN1, adjust electron transport and functions in photoprotective mechanisms (Figs. 5 and 7; Carraretto et al. 2013; Kunz et al. 2014; Duan et al. 2016; Herdean et al. 2016). The knockout of Best1/VCCN1, which leads to an influx of Cl^- into the lumen, resulted in disturbance of the *pmf* components, resulting in a decreased rate of NPQ induction (Duan et al. 2016; Herdean et al. 2016). These data suggest that a Cl^- influx into the lumen would fine-tune *pmf* and allow the plant to adjust photosynthesis to variable light. On the other hand, TPK3 effluxes K^+

from the thylakoid lumen to the stroma and partially dissipates $\Delta\Psi$ to allow more H+ to enter the lumen and thus enables a significant ΔpH to be formed, thus balancing photoprotection and photochemical efficiency (Carraretto et al. 2013). Moreover, KEA3 effluxes H+ with the counter influx of K+, exchanging $\Delta\psi$ for ΔpH, which is critical for photosynthetic acclimation after transitions from high to low light (Kunz et al. 2014; Armbruster et al. 2014). The activity of KEA3 accelerates the down-regulation of pH-dependent NPQ after transitions to low light, leading to the faster recovery of high photosystem II quantum efficiency and increased CO_2 assimilation. The overexpression of KEA3 accelerates the relaxation of photoprotective energy-dependent quenching after transitions from high to low light in *Arabidopsis* and tobacco (Armbruster et al. 2016). Thus, the KEA3 function is critical in terms of realizing high photosynthetic efficiency under fluctuating light. Taken together, these findings underscore the potential for accelerating NPQ relaxation once light intensity is decreased so as not to decrease the efficiency of light energy use under light-limiting conditions in improving photosynthetic efficiency under fluctuating light in field conditions (Fig. 7).

7.2 Activation of Calvin-Cycle Enzymes, Especially Rubisco

Rubisco must be activated by Rubisco activase to catalyze CO_2 assimilation in the Calvin–Benson cycle (Fig. 7). A positive relationship has been observed between Rubisco activase content and the speed of the photosynthetic induction response in *A. thaliana* (Mott et al. 1997), tobacco (Hammond et al. 1998; Yamori and von Caemmerer 2009), and rice (Masumoto et al. 2012; Yamori et al. 2012). Thus, it is considered that the Rubisco activation state could be a limiting factor for the induction response to a sudden increase in light intensity. In most species, Rubisco activase is present in two isoforms: redox-regulated α-isoform and redox-insensitive β-isoform (Portis Jr 2003). In transgenic *Arabidopsis* plants containing only the β-isoform, photosynthetic induction after a transition from low to high light was faster than in the wild type, as Rubisco activase activity was constitutively high and independent of irradiance (Carmo-Silva and Salvucci 2013; Kaiser et al. 2016). Furthermore, the overexpression of β-isoform from maize in rice led to an improvement in photosynthetic induction via the rapid regulation of the Rubisco activation state by Rubisco activase following an increase in light intensity and/or the maintenance of a high Rubisco activation state under low light (Yamori et al. 2012). Taken together, modifying the concentration of Rubisco activase and its composition could be used to improve photosynthetic performance and plant growth under fluctuating light conditions.

Thioredoxins are ubiquitous enzymes in chloroplasts, and the thioredoxin systems are responsible for the light-induced activation of enzymes in the Calvin–Benson cycle, including GAPDH, FBPase, SBPase, and PRK (Thormählen et al. 2017); ATP synthesis (Hisabori et al. 2013); malate-oxaloacetate shuttle

(Miginiac-Maslow et al. 2000); and starch metabolism (Thormählen et al. 2013). There are two plastid thioredoxins systems: (1) the ferredoxin-thioredoxin system, which consists of ferredoxin-thioredoxin reductase (FTR) and multiple thioredoxins, and (2) the NADPH-dependent thioredoxin reductase (NTRC) system, which contains a complete thioredoxin system (Fig. 7). Recent reports focusing on the overexpression of chloroplast thioredoxin components in plants support the concept of the high impact of thioredoxins on plant fitness. Transgenic tobacco lines overexpressing thioredoxin f, one of the thioredoxin families, showed a large increase in plant biomass and starch content, which was further stimulated by an increase in light intensity (Sanz-Barrio et al. 2013). The overexpression of the endogenous NTRC gene in *Arabidopsis* also increased plant growth under moderate light intensity (Toivola et al. 2013). Furthermore, a recent study showed that both ferredoxin-dependent thioredoxin m, one of the thioredoxin families, and NADPH-dependent NTRC are indispensable for photosynthetic acclimation in fluctuating light intensities (Nikkanen et al. 2016; Thormählen et al. 2017). Thus, it is highly possible that thioredoxin-mediated redox regulation allows the activation state of these enzymes to be modulated in response to fluctuating light in field conditions.

7.3 CO_2 Diffusion into the Chloroplast

The diffusion of CO_2 to the Rubisco catalytic sites in the chloroplast is mediated by both stomatal and mesophyll conductance (Fig. 7). Under naturally fluctuating environmental conditions, stomatal responses are much slower than photosynthetic responses. Manipulating stomatal conductance so that it responds more quickly to irradiance could greatly enhance photosynthesis and water-use efficiency in fluctuating irradiance (Lawson and Blatt 2014; Vialet-Chabrand et al. 2017). Removal of the stomatal limitation could increase photosynthetic induction in *aba*2-1 *Arabidopsis* mutant, which impaired ABA synthesis and thus showed constitutively high stomatal conductance (Kaiser et al. 2016). Moreover, *SLAC1*-deficient rice mutant, which knocked out an anion channel protein in the plasma membrane of stomatal guard cells, constitutively opened stomata and contributed to higher photosynthetic rates more than the wild type in naturally fluctuating light (Yamori et al. 2020). Papanatsiou et al. (2019) induced a synthetic, light-gated K^+ channel in guard cells in *Arabidopsis* and succeeded in facilitating stomatal opening under light exposure and closing after irradiation, leading to greater plant growth in fluctuating light. Furthermore, several *Arabidopsis* mutants with stay-open stomata and the *PATROL1* (proton ATPase translocation control 1) overexpression *Arabidopsis* line with faster stomatal opening responses exhibited higher photosynthetic rates and plant growth in fluctuating light than the wild type, whereas those lines showed similar photosynthetic rates and plant growth in constant light (Shimadzu et al. 2019; Kimura et al. 2020). Taken together, enhancing stomatal conductance could result in better use of plant photosynthetic capacity in naturally fluctuating light.

In addition to stomatal conductance, mesophyll conductance could place a large diffusional limitation on photosynthesis. The extent to which mesophyll conductance limits photosynthesis under fluctuating light is largely unknown, although it has been reported that mesophyll conductance could impose a major limitation to photosynthesis during the steady state. Mesophyll conductance can vary within minutes, and is affected by changes in irradiance, CO_2, and temperature (Flexas et al. 2007, 2008, 2012; Tazoe et al. 2011; Tholen et al. 2008; Yamori et al. 2006a), making it a potentially important process. Recently, we succeeded to characterize induction both of mesophyll conductance and stomatal conductance after a step change in light from darkness to high or low light and showed that mesophyll conductance would impose a smaller limitation to photosynthesis under fluctuating light conditions, but both of mesophyll conductance and stomatal conductance would contribute to the limitation of photosynthesis during induction (Sakoda et al. 2021). Relevant factors that might contribute to variations in mesophyll conductance are carbonic anhydrase and aquaporins.

8   Future Prospects

The present rate of increase in crop yields is insufficient to keep pace with the rapid increase in the global population. Thus, the development of crops with higher yield by improving photosynthesis is essential if we are to meet future food and energy demands. Therefore, suitable approaches must be explored for generating more efficient plants with higher yield. Enhancement of leaf photosynthetic capacity would provide one attractive way of achieving increases in crop yield since plant growth depends largely on photosynthesis. In this review, we have highlighted crucial targets that could be manipulated to enhance crop productivity (Fig. 7). To date, research into finding new ways to increase crop yield has focused on improving steady-state photosynthesis. However, leaves in natural plant canopies experience a highly variable light environment over the course of a day. Thus, the improvements in photosynthesis and yield observed in model plants grown in constant growth chambers may not be completely transferrable to crop species under field conditions. Therefore, an understanding of the key factors operating in natural environments and responsible for increases in yield is essential if we are to achieve the maximum yield potential.

Furthermore, improving photosynthesis to increase food production ultimately means maximizing the photosynthetic efficiency of the crop canopy rather than that of an individual plant. One approach would be to alter the plant architecture and biochemistry and thus distribute irradiation more evenly throughout the canopy in order to achieve the highest conversion efficiency of solar radiation to biomass. Recent genome editing technologies have been progressing and they will enable easier and more precise manipulation of the photosynthesis process in crops. Our understanding of photosynthesis will help us to achieve our goal of sustainable food production.

References

Ainsworth EA, Leakey ADB, Ort DR, Long SP (2008) FACE-ing the facts: inconsistencies and interdependence among field, chamber and modeling studies of elevated CO_2 impacts on crop yield and food supply. New Phytol 179:5–9

Alexandratos N, Bruinsma J (2012) World agriculture towards 2030/2050: the 2012 revision. ESA Working paper No. 12-03. FAO, Rome

Allahverdiyeva Y, Mustila H, Ermakova M, Bersanini L, Richaud P, Ajlani G, Battchikova N, Cournac L, Aro EM (2013) Flavodiiron proteins Flv1 and Flv3 enable cyanobacterial growth and photosynthesis under fluctuating light. Proc Natl Acad Sci U S A 110:4111–4116

Armbruster U, Carrillo LR, Venema K, Pavlovic L, Schmidtmann E, Kornfeld A, Jahns P, Berry JA, Kramer DM, Jonikas MC (2014) Ion antiport accelerates photosynthetic acclimation in fluctuating light environments. Nat Commun 5:5439

Armbruster U, Leonelli L, Correa Galvis V, Strand D, Quinn EH, Jonikas MC, Niyogi KK (2016) Regulation and levels of the thylakoid K^+/H^+ antiporter KEA3 shape the dynamic response of photosynthesis in fluctuating light. Plant Cell Physiol 57:1557–1567

Ballantyne AP, Alden CB, Miller JB, Tans PP, White JWC (2012) Increase in observed net carbon dioxide uptake by land and oceans during the last 50 years. Nature 488:70–72

Biswal AK, Pattanayak GK, Pandey SS, Leelavathi S, Reddy VS, Govindjee, Tripathy BC (2012) Light intensity-dependent modulation of chlorophyll b biosynthesis and photosynthesis by overexpression of chlorophyllide *a* oxygenase in tobacco. Plant Physiol 159:433–449

Burkey KO (1994) Genetic variation of photosynthetic electron transport in barley: identification of plastocyanin as a potential limiting factor. Plant Sci 98:177–187

Burkey KO, Gizlice Z, Carter TE (1996) Genetic variation in soybean photosynthetic electron transport capacity is related to plastocyanin concentration in the chloroplast. Photosynth Res 49:141–149

Busch FA, Sage TL, Cousins AB, Sage RF (2013) C_3 plants enhance rates of photosynthesis by reassimilating photorespired and respired CO_2. Plant Cell Environ 36:200–212

Carmo-Silva AE, Salvucci ME (2012) The temperature response of CO_2 assimilation, photochemical activities and RuBisCO activation in *Camelina sativa*, a potential bioenergy crop with limited capacity for acclimation to heat stress. Planta 236:1433–1445

Carmo-Silva AE, Salvucci ME (2013) The regulatory properties of RuBisCO activase differ among species and affect photosynthetic induction during light transitions. Plant Physiol 161:1645–1655

Carraretto L, Formentin E, Teardo E, Checchetto V, Tomizioli M, Morosinotto T, Giacometti GM, Finazzi G, Szabó I (2013) A thylakoid-located two-pore K+ channel controls photosynthetic light utilization in plants. Science 342:114–118

Carvalho J, Madgwick P, Powers S, Keys A, Lea P, Parry M (2012) An engineered pathway for glyoxylate metabolism in tobacco plants aimed to avoid the release of ammonia in photorespiration. BMC Biotechnol 11:111

Chai MF, Chen QJ, An R, Chen YM, Chen J, Wang XC (2005) NADK2, an *Arabidopsis* chloroplastic NAD kinase, plays a vital role in both chlorophyll synthesis and chloroplast protection. Plant Mol Biol 59:553–564

Chen M, Blankenship RE (2011) Expanding the solar spectrum used by photosynthesis. Trends Plant Sci 16:427–431

Chen M, Schliep M, Willows RD, Cai Z-L, Neilan BA, Scheer H (2010) A red-shifted chlorophyll. Science 329:1318–1319

Chida H, Nakazawa A, Akazaki H, Hirano T, Suruga K, Ogawa M, Satoh T, Kadokura K, Yamada S, Hakamata W et al (2007) Expression of the algal cytochrome c_6 gene in *Arabidopsis* enhances photosynthesis and growth. Plant Cell Physiol 48:948–957

Covshoff S, Hibberd JM (2012) Integrating C_4 photosynthesis into C_3 crops to increase yield potential. Curr Opin Biotechnol 23:209–214

Crafts-Brandner SJ, Salvucci ME (2000) Rubisco activase constrains the photosynthetic potential of leaves at high temperature and CO_2. Proc Natl Acad Sci U S A 97:13430–13435

Ding F, Wang M, Zhang S, Ai X (2016) Changes in SBPase activity influence photosynthetic capacity, growth, and tolerance to chilling stress in transgenic tomato plants. Sci Rep 32741

Driever SM, Simkin AJ, Alotaibi S, Fisk SJ, Madgwick PJ, Sparks CA, Jones HD, Lawson T, Parry MAJ, Raines CA (2017) Increased SBPase activity improves photosynthesis and grain yield in wheat grown in greenhouse conditions. Philos Trans R Soc Lond B Biol Sci 372:20160384

Duan Z, Kong F, Zhang L, Li W, Zhang J, Peng L (2016) A bestrophin-like protein modulates the proton motive force across the thylakoid membrane in *Arabidopsis*. J Integr Plant Biol 58:848–858

Evans JR (1989) Photosynthesis and nitrogen relationships in leaves of C_3 plants. Oecologia 78:9–19

Evans JR, Kaldenhoff R, Genty B, Terashima I (2009) Resistances along the CO_2 diffusion pathway inside leaves. J Exp Bot 60:2235–2248

Farquhar GD, von Caemmerer S, Berry JA (1980) A biochemical model of photosynthetic CO_2 assimilation in leaves of C3 species. Planta 149:78–90

Finazzi G, Petroutsos D, Tomizioli M, Flori S, Sautron E, Villanova V, Rolland N, Seigneurin-Berny D (2015) Ions channels/transporters and chloroplast regulation. Cell Calcium 58:86–97

Flexas J, Ribas-Carbó M, Hanson DT, Bota J, Otto B, Cifre J, McDowell N, Medrano H, Kaldenhoff R (2006) Tobacco aquaporin NtAQP1 is involved in mesophyll conductance to CO_2 *in vivo*. Plant J 48:427–439

Flexas J, Diaz-Espejo A, Galmés J, Kaldenhoff R, Medrano H, Ribas-Carbo M (2007) Rapid variations of mesophyll conductance in response to changes in CO_2 concentration around leaves. Plant Cell Environ 30:1284–1298

Flexas J, Ribas-Carbó M, Diaz-Espejo A, Galmés J, Medrano H (2008) Mesophyll conductance to CO_2: current knowledge and future prospects. Plant Cell Environ 31:602–621

Flexas J, Barbour MM, Brendel O, Cabrera HM, Carriquí M, Diaz-Espejo A, Douthe C, Dreyer E, Ferrio JP, Gago J, Gallé A, Galmés J, Kodama N, Medrano H, Niinemets Ü, Peguero-Pina JJ, Pou A, Ribas-Carbó M, Tomás M, Tosens T, Warren CR, Gallé A (2012) Mesophyll diffusion conductance to CO_2: an unappreciated central player in photosynthesis. Plant Sci 193–194:70–84

Furbank RT, Quick WP, Sirault XR (2015) Improving photosynthesis and yield potential in cereal crops by targeted genetic manipulation: prospects, progress and challenges. Field Crops Res 182:19–29

Gotoh E, Suetsugu N, Yamori W, Ishishita K, Kiyabu R, Fukuda M, Higa T, Shirouchi B, Wada M (2018) Chloroplast accumulation response enhances leaf photosynthesis and plant biomass production. Plant Physiol 178:1358–1369

Gu J, Yin X, Stomph TJ, Struik PC (2014) Can exploiting natural genetic variation in leaf photosynthesis contribute to increasing rice productivity? A simulation analysis. Plant Cell Environ 37:22–34

Haake V, Zrenner R, Sonnewald U, Stitt M (1998) A moderate decrease of plastid aldolase activity inhibits photosynthesis, alters the levels of sugars and starch, and inhibits growth of potato plants. Plant J 14:147–157

Haake V, Geiger M, Walch-Liu P, Engels C, Zrenner R, Stitt M (1999) Changes in aldolase activity in wild-type potato plants are important for acclimation to growth irradiance and carbon dioxide concentration, because plastid aldolase exerts control over the ambient rate of photosynthesis across a range of growth conditions. Plant J 17:479–489

Hajirezaei M-R, Peisker M, Tschiersch H, Palatnik JF, Valle EM, Carrillo N, Sonnewald U (2002) Small changes in the activity of chloroplastic NADP$^+$-dependent ferredoxin oxidoreductase lead to impaired plant growth and restrict photosynthetic activity of transgenic tobacco plants. Plant J 29:281–293

Hammond ET, Andrews TJ, Mott KA, Woodrow IE (1998) Regulation of Rubisco activation in antisense plants of tobacco containing reduced levels of Rubisco activase. Plant J 14:101–110

Hanba YT, Shibasaka M, Hayashi Y, Hayakawa T, Kasamo K, Terashima I, Katsuhara M (2004) Overexpression of the barley aquaporin HvPIP2;1 increases internal CO_2 conductance and CO_2 assimilation in the leaves of transgenic rice plants. Plant Cell Physiol 45:521–529

Harrison EP, Willingham NM, Lloyd JC, Raines CA (1998) Reduced sedoheptulose-1,7-bisphosphatase levels in transgenic tobacco lead to decreased photosynthetic capacity and altered carbohydrate accumulation. Planta 204:27–36

Harrison EP, Olcer H, Lloyd JC, Long SP, Raines CA (2001) Small decreases in SBPase cause a linear decline in the apparent RuBP regeneration rate, but do not affect Rubisco carboxylation capacity. J Exp Bot 52:1779–1784

Hatano-Iwasaki A, Ogawa K (2012) Biomass production is promoted by increasing an aldolase undergoing glutathionylation in *Arabidopsis thaliana*. Int J Dev Biol 6:1–8

Hay WT, Bihmidine S, Mutlu N, Le Hoang K, Awada T, Weeks DP, Clemente TE, Long SP (2017) Enhancing soybean photosynthetic CO_2 assimilation using a cyanobacterial membrane protein, *ictB*. J Plant Physiol 212:58–68

Heckwolf M, Pater D, Hanson DT, Kaldenhoff R (2011) The *Arabidopsis thaliana* aquaporin AtPIP1;2 is a physiologically relevant CO_2 transport facilitator. Plant J 67:795–804

Herdean A, Teardo E, Nilsson AK, Pfeil BE, Johansson ON, Ünnep R, Nagy G, Zsiros O, Dana S, Solymosi K, Garab G (2016) A voltage-dependent chloride channel fine-tunes photosynthesis in plants. Nat Commun 7:11654

Hisabori T, Sunamura E, Kim Y, Konno H (2013) The chloroplast ATP synthase features the characteristic redox regulation machinery. Antioxid Redox Signal 19:1846–1854

IPCC (2013) Annex II: climate system scenario tables. In: Prather M, Flato G, Friedlingstein P et al (eds) Climate change 2013: the physical science basis. Contribution of Working Group I to the fifth assessment report of the Intergovernmental Panel on Climate Change. Cambridge University Press, Cambridge

Ishikawa C, Hatanaka T, Misoo S, Miyake C, Fukayama H (2011) Functional incorporation of sorghum small subunit increases the catalytic turnover rate of RuBisCO in transgenic rice. Plant Physiol 156:1603–1611

Kaiser E, Morales A, Harbinson J, Heuvelink E, Prinzenberg AE, Marcelis LFM (2016) Metabolic and diffusional limitations of photosynthesis in fluctuating irradiance in *Arabidopsis thaliana*. Sci Rep 6:31252

Kaldenhoff R (2012) Mechanisms underlying CO_2 diffusion in leaves. Curr Opin Plant Biol 15:276–281

Kawai S, Murata K (2008) Structure and function of NAD kinase and NADP phosphatase: key enzymes that regulate the intracellular balance of NAD(H) and NADP(H). Biosci Biotechnol Biochem 72:919–930

Kebeish R, Niessen M, Thiruveedhi K, Bari R, Hirsch HJ, Rosenkranz R, Stäbler N, Schönfeld B, Kreuzaler F, Peterhänsel C (2007) Chloroplastic photorespiratory bypass increases photosynthesis and biomass production in *Arabidopsis thaliana*. Nat Biotechnol 25:593–599

Kimura H, Hashimoto-Sugimoto M, Iba K, Terashima I, Yamori W (2020) Improved stomatal opening enhances photosynthetic rate and biomass production in fluctuating light. J Exp Bot 71:2339–2350

Köhler IH, Ruiz-Vera UM, VanLoocke A, Thomey ML, Clemente T, Long SP, Ort DR, Bernacchi CJ (2016) Expression of cyanobacterial FBP/SBPase in soybean prevents yield depression under future climate conditions. J Exp Bot 68:715–726

Kossmann J, Sonnewald U, Willmitzer L (1994) Reduction of the chloroplastic fructose-1,6-bisphosphatase in transgenic potato plants impairs photosynthesis and plant growth. Plant J 6:637–650

Kozaki A, Takeba G (1996) Photorespiration protects C_3 plants from photooxidation. Nature 384:557

Kramer DM, Cruz JA, Kanazawa A (2003) Balancing the central roles of the thylakoid proton gradient. Trends Plant Sci 8:27–32

Kromdijk J, Głowacka K, Leonelli L, Gabilly ST, Iwai M, Niyogi KK, Long SP (2016) Improving photosynthesis and crop productivity by accelerating recovery from photoprotection. Science 354:857–861

Kumar A, Li C, Portis AR Jr (2009) *Arabidopsis thaliana* expressing a thermostable chimeric Rubisco activase exhibits enhanced growth and higher rates of photosynthesis at moderately high temperatures. Photosynth Res 100:143–153

Kunz HH, Gierth M, Herdean A, Satoh-Cruz M, Kramer DM, Spetea C, Schroeder JI (2014) Plastidial transporters KEA1, -2, and -3 are essential for chloroplast osmoregulation, integrity, and pH regulation in *Arabidopsis*. Proc Natl Acad Sci U S A 111:7480–7485

Kurek I, Chang TK, Bertain SM, Madrigal A, Liu L, Lassner MW, Zhu G (2007) Enhanced thermostability of *Arabidopsis* rubisco activase improves photosynthesis and growth rates under moderate heat stress. Plant Cell 19:3230–3241

Kusumi K, Hirotsuka S, Kumamaru T, Iba K (2012) Increased leaf photosynthesis caused by elevated stomatal conductance in a rice mutant deficient in SLAC1, a guard cell anion channel protein. J Exp Bot 63:5635–5644

Lawson T, Blatt MR (2014) Stomatal size, speed, and responsiveness impact on photosynthesis and water use efficiency. Plant Physiol 164:1556–1570

Lawson T, Bryant B, Lefebvre S, Lloyd JC, Raines CA (2006) Decreased SBPase activity alters growth and development in transgenic tobacco plants. Plant Cell Environ 29:48–58

Lefebvre S, Lawson T, Zakhleniuk OV, Lloyd JC, Raines CA, Fryer M (2005) Increased sedoheptulose-1,7-bisphosphatase activity in transgenic tobacco plants stimulates photosynthesis and growth from an early stage in development. Plant Physiol 138:451–460

Lieman-Hurwitz J, Rachmilevitch S, Mittler R, Marcus Y, Kaplan A (2003) Enhanced photosynthesis and growth of transgenic plants that express ictB, a gene involved in HCO_3^- accumulation in cyanobacteria. Plant Biotechnol J 1:43–50

Lin MT, Occhialini A, Andralojc PJ, Parry MAJ, Hanson MR (2014) A faster Rubisco with potential to increase photosynthesis in crops. Nature 513:547–550

Long BM, Hee WY, Sharwood RE, Rae BD, Kaines S, Lim YL, Nguyen ND, Massey B, Bala S, von Caemmerer S, Badger MR (2018) Carboxysome encapsulation of the CO_2-fixing enzyme Rubisco in tobacco chloroplasts. Nat Commun 9:3570

Lopez-Calcagno PE, Fisk S, Brown KL, Bull SE, South PF, Raines CA (2018) Overexpressing the H-protein of the glycine cleavage system increases biomass yield in glasshouse and field grown transgenic tobacco plants. Plant Biotechnol J 17(1):141–151. https://doi.org/10.1111/pbi.12953

Maier A, Fahnenstich H, von Caemmerer S, Engqvist MKM, Weber APM, Fluegge U-I, Maurino VG (2012) Transgenic introduction of a glycolate oxidative cycle into *A. thaliana* chloroplasts leads to growth improvement. Front. Plant Sci 3:38

Makino A, Sato T, Nakano H, Mae T (1997) Leaf photosynthesis, plant growth and nitrogen allocation in rice under different irradiances. Planta 203:390–398

Masumoto C, Fukayama H, Hatanaka T, Uchida N (2012) Photosynthetic characteristics of antisense transgenic rice expressing reduced levels of Rubisco activase. Plant Prod Sci 15:174–182

Miginiac-Maslow M, Johansson K, Ruelland E, Issakidis-Bourguet E, Schepens I, Goyer A, Lemaire-Chamley M, Jacquot J-P, Le Maréchal P, Decottignies P (2000) Light-activation of NADP-malate dehydrogenase: a highly controlled process for an optimized function. Physiol Plant 110:322–329

Miyagawa Y, Tamoi M, Shigeoka S (2001) Overexpression of a cyanobacterial fructose-1,6-/sedoheptulose-1,7-bisphosphatase in tobacco enhances photosynthesis and growth. Nat Biotechnol 19:965–969

Miyashita H, Ikemoto H, Kurano N, Adachi K, Chihara M, Miyachi S (1996) Chlorophyll *d* as a major pigment. Nature 383:402–402

Morris PF, Layzell DB, Canvin DT (1988) Ammonia production and assimilation in glutamate synthase mutants of *Arabidopsis thaliana*. Plant Physiol 87:148–154

Mott KA, Snyder GW, Woodrow IE (1997) Kinetics of Rubisco activation as determined from gas-exchange measurements in antisense plants of *Arabidopsis thaliana* containing reduced levels of Rubisco activase. Aust J Plant Physiol 24:811–818

Mueller-Cajar O, Stotz M, Wendler P, Hartl FU, Bracher A, Hayer-Hartl M (2011) Structure and function of the AAA+ protein CbbX, a red-type Rubisco activase. Nature 479:194–199

Murchie EH, Niyogi KK (2011) Manipulation of photoprotection to improve plant photosynthesis. Plant Physiol 155:86–92

Naumburg E, Ellsworth DS (2002) Short-term light and leaf photosynthetic dynamics affect estimates of daily understory photosynthesis in four tree species. Tree Physiol 22:393–401

Nikkanen L, Toivola J, Rintamäki E (2016) Crosstalk between chloroplast thioredoxin systems in regulation of photosynthesis. Plant Cell Environ 39:1691–1705

Ölcer H, Lloyd JC, Raines CA (2001) Photosynthetic capacity is differentially affected by reductions in sedoheptulose-1,7-bisphosphatase activity during leaf development in transgenic tobacco plants. Plant Physiol 125:982–989

Papanatsiou M, Petersen J, Henderson L, Wang Y, Christie JM, Blatt MR (2019) Optogenetic manipulation of stomatal kinetics improves carbon assimilation, water use, and growth. Science 363:1456–1459

Parry MAJ, Keys AJ, Madgwick PJ, Carmo-Silva AE, Andralojc PJ (2008) Rubisco regulation: a role for inhibitors. J Exp Bot 59:1569–1580

Pearcy RW (1990) Sunflecks and photosynthesis in plant canopies. Annu Rev Plant Physiol Plant Mol Biol 41:421–453

Pesaresi P, Scharfenberg M, Weigel M, Granlund I, Schroder WP, Finazzi G, Rappaport F, Masiero S, Furini A, Jahns P, Leister D (2009) Mutants, overexpressors, and interactors of *Arabidopsis* plastocyanin isoforms: revised roles of plastocyanin in photosynthetic electron flow and thylakoid redox state. Mol Plant 2:236–248

Peterhansel C, Blume C, Offermann S (2013) Photorespiratory bypasses: how can they work? J Exp Bot 64:709–715

Portis AR Jr (2003) Rubisco activase: Rubisco's catalytic chaperone. Photosynth Res 75:11–27

Pribil M, Pesaresi P, Hertle A, Barbato R, Leister D (2010) Role of plastid protein phosphatase TAP38 in LHCII dephosphorylation and thylakoid electron flow. PLoS Biol 8:e1000288

Price GD, von Caemmerer S, Evans JR, Siebke K, Anderson JM, Badger MR (1998) Photosynthesis is strongly reduced by antisense suppression of chloroplastic cytochrome *bf* complex in transgenic tobacco. Funct Plant Biol 25:445–452

Price GD, Badger MR, von Caemmerer S (2011) The prospect of using cyanobacterial bicarbonate transporters to improve leaf photosynthesis in C_3 crop plants. Plant Physiol 155:20–26

Price GD, Pengelly JJL, Forster B, Du J, Whitney SM, von Caemmerer S, Badger MR, Howitt SM, Evans JR (2013) The cyanobacterial CCM as a source of genes for improving photosynthetic CO_2 fixation in crop species. J Exp Bot 64:753–768

Raines CA (2003) The Calvin cycle revisited. Photosynth Res 75:1–10

Raines CA, Paul MJ (2006) Products of leaf primary carbon metabolism modulate the developmental programme determining plant morphology. J Exp Bot 57:1857–1862

Raines CA, Lloyd JC, Dyer TA (1999) New insights into the structure and function of sedoheptulose-1,7-bisphosphatase; an important but neglected Calvin cycle enzyme. J Exp Bot 50:1–8

Richards RA (2000) Selectable traits to increase crop photosynthesis and yield of grain crops. J Exp Bot 51:447–458

Rodriguez RE, Lodeyro A, Poli HO, Zurbriggen M, Palatnik JF, Tognetti VB, Tschiersch H, Hajirezaei MR, Valle EM, Carrillo N (2007) Transgenic tobacco plants overexpressing chloroplastic ferredoxin-NADP(H) reductase display normal rates of photosynthesis and increased tolerance to oxidative stress. Plant Physiol 143:639–649

Rojas-González JA, Soto-Súarez M, García-Díaz Á, Romero-Puertas MC, Sandalio LM, Mérida Á, Thormählen I et al (2015) Disruption of both chloroplastic and cytosolic FBPase genes results in a dwarf phenotype and important starch and metabolite changes in *Arabidopsis thaliana*. J Exp Bot 66:2673–2689

Rosenthal D, Locke A, Khozaei M, Raines C, Long S, Ort D (2011) Over-expressing the C_3 photosynthesis cycle enzyme sedoheptulose-1-7 bisphosphatase improves photosynthetic carbon gain and yield under fully open air CO_2 fumigation (FACE). BMC Plant Biol 11:123

Sage RF, Sage TL, Kocacinar F (2012) Photorespiration and the evolution of C_4 photosynthesis. Annu Rev Plant Biol 63:19–47

Sakoda K, Yamori W, Groszmann M, Evans JR (2021) Stomatal, mesophyll conductance, and biochemical limitations to photosynthesis during induction. Plant Physiology

Sahrawy M, Avila C, Chueca A, Canovas FM, Lopez-Gorge J (2004) Increased sucrose level and altered nitrogen metabolism in *Arabidopsis thaliana* transgenic plants expressing antisense chloroplastic fructose-1,6-bisphosphatase. J Exp Bot 55:2495–2503

Salvucci ME, Crafts-Brandner SJ (2004a) Relationship between the heat tolerance of photosynthesis and the thermal stability of Rubisco activase in plants from contrasting thermal environments. Plant Physiol 134:1460–1470

Salvucci ME, Crafts-Brandner SJ (2004b) Inhibition of photosynthesis by heat stress: the activation state of Rubisco as a limiting factor in photosynthesis. Physiol Plant 120:179–186

Salvucci ME, DeRidder BP, Portis AR (2006) Effect of activase level and isoform on the thermotolerance of photosynthesis in *Arabidopsis*. J Exp Bot 57:3793–3799

Sanz-Barrio R, Corral-Martinez P, Ancin M, Segui-Simarro JM, Farran I (2013) Overexpression of plastidial thioredoxin f leads to enhanced starch accumulation in tobacco leaves. Plant Biotechnol J 11:618–627

Scafaro AP, Atwell BJ, Muylaert S, Van Reusel B, Alguacil Ruiz G, Van Rie J, Gallé A (2018) A thermotolerant variant of Rubisco activase from a wild relative improves growth and seed yield in rice under heat stress. Front Plant Sci 9:1663

Schöttler MA, Kirchhoff H, Weis E (2004) The role of plastocyanin in the adjustment of the photosynthetic electron transport to the carbon metabolism in tobacco. Plant Physiol 136:4265–4274

Shimadzu S, Seo M, Terashima I, Yamori W (2019) Whole irradiated plant leaves showed faster photosynthetic induction than individually irradiated leaves via improved stomatal opening. Front Plant Sci 10:1512

Simkin AJ, McAusland L, Lawson T, Raines CA (2017) Over-expression of the Rieske FeS protein increases electron transport rates and biomass yield. Plant Physiol 175:134–145

Sinclair TR, Purcell LC, Sneller CH (2004) Crop transformation and the challenge to increase yield potential. Trends Plant Sci 9:70–75

Spreitzer RJ, Salvucci ME (2002) Rubisco: structure, regulatory interactions, and possibilities for a better enzyme. Annu Rev Plant Biol 53:449–475

Stotz M, Mueller-Cajar O, Ciniawsky S, Wendler P, Hartl FU, Bracher A, Hayer-Hartl M (2011) Structure of green-type RuBisCO activase from tobacco. Nat Struct Mol Biol 18:1366–1370

Takahara K, Kasajima I, Takahashi H, Hashida SN, Itami T, Onodera H, Toki S, Yanagisawa S, Kawai-Yamada M, Uchimiya H (2010) Metabolome and photochemical analysis of rice plants overexpressing *Arabidopsis* NAD kinase gene. Plant Physiol 152:1863–1873

Tamoi M, Nagaoka M, Miyagawa Y, Shigeoka S (2006) Contribution of fructose-1,6-bisphosphatase and sedoheptulose-1,7-bisphosphatase to the photosynthetic rate and carbon flow in the Calvin cycle in transgenic plants. Plant Cell Physiol 47:380–390

Tanaka Y, Sugano SS, Shimada T, Hara-Nishimura I (2013) Enhancement of leaf photosynthetic capacity through increased stomatal density in *Arabidopsis*. New Phytol 198:757–764

Tanaka Y, Adachi S, Yamori W (2019) Natural genetic variation of the photosynthetic induction response to fluctuating light environment. Curr Opin Plant Biol 49:52–59

Taylor SH, Long SP (2017) Slow induction of photosynthesis on shade to sun transitions in wheat may cost at least 21% of productivity. Philos Trans R Soc Lond B Biol Sci 372:20160543

Tazoe Y, von Caemmerer S, Estavillo GM, Evans JR (2011) Using tunable diode laser spectroscopy to measure carbon isotope discrimination and mesophyll conductance to CO_2 diffusion dynamically at different CO_2 concentrations. Plant Cell Environ 344:580–591

Terashima I, Hanba YT, Tholen D, Niinemets Ü (2011) Leaf functional anatomy in relation to photosynthesis. Plant Physiol 155:108–116

Tholen D, Zhu XG (2011) The mechanistic basis of internal conductance: a theoretical analysis of mesophyll cell photosynthesis and CO_2 diffusion. Plant Physiol 156:90–105

Tholen D, Boom C, Noguchi K, Ueda S, Katase T, Terashima I (2008) The chloroplast avoidance response decreases internal conductance to CO_2 diffusion in *Arabidopsis thaliana* leaves. Plant Cell Environ 31:1688–1700

Thormählen I, Ruber J, von Roepenack-Lahaye E, Ehrlich S, Massot V, Huemmer C, Tezycka J, Issakidis-Bourguet E, Geigenberger P (2013) Inactivation of thioredoxin f1 leads to decreased light activation of ADP-glucose pyrophosphorylase and altered diurnal starch turnover in leaves of *Arabidopsis* plants. Plant Cell Environ 36:16–29

Thormählen I, Zupok A, Rescher J, Leger J, Weissenberger S, Groysman J, Orwat A, Chatel-Innocenti G, Issakidis-Bourguet E, Armbruster U et al (2017) Thioredoxins play a crucial role in dynamic acclimation of photosynthesis in fluctuating light. Mol Plant 10:168–182

Tikkanen M, Grieco M, Nurmi M, Rantala M, Suorsa M, Aro EM (2012) Regulation of the photosynthetic apparatus under fluctuating growth light. Philos Trans R Soc Lond B Biol Sci 367:3486–3493

Tilman D, Balzer C, Hill J, Befort BL (2011) Global food demand and the sustainable intensification of agriculture. Proc Natl Acad Sci U S A 108:20260–20264

Timm S, Florian A, Arrivault S, Stitt M, Fernie AR, Bauwe H (2012) Glycine decarboxylase controls photosynthesis and plant growth. FEBS Lett 586:3692–3697

Timm S, Wittmiss M, Gamlien S, Ewald R, Florian A, Frank M, Wirtz M et al (2015) Mitochondrial dihydrolipoyl dehydrogenase activity shapes photosynthesis and photorespiration of *Arabidopsis thaliana*. Plant Cell 27:1968–1984

Timm S, Florian A, Fernie AR, Bauwe H (2016) The regulatory interplay between photorespiration and photosynthesis. J Exp Bot 67:2923–2929

Toivola J, Nikkanen L, Dahlström KM, Salminen TA, Lepistö A, Vignols F, Rintamäki E (2013) Overexpression of chloroplast NADPH-dependent thioredoxin reductase in *Arabidopsis* enhances leaf growth and elucidates in vivo function of reductase and thioredoxin domains. Front Plant Sci 4:389

Uehlein N, Sperling H, Heckwolf M, Kaldenhoff R (2012) The *Arabidopsis* aquaporin PIP1;2 rules cellular CO_2 uptake. Plant Cell Environ 35:1077–1083

Uematsu K, Suzuki N, Iwamae T, Inui M, Yukawa H (2012) Increased fructose 1,6-bisphosphate aldolase in plastids enhances growth and photosynthesis of tobacco plants. J Exp Bot 63:3001–3009

Vialet-Chabrand SRM, Matthews JSA, McAusland L, Blatt MR, Griffiths H, Lawson T (2017) Temporal dynamics of stomatal behavior: modeling and implications for photosynthesis and water use. Plant Physiol 174:603–613

von Caemmerer S, Quick WP, Furbank RT (2012) The development of C_4 rice: current progress and future challenges. Science 336:1671–1672

Wada S, Yamamoto H, Suzuki Y, Yamori W, Shikanai T, Makino A (2018) Flavodiiron protein substitutes for cyclic electron flow without competing CO_2 assimilation in rice. Plant Physiol 176:1509–1518

Walker BJ, VanLoocke A, Bernacchi CJ, Ort DR (2016) The costs of photorespiration to food production now and in the future. Annu Rev Plant Biol 67:107–129

Whitney SM, Sharwood RE, Orr D, White SJ, Alonso H, Galmés J (2011) Isoleucine 309 acts as a C_4 catalytic switch that increases ribulose-1,5-bisphosphate carboxylase/oxygenase (rubisco) carboxylation rate in *Flaveria*. Proc Natl Acad Sci U S A 108:14688–14693

Wigge B, Krömer S, Gardeström P (1993) The redox levels and subcellular distribution of pyridine nucleotides in illuminated barley leaf protoplasts studied by rapid fractionation. Physiol Plant 88:10–18

Wright IJ, Reich PB, Westoby M, Ackerly DD, Baruch Z et al (2004) The worldwide leaf economics spectrum. Nature 428:821–827

Yamamoto H, Kato H, Shinzaki Y, Horiguchi S, Shikanai T, Hase T, Endo T, Nishioka M, Makino A, Tomizawa K, Miyake C (2006) Ferredoxin limits cyclic electron flow around PSI (CEF-PSI) in higher plants: stimulation of CEF-PSI enhances non-photochemical quenching of Chl fluorescence in transplastomic tobacco. Plant Cell Physiol 47:1355–1371

Yamamoto H, Takahashi S, Badger MR, Shikanai T (2016) Artificial remodelling of alternative electron flow by flavodiiron proteins in *Arabidopsis*. Nat Plants 2:16012

Yamori W (2016) Photosynthetic response to fluctuating environments and photoprotective strategies under abiotic stress. J Plant Res 129:379–395

Yamori W, Shikanai T (2016) Physiological functions of cyclic electron transport around photosystem I in sustaining photosynthesis and plant growth. Annu Rev Plant Biol 67:81–106

Yamori W, von Caemmerer S (2009) Effect of Rubisco activase deficiency on the temperature response of CO_2 assimilation rate and Rubisco activation state: insights from transgenic tobacco with reduced amounts of Rubisco activase. Plant Physiol 151:2073–2082

Yamori W, Noguchi K, Hanba YT, Terashima I (2006a) Effects of internal conductance on the temperature dependence of the photosynthetic rate in spinach leaves from contrasting growth temperatures. Plant Cell Physiol 47:1069–1080

Yamori W, Suzuki K, Noguchi K, Nakai M, Terashima I (2006b) Effects of Rubisco kinetics and Rubisco activation state on the temperature dependence of the photosynthetic rate in spinach leaves from contrasting growth temperatures. Plant Cell Environ 29:1659–1670

Yamori W, Takahashi S, Makino A, Price GD, Badger MR, von Caemmerer S (2011) The roles of ATP synthase and the cytochrome b_6/f complexes in limiting chloroplast electron transport and determining photosynthetic capacity. Plant Physiol 155:956–962

Yamori W, Masumoto C, Fukayama H, Makino A (2012) Rubisco activase is a key regulator of non-steady-state photosynthesis at any leaf temperature and, to a lesser extent, of steady-state photosynthesis at high temperature. Plant J 71:871–880

Yamori W, Hikosaka K, Way DA (2014) Temperature response of photosynthesis in C_3, C_4 and CAM plants: temperature acclimation and temperature adaptation. Photosynth Res 119:101–117

Yamori W, Kondo E, Sugiura D, Terashima I, Suzuki Y, Makino A (2016a) Enhanced leaf photosynthesis as a target to increase grain yield: insights from transgenic rice lines with variable Rieske FeS protein content in the cytochrome $b6/f$ complex. Plant Cell Environ 39:80–87

Yamori W, Irving LJ, Adachi S, Busch FA (2016b) Strategies for optimizing photosynthesis with biotechnology to improve crop yield. In: Pessarakli M (ed) Handbook of photosynthesis, 3rd edn. Taylor & Francis, Boca Raton, pp 741–759

Yamori W, Makino A, Shikanai T (2016c) A physiological role of cyclic electron transport around photosystem I in sustaining photosynthesis under fluctuating light in rice. Sci Rep 6:20147

Yamori W, Kusumi K, Iba K, Terashima I (2020) Increased stomatal conductance induces rapid changes to photosynthetic rate in response to naturally fluctuating light conditions in rice. Plant Cell Environ 43:1230–1240

Green Super Rice (GSR) Traits: Breeding and Genetics for Multiple Biotic and Abiotic Stress Tolerance in Rice

Jauhar Ali, Mahender Anumalla, Varunseelan Murugaiyan, and Zhikang Li

Abstract The frequent fluctuations in global climate variability (GCV), decreases in farmland and irrigation water, soil degradation and erosion, and increasing fertilizer costs are the significant factors in declining rice productivity, mainly in Asia and Africa. Under GCV scenarios, it is a challenging task to meet the rice food demand of the growing population. Identifying green traits (tolerance of biotic and abiotic stresses, nutrient-use efficiency, and nutritional grain quality) and stacking them in high-yielding elite genetic backgrounds is one promising approach to increase rice productivity. To this end, the Green Super Rice (GSR) breeding strategy helps to pool multi-stress-tolerance traits by stringent selection processes and to develop superior GSR cultivars within a short span of 4–5 years. In the crossing and selection process of GSR breeding, selective introgression lines (SILs) derived from sets of early backcross BC_1F_2 bulk populations through both target traits and non-target traits were selected. Genotyping of SILs with high-density SNP markers leads to the identification of a large number of SNP markers linked with the target green traits. The identified SILs with superior trait combinations were used for designed QTL pyramiding to combine different target green traits. The GSR breeding strategy also focused on nutrient- and water-use efficiency besides environment-friendly green features primarily to increase grain yield and income returns for resource-poor farmers. In this chapter, we have highlighted the GSR breeding strategy and QTL introgression of green traits in rice. This breeding strategy has successfully dissected many complex traits and also released several multi-stress-tolerant varieties with high grain yield and productivity in the target regions of Asia and Africa.

J. Ali (✉) · M. Anumalla
Rice Breeding Platform, International Rice Research Institute (IRRI),
Los Baños, Laguna, Philippines
e-mail: J.Ali@irri.org

V. Murugaiyan
Rice Breeding Platform, International Rice Research Institute (IRRI),
Los Baños, Laguna, Philippines

Plant Nutrition, Institute of Crop Sciences and Resource Conservation (INRES), University of Bonn, Bonn, Germany

Z. Li
National Key Facility for Crop Gene Resources and Genetic Improvement, Institute of Crop Science, Chinese Academy of Agricultural Sciences (CAAS), Beijing, China

© The Author(s) 2021
J. Ali, S. H. Wani (eds.), *Rice Improvement*,
https://doi.org/10.1007/978-3-030-66530-2_3

59

Keywords Green traits · Molecular breeding strategies · QTLs and genes ·
Multiple-stress tolerance · Green super rice varieties

1 Introduction

Food security is a global challenge for plant researchers to increase crop productivity, especially under changing climatic conditions. Rice (*Oryza sativa* L.) is one of the staple food crops for more than half of the world population. More than 95% of global rice is produced and consumed by the top 10 rice-producing countries (Fig. 1). China consumes about 143 million metric tons (MMT), followed by India (103 MMT), with these being the two most populated countries (https://www. statista.com). The rapid growth of population is estimated to reach 9.7 billion by 2050, and this would have a direct bearing on the demand side of global food (United Nations 2019). Worldwide, more than 800 million people are affected by malnutrition, thus hindering sustainable development programs, and food demand is expected to increase by 59–98% by 2050 (Elferink and Schierhorn 2016). Apart from this great challenge, increased global climate variability (GCV) (abiotic stresses: drought, salinity, low/high temperature, submergence/flooding; and biotic stresses: blast, bacterial leaf blight, brown planthopper, etc.) and decreasing natural resources (NRS) (e.g., decreased availability of irrigation water, labor scarcity, arable land reduction, and soil nutrient deficiency and toxicity) are the foremost factors that slow the pace toward food security for the global population. As per the prediction of the Intergovernmental Panel on Climate Change (IPCC), Earth's global surface temperature is expected to surge by 1.4–5.8 °C by 2100 (Fahad et al. 2019),

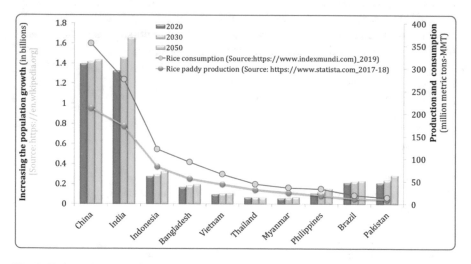

Fig. 1 Estimated growth of population in the top 10 rice-producing countries

thereby resulting in a decrease in precipitation in the subtropics and a possible rise in the frequent occurrences of extreme climatic events. However, these two major factors, GCV and NRS, would have direct or indirect effects on crop growth and significantly decrease crop yield (Pandey et al. 2017). Accordingly, more than 90% of arable lands are prone to one or more than two combinations of stresses, which cause up to 70% yield losses in the major food crops (Ray et al. 2015; Fahad et al. 2017; Tigchelaar et al. 2018). To provide nutrition and food security by 2050, it has been projected that crop yield must be increased by 1%, 0.9%, and 1.6% annually for rice, wheat, and maize, respectively (Fischer and Edmeades 2010). Thus, current grain yield levels are insufficient to sustain the rapid growth of the human population. The production of the principal crops (rice, maize, and wheat) has to increase by ~42%, ~67%, and ~38%, respectively, by 2050 to meet this demand (Ray et al. 2013). As compared to cereals, rice is the only major staple food crop to have more nutritional and health benefits for increasing rice consumption; it also has sensitivity to the major biotic and abiotic stress factors. Therefore, to overcome the challenges of growing GCV and decreasing NRS, the development of superior rice cultivars with higher grain yield and multiple-stress tolerance is necessary to provide a kind of crop insurance besides increasing the income of poor farmers across mainly Asian and African countries.

Multiple abiotic stresses greatly influence crop productivity and have adverse effects on plant growth and development (Fahad et al. 2017; Cohen and Leach 2019). For example, drought is a major constraint in rice production and approximately 42 million hectares of rice in Asia suffer significant yield losses from varying amounts of drought stress at the different crop growth stages (Saikumar et al. 2016; Sandhu and Kumar 2017; Mukamuhirwa et al. 2019). Besides drought, soil salinity and alkalinity are also devastating factors that cause significant yield losses of rice crops at the vegetative and reproductive stages. Globally, about 20% of the cultivable land and 33% of irrigated agricultural areas are afflicted by salinity stress (Shrivastava and Kumar 2015). Thus, it is crucial to identify stress-tolerant rice varieties to maintain productivity and meet global food security needs. Beyond abiotic stresses, sustainable rice production has become one of the top priorities in developing and adopting eco-friendly rice varieties with low input-use efficiency. In the majority of the developed countries, farmers have been using considerably more fertilizer to increase their crop yield. According to Cheng et al. (2007), rice production in China has to be increased by 14% by 2030 to meet the food requirements of its rapidly growing population. In Jiangsu Province, farmers are applying 300–350 kg N/ha to achieve high yield for rice crops (Sui et al. 2013), which is 90% more than the global average N application (Chen et al. 2014a) . The excess amounts of nutrient fertilizer also cause the accumulation of higher salt concentration in the soil. This further decreases water and nutrient absorption, leading to several changes in physiological processes as leaf dehydration decreases photosynthesis efficiency, decreases grain micronutrients, and also causes severe environmental pollution (Kreuzwieser and Gessler 2010; Panda et al. 2012; Guo et al. 2017; Ali et al. 2018b). Even in China, the excessive amounts of fertilizer application in paddy lands to boost yield are no longer a viable option, especially under increasing fertilizer costs. This is further driving the search for rice cultivars with high nutrient-use efficiency that are required for sustainable rice production and increasing grain yield under optimal and suboptimal rates of fertilizer dosage. Thus, it is essential to develop rice varieties with high nutrient-use efficiency to maximize yield and productivity with the best agronomic management options such as judicious stage-specific dosages of fertilizer application. Further, it is essential to develop high-yielding, multiple-stress-tolerant rice varieties with combinations of several green traits such as tolerance of/resistance to drought, salinity, high/low temperature, flooding, blast, bacterial blight, tungro, brown planthopper, and stem borer, along with water-use efficiency. These multi-stress-tolerant varieties also need to meet market segment requirements such as duration, grain shape, and quality preferences. Recently, the incorporation of desirable nutrients for improved grain quality such as iron and zinc in multi-stress-tolerant cultivars has become necessary in rice improvement programs to ensure food security and overcome hidden hunger (Ali et al. 2020; Yu et al. 2020).

2 Green Super Rice

During the past two decades, breeders and biotechnologists have been working on various biotic and abiotic stress tolerances with a focus on increasing crop yield. This is possible by modifying the plant architecture and introgressing the target traits into the desired background through conventional and marker-assisted breeding approaches (Mehta et al. 2019; Oladosu et al. 2019; Gautam et al. 2020; Muthu et al. 2020). Complex abiotic stress tolerances such as drought and salinity tolerance are polygenic, and may have a negative association with grain yield components and might also interact with genetic and physiological mechanisms of similar traits (Guojun et al. 2009). Also, choosing the right donors for different target traits is challenging. In most cases, the abiotic stress-tolerant rice varieties are developed by crossing with tolerant landraces. However, concern exists regarding the difficulty in breaking undesirable linkages through conventional breeding approaches, which often represent a time-consuming process for selecting and fixing desirable lines, with limited success (Zhou et al. 2010; Wang et al. 2013). These developed tolerant rice varieties are low to moderate yielding under irrigated and rainfed conditions. In light of this apprehension, the Green Super Rice (GSR) breeding strategy began in 2008 at IRRI to efficiently develop lines with multiple-stress tolerance and more nutrient- and water-use efficiency with high genetic gains for various targeted ecosystems. The development of superior rice varieties with the GSR breeding approach under decreased rates of fertilizer, pesticide, and irrigation water, grown in marginal environments and producing fewer greenhouse gas emissions, and with improved grain nutrient elements (Zhang 2007; Wing et al. 2018), is described in Fig. 2. GSR breeding aims to develop stable high-yielding GSR rice varieties with several green traits suitable to be grown under lower input conditions in irrigated and rainfed areas of Asian and African countries.

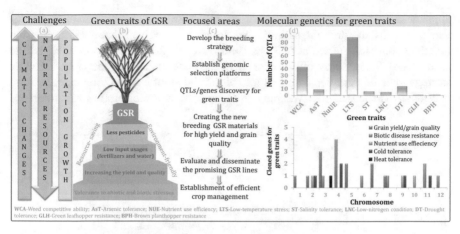

Fig. 2 A schematic representation of the Green Super Rice (GSR) breeding program for the development of green traits and increasing grain yield with improved grain quality. (**a**) The drastic changes in climatic variations, degradation of natural resources, and increasing global population are the major driving forces for increasing food demand by about 42% by 2050. (**b**) To ensure food security against these challenges, the GSR concept was proposed in 2005, and its major focus was to develop novel rice varieties with green traits, which included tolerance of multiple stresses (biotic and abiotic), high nutrient-use efficiency, fewer inputs of fertilizer, and water-saving. The combination of green traits and the GSR breeding concept provided an environment-friendly and natural resource–saving dimension. (**c**) The major target of GSR breeding is to develop new GSR cultivars through the integration of an advanced genomics platform for determining the genomic regions for green traits and further disseminating promising cultivars for specific target regions to improve the income of resource-poor farmers. (**d**) As of now, many QTLs have been identified using the GSR breeding strategy, and several genes have been functionally characterized and cloned for increasing the tolerance of abiotic stresses (Ali et al. 2020; Yu et al. 2020). These identified QTLs and genes provide abundant genetic information for the development of novel GSR varieties

Integrating advanced genomics and stringent phenotypic selection in the GSR breeding strategy helped to identify promising rice varieties besides understanding the molecular genetics and physiological mechanisms underlying trait expression (Wing et al. 2018; Ali et al. 2020; Yu et al. 2020). This breeding strategy played a significant role in the quick introgression of the desired target traits with the highest precision and with less or no genetic drag.

2.1 GSR Breeding and Population Development

The concept of GSR and its breeding strategy followed a systematic breeding effort and advanced genotyping approaches for developing varieties with significantly improved tolerance of multiple biotic and abiotic stresses, along with improved nutrient- and water-use efficiency. A schematic view of the backcross GSR breeding technology for developing early backcross-selective introgression lines (EB-SILs)

appears in Fig. 3. The development of several promising stress-tolerant rice cultivars that withstand drought, salinity, and flooding through a selective introgression
breeding approach started at IRRI in 1998 (Li et al. 2005; Ali et al. 2006; Lafitte
et al. 2006). These early efforts helped in the development of the GSR breeding
strategy to breed varieties with multiple biotic and abiotic stress tolerance. Starting
in 2009, IRRI focused on developing cultivars with multiple abiotic stress tolerance

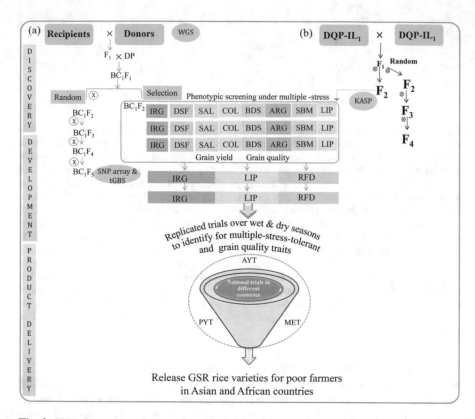

Fig. 3 IRRI-Green Super Rice early backcross breeding strategy to develop rice varieties with
multiple-stress tolerance, resource-use efficiency, and high yield with market-required grain quality. (**a**) The selected donors were crossed with HHZ and WTR-1 and later selfed to create 16 BC$_1$F$_2$
bulk populations. These backcrossed lines were used in three rounds of phenotypic selection under
different environments and selected lines with better performance were evaluated in the different
targeted ecosystems. (**b**) Identified promising trait-specific selective introgression lines with more
than one target green trait were used for the designed QTL pyramiding to combine various desirable traits into one single line. The blue-colored circles indicated in the schematic diagram were
used to show the different genotyping technologies used to identify genomic regions associated
with green traits in GSR breeding populations. (*IRG*, irrigated conditions; *DSF*, drought stress in
field conditions; *BDS*, biotic-stress phenotypic screening; *RFD*, rainfed environment; *SAL*, salinity
at the seedling stage at 18 d/sm (deciSiemens per metre); *COL*, cold stress; *LIP*, low inputs (without any fertilizer, pesticide, or herbicide); *ARG*, anaerobic germination (direct-seeded and immediately submerged in water for 21 days and maintained at 10-cm water depth); *SBM*, submergence
for 21 days at 14 days of seedling stage

through an innovative GSR breeding strategy to benefit poor and smallholder farmers (Ali et al. 2012). The concept of a GSR breeding program primarily involved two fundamental steps: first, developing superior EB-SILs; second, identifying suitable SILs to develop designed QTL pyramiding (DQP) lines. These two approaches helped to develop GSR varieties with multiple biotic and abiotic stress tolerance without compromising on grain yield, grain appearance, and cooking quality parameters.

The GSR breeding program at IRRI initially focused on identifying and exploring genetic variation and further selecting ~500 elite EB-SILs in an elite and widely adaptable Chinese variety, Huang-Hua-Zhan (HHZ), background with 16 donors from the mini-core collection of rice germplasm and a part of the 3000 Rice Genomes Project. These parental varieties underwent whole-genome sequencing to identify genomic variations. In the IRRI GSR breeding program, two additional elite recipients, Weed Tolerant Rice-1 (WTR-1) and TME80518, were used for developing EB-SILs (Ali et al. 2012, 2013b, 2018a). In this GSR breeding program, the BC_1F_2 populations were derived from crosses between the recipients (HHZ, WTR-1, and TME80518) and the 16 donors at IRRI. The F_1s derived from the HHZ background were again backcrossed with HHZ and subsequently selfed, and the progenies bulked to generate a total of 16 BC_1F_2 bulk populations. These bulk populations were screened over three rounds of selection for different abiotic stresses (drought, salinity, submergence, and low-input fertilizer) and biotic stresses (blast, bacterial leaf blight, and tungro), and in normal irrigated conditions. This led to the identification of a total of 1333 trait-specific SILs from the HHZ background, 2232 SILs from the WTR-1 background, and 1408 SILs from the TME80518 background.

The developed SILs were phenotyped across different stress and non-stress conditions and genotyped with SSR and high-density SNP markers. The phenotypic and marker data generated using SSR and SNP analysis allowed us to identify the donor QTLs associated with target abiotic and biotic stress tolerance/resistance and yield in the SILs. Then, the best SILs were selected as parents based on both their superior phenotypes and complementary donor alleles for designed QTL pyramiding (DQP) to develop better GSR varieties that combine various target traits from two or more donors. These promising SILs showed high tolerance vis-à-vis the respective tolerant checks IR74371-70-1-1, FL478, IR49830, and PSBRc82 under normal and stress conditions. Through these DQP approaches, a total of 2023 pyramiding lines (PDLs) from the HHZ background and 661 PDLs from the WTR-1 background were developed and found superior to the checks for all of the traits studied (Ali ct al. 2013a, 2020; Li and Ali 2017). The selected 564 EB-SILs derived from the genetic background of WTR-1 as recipient parents and 11 donor parents were sequenced using tunable genotyping-by-sequencing (tGBS) technology. Table 1 provides the list of donors and recipients used. These SILs developed through early backcross breeding were more advantageous in terms of high allelic diversity, which comprised multiple donor parents. As compared with the biparental mapping population, these SILs provide more precise QTL detection and fine-mapping of candidate genes for biotic and abiotic stress tolerance (Ali et al. 2017, 2018a). A total of 102 loci were identified from the 11 populations, and they contain

Table 1 List of donors used in the GSR breeding program

S. no.	Parent	Rice variety	Origin	Key features
1	DP	OM1723	Vietnam	Long panicle, salinity and drought tolerance
2	DP	Phalguna	India	Fine-grain type, resistance to blast and gall midge disease
3	DP	IR50	Philippines	Resistance to insects and diseases, superior grain quality
4	DP	IR64	Philippines	High yield, tolerance of lodging, blast resistance, long slender grain
5	DP	Teqing	China	High yield, tolerance of nitrogen deficiency and zinc deficiency
6	DP	PSBRc66	Philippines	High amylose content
7	DP	CDR22	India	Strong restoring ability and high combining ability, resistance to blast
8	DP	PSBRc28	Philippine	High yield, resistance to blast, moderate resistance to brown planthopper, bacterial leaf blight, green leafhopper, and stem borer
9	DP	Yue-Xiang-Zhan	China	Wide adaptation and high harvest index
10	DP	Khazar	Iran	Tolerance of salinity, zinc deficiency, and anaerobic germination
11	DP	OM1706	Vietnam	Tolerance of salinity, anaerobic germination, and submergence, resistance to brown planthopper
12	DP	IRAT352	CIAT	Tolerance of low pH and aluminum toxicity
13	DP	Zhong 413	China	High yield, wide compatibility, restoring ability
14	DP	R644	China	Resistance to brown planthopper
15	DP	IR58025B	Philippines	Popular wild-abortive elite maintainer
16	DP	Bg304	Sri Lanka	Tolerance of salinity and zinc deficiency, high yield of red rice, resistance to glyphosate and gall midge, blast, and bacterial leaf blight diseases
17	RP	Huang-Hua-Zhan	China	Chinese *indica* variety with high yield and wide adaptability, superior grain quality, moderate tolerance of salt and drought stresses, long panicle
18	DP	Haoannong	China	Long panicle, tolerance of low-temperature stress
19	DP	Cheng-Hui 448	China	Excellent restorer line, salinity tolerance
20	DP	Feng-Ai-Zan	China	High yield
21	DP	Y-134	China	Tolerance of submergence and anaerobic germination, resistance to brown planthopper
22	DP	Zhong 413	China	High yield, wide compatibility, restoring ability
23	DP	Khazar	Iran	High yield, blast resistance, moderate resistance to stem borer, mild aroma
24	DP	BG 300	Sri Lanka	Drought tolerance, resistance to brown planthopper, gall midge, blast, and bacterial leaf blight
25	DP	OM 997	Vietnam	Drought tolerance, resistance to blast
26	DP	Basmati-385	Pakistan	Tolerance of zinc deficiency

(continued)

Table 1 (continued)

S. no.	Parent	Rice variety	Origin	Key features
27	DP	M 401	United States	Premium grain quality, large kernel, tolerance of low nitrogen, good yield, late maturity
28	DP	X 21	Vietnam	High yield, disease resistance
29	RP	Weed Tolerant Rice-1	China	High yield, widely adapted rice variety

120 deleterious SNPs, with a range of one to four SNPs per loci. These loci can substitute the amino acid, and this leads to changes in the functional proteins in either positive or negative regulations (Ali et al. 2018a). Interestingly, the donors of SILs have a different introgression frequency of alleles in the recipient genomic regions and are mainly attributed to the severe selection pressure under abiotic stress conditions. With the availability of tGBS marker information with the GSR breeding approach, several genetic loci have been identified through the various linkage-based mapping methods. This innovative approach assured the development of new rice varieties that were tolerant of multiple stresses.

3 Genetics of Green Traits

In the GSR breeding program, the critical genetic determinants of various abiotic stress tolerances can be identified through QTL mapping approaches. As of now, the details of the identified QTLs and breeding materials are listed in Table 2. In many instances, advancement in the identification of stress-tolerance genes and QTLs that determine the trait has led to insights into the specific physiological and molecular mechanisms in response to various biotic and abiotic stresses (Wing et al. 2018; Yu et al. 2020). The highlights of the green traits associated with genetic information that confers stress tolerance and adaptive strategies provide valuable information for crop improvement.

3.1 *Drought Tolerance*

Drought is a complex trait. Understanding its mechanisms in terms of plant-water relationship traits, molecular breeding strategies, and dissecting the molecular genetics of QTLs and deployment in the breeding pipeline are the key steps for the development of drought-tolerant rice varieties. For drought improvement, it is essential to use advanced genomics and omics platforms that provide an opportunity for the mining of trait-specific allele functions. Identifying and developing drought-tolerant rice cultivars mainly depends on two major factors: prospecting for donors and effective

Table 2 List of QTLs and putative candidate genes for green traits in the Green Super Rice breeding program

Traits	Genotyping	Parents (RP/DP)	Breeding strategy	Polymorphic SNPs	Number of QTLs	Putative candidate genes	Chromosomes	Reference
Weed-competitive ability	SNP array	WTR-1/Y-134	EB-SILs	677	43	13	1, 2, 3, 5, 6, 7, 9, 10, 11, and 12	Dimaano et al. (2020)
Arsenic tolerance	SNP array	WTR-1/Haoannong	EB-SILs	704	9	25	1, 2, 5,6, 8, and 9	Murugaiyan et al. (2019)
Nutrient-use efficiency	tGBS	WTR-1/Haoannong	EB-SILs	1174	13	90	1, 2, 3, 4, 5, 9, 10, and 12	Mahender et al. (2019)
		WTR-1/Cheng-Hui 448	EB-SILs	1110	4	30	4, 5, 6, and 8	
		WTR-1/Zhong 413	EB-SILs	834	2	–	1,11	
Nutrient-use efficiency	SNP array	WTR-1/Haoannong	EB-SILs	704	49	–	1, 2, 3, 4, 5, 6, 7, 8, 9, 10, 11, and 12	Jewel et al. (2019)
Low-temperature stress (LTS) tolerance	SNP array	WTR-1/Haoannong	EB-SILs	704	82	16	1, 2, 3, 4, 5, 6, 7, 8, 9, 10, 11, and 12	Najeeb et al. (2020)
Salt tolerance	tGBS	WTR-1/Khazar and BG 300/WTR-1	DQP	9244	6	87	1, 2, and 4	Pang et al. (2017b)
Brown planthopper resistance	SNP array	Huang-Huan-Zhan/Khazar	EB-SILs	702	1	71	1	Balachiranjeevi et al. (2019)
Low-nitrogen	KASP and WGS	HHZ/Teqing, CDR22, and OM1723	DQP	3162	7	–	1, 2, and 3	Feng et al. (2018)
Drought tolerance	KASP	HHZ/Teqing, CDR22, and OM1723	DQP	3162	9	–	2, 3, 5, 6, 8, and 12	Feng et al. (2018)

SNP array single nucleotide polymorphism array, *WGA* whole-genome sequence, *DQP* designed QTL pyramiding, *EB-SILs* early backcross-derived selective introgression lines, *KASP* competitive allele-specific PCR phenotypic evaluation methods required for crossing programs and a successful selection scheme. However, improving drought tolerance in rice varieties has had slow progress due to the complexity of the trait and some associated undesirable characteristics, including low grain yield and a lack of preferable nutritional grain quality traits (Swamy and Kumar 2013; Kumar et al. 2014). With pre-breeding activity and marker-assisted breeding approaches, several drought-tolerant lines have been developed and further used in breeding programs for improving lines with multiple-stress tolerance (Singh et al. 2016; Ali et al. 2017; Pang et al. 2017a; Gautam et al. 2020; Kumar et al. 2020). However, improving yield potential under drought environments is limited due to the intensity, duration, and timing of drought stress. Important are the establishment of efficient phenotypic screening protocols and using precise phenotypic selection criteria (Ouk et al. 2006; Oladosu et al. 2019). From the recent developments in various molecular breeding strategies, precise high-throughput phenotypic technologies have been identified for promising traits such as grain yield, and morpho-physiological component traits have shown moderate to high heritability under drought stress conditions (Tuberosa 2012; Kumar et al. 2014; Swain et al. 2014; Ali et al. 2017; Sahebi et al. 2018; Yadav et al. 2019). Several researchers have used direct selection of GY, along with considering other secondary traits using molecular and genomic approaches (Oladosu et al. 2019). The results led to the identification of the QTLs and genes that are associated with these traits. Discovering these QTLs/genes that are responsible for tolerance traits is essential for developing crops with tolerance through marker-assisted selection or genetic engineering approaches. To this end were identified the major-effect drought tolerance QTLs from drought-tolerant rice varieties Apo ($qDTY_{1.1}$, $qDTY_{2.1}$, $qDTY_{3.1}$, and $qDTY_{6.1}$), Way Rarem ($qDTY_{12.1}$), Nagina 22 and Dhagaddeshi ($qDTY_{1.1}$), and Vandana ($qDTY_{2.3}$ and $qDTY_{3.2}$) (Bernier et al. 2007; Venuprasad et al. 2009; Vikram et al. 2011, 2015; Ghimire et al. 2012; Dixit et al. 2014). With the marker-assisted introgression approach, grain yield QTLs with major and consistent effects ($qDTY_{1.1}$, $qDTY_{2.1}$, qDTY2.2, $qDTY_{3.1}$, $qDTY_{3.2}$, $qDTY_{6.1}$, and $qDTY_{12.1}$) on chromosomes 1, 2, 3, 6, and 12 have been validated, deployed into drought breeding programs, and successfully improved drought-susceptible rice mega-varieties (Kumar et al. 2014, 2017b, 2020; Singh et al. 2016; Sandhu and Kumar 2017). However, the locally adapted high-yielding commercial mega-varieties show a sensitive reaction to other biotic and abiotic stresses. These improved tolerant rice varieties have not been able to show a similar yield performance in varied agroecosystems. Introgression of these major QTLs/genes in a different combination using a marker-assisted selection approach provides an opportunity to increase desirable phenotypic traits and thereby improve grain yield. Still, pyramiding of these major QTLs/genes has not been fully exploited due to G × E interaction of this complex trait epistasis, and pleiotropy could have negative/positive effects on the trait of interest (Li 1998; Yano et al. 2003; Xu and Crouch 2008). As a consequence, it has been tough to make significant genetic improvements in grain yield under drought stress through conventional breeding methods. Therefore, the development of large-scale phenotypic screening, marker-assisted breeding strategies, and advanced genotyping technologies has significantly accelerated varietal improvement programs on drought tolerance. Further, the identification of QTLs/genes affecting grain yield could result in yield improvement and stability under drought stress. In comparison to the classical breeding and genetic mapping approaches, the GSR breeding strategy used SILs and designed QTL pyramiding approaches in the development of multiple-stress-tolerant rice varieties for improving grain yield under stress conditions.

The GSR breeding strategy has three significant advantages: first, the selected few SILs require lesser costs in both genotyping and phenotyping; second, the strategy significantly increased the power in detecting the genomic regions of QTLs/genes; third, the selected lines mostly carried the beneficial alleles of QTLs (Ali et al. 2017). Based on the DQP approach, Feng et al. (2018) identified a total of nine drought tolerance QTLs on chromosomes 2, 3, 5, 6, 8, and 12 using a segregation distortion approach. These QTLs were detected from the three different genetic backgrounds of trait-specific SILs, which were generated from the 63 SILs from the cross HHZ × Teqing, 68 SILs from the cross HHZ × CDR22, and 75 SILs from the cross HHZ × OM1723 (Ali et al. 2017). The major significant QTLs on chromosomes 3, 5, and 8 were closely associated with the earlier reports of grain yield QTLs and seed fertility QTLs under drought stress. Similarly, researchers exploited the genotypic information on EB-SILs derived from the 11 donors crossed with recurrent parent WTR-1, which comprised a total of 564 diverse SILs used for identifying synonymous and non-synonymous deleterious polymorphic SNPs (Ali et al. 2018b). Of these, the significant locus *Os01g01689* on chromosome 1 possessed a G/A deleterious SNP altering an amino acid from *Ala* to *Thr* in the background of Haoannong and Y-134. This locus was associated with abiotic stress tolerance. This innovative breeding strategy addresses the GCV challenge and allows poor farmers to benefit from the use of stress-tolerant and high-yielding rice varieties under limited resources. Therefore, genotyping the EB-SILs and DQP populations helped to identify several promising genomic regions for complex traits and assisted in molecular marker development and genomic applications in molecular breeding programs.

3.2   Salinity Tolerance

More than 20% of global agricultural land is affected by salinity, especially in the coastal regions of South and Southeast Asian countries. Salinity is mainly caused by the drastic changes in the climatic events that are significantly associated with increases in salinity in the soil. During crop growth stages, the accumulated Na^+ and Cl^- concentrations in plants lead to inhibiting cell expansion and photosynthetic activity, followed by changes in physiological and molecular pathways in roots for the uptake of water and nutrients, leading to cytotoxic effects. Mainly, the accumulation of these elements can be processed through osmotic and ionic stresses, which can activate the formation of reactive oxygen species (ROS) and cause leaf damage or plant death. Thus, developing rice varieties with salinity tolerance is a prominent approach to resolve the salinity problem. However, progress in developing salinity-tolerant rice varieties has been slow because of the complexity of the salinity tolerance trait, and this has involved multiple physiological and biochemical pathways.

As of now, several researchers had worked on the complex traits using advanced genomic-assisted breeding tools in traditional molecular breeding strategies and sequencing at the whole-genome level. In the past two decades, >700 QTLs and 200 genes have been reported for salinity tolerance at different growth stages (Rahman et al. 2017; van Oort 2018) (http://archive.gramene.org/). Most of these genes and QTLs are associated with different traits such as salt injury score, shoot and root growth, fresh and dry weight of shoot and root, Na^+ and K^+ content in shoot and root, grain yield, and chlorophyll content. The genomic regions of these QTLs and genes are involved in the different physiological and molecular mechanisms such as ionic equilibrium, osmotic adjustment, transcription regulation, and signaling pathway (Molla et al. 2015; Pradhan et al. 2015; Gimhani et al. 2016; Kumar et al. 2017a). However, the identified genomic regions have a more significant interval gap because they employ a low density of molecular markers and possibly because of epistatic QTLs and environmental interaction. It's possible to have hundreds of genes in those QTL positions, which makes it difficult to track the genetic pathways.

The recent development of genotyping technologies provided robust genotyping information and can remarkably reduce chromosomal intervals and also help in identifying accurately predicted gene functions related to the target traits. Dissecting these complex traits and identifying superior tolerant lines, the GSR breeding program began an integrated molecular breeding strategy to identify genome-wide trait-specific introgression lines (ILs) through the DQP approach. Using high-density SNP markers of tGBS technology, three significant QTLs, *qSES2*, *qSES4* (*qChlo4*), and *qChlo1*, were mapped, and this further narrowed down the list of genes through the use of grandparent genotypic information (Pang et al. 2017b). As a result of this approach, 13, 34, and 40 candidate genes were identified in the QTLs regions, respectively. Pang et al. (2017b) carried out functional analyses of the candidate genes for the salinity-tolerance QTLs to infer two, five, and six genes as the most likely candidates of *qSES2*, *qSES4* (*qChlo4*), and *qChlo1*, respectively. This combination of high-density SNP markers with parent and grandparent information could be a potential approach for identifying the right genes for desired target traits in rice. With this strategy, several high-yielding, multi-stress-tolerant rice varieties such as NSIC Rc480, NSIC Rc534, NSIC Rc 390, NSIC Rc 392, NSIC Rc 554, and NSIC Rc 556 were developed and released for cultivation under saline conditions (Guan et al. 2010; Ali et al. 2017; Yu et al. 2020).

3.3 Submergence Tolerance

Submergence is an essential factor that limits rice yield over 15 million ha in rainfed lowland regions of Asia (Septiningsih et al. 2009). Rice is known to adapt well to flooded conditions, but most rice cultivars become vulnerable upon complete submergence. Complete submergence of rice plants, particularly under turbid water, causes severe damage through decreased respiration and photosynthesis (Ella et al. 2003; Bailey-Serres et al. 2010). Submergence tolerance is an important trait that could help rice plants to overcome flooding with a water head of 1 m for 14 days at the active tillering stage. The genetics of submergence tolerance from FR13A showed a high heritability and was governed by both major genes and some quantitative trait loci (QTLs) (Suprihatno and Coffman 1981; Nandi et al. 1997; Sripongpangkul et al. 2000; Toojinda et al. 2003). Xu et al. (2006) used FR13A for cloning *Sub1A*, a major QTL on chromosome 9, which is a transcription factor concerning ethylene responsiveness. *Sub1A-1*, one of the alleles at *Sub1A*, was found to be essential for submrgence tolerance in several semi-dwarf varieties (Septiningsih et al. 2009). Ali et al. (2006) demonstrated that progenies with submergence tolerance were identified consistently in most BC populations derived from sensitive parents. This suggested the presence of hidden genetic diversity for submergence tolerance in the primary gene pool of rice. Wang et al. (2015) determined the genetic basis of submergence tolerance in rice and facilitated the simultaneous improvement of submergence tolerance in rice. In their study, they characterized the genome-wide responses of 162 SILs with submergence tolerance from 12 BC populations using SSR markers that helped in the dissection of the hidden diversity and transgressive segregation. Their results provided insights into the genetic basis of submergence tolerance of rice and demonstrated a novel strategy for simultaneous improvement and genetic dissection of complex traits using the approach of selective introgression (Li et al. 2005). The genome-wide responses of donor alleles to strong phenotypic selection for submergence tolerance can be understood with three key findings from Wang et al. (2015). First, they found significant over-introgression of the donor alleles at 295 loci in 167 functional genetic units (FGUs) across the rice genome. Second, they observed significantly increased homozygosity or "loss of heterozygosity" genome-wide. Third, pronounced non-random associations between or among the detected submergence tolerance loci led to the discovery of putative genetic networks (multi-loci structures) underlying submergence tolerance in rice. Further, their results suggested that submergence tolerance of rice is controlled by large numbers of loci involved in multiple positively regulated signaling pathways (Wang et al. 2015). It is essential here to understand that the restoration of one or more of these broken pathways in the BC progenies by genetic complementation from the introgressed functional donor alleles at submergence tolerance loci provided an appropriate explanation for the transgressive segregation of submergence tolerance and this could be extended to other complex traits in rice. The GSR breeding strategy developed several promising salinity-tolerant materials such as NSIC Rc480, GSR 5, and GSR11. Yorobe et al. (2016) found that the mean difference in net farm income between GSR and non-GSR varieties was positive and significant and gave farmers an income advantage of USD 230.90/ha for GSR variety users. In their study, Yorobe et al. (2016) found that, with a high occurrence of flooding in the wet season, the use of GSR varieties assured rice farmers of a positive net farm income.

3.4 Nutrient-Use Efficiency

Nutrient-use efficiency (NUE) is one of the most critical traits for increasing yield and productivity by using nutrients such as nitrogen (N), phosphorus (P), and potassium (K) to increase NUE. During the Green Revolution (GR) period in the 1960s and post-GR, yield increased significantly, and this was primarily achieved with a higher input dosage of fertilizer, pesticide, and water (Ali et al. 2018b). Recently, Hawkesford and Griffiths (2019) mentioned that only 33% of global N is recovered from harvested grain, while the remaining N is a significant pollutant and a colossal waste of resources. A higher amount of fertilizer application can lead to a substantial imbalance in nutrient availability in the soil, increase the risk of pests and diseases, and not be cost-effective for poor farmers (Chen et al. 2014a, b; Rahman and Zhang 2018). Regarding environmental safety and the use of low inputs in farmers' fields, the GSR breeding team started a program at IRRI to identify nutrient-use-efficient breeding lines that could obtain higher grain yield under integrated nutrient management techniques. This was the first initiative in the NUE breeding program at IRRI (Jewel et al. 2018) and it was reported as a unique and systematic breeding approach through the selection of SILs with higher NUE through the early backcross breeding program. SILs were selected in four consecutive seasons under different nutrient fertilizer combinations of N, P, and K dosages. Five promising SILs (Nue-115, Nue-114, Nue-112, Nue-229, and Nue-230) were identified with higher grain yield and nutrient-use efficiency. These SILs could provide valuable information for rice breeding programs. The genetics of NUE (Mahender et al. 2019) identified a total of 19 QTLs that were associated with three agronomic traits by using tGBS technology. These major QTLs were located on chromosomes 2, 5, 8, 9, and 12. The genomic regions of these QTLs were co-localized with earlier reports of low nitrogen and phosphorus conditions. Importantly, in silico analysis of these QTL positions suggested that several key candidate genes played a major role in the various molecular and physiological pathways in response to abiotic stress tolerance and also maintenance of the homeostasis mechanism under low-input conditions. Similarly, Jewel et al. (2019) detected a total of 49 main-effect QTLs under six nutrient conditions. These QTLs explained a phenotypic variance range from 20.25 to 34.68%. They were located on all 12 chromosomes, except on chromosomes 7, 11, and 12. Among them, four hotspot QTLs were identified on chromosomes 3, 5, 9, and 11. Interestingly, 22 QTLs for partial factor productivity and four QTLs for agronomic efficiency were detected as novel under –P and 75 N conditions. Several genes and transporters were located in the interval regions of these QTLs, and they were involved in nutrient uptake and transporting mechanisms from soil to plants. Therefore, the hotspot regions of QTLs and genes may offer significant value for marker-assisted selection and pyramiding of multiple QTLs for improving NUE in rice.

3.5   Weed-Competitive Ability Traits

Threats such as GCV, rising labor shortages, decreasing arable land, and the increasing prices of fertilizer and pesticide are the major contributors to the decrease in rice production (Singh et al. 2013). To overcome these constraints, shifting from the manual rice transplanting system to direct-seeded rice (DSR) is the most promising approach to improve rice sustainability (Chauhan and Abugho 2013; Mahender et al. 2015). It has numerous benefits, mainly in decreasing water use by 35–75%, decreasing labor demand, shortening crop duration, mitigating methane gas emissions, and lowering the cost of cultivation (Mahender et al. 2015; Dimaano et al. 2017). However, vigorous growth of weeds is one of the most complicated biological constraints in the DSR system to attaining optimal grain yield (Antralina et al. 2015; Chauhan et al. 2015a, b; Jabran and Chauhan 2015). Several options such as tillage operations and herbicide application are available to control weeds, but these are laborious and costly (Rahman et al. 2012). Globally, more than USD 100 billion are lost annually because of weed control (Appleby et al. 2002). Therefore, developing a breeding strategy for weed-competitive (WC) rice varieties is a critical solution for decreasing tillage operations, hand weeding, and herbicide inputs in the DSR system. WC ability is a complex trait, and it interacts with several agro-morphological traits (Chauhan et al. 2015a; Raj and Syriac 2017). Thus, it is essential to understand the trait interactions and mechanisms that confer WC ability, and this could be useful for speeding up the breeding activities for developing WC rice cultivars. Recently, as a part of the GSR breeding program at IRRI, Chauhan et al. (2015b) and Dimaano et al. (2017) followed a systematic breeding effort to identify WC ability traits related to early seed germination (ESG) and early seedling vigor (ESV). The breeding materials were developed from the four early generations of backcross populations derived from one common recipient parent, Weed Tolerant Rice-1 (WTR-1), and four donors, Y134, Zhong 143, Khazar, and Cheng Hui-448. These SILs were evaluated over three rounds of selection in upland weed-free, upland weedy, and lowland weedy conditions. Among the total SILs, five (G-6-L2-WL-3, G-6-RF6-WL-3, G-6-L15-WU-1, G-6-Y16-WL-2, and G-6-L6-WU-3) were found to be promising in lowland weedy conditions, whereas four SILs (G-6-Y7-WL-3, G-6-Y6-WU-3, G-6-Y3-WL-3, and G-6-Y8-WU-1) were found to have the highest grain yield under upland weedy conditions (Dimaano et al. 2017). The primary requirements for the DSR system to be successful are the following: uniformity and speed of germination rate and early seedling growth are significantly associated with robust and vigorous crop growth and a better crop establishment, which can influence WC ability (Cui et al. 2002; Foolad et al. 2007; Diwan et al. 2013; Dang et al. 2014). These traits can provide support for traits attributed to efficient root growth that can help in the absorption of more nutrients (Farooq et al. 2011; Matsushima and Sakagami 2013; Singh et al. 2015; Khan et al. 2016). For the molecular genetics of WC rice cultivars, a total of 43 QTLs were mapped on all 12 chromosomes, except on chromosomes 4 and 8, by using 677 high-quality SNP markers (Dimaano et al. 2020). Interestingly, 29 novel genetic loci were associated with ESG and ESV traits on chromosomes 1, 3, 5, 6, 7, 10, 11, and 12. The hotspot regions of chromosomes 11 and 12 were associated with multiple traits (Dimaano et al. 2020). Many of these QTLs were co-localized in previous reports, which are related to germination rate, germination index, germination percentage, and germination time in different genetic backgrounds of mapping populations (Mahender et al. 2015). In addition, some critical genes located in the co-localized hotspot regions can influence the regulation of various physiological functions such as chloroplast development, photosynthesis, hybrid sterility, seed development, and seedling lethality during plant growth stages (Gothandam et al. 2005; Matthus et al. 2015; Sharma and Pandey 2016; Yu et al. 2016). Therefore, the hotspot regions with co-localized QTLs and genes may have a more significant role in the improvement of weed-competitiveness, mainly in African and Asian countries, to decrease rice production costs.

3.6   Low-Temperature Stress Tolerance at Different Crop Growth Stages

Rice is one of the most sensitive among the cereal crops to low-temperature stress (LTS)/cold stress (CS), mainly at the germination to booting stage, which can cause a significant yield decrease due to reduced germination rate and seedling growth, high spikelet sterility, delay in flowering, and lower grain filling (Ranawake et al. 2014; Schläppi et al. 2017; Shakiba et al. 2017; Najeeb et al. 2020). The drastic changes in GCV in rice-growing areas mainly in the tropical, subtropical, and temperate regions had severe effects from low-temperature stress during the critical stages of seedling growth and pollen abortion at the booting stage, leading to grain yield decreases (Ye et al. 2009; Jena et al. 2010; Sun et al. 2018). LTS mainly inhibits sugar accumulation in the pollen and further leads to male sterility, which is regulated by the invertase enzyme through the hormone abscisic acid (ABA) pathways that transport sugar to the tapetum. The results of these mechanisms decrease invertase amounts in susceptible rice varieties, leading to lower pollination (Oliver et al. 2007). This situation warrants the development of LTS-tolerant varieties through a systematic breeding effort to increase rice production in 25 countries, including the major rice-producing countries (Cruz et al. 2013; Zhang et al. 2017). Therefore, to minimize yield losses under LTS, particularly in cold-affected regions, it is crucial to identify potential donors to improve LTS tolerance. LTS tolerance breeding for widely adapted and high-yielding rice cultivars is needed to meet future food demand worldwide.

Information on the genomic regions of QTLs and genes governing tolerance of LTS is limited for different growth stage-specific traits in rice due to its complex nature, and this significantly influences QTL × environment interactions. Up to now, several studies have been reported for QTL mapping for LTS. Recently, Najeeb et al. (2020) reviewed the genetics of LTS tolerance QTLs in rice. A total of 239 and 339

QTLs were identified on 12 chromosomes using genome-wide association studies and biparental mapping populations, respectively (http://archive.gramene.org). However, mapping of the sensitive stage, especially at the reproductive phase of a complex trait, involves multiple genes, environment interactions, and difficulties in phenotypic screening. Despite these large numbers of QTLs for LTS, a few of them were studied for fine-mapping and cloning. Based on the public domain and rice database, a total of 38 candidate genes were functionally characterized for LTS tolerance. Importantly, four genes on chromosome 1 (*OsCOIN, OsGSK1, OsGH3-2,* and *OsMYB3R-2*), two genes on chromosome 6 (*OsiSAP8* and *OsbZIP52*,) and a single gene on chromosome 4 (*OsCAF1B*) and chromosome 11 (*OsAsr1*) were significantly associated with LTS tolerance in the seedling and reproductive stage in rice. In contrast, breeding strategies for LTS tolerance remained a slow process by conventional methods because most of the QTL mapping methods could not explain G × E interaction well and appropriate statistical and advanced breeding strategies were lacking. Improvement of LTS toler-ance at the reproductive stage is needed through a selective introgression method (Liang et al. 2018). A population was derived from BC_2F_4 onward using five donors and a *japonica* (Geng) recipient parent that were screened over three rounds to select under LTS conditions at the reproductive stage. This approach helped in dissecting the complex trait of LTS tolerance and developed trait-specific introgression lines. A total of 17 QTLs were identified using five different populations of EB-SIL derived from five different donors into a common recipient parent. Further, multi-locus probability tests and linkage disequilibrium results showed that a total of 46 functional genetic units were distributed across the rice genome for cold tolerance. Studies showed the presence of strong epistasis and power of the statistical approach for the development of selective introgressions for simultaneous improvement and genetic dissection of complex traits. Zhu et al. (2015) used an inter-connected breeding (IB) population comprising 497 SILs derived from eight BC families with the same recipient parent for identifying and fine-mapping QTLs for LTS tolerance at the booting stage. A total of 41,754 high-quality SNPs were obtained through the re-sequencing of the IB popu-lation. Phenotyping was conducted under field conditions in 2 years and three loca-tions. Association analysis identified six QTLs for LTS tolerance on chromosomes 3, 4, and 12. The stably expressed QTL *qCT-3-2* was fine-mapped and narrowed down to approximately 192.9 kb on the reference genome (Zhu et al. 2015). The QTL *qCT-3-2* is essential for developing varieties with LTS tolerance at the booting stage, which are in high demand in temperate and high-altitude rice production regions. GWAS applied to an IB population allowed better integration of gene discovery and breeding. QTLs can be mapped in high resolution and quickly used for breeding.

3.7   Grain Quality

Improvement of grain quality traits (grain appearance, cooking and eating quality) is a major concern in rice breeding programs. Earlier rice breeding efforts over the past few decades primarily focused on increasing grain yield potential as the primary target in the major Asian rice-growing countries, and most of the high-yielding and popular rice varieties had poor grain quality traits (Custodio et al. 2019). Approximately one-third of the global population is suffering from nutritional deficiency, mainly caused by inferior grain quality traits such as low protein content and lack of vitamins and minerals (Balyan et al. 2013). Therefore, to meet the global food demand, it is essential to keep nutritional quality traits integrated well in our breeding programs. Consumer acceptance and their preferred grain traits are major factors that influence different markets across the globe. For instance, grain appearance traits such as long and slender grain shape are mostly preferred in countries such as India, Vietnam, the United States, and most Asian countries, whereas northern China, Korea, and Japan prefer short and round rice grains (Unnevehr et al. 1985; Cuevas et al. 2016). Improvement in grain quality is for a complex trait involving many traits such as milling efficiency, grain appearance (size and shape), and cooking and eating quality (apparent amylose content, gelatinization temperature, gel consistency). Many of these grain quality component traits are highly complex and quantitative in nature and are significantly influenced by environmental factors. In a global view, grain shape traits need to be understood, matching market and consumer preferences and needing to be adequately addressed in breeding programs for different target regions. Several researchers have identified many critical genetic loci and candidate genes for grain quality traits (Yun et al. 2014; Balakrishnan et al. 2016; Pang et al. 2016; Wang et al. 2017; Mogga et al. 2018; Kavurikalpana and Shashidhara 2018; Wing et al. 2018; Bazrkar-khatibani et al. 2019; Calayugan et al. 2020; Zhang et al. 2020). Mahender et al. (2016) reviewed grain nutritional traits and associated genetic information on QTLs and genes in rice. More than 400 QTLs have been reported for these grain quality traits, including grain appearance and cooking and nutritional properties (Mahender et al. 2016; Bazrkar-khatibani et al. 2019).

As of now, 28 major genes are involved in controlling grain shape and 65 genes are associated with eating quality traits in rice (http://qtaro.abr.affrc.go.jp/). Importantly, the natural variation of eating quality genes such as *ALK* and *WAXY* on chromosome 6, *Badh*-2 on chromosome 8, and *LOX*-3 on chromosome 3 regulates cooking quality traits and encodes the strong component in rice fragrance (Chen et al. 2008; Shirasawa et al. 2008; Gao et al. 2011; Venu et al. 2011; Wang et al. 2017). Similarly, overexpression of the six genes (*OASA2, BiP, OASA1D, OsAAT1, OsAAT2, OsISA1*, and *OsISA2*) on chromosomes 1, 2, 3, 5, and 8 significantly increased amino acid content and seed storage protein content in rice grain (Zhou et al. 2009; Saika et al. 2011; Utsumi et al. 2011). Recent successful GSR breeding strategies help to a great extent to understand the various green traits that include grain quality and biotic and abiotic stress tolerance genes that have been cloned (Yu et al. 2020). These include the critical genes *OsAAP6* on chromosome 1, *OsGRF4* on chromosome 2, three genes (*GNP1, lgy3*, and *GL3*) on chromosome 3, two genes (*Chalk5* and *GW5*) on chromosome 5, two genes (*W7* and *OsSPL13*) on chromosome 7, and a single gene (*OsOTUB1*) on chromosome 8 associated with grain quality traits, and some of these genes are significantly associated with yield-attributed traits and NUE traits (Li et al. 2018). Recently, Mahender et al. (2019)

identified 19 QTLs for agro-morphological traits under different dosages of nutrient fertilizer. Interestingly, one of the major QTLs for leaf chlorophyll content on chromosome 2 and in the same genomic region co-localized with the nitrate transporter *OsNPF7.2* and *GROWTH-REGULATOR FACTOR 4* (*OsGRF4*). These hotspot genomic regions play a vital role in the interaction of GRF4 and DELLA proteins that are involved in multiple gene regulation related to grain quality traits and that also maintain homeostatic coordination of nitrogen and carbon metabolism (Serrano-Mislata et al. 2017; Li et al. 2018; Xing et al. 2018).

3.8   Biotic Stress Tolerance

Rice grain yield declined globally by more than 52% because of various biotic stresses such as insect pests (brown planthopper, green rice leafhopper, and yellow stem borer) and diseases (bacterial leaf blight, blast, sheath blight, and tungro). Yield was severely affected in the rice-growing countries of Asia and Africa (Van Oort and Zwart 2018). These biotic stresses significantly cause annual crop losses that threaten global food security. According to the estimation of Roy-Barman and Chattoo (2005), global yield losses annually are equivalent to the quantity required to feed 60 million people. Worldwide, fungal disease alone is estimated at 14% of the annual rice yield decline. Hence, significant changes in GCV lead to increased extreme weather patterns and this combined with increasing air temperatures are the leading causes of the spread of disease into different areas, which has also differed from region to region and from one agroecology to another (Anderson et al. 2004; Agrios 2005). Therefore, exploiting the diverse resources of rice germplasm and understanding the host-plant resistance mechanism at molecular genetics and cellular levels will provide a viable option to manage this disease vis-à-vis the application of various pesticides. For better management of biotic stresses, integrative strategies are required for the selection of resistant rice varieties, and identifying QTLs and genes to understand the genetics of resistance from advanced genotyping technologies and pathogen races. This will deliver valuable information for the development of future climate-smart rice varieties.

Recent progress in QTL and gene/allele identification technologies, marker-assisted selection (MAS), and advanced genomic techniques has been used to develop disease and insect resistance in rice. The molecular genetics of the resistance to the various pathogens has been well documented through conventional or molecular marker-assisted breeding strategies. Therefore, the use of this disease resistance and insect resistance in genomic regions of QTLs and genes provides the most cost-effective and prominent approach for decreasing pesticide use. To date, more than 100 genes for blast resistance, 40 genes for bacterial blight resistance, 34 genes for brown planthopper resistance, 11 genes for general insect resistance, and 6 genes for sheath blight resistance have been identified on different chromosomes (http://qtaro.abr.affrc.go.jp/). Further, some of these genes have been isolated and functionally characterized (http://qtaro.abr.affrc.go.jp/). However, the two major diseases (blast and BLB) frequently affect crop yield and decrease it in most rice-growing regions worldwide. These cloned genes are providing valuable sources to understand the interaction between the disease and host, pathogen mechanism, and, further, as a means to enhance resistance mechanisms in the various backgrounds using marker-assisted breeding and genetic engineering approaches. However, breakdown of the resistance of a single gene has also been identified after 2 or 3 years. This indicates changes in the frequency of pathotypes or the emergence of new ones through mutations and other mechanisms. Therefore, to develop broad-spectrum resistance in rice varieties by stacking multiple resistance genes or QTLs into a single rice variety seeks to provide resistance to a wide range of races as compared to one or two gene combinations.

Pyramiding of multiple resistance genes is a useful molecular breeding strategy for the expression simultaneously of more than one gene to achieve durable resistance against desired target diseases, and it also needs to prevent or delay the breakdown of resistance (Shinada et al. 2014; Feng et al. 2018; Kumar et al. 2018; Liu et al. 2020). Ashkani et al. (2015) have reviewed different successful molecular breeding schemes and gene pyramiding strategies to improve biotic disease resistance genes in rice. Notably, Ji et al. (2016) used a gene pyramiding strategy to develop restorer lines with resistance to multiple diseases: blast (*Pita, Pi1,* and *Pi2*), bacterial blight (*Xa23* and *xa5*), and brown planthopper (*Bph3*). Therefore, restorer lines are useful in hybrid rice breeding programs. In the GSR breeding program, several biotic disease-resistance QTLs and cloned genes have been used in developing novel GSR breeding materials. *Pi2, Xa23, Bph14,* and *Bph15* genes have been mainly introgressed into different GSR lines. Recently, a novel gene, *Bph38*(t), identified on chromosome 1 explained phenotypic variation of 35.9% in a backcross population derived from a cross of HHZ and Khazar (Balachiranjeevi et al. 2019). The development of molecular mapping and functional genomics of insect resistance revealed that most of the BPH resistance genes were clustered together in specific regions on different chromosomes (3, 4, 6, and 12), except for *bph5, bph8,* and *BPH22(*t) to *BPH24(t)*. For example, 8 genes were located in the 19.1–24.4 Mb region on chromosome 12 and 12 genes in the 4.1–8.9 Mb region on chromosome 4 (Du et al. 2020). These clustered genes provide a valuable resource for identifying specific alleles and interaction of the different genes that are involved in the molecular and physiological pathways for the insect resistance mechanism in rice.

4 Molecular Genetics and Breeding Strategies to Combine Multiple Stresses

To develop rice varieties with multiple stress tolerance along with superior grain yield and quality, breeders are exploring different breeding strategies. However, limited progress has been made in this direction through conventional breeding approaches, which are lengthy and laborious processes. Recently available genotyping technologies such as SNP chips, genotyping by sequencing (GBS), and whole-genome sequencing (WGS) have led to the identification of major-effect QTLs associated with the complex abiotic stresses such as drought, salinity, and flooding in rice (Guo et al. 2014; Wang et al. 2017; Ali et al. 2018b; Le Nguyen et al. 2019; Yadav et al. 2019). The identification of molecular markers tightly associated with a trait is valuable for marker-assisted selection. The marker-assisted breeding efforts that began at IRRI led to the identification of a major QTL, *Sub1*, derived from FR13A, an Indian landrace; *qDTY$_{1.1}$* and *qDTY$_{2.1}$* from drought-tolerant genotype Apo; *qDTY$_{12.1}$* from Way Rarem; and *Saltol* identified from Pokkali, which showed large effects across different genetic backgrounds. Deployment of these QTLs in molecular breeding strategies paved the way for the development and release of rice varieties such as Swarna-Sub1, Samba Mahsuri-Sub1, Swarna-Sub1, DRR Dhan-50, and CR Dhan 802, which are suitable for different ecosystems (Bhandari et al. 2019). Most of the rice-growing areas in the rainfed environment in South and Southeast Asia are frequently affected by multiple abiotic stresses even within the same cropping season near the coastal areas. Therefore, developing new breeding materials that could tolerate multiple stresses and also provide higher grain yield is essential for global food security.

Over the past decade, there have been only a few significant reports on pyramiding QTLs for stacking multiple traits in the different backgrounds of popular high-yielding rice varieties through MAS approaches. The identified superior and stable lines across different environments and having acceptable grain quality traits are promoted for release in different countries. Pyramiding of the major biotic and abiotic stress-tolerance QTL combinations has been used in different breeding strategies for developing rice varieties with multiple-stress tolerance (Zhu et al. 2015; Dixit et al. 2017; Pang et al. 2017b; Feng et al. 2018; Kumar et al. 2018). Recently, Muthu et al. (2020) developed an improved White Ponni, a popular high-yielding rice variety with significant tolerance against drought, salinity, and submergence by introgressing major-effect QTLs (*qDTY$_{1.1}$*, *qDTY$_{2.1}$*, *Saltol*, and *Sub1*) through a marker-assisted backcross breeding approach. However, the innovative Green Super Rice breeding strategy successfully demonstrated that a high genetic diversity exists within the primary gene pool for improving multiple-stress-tolerance traits, especially for rainfed environments (Ali et al. 2006, 2017). Stringent simultaneous phenotypic selection by screening under multiple abiotic stresses in the early generations has a major advantage in developing trait-specific backcross inbred lines with significantly improved tolerance. This provides an opportunity for the discovery of genes/QTLs underlying the target and non-target traits. With the GSR breeding strategy within the span of 7 years, a total of 27 IRRI-bred GSR varieties were released, and more than 104 rice varieties have been nominated for national cooperative yield trials from three recipient parents and 16 donors. These varieties are now being cultivated on more than 2.7 million ha on a seed distribution basis alone for farmers in Asia and Africa (Ali et al. 2017; Feng et al. 2018; Yu et al. 2020). The integration of advanced genotyping technology such as SNP genotyping array and tGBS in the GSR breeding program provides a high-quality SNP calling accuracy with a low percentage of missing rates across populations. This genotyping information from SNPs is an excellent source for understanding the genetics of green traits and can be used for dissecting many complex traits by using marker-trait association studies (Wing et al. 2018; Feng et al. 2018). Among the various breeding populations, the early backcross-selective introgression lines (EB-SILS) and designed QTL pyramiding (DQP) approach have proven to be an effective strategy for dissecting complex traits such as drought, salinity, and low-temperature stress tolerance; arsenic toxicity tolerance; and nutrient-use efficiency (Dimaano et al. 2017; Pang et al. 2017b; Ali et al. 2018b; Feng et al. 2018; Mahender et al. 2019; Murugaiyan et al. 2019; Najeeb et al. 2020). So far, more than 3200 genes have been listed in Oryzabase, and they were associated with a wide range of stress-tolerance mechanisms in biotic and abiotic stresses (https://shigen.nig.ac.jp/rice/oryzabase/). More than 1800 genes were functionally characterized and deposited in the Q-TARO (QTL Annotation Rice Online) database (http://qtaro.abr.affrc.go.jp/). Of these, cloned rice genes associated with tolerance of multiple biotic stresses (diseases and insect pests) and abiotic stresses (drought, salinity, flooding, anaerobic germination, low nutrient-use efficiency) and also grain quality traits (for eating) are a vital concern in GSR breeding. The availability of rice genome annotation and functionally characterized genes from these two databases are highlighted in Fig. 4. The main focus of green traits involves the genes governing them: 104 for drought tolerance, 95 for salinity tolerance, 60 for low-input tolerance, 52 for cold tolerance, and 8 for submergence tolerance. These are also referred to as green genes, and they mostly represent resource-saving and environment-friendly approaches. As a result of the molecular genetics of green traits in the GSR breeding program, a total of 225 QTLs and 332 candidate genes have been identified on 12 chromosomes, and, trait-wise, each is explained in Table 2 and Fig. 2.

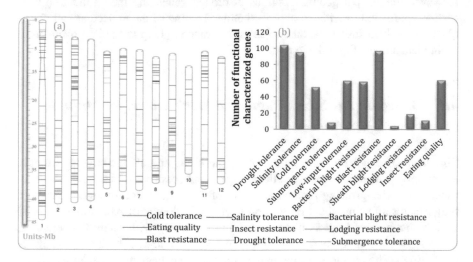

Fig. 4 Distribution of cloned genes in rice. (**a**) The important genes associated with green traits were highlighted across the 12 chromosomes using a Phenogram plot. (**b**) Traits were associated with the number of genes functionally characterized and these genes provide a valuable resource for understanding the complex nature of stress tolerance and adaptive traits. (*Source*: https://orygenesdb.cirad.fr/data.html and http://qtaro.abr.affrc.go.jp/)

In the genomic region of these QTL positions, a large number of candidate genes were identified through the in silico database and many of them have known functions related to green traits. Also, a few of them have unknown functions. The comprehensive literature survey and rice genomic database showed that genes such as *OsCOIN, OsDREB2A, OsGSK1,* and *OsDREB1F* on chromosome 1; *OsGS1;2, OsMYB2,* and *ZFP182* on chromosome 3; *OsTPS1* on chromosome 5; *OsDREB1C* and *OsiSAP8* on chromosome 6; *OsDREB1B* and *OsDREB1A* on chromosome 9; and *OsNAC5* on chromosome 11 are responsible for the multiple stress-tolerance mechanism. These genes act in ABA signaling pathways, hormonal regulation, accumulation of sugar and other compatible solutes such as proline, and also many other developmental and physiological processes involved in the regulation of the multiple stress-tolerance mechanism. Most of the multiple stress-tolerance gene expression is induced by ABA, and it depends on the presence of a *cis*-acting element referred to as ABA-responsive element (ABRE). These ABA-independent and dependent signaling pathways are involved in stress responses including drought, cold, heat, and cold. However, certain genomic regions on chromosome 1 (33.16–33.97 Mb; 40.15–41.90 Mb), chromosome 3 (11.07–11.75 Mb; 35.01–35.56 Mb), and chromosome 9 (21.13–21.98 Mb) played a significant role in the biotic and abiotic stress-tolerance pathways. They mainly played a role in major ABA signaling pathways and other stress signal transduction mechanisms in regulating stress tolerance and crosstalk between the other transcription factors to enhance gene regulation against multiple stresses. These overlapping stress-tolerance genes and transcription factors regulating similar stress-tolerance signaling pathways exist besides the crosstalk among the biotic and abiotic stresses. However, the hotspot genomic regions of cloned genes and their validation with earlier reports will provide an improved understanding of the molecular and physiological mechanisms in response to stress tolerance to help in the improvement of grain yield and quality traits in rice.

4.1 Dissecting the Stress-Regulated Mechanisms for Multiple Stress Tolerance

Simultaneous exposure to a single or multiple biotic and abiotic stresses strongly affects crop production, mainly in the rice-growing countries of Asia and Africa. In most of the rice-growing regions in the rainfed environment, drought stress is a major factor affecting about 23 million hectares in Southeast Asia and, combined with biotic stresses such as diseases and insects, along with rising temperatures, could further significantly decrease grain yield (Aghamolki et al. 2014; Bahuguna et al. 2018). The response of the plant's tolerance mechanism and adaptive strategies toward multiple stresses are significantly limited. It is crucial to understand this at the genomics and metabolic levels because of non-additive interactions, extensive overlaps, and crosstalk between stress-response signaling pathways, and also the interactions between transcription factors (TFs) and *cis*-elements on the promoters of target genes (Kissoudis et al. 2014; Verma and Deepti 2016). Generally, the combination of biotic and abiotic stress impacts depends on two factors: the host tolerance/susceptibility mechanism and the interaction of plant-microbial reactions, which influence stress responses to plants. However, a few common morphophysiological traits such as leaf wilting, tiller number, harvest index, root growth pattern, and chlorophyll content have been identified in the case of both drought and bacterial infections. They further decrease the photosynthetic machinery and diminish grain yield (Pandey et al. 2017). Importantly, the central role of hormonal balance and interaction is the critical controller of genes that play a significant role in regulating the stress-tolerance genes that are involved in different molecular and physiological mechanisms for multiple stresses. Hormonal interaction is regulated by ethylene, ABA, salicylic acid, jasmonic acid, cytokinin, and brassinosteroid. They, in turn, regulate several growth development traits. They also connect to the multiple stress signaling pathways to regulate stress-responsive gene expression (Vemanna et al. 2019). The receptor of these hormones regulates the various TF families such as NAC, AP2/ERF, bZIP, and MYC, which have altered the stress response to biotic and abiotic stresses (Huang et al. 2011; Li et al. 2018). However, integration of the omics approach has provided more profound insights into the molecular mechanisms for a better understanding of the multiple stress-responsive candidate genes. Further, this helped in the functional characterization of each gene that is involved in the stress signaling and tolerance mechanisms for abiotic and biotic stress adaptation.

Recently, Vemanna et al. (2019) reviewed comprehensive information on crosstalk signaling and multiple stress-tolerance mechanisms in rice. The overlapping stress-tolerance mechanisms and transcriptional responses in combined stresses are quite complex and interact with several biological processes. There are few reports on a meta-analysis using transcriptome data, and they found certain common genomic regions shared with multiple or individual stress tolerances. This indicates that, in response to stress tolerance, several hormonal signaling pathways are overlapping and cross-talking. Interestingly, Zhang et al. (2016) identified a total of 178 genes that were commonly expressed in drought and bacterial pathogen infections from the transcriptome analysis. However, some genes play specific functional and opposite roles in biotic and abiotic stresses. For example, one of the CDPK family proteins, *OsCPK12*, regulates drought, salinity, and cold stress tolerance, but it is a negative regulator for blast resistance (Asano et al. 2012; Fang et al. 2019). Similarly, the WRKY transcription factor family of WRKY71 is responsible for increasing the tolerance of bacterial infections, whereas the overexpression of WRKY45 showed susceptibility. Universal stress-tolerance proteins such as ABC transporters play a vital role in the development of stress tolerance in both biotic and abiotic stresses.

In the GSR breeding strategy, Murugaiyan et al. (2019) identified a robust QTL for arsenic tolerance on chromosome 1 and found a multi-drug resistance-associated protein (*OsMRP2*), which belongs to the subfamily of ABC transporters. It is mainly involved in the vacuolar sequestration of toxic metabolites (Brunetti et al. 2015). In the GSR breeding program, we primarily focused on dissecting complex abiotic traits. The rice crop is susceptible to cold stress, which adversely affects the crop at various growth stages, causing significant yield decreases mainly in temperate, tropical, and subtropical rice-growing regions. Similary, Najeeb et al. (2020) identified low-temperature stress-tolerance QTLs and found some promising genomic regions that were involved in the multiple stress-tolerance mechanism in rice. For instance, the major QTL on chromosome 5 had a possible candidate gene, $Os05g49970$, encoding translation initiation factor-2 (eIF2). These genes are involved in several cellular and metabolic processes in the early growth developmental stages, hormonal signaling pathways in plant defense mechanisms, and tolerance of various abiotic stresses (Martínez-Silva et al. 2012; Mutuku et al. 2015). Likewise, on chromosome 6, two genes ($Os06g17220$ and $Os06g48300$) were involved in major ABA-dependent signaling pathways and were responsible for sucrose-synthetic activity during anoxia conditions (Lasanthi-Kudahettige et al. 2007; Bhatnagar et al. 2017). Zhang et al. (2011) had attempted to integrate contemporary knowledge of signal transduction pathways with the principles of quantitative and population genetics to illustrate the genetic networks underlying complex traits using a model established upon the one-way functional dependency of downstream genes on upstream regulators depicting the principle of hierarchy. The mutual functional dependence among related genes was determined as the functional genetic units (FGUs). Interestingly, both simulated and real data suggested that complementary epistasis contributes significantly to quantitative trait variation and obscures the phenotypic effects of many "downstream" loci in pathways. Downstream FGUs were more vulnerable to loss of function than their upstream regulators; however, this vulnerability was compensated by different FGUs of similar functions (Zhang et al. 2011). Dissecting the complex trait of nutrient-use efficiency under low-input conditions (Mahender et al. 2019) identified a key regulator as F-box protein and calcium-dependent protein kinase on chromosome 2. These genomic regions played a major role in carbon and nitrogen metabolism and also maintained the homeostatic coordination of other enzymatic activities. The integrated approaches of omics and gene network analysis could play a crucial role in understanding stress tolerance in combined or individual stress-tolerance mechanisms in rice. These strategies need to be assessed more to gain deeper insights into dissecting the complexity and expanding the knowledge on each specific role of stress-responsive genes and transcription factors. Placing all the outputs from these strategies together could help to understand the unique and shared molecular and physiological pathways and possibly increase the adaptation to multiple abiotic and biotic stress factors.

4.2 Breeding Products Combining Tolerance of Multiple Stresses

Maintaining genetic diversity in breeding programs is a critical component, and it provides an excellent opportunity for breeders to develop novel and improved cultivars with desirable target traits in their breeding program. The GSR breeding program contains two interlinked breeding strategies: selective introgression lines

(SILs) and the designed QTL pyramiding (DQP) strategy. First is to develop multi-trait-specific SILs from several BC_1F_2 bulk populations derived from a few highly adapted recipient parents and 10–15 donor parents with three rounds of multiple-stress screening. This results in the development of EB-SILs, and these high-yielding, multi-stress-tolerant BC_1F_5 SILs then undergo two seasons of preliminary yield trials and advanced yield trials to identify superior lines simultaneously across drought, low-input, and irrigated conditions. The superior SILs that outyield the standard checks under different conditions are shortlisted for global multi-location trials. Superior multi-trait-specific SILs are crossed with another IL with complementary traits from within the same recipient/donor combination or different donors. Such a designed cross based on genotypic and phenotypic information is referred to as a DQP approach. This has played a vital role in increasing grain yield, improving tolerance of abiotic and biotic stresses, and using fewer inputs such as fertilizer and pesticide. This breeding strategy began in 2008 at IRRI and was used to create breeding materials that involved 500 elite rice varieties that belong to the mini-core collection and are part of the 3000 Rice Genomes Project (Ali et al. 2012). Among these, only three recipient parents, Huanghuazhan (HHZ), Weed Tolerant Rice-1 (WTR-1), and TME80518, and 16 donors were fully used for the EBBP and DQP approach to identify promising GSR materials with multiple abiotic and biotic stress tolerance (Ali et al. 2012, 2013b, 2018a) (Fig. 3). The BC_1F_2 populations underwent three rounds of precise phenotypic screening under drought, salinity, submergence, low-chemical-input, and normal irrigated conditions. This led to identifying 845 trait-specific SILs that outperformed the tolerant checks and these were further evaluated under multi-environment locations in Asian and African countries. As of now, a total of 66 GSR breeding materials have been registered in China and 59 GSR breeding materials have been released across Asian and African countries (Fig. 5). More than 90 GSR rice varieties were nominated in national cooperative yield trials from 2016 to 2018. The released varieties showed consistency in higher grain yield under low inputs (fertilizer and pesticide) and tolerance of multiple biotic and abiotic stresses. For instance, Yorobe et al. (2016) assessed the impact of GSR varieties and evaluated the income of farmer-users per hectare. Based on the survey data and fixed-effect model approach, farmer income increased significantly compared with that of farmers using conventional inbred varieties (non-GSR materials) with the frequent occurrence of flood and submergence conditions (Yorobe et al. 2014; Kodama et al. 2019). In addition, the net farm income advantage with GSR is quite high under increasing percentiles of rainfall. Several GSR varieties outperformed the local checks (Table 2). One GSR line (IRIS 179-880151) in data from two seasons for 2 years showed a 10% yield advantage vis-à-vis the local check variety and also had tolerance of drought, salinity, and submergence (Ali et al. 2013a). Similarly, multiple-stress-tolerant lines such as GSR IR1-12-D10-S1-D1, GSR IR1-5-S10-D1-D1, and GSR IR1-8-S12-Y2-D1 performed well under different environments and had a higher grain yield advantage of 25% to 40% over the drought-tolerant checks (Marcaida III et al. 2014). Another recent example of the potential impact of GSR varietal performance is found in the sub-Saharan African region of Mozambique, which is one of the major rice

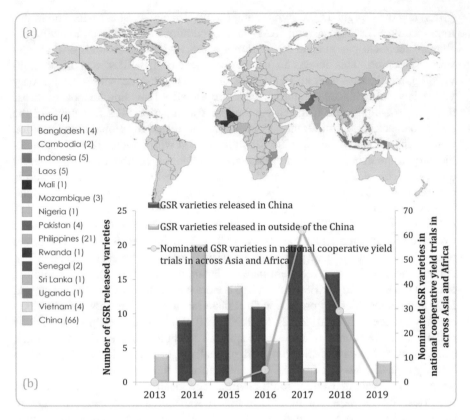

Fig. 5 Achievements of GSR breeding program and released GSR varieties across the globe. (**a**) Map showing the number of released GSR varieties highlighted country-wise. (**b**) In a span of 6 years, the total number of GSR varieties released and nominated in national cooperative yield trials across Asia and Africa

exporters. Kodama (2019) conducted a farm-level survey in three regions (Gaza, Sofala, and Nampula) and used an endogenous switching regression (ESR) model to assess GSR varieties among smallholder rice farmers and smallholder rice producers. Interestingly, smallholders carry out about 90% of the rice production in this region, and average yield is 1.0–1.2 t/ha in rainfed systems and 2.8–3.5 t/ha in normal irrigated conditions. This region is more prone to abiotic stresses, and mainly to the variability and duration of rainfall. With the adoption of GSR varieties, Kodama (2019) noticed a significant yield advantage, about ten times higher for smallholder farmers than for smallholder rice producers adopting non-GSR varieties. This indicates the strong and positive yield advantage and tolerance of multiple stresses across Asian and African countries that benefit poor farmers' income. This provides a great opportunity for GSR varieties to provide more rice and help alleviate poverty across the major rice-growing countries.

4.3   Development of Rice Hybrids with Multiple-Stress Tolerance

Hybrid rice breeding is one of the major core components for ensuring global food security. From a historical perspective, two milestones have been reached in enhancing grain yield. The first was by developing semi-dwarf rice varieties through the incorporation of the semi-dwarf (*sd1*) gene in rice breeding at IRRI in the early 1960s and the second was by the use of heterosis in hybrid rice (Yuan 2017). The potential of these two strategies was shown by rice grain yield increasing by about 30% with the semi-dwarf gene and a 15–20% yield increase with the use of heterosis in the major rice-producing areas of China (Peng et al. 2009). Currently, there are three approaches for increasing hybrid vigor: the cytoplasmic male sterility/fertility restoration (*CMS/Rf*) locus, environment-sensitive genic male sterility, and apomixis (Birchler et al. 2006; Bar-Zvi et al. 2017; Xie et al. 2019). At IRRI, we began developing climate-smart rice hybrids by adopting EB-SILs and using the DQP approach in separate restorer and maintainer backgrounds (Ali et al. 2020). The promising rice varieties that are tolerant of multiple stresses and possess higher grain yield and superior grain quality traits were considered for parental selection based on the fertility restoration locus and possession of desirable floral characteristics. Based on genotyping results, the GSR lines positive for fertility restoration loci *Rf3* and *Rf4* will be considered as restorers and those entirely negative for those loci as maintainers. More than 100 restorer lines with a strong general combining ability, tolerance of multiple stresses, and having preferable grain quality have been bred from using the promising lines from the GSR breeding program. These climate-smart parental lines are essential for complementing the introgression of QTLs and genes with green traits into the F_1s. Hybrid rice-related traits governing wide compatibility, floral traits that promote outcrossing, and seed reproducibility are important for parental line-breeding besides addressing the market requirements of the target regions. Further, by selecting the maximum genetic distance between the parental lines from diverse heterotic pools, this helped to identify promising high-yielding climate-resilient hybrid rice combinations. Recently, genomic hybrid breeding has emerged using whole-genome markers to predict future hybrids. Only the superior predicted hybrids are then field-evaluated and later released as new hybrid cultivars based on their actual performance in the target conditions. This approach offers an opportunity to select truly superior hybrids at limited cost. Cui et al. (2019) used genomic best linear unbiased prediction to predict hybrid performance using an existing rice population of 1495 hybrids. Further, a replicated ten-fold cross-validation showed that the prediction ability on ten agronomic traits ranged from 0.35 to 0.92 (Cui et al. 2019). By keeping this population of 1495 hybrids, it can be used to predict hybrids from seemingly unrelated parents. Machine learning algorithms and artificial intelligence are also being employed for predicting the best hybrid combinations at IRRI using genomic and historical hybrid yield trial data.

5   Conclusions

Identifying the promising genomic regions of QTLs and genes that are associated with the green traits that are involved in multiple-stress tolerance mechanisms will be adapted to increase rice yield in the future. The Green Super Rice breeding strategy using EB-SILs and DQP, along with combined advanced genomic technologies, significantly improved the development of rice varieties with multiple-stress tolerance. This breeding strategy also helped to identify promising QTLs and genes that were associated with the green traits related to higher grain yield, tolerance of abiotic and biotic stresses, low-input use of fertilizer and pesticide, and superior grain quality. These resource-saving and environment-friendly features have been well exploited under the GSR breeding strategy. The stress-tolerance mechanism of these combined or individual stresses that triggers a network of signaling pathways at the plant cell level needs to be studied in detail. The crosstalk of these biotic and abiotic stresses could reveal similar pathways for maintaining the cellular mechanism for tolerance. Omics approaches should occupy the core component in breeding and trait development activities for identifying multiple-stress tolerance, and this would remain the primary target trait in the next generation of crop breeding programs.

References

Aghamolki MTK, Yusop MK, Oad FC et al (2014) Heat stress effects on yield parameters of selected rice cultivars at reproductive growth stages. J Food Agric Environ 12:741–746

Agrios GN (2005) Plant pathology, 5th edn. Elsevier Academic Press, Burlington, pp 79–103

Ali AJ, Xu JL, Ismail AM et al (2006) Hidden diversity for abiotic and biotic stress tolerances in the primary gene pool of rice revealed by a large backcross breeding program. Field Crops Res 97:66–76

Ali J, Xu JL, Gao YM et al (2012) Green super rice (GSR) technology: an innovative breeding strategy-achievements & advances. In: The 12th SABRAO congress on plant breeding towards 2025: challenges in a rapidly changing world. Chiang Mai, Thailand. pp 16–17

Ali J, Xu JL, Gao YM et al (2013a) Breeding for yield potential and enhanced productivity across different rice ecologies through green super rice (GSR) breeding strategy. In: Muralidharan K, Siddiq EA (eds) International dialogue on perception and prospects of designer rice. Society for Advancement of Rice Research, Hyderabad, pp 60–68

Ali N, Paul S, Gayen D et al (2013b) RNAi mediated down regulation of *myo*-inositol-3-phosphate synthase to generate low phytate rice. Rice 6:12. https://doi.org/10.1186/1939-8433-6-12

Ali J, Xu J-L, Gao Y-M et al (2017) Harnessing the hidden genetic diversity for improving multiple abiotic stress tolerance in rice (*Oryza sativa* L.). PLoS One 12:e0172515. https://doi.org/10.1371/journal.pone.0172515

Ali J, Aslam UM, Tariq R et al (2018a) Exploiting the genomic diversity of rice (*Oryza sativa* L.): SNP-typing in 11 early-backcross introgression-breeding populations. Front Plant Sci 9:1–10. https://doi.org/10.3389/fpls.2018.00849

Ali J, Jewel ZA, Mahender A et al (2018b) Molecular genetics and breeding for nutrient use efficiency in rice. Int J Mol Sci 19:1762. https://doi.org/10.3390/ijms19061762

Ali J, Mahender A, Prahalada GD et al (2020) Genomics-assisted breeding of climate-smart inbred and hybrid rice varieties. In: Kole C (ed) Genomic designing of climate-smart cereal crops. Springer, Cham, Springer Nature Switzerland AG, pp 1–43. https://doi.org/10.1007/978-3-319-93381-8_1

Anderson PK, Cunningham AA, Patel NG et al (2004) Emerging infectious diseases of plants: pathogen pollution, climate change and agrotechnology drivers. Trends Ecol Evol 19:535–544

Antralina M, Istina IN, Simarmata T (2015) Effect of difference weed control methods to yield of lowland rice in the SOBARI. Procedia Food Sci 3:323–329

Appleby AP, Müller F, Carpy S (2002) Weed control. Ullmann's encyclopedia of industrial chemistry. Wiley, Weinheim

Asano T, Hayashi N, Kobayashi M et al (2012) A rice calcium-dependent protein kinase OsCPK12 oppositely modulates salt-stress tolerance and blast disease resistance. Plant J 69:26–36

Ashkani S, Rafii MY, Shabanimofrad M et al (2015) Molecular breeding strategy and challenges towards improvement of blast disease resistance in rice crop. Front Plant Sci 6:886

Bahuguna RN, Gupta P, Bagri J et al (2018) Forward and reverse genetics approaches for combined stress tolerance in rice. Indian J Plant Physiol 23:630–646

Bailey-Serres J, Fukao T, Ronald P et al (2010) Submergence tolerant rice: *SUB1*'s journey from landrace to modern cultivar. Rice 3:138–147

Balachiranjeevi CH, Prahalada GD, Mahender A et al (2019) Identification of a novel locus, *BPH38* (t), conferring resistance to brown planthopper (*Nilaparvata lugens* Stal.) using early backcross population in rice (*Oryza sativa* L.). Euphytica 215:185

Balakrishnan D, Subrahmanyam D, Badri J et al (2016) Genotype × environment interactions of yield traits in backcross introgression lines derived from *Oryza sativa* cv. Swarna/*Oryza nivara*. Front Plant Sci 7:1–19. https://doi.org/10.3389/fpls.2016.01530

Balyan HS, Gupta PK, Kumar S et al (2013) Genetic improvement of grain protein content and other health-related constituents of wheat grain. Plant Breed 132:446–457

Bar-Zvi D, Lupo O, Levy AA, Barkai N (2017) Hybrid vigor: the best of both parents, or a genomic clash? Curr Opin Syst Biol 6:22–27

Bazrkar-Khatibani L, Fakheri BA, Hosseini-Chaleshtori M et al (2019) Genetic mapping and validation of quantitative trait loci (QTL) for the grain appearance and quality traits in rice (*Oryza sativa* L.) by using recombinant inbred line (RIL) population. Int J Genomics 2019:3160275. https://doi.org/10.1155/2019/3160275

Bernier J, Kumar A, Ramaiah V et al (2007) A large effect QTL for grain yield under reproductive-stage drought stress in upland rice. Crop Sci 47:507–516. https://doi.org/10.2135/cropsci2006.07.0495

Bhandari A, Jayaswal P, Yadav N et al (2019) Genomics-assisted backcross breeding for infusing climate resilience in high-yielding green revolution varieties of rice. Indian J Genet 79(1 Suppl):160–170

Bhatnagar N, Min M-K, Choi E-H et al (2017) The protein phosphatase 2C clade A protein *OsPP2C51* positively regulates seed germination by directly inactivating *OsbZIP10*. Plant Mol Biol 93:389–401

Birchler JA, Yao H, Chudalayandi S (2006) Unraveling the genetic basis of hybrid vigor. Proc Natl Acad Sci U S A 103:12957–12958

Brunetti P, Zanella L, De Paolis A et al (2015) Cadmium-inducible expression of the ABC-type transporter AtABCC3 increases phytochelatin-mediated cadmium tolerance in *Arabidopsis*. J Exp Bot 66(13):3815–3829. https://doi.org/10.1093/jxb/erv185

Calayugan MIC, Formantes AK, Amparado A et al (2020) Genetic analysis of agronomic traits and grain iron and zinc concentrations in a doubled haploid population of rice (*Oryza sativa* L.). Sci Rep 10:1–14

Chauhan BS, Abugho SB (2013) Weed management in mechanized-sown, zero-till dry-seeded rice. Weed Technol 27:28–33

Chauhan BS, Awan TH, Abugho SB, Evengelista G (2015a) Effect of crop establishment methods and weed control treatments on weed management, and rice yield. Field Crops Res 172:72–84

Chauhan BS, Opeña J, Ali J et al (2015b) Response of 10 elite "Green Super Rice" genotypes to weed infestation in aerobic rice systems. Plant Prod Sci 18:228–233. https://doi.org/10.1626/pps.18.228

Chen S, Yang Y, Shi W et al (2008) *Badh2*, encoding betaine aldehyde dehydrogenase, inhibits the biosynthesis of 2-acetyl-1-pyrroline, a major component in rice fragrance. Plant Cell 20(7):1850–1861

Chen X, Cui Z, Fan M et al (2014a) Producing more grain with lower environmental costs. Nature 514:486–489

Chen L, Lin L, Cai G et al (2014b) Identification of nitrogen, phosphorus, and potassium deficiencies in rice based on static scanning technology and hierarchical identification method. PLoS One 9:e113200. https://doi.org/10.1371/journal.pone.0113200

Cheng SH, Zhuang JY, Fan YY et al (2007) Progress in research and development on hybrid rice: a super-domesticate in China. Ann Bot 100:959–966

Cohen SP, Leach JE (2019) Abiotic and biotic stresses induce a core transcriptome response in rice. Sci Rep 9:1–11

Cuevas RP, Pede VO, McKinley J et al (2016) Rice grain quality and consumer preferences: a case study of two rural towns in the Philippines. PLoS One 11:e0150345

Cui K, Peng S, Xing Y et al (2002) Molecular dissection of seedling-vigor and associated physiological traits in rice. Theor Appl Genet 105:745–753

Cui Y, Li R, Li G et al (2019) Hybrid breeding of rice via genomic selection. Plant Biotechnol J 18:57–67

Custodio MC, Cuevas RP, Ynion J et al (2019) Rice quality: how is it defined by consumers, industry, food scientists, and geneticists? Trends Food Sci Technol 92:122–137

da Cruz RP, Sperotto RA, Cargnelutti D et al (2013) Avoiding damage and achieving cold tolerance in rice plants. Food Energy Secur 2:96–119. https://doi.org/10.1002/fes3.25

Dang X, Thi TGT, Dong G et al (2014) Genetic diversity and association mapping of seed vigor in rice (*Oryza sativa* L.). Planta 239:1309–1319

Dimaano NGB, Ali J, Cruz PCSS et al (2017) Performance of newly developed weed-competitive rice cultivars under lowland and upland weedy conditions. Weed Sci 65:798–817. https://doi.org/10.1017/wsc.2017.57

Dimaano NGB, Ali J, Mahender A et al (2020) Identification of quantitative trait loci governing early germination and seedling vigor traits related to weed competitive ability in rice. Euphytica 216:159. https://doi.org/10.1007/s10681-020-02694-8

Diwan J, Channbyregowda M, Shenoy V et al (2013) Molecular mapping of early vigour related QTLs in rice. Res J Biol 1:24–30

Dixit S, Singh A, Cruz MTS et al (2014) Multiple major QTL lead to stable yield performance of rice cultivars across varying drought intensities. BMC Genet 15:16

Dixit S, Singh A, Sandhu N et al (2017) Combining drought and submergence tolerance in rice: marker-assisted breeding and QTL combination effects. Mol Breed 37:143

Du B, Chen R, Guo J, He G (2020) Current understanding of the genomic, genetic, and molecular control of insect resistance in rice. Mol Breed 40:24

Elferink M, Schierhorn F (2016) Global demand for food is rising. Can we meet it? Harv Bus Rev 7:2016

Ella ES, Kawano N, Osamu H (2003) Importance of active oxygen scavenging system in the recovery of rice seedlings after submergence. Plant Sci 165:85–93

Fahad S, Bajwa AA, Nazir U et al (2017) Crop production under drought and heat stress: plant responses and management options. Front Plant Sci 8:1147

Fahad S, Adnan M, Noor M et al (2019) Major constraints for global rice production. In: Advances in rice research for abiotic stress tolerance. Elsevier, Amsterdam, pp 1–22

Fang WB, Yong CL, Hua CS et al (2019) Impaired function of the calcium-dependent protein kinase, OsCPK12, leads to early senescence in rice (*Oryza sativa* L.). Front Plant Sci 10:52

Farooq M, Siddique KHM, Rehman H et al (2011) Rice direct seeding: experiences, challenges and opportunities. Soil Tillage Res 111:87–98

Feng B, Chen K, Cui Y et al (2018) Genetic dissection and simultaneous improvement of drought and low nitrogen tolerances by designed QTL pyramiding in rice. Front Plant Sci 9:306. https://doi.org/10.3389/fpls.2018.00306

Fischer RA, Edmeades GO (2010) Breeding and cereal yield progress. Crop Sci 50:S-85

Foolad MR, Subbiah P, Zhang L (2007) Common QTL affect the rate of tomato seed germination under different stress and nonstress conditions. Int J Plant Genomics 2007:97386. https://doi.org/10.1155/2007/97386

Gao Z, Zeng D, Cheng F et al (2011) ALK, the key gene for gelatinization temperature, is a modifier gene for gel consistency in rice. J Integr Plant Biol 53:756–765

Gautam T, Saripalli G, Kumar A et al (2020) Introgression of a drought insensitive grain yield QTL for improvement of four Indian bread wheat cultivars using marker assisted breeding without background selection. J Plant Biochem Biotechnol 29:1–12. https://doi.org/10.1007/s13562-020-00553-0

Ghimire KH, Quiatchon LA, Vikram P et al (2012) Identification and mapping of a QTL (*qDTY1.1*) with a consistent effect on grain yield under drought. Field Crops Res 131:88–96

Gimhani DR, Gregorio GB, Kottearachchi NS, Samarasinghe WLG (2016) SNP-based discovery of salinity-tolerant QTLs in a bi-parental population of rice (*Oryza sativa*). Mol Gen Genomics 291:2081–2099

Gothandam KM, Kim E-S, Cho H, Chung Y-Y (2005) OsPPR1, a pentatricopeptide repeat protein of rice, is essential for the chloroplast biogenesis. Plant Mol Biol 58:421–433

Guan YS, Serraj R, Liu SH et al (2010) Simultaneously improving yield under drought stress and non-stress conditions: a case study of rice (*Oryza sativa* L.). J Exp Bot 61:4145–4156. https://doi.org/10.1093/jxb/erq212

Guo L, Gao Z, Qian Q (2014) Application of resequencing to rice genomics, functional genomics and evolutionary analysis. Rice 7:4. https://doi.org/10.1186/s12284-014-0004-7

Guo J, Hu X, Gao L et al (2017) The rice production practices of high yield and high nitrogen use efficiency in Jiangsu, China. Sci Rep 7:2101

Guojun P, Shuqiang C, Chengyan S et al (2009) Study on relationship between resistance to blast and yield traits in early japonica rice in cold region [J]. Chin Agric Sci Bull 19

Hawkesford MJ, Griffiths S (2019) Exploiting genetic variation in nitrogen use efficiency for cereal crop improvement. Curr Opin Plant Biol 49:35–42

Huang XS, Luo T, Fu XZ et al (2011) Cloning and molecular characterization of a mitogen-activated protein kinase gene from *Poncirus trifoliata* whose ectopic expression confers dehydration/drought tolerance in transgenic tobacco. J Exp Bot 62:5191–5206. https://doi.org/10.1093/jxb/err229

Jabran K, Chauhan BS (2015) Weed management in aerobic rice systems. Crop Prot 78:151–163

Jena KK, Kim SM, Suh JP, Kim YG (2010) Development of cold-tolerant breeding lines using QTL analysis in rice. Second Africa Rice Congr 22–26

Jewel Z, Ali J, Pang Y et al (2018) Developing Green Super Rice varieties with high nutrient use efficiency by phenotypic selection under varied nutrient conditions. Crop J 7(3):368–377. https://doi.org/10.20944/preprints201807.0216.v1

Jewel ZA, Ali J, Mahender A et al (2019) Identification of quantitative trait loci associated with nutrient use efficiency traits, using SNP markers in an early backcross population of rice (*Oryza sativa* L.). Int J Mol Sci 20:900. https://doi.org/10.3390/ijms20040900

Ji Z, Yang S, Zeng Y et al (2016) Pyramiding blast, bacterial blight and brown planthopper resistance genes in rice restorer lines. J Integr Agric 15:1432–1440

Kavurikalpana TS, Shashidhara N (2018) Validation of molecular markers linked to grain quality traits in rice (*Oryza sativa* L.). Int J Curr Microbiol Appl Sci 7:1897–1902

Khan FA, Narayan S, Bhat SA, Maqbool R (2016) Vermipriming: a noble technology for seed invigouration in rice (*Oryza sativa* L.). SKUAST J Res 18:124–129

Kissoudis C, van de Wiel C, Visser RGF, van der Linden G (2014) Enhancing crop resilience to combined abiotic and biotic stress through the dissection of physiological and molecular cross-talk. Front Plant Sci 5:207

Kodama W, Pede VO, Mishra AK, Cabrera ER (2019) Assessing the benefits of Green Super Rice in Sub-Saharan Africa: evidence from Mozambique. Selected paper prepared for presentation at the 2019 Agricultural & Applied Economics Association Annual Meeting, Atlanta, Georgia, July 21–23. 23 p

Kreuzwieser J, Gessler A (2010) Global climate change and tree nutrition: influence of water availability. Tree Physiol 30:1221–1234

Kumar A, Dixit S, Ram T et al (2014) Breeding high-yielding drought-tolerant rice: genetic variations and conventional and molecular approaches. J Exp Bot 65:6265–6278. https://doi.org/10.1093/jxb/eru363

Kumar A, Sandhu N, Yadav S et al (2017a) Rice varietal development to meet future challenges. In: The future rice strategy for India. Elsevier, Amsterdam, pp 161–220

Kumar M, Gho Y-S, Jung K-H, Kim S-R (2017b) Genome-wide identification and analysis of genes, conserved between japonica and indica rice cultivars, that respond to low-temperature stress at the vegetative growth stage. Front Plant Sci 8:1120

Kumar A, Sandhu N, Dixit S et al (2018) Marker-assisted selection strategy to pyramid two or more QTLs for quantitative trait-grain yield under drought. Rice (N Y) 11:35

Kumar A, Sandhu N, Venkateshwarlu C et al (2020) Development of introgression lines in high yielding, semi-dwarf genetic backgrounds to enable improvement of modern rice varieties for tolerance to multiple abiotic stresses free from undesirable linkage drag. Sci Rep 10:1–13

Lafitte HR, Li ZK, Vijayakumar CHM et al (2006) Improvement of rice drought tolerance through backcross breeding: evaluation of donors and selection in drought nurseries. Field Crops Res 97:77–86. https://doi.org/10.1016/j.fcr.2005.08.017

Lasanthi-Kudahettige R, Magneschi L, Loreti E et al (2007) Transcript profiling of the anoxic rice coleoptile. Plant Physiol 144:218–231

Le Nguyen K, Grondin A, Courtois B, Gantet P (2019) Next-generation sequencing accelerates crop gene discovery. Trends Plant Sci 24:263–274

Li Z (1998) Molecular analysis of epistasis affecting complex traits. In: Molecular dissection of complex traits. CRC Press, Boca Raton, pp 119–130

Li Z, Ali J (2017) Breeding green super rice (GSR) varieties for sustainable rice cultivation. In: Sasaki T (ed) Achieving sustainable cultivation of rice, vol 1 edn. Burleigh Dodds Science Publishing, Cambridge, pp 131–152

Li ZK, Fu BY, Gao YM et al (2005) Genome-wide introgression lines and their use in genetic and molecular dissection of complex phenotypes in rice (Oryza sativa L.). Plant Mol Biol 59:33–52. https://doi.org/10.1007/s11103-005-8519-3

Li S, Tian Y, Wu K et al (2018) Modulating plant growth–metabolism coordination for sustainable agriculture. Nature 560:595. https://doi.org/10.1038/s41586-018-0415-5

Liang Y, Meng L, Lin X et al (2018) QTL and QTL networks for cold tolerance at the reproductive stage detected using selective introgression in rice. PLoS One 13:e0200846

Liu C, Ding S, Zhang A et al (2020) Development of nutritious rice with high zinc/selenium and low cadmium in grains through QTL pyramiding. J Integr Plant Biol 62:349–359

Mahender A, Anandan A, Pradhan SK (2015) Early seedling vigour, an imperative trait for direct-seeded rice: an overview on physio-morphological parameters and molecular markers. Planta 241:1027–1050

Mahender A, Anandan A, Pradhan SK, Pandit E (2016) Rice grain nutritional traits and their enhancement using relevant genes and QTLs through advanced approaches. Springerplus 5(1):2086. https://doi.org/10.1186/s40064-016-3744-6

Mahender A, Ali J, Prahalada GD et al (2019) Genetic dissection of developmental responses of agro-morphological traits under different doses of nutrient fertilizers using high-density SNP markers. PLoS One 14(7):e0220066

Marcaida M III, Li T, Angeles O et al (2014) Biomass accumulation and partitioning of newly developed Green Super Rice (GSR) cultivars under drought stress during the reproductive stage. Field Crops Res 162:30–38. https://doi.org/10.1016/j.fcr.2014.03.013

Martínez-Silva AV, Aguirre-Martínez C, Flores-Tinoco CE et al (2012) Translation initiation factor AteIF(iso)4E is involved in selective mRNA translation in *Arabidopsis thaliana* seedlings. PLoS One 7(2):e31606. https://doi.org/10.1371/journal.pone.0031606

Matsushima K-I, Sakagami J-I (2013) Effects of seed hydropriming on germination and seedling vigor during emergence of rice under different soil moisture conditions. Am J Plant Sci 4:1584. https://doi.org/10.4236/ajps.2013.48191

Matthus E, Wu LB, Ueda Y et al (2015) Loci, genes, and mechanisms associated with tolerance to ferrous iron toxicity in rice (*Oryza sativa* L.). Theor Appl Genet 128:2085–2098. https://doi.org/10.1007/s00122-015-2569-y

Mehta S, Singh B, Dhakate P et al (2019) Rice, marker-assisted breeding, and disease resistance. In: Wani SH (ed) Disease resistance in crop plants. Springer, Cham, pp 83–111

Mogga M, Sibiya J, Shimelis H et al (2018) Diversity analysis and genome-wide association studies of grain shape and eating quality traits in rice (*Oryza sativa* L.) using DArT markers. PLoS One 13:e0198012

Molla KA, Debnath AB, Ganie SA, Mondal TK (2015) Identification and analysis of novel salt responsive candidate gene based SSRs (cgSSRs) from rice (*Oryza sativa* L.). BMC Plant Biol 15:122

Mukamuhirwa A, Persson Hovmalm H, Bolinsson H et al (2019) Concurrent drought and temperature stress in rice—a possible result of the predicted climate change: effects on yield attributes, eating characteristics, and health promoting compounds. Int J Environ Res Public Health 16:1043

Murugaiyan V, Ali J, Mahender A et al (2019) Mapping of genomic regions associated with arsenic toxicity stress in a backcross breeding populations of rice (*Oryza sativa* L.). Rice 12:61

Muthu V, Abbai R, Nallathambi J et al (2020) Pyramiding QTLs controlling tolerance against drought, salinity, and submergence in rice through marker assisted breeding. PLoS One 15:e0227421

Mutuku JM, Yoshida S, Shimizu T et al (2015) The WRKY45-dependent signaling pathway is required for resistance against *Striga hermonthica* parasitism. Plant Physiol 168:1152–1163

Najeeb S, Ali J, Mahender A et al (2020) Identification of main-effect quantitative trait loci (QTLs) for low-temperature stress tolerance germination- and early seedling vigor-related traits in rice (*Oryza sativa* L.). Mol Breed 40:10

Nandi S, Subudhi PK, Senadhira D et al (1997) Mapping QTLs for submergence tolerance in rice by AFLP analysis and selective genotyping. Mol Gen Genet 255(1):1–8

Oladosu Y, Rafii MY, Samuel C et al (2019) Drought resistance in rice from conventional to molecular breeding: a review. Int J Mol Sci 20:3519. https://doi.org/10.3390/ijms20143519

Oliver SN, Dennis ES, Dolferus R (2007) ABA regulates apoplastic sugar transport and is a potential signal for cold-induced pollen sterility in rice. Plant Cell Physiol 48:1319–1330

Ouk M, Basnayake J, Tsubo M et al (2006) Use of drought response index for identification of drought tolerant genotypes in rainfed lowland rice. Field Crops Res 99:48–58

Panda BB, Sharma S, Mohapatra PK, Das A (2012) Application of excess nitrogen, phosphorus, and potassium fertilizers leads to lowering of grain iron content in high-yielding tropical rice. Commun Soil Sci Plant Anal 43:2590–2602

Pandey P, Irulappan V, Bagavathiannan MV, Senthil-Kumar M (2017) Impact of combined abiotic and biotic stresses on plant growth and avenues for crop improvement by exploiting physio-morphological traits. Front Plant Sci 8:537

Pang Y, Ali J, Wang X et al (2016) Relationship of rice grain amylose, gelatinization temperature and pasting properties for breeding better eating and cooking quality of rice varieties. PLoS One 11:e0168483

Pang Y, Chen K, Wang X et al (2017a) Recurrent selection breeding by dominant male sterility for multiple abiotic stresses tolerant rice cultivars. Euphytica 213:268. https://doi.org/10.1007/s10681-017-2055-5

Pang Y, Chen K, Wang X et al (2017b) Simultaneous improvement and genetic dissection of salt tolerance of rice (*Oryza sativa* L.) by designed QTL pyramiding. Front Plant Sci 8:1275. https://doi.org/10.3389/fpls.2017.01275

Peng S, Tang Q, Zou Y (2009) Current status and challenges of rice production in China. Plant Prod Sci 12:3–8

Pradhan A, Naik N, Sahoo KK (2015) RNAi mediated drought and salinity stress tolerance in plants. Am J Plant Sci 6:1990

Rahman KM, Zhang D (2018) Effects of fertilizer broadcasting on the excessive use of inorganic fertilizers and environmental sustainability. Sustainability 10:759

Rahman M, Juraimi AS, Suria J et al (2012) Response of weed flora to different herbicides in aerobic rice system. Sci Res Essays 7:12–23

Rahman MA, Bimpong IK, Bizimana JB et al (2017) Mapping QTLs using a novel source of salinity tolerance from Hasawi and their interaction with environments in rice. Rice 10(1):47. https://doi.org/10.1186/s12284-017-0186-x

Raj SK, Syriac EK (2017) Weed management in direct seeded rice: a review. Agric Rev 38:41–50. https://doi.org/10.18805/ag.v0iOF.7307

Ranawake AL, Manangkil OE, Yoshida S et al (2014) Mapping QTLs for cold tolerance at germination and the early seedling stage in rice (*Oryza sativa* L.). Biotechnol Biotechnol Equip 28:989–998. https://doi.org/10.1080/13102818.2014.978539

Ray DK, Mueller ND, West PC, Foley JA (2013) Yield trends are insufficient to double global crop production by 2050. PLoS One 8:e66428

Ray DK, Gerber JS, MacDonald GK, West PC (2015) Climate variation explains a third of global crop yield variability. Nat Commun 6:1–9

Roy-Barman S, Chattoo BB (2005) Rice blast fungus sequenced. Curr Sci 89:930

Sahebi M, Hanafi MM, Rafii MY et al (2018) Improvement of drought tolerance in rice (*Oryza sativa* L.): genetics, genomic tools, and the WRKY gene family. Biomed Res Int 2018:3158474

Saika H, Oikawa A, Matsuda F et al (2011) Application of gene targeting to designed mutation breeding of high-tryptophan rice. Plant Physiol 156:1269–1277

Saikumar S, Varma CMK, Saiharini A et al (2016) Grain yield responses to varied level of moisture stress at reproductive stage in an interspecific population derived from Swarna/*O. glaberrima* introgression line. NJAS Wageningen J Life Sci 78:111–122

Sandhu N, Kumar A (2017) Bridging the rice yield gaps under drought: QTLs, genes, and their use in breeding programs. Agronomy 7:27. https://doi.org/10.3390/agronomy7020027

Schläppi MR, Jackson AK, Eizenga GC et al (2017) Assessment of five chilling tolerance traits and GWAS mapping in rice using the USDA mini-core collection. Front Plant Sci 8:957. https://doi.org/10.3389/fpls.2017.00957

Septiningsih EM, Pamplona AM, Sanchez DL et al (2009) Development of submergence-tolerant rice cultivars: the *Sub1* locus and beyond. Ann Bot 103(2):151–160

Serrano-Mislata A, Bencivenga S, Bush M et al (2017) DELLA genes restrict inflorescence meristem function independently of plant height. Nat Plants 3:749

Shakiba E, Edwards JD, Jodari F et al (2017) Genetic architecture of cold tolerance in rice (*Oryza sativa*) determined through high resolution genome-wide analysis. PLoS One 12:1–22. https://doi.org/10.1371/journal.pone.0172133

Sharma M, Pandey GK (2016) Expansion and function of repeat domain proteins during stress and development in plants. Front Plant Sci 6:1218

Shinada H, Iwata N, Sato T, Fujino K (2014) QTL pyramiding for improving of cold tolerance at fertilization stage in rice. Breed Sci 63:483–488. https://doi.org/10.1270/jsbbs.63.483

Shirasawa K, Takeuchi Y, Ebitani T et al (2008) Identification of gene for rice (*Oryza sativa*) seed lipoxygenase-3 involved in the generation of stale flavor and development of SNP markers for lipoxygenase-3 deficiency. Breed Sci 58(2):169–176

Shrivastava P, Kumar R (2015) Soil salinity: a serious environmental issue and plant growth promoting bacteria as one of the tools for its alleviation. Saudi J Biol Sci 22:123–131

Singh K, Kumar V, Saharawat YS et al (2013) Weedy rice: an emerging threat for direct-seeded rice production systems in India. J Rice Res 1:1–6. (open access)

Singh H, Jassal RK, Kang JS et al (2015) Seed priming techniques in field crops: a review. Agric Rev 36:251–264

Singh R, Singh Y, Xalaxo S et al (2016) From QTL to variety-harnessing the benefits of QTLs for drought, flood and salt tolerance in mega rice varieties of India through a multi-institutional network. Plant Sci 242:278–287

Sripongpangkul K, Posa GBT, Senadhira DW et al (2000) Genes/QTLs affecting flood tolerance in rice. Theor Appl Genet 101(7):1074–1081

Sui B, Feng X, Tian G et al (2013) Optimizing nitrogen supply increases rice yield and nitrogen use efficiency by regulating yield formation factors. Field Crops Res 150:99–107

Sun J, Yang L, Wang J et al (2018) Identification of a cold-tolerant locus in rice (*Oryza sativa* L.) using bulked segregant analysis with a next-generation sequencing strategy. Rice 11:24

Suprihatno B, Coffman WR (1981) Inheritance of submergence tolerance in rice (*Oryza sativa* L.). SABRAO J 13(2):98–108

Swain P, Anumalla M, Prusty S et al (2014) Characterization of some Indian native land race rice accessions for drought tolerance at seedling stage. Aust J Crop Sci 8:324–331

Swamy BPM, Kumar A (2013) Genomics-based precision breeding approaches to improve drought tolerance in rice. Biotechnol Adv 31:1308–1318. https://doi.org/10.1016/j.quaint.2017.02.033

Tigchelaar M, Battisti DS, Naylor RL, Ray DK (2018) Future warming increases probability of globally synchronized maize production shocks. Proc Natl Acad Sci U S A 115:6644–6649

Toojinda T, Siangliw M, Tragoonrung S et al (2003) Molecular genetics of submergence tolerance in rice: QTL analysis of key traits. Ann Bot 91(2):243–253

Tuberosa R (2012) Phenotyping for drought tolerance of crops in the genomics era. Front Physiol 3:347. https://doi.org/10.3389/fphys.2012.00347

United Nations (2019) World population prospects 2019: highlights. United Nations Department of Economic and Social Affairs, New York

Unnevehr LJ, Juliano BO, Perez CM (1985) Consumer demand for rice grain quality in Southeast Asia. International rice research conference, International Rice Research Institute, pp 15–23

Utsumi Y, Utsumi C, Sawada T et al (2011) Functional diversity of isoamylase oligomers: the ISA1 homo-oligomer is essential for amylopectin biosynthesis in rice endosperm. Plant Physiol 156:61–77

Van Oort PAJJ (2018) Mapping abiotic stresses for rice in Africa: drought, cold, iron toxicity, salinity and sodicity. Field Crops Res 219:55–75. https://doi.org/10.1016/j.fcr.2018.01.016

Van Oort PA, Zwart SJ (2018) Impacts of climate change on rice production in Africa and causes of simulated yield changes. Glob Chang Biol 24(3):1029–1045

Vemanna RS, Bakade R, Bharti P et al (2019) Cross-talk signaling in rice during combined drought and bacterial blight stress. Front Plant Sci 10:193

Venu RC, Sreerekha MV, Nobuta K et al (2011) Deep sequencing reveals the complex and coordinated transcriptional regulation of genes related to grain quality in rice cultivars. BMC Genomics 12:190. https://doi.org/10.1111/j.1551-2916.2005.00910.x

Venuprasad R, Dalid CO, Del Valle M et al (2009) Identification and characterization of large-effect quantitative trait loci for grain yield under lowland drought stress in rice using bulk-segregant analysis. Theor Appl Genet 120:177–190

Verma AK, Deepti S (2016) Abiotic stress and crop improvement: current scenario. Adv Plants Agric Res 4:149

Vikram P, Swamy BPM, Dixit S et al (2011) *qDTY 1.1*, a major QTL for rice grain yield under reproductive-stage drought stress with a consistent effect in multiple elite genetic backgrounds. BMC Genet 12:89

Vikram P, Swamy BPM, Dixit S et al (2015) Drought susceptibility of modern rice varieties: an effect of linkage of drought tolerance with undesirable traits. Sci Rep 5:1–18. https://doi.org/10.1063/1.4791353

Wang Y, Zhang L, Nafisah A et al (2013) Selection efficiencies for improving drought/salt tolerances and yield using introgression breeding in rice (*Oryza sativa* L.). Crop J 1:134–142. https://doi.org/10.1016/j.cj.2013.07.006

Wang W, Fu B, Ali J et al (2015) Genome-wide responses to selection and genetic networks underlying submergence tolerance in rice. Plant Genome 8(2):1–13

Wang X, Pang Y, Zhang J et al (2017) Genome-wide and gene-based association mapping for rice eating and cooking characteristics and protein content. Sci Rep 7:1–10. https://doi.org/10.1038/s41598-017-17347-5

Wing RA, Purugganan MD, Zhang Q (2018) The rice genome revolution: from an ancient grain to Green Super Rice. Nat Rev Genet 19:505–517. https://doi.org/10.1038/s41576-018-0024-z

Xie Y, Shen R, Chen L, Liu Y-G (2019) Molecular mechanisms of hybrid sterility in rice. Sci China Life Sci 62(6):737–743

Xing Y, Guo S, Chen X et al (2018) Nitrogen metabolism is affected in the nitrogen-deficient rice mutant esl4 with a calcium-dependent protein kinase gene mutation. Plant Cell Physiol 59:2512–2525

Xu Y, Crouch JH (2008) Marker-assisted selection in plant breeding: from publications to practice. Crop Sci 48:391–407

Xu K, Xu X, Fukao T et al (2006) Sub1A is an ethylene-response-factor-like gene that confers submergence tolerance to rice. Nature 442(7103):705–708

Yadav S, Sandhu N, Singh VK et al (2019) Genotyping-by-sequencing based QTL mapping for rice grain yield under reproductive stage drought stress tolerance. Sci Rep 9:1–12

Yano M, Lin HX, Takeuchi Y et al (2003) Marker-assisted dissection and pyramiding of complex traits in rice. In: Mew TW et al (eds) Rice Science: Innovations and Impact for Livelihood. IRRI, pp 257–263

Ye C, Fukai S, Godwin I et al (2009) Cold tolerance in rice varieties at different growth stages. Crop Pasture Sci 60:328–338

Yorobe J, Pede V, Rejesus R et al (2014) Yield and income effects of the Green Super Rice (GSR) varieties: evidence from a fixed-effects model in the Philippines. Selected paper prepared for presentation at the Agricultural & Applied Economics Association's 2014 Annual Meeting, Minneapolis, MN, July 27–29, 2014. 31 p. https://doi.org/10.1063/1.1609251

Yorobe JM, Ali J, Pede VO et al (2016) Yield and income effects of rice varieties with tolerance of multiple abiotic stresses: the case of green super rice (GSR) and flooding in the Philippines. Agric Econ 47:1–11

Yu Y, Zhao Z, Shi Y et al (2016) Hybrid sterility in rice (*Oryza sativa* L.) involves the tetratrico-peptide repeat domain containing protein. Genetics 203:1439–1451

Yu S, Ali J, Zhang C et al (2020) Genomic breeding of Green Super Rice varieties and their deployment in Asia and Africa. Theor Appl Genet 133:1427–1442

Yuan LP (2017) Progress in super-hybrid rice breeding. Crop J 5:100W102

Yun BB-W, Kim M-GM, Handoyo T, Kim KK-M (2014) Analysis of rice grain quality-associated quantitative trait loci by using genetic mapping. Am J Plant Sci 5:1125. https://doi.org/10.4236/ajps.2014.59125

Zhang Q (2007) Strategies for developing Green Super Rice. Proc Natl Acad Sci U S A 104:16402–16409. https://doi.org/10.1073/pnas.0708013104

Zhang F, Zhai HQ, Paterson AH et al (2011) Dissecting genetic networks underlying complex phenotypes: the theoretical framework. PLoS One 6(1):e14541

Zhang F, Zhang F, Huang L et al (2016) Overlap between signaling pathways responsive to *Xanthomonas oryzae* pv. *oryzae* infection and drought stress in rice introgression line revealed by RNA-seq. J Plant Growth Regul 35:345–356. https://doi.org/10.1007/s00344-015-9538-1

Zhang Z, Li JJJ, Pan Y et al (2017) Natural variation in CTB4a enhances rice adaptation to cold habitats. Nat Commun 8:1–13. https://doi.org/10.1007/BF01337500

Zhang J, Guo T, Yang J et al (2020) QTL mapping and haplotype analysis revealed candidate genes for grain thickness in rice (*Oryza sativa* L.). Mol Breed 40:1–12

Zhou Y, Cai H, Xiao J et al (2009) Over-expression of aspartate aminotransferase genes in rice resulted in altered nitrogen metabolism and increased amino acid content in seeds. Theor Appl Genet 118:1381–1390

Zhou Z, Li H, Sun Y et al (2010) Effect of selection for high yield, drought and salinity tolerances on yield-related traits in rice (*Oryza sativa* L.). Acta Agron Sin 36:1725–1735

Zhu Y, Chen K, Mi X et al (2015) Identification and fine mapping of a stably expressed QTL for cold tolerance at the booting stage using an interconnected breeding population in rice. PLoS One 10:e0145704. https://doi.org/10.1371/journal.pone.0145704

Advances in Two-Line Heterosis Breeding in Rice via the Temperature-Sensitive Genetic Male Sterility System

Jauhar Ali, Madonna Dela Paz, and Christian John Robiso

Abstract Hybrid rice technology is a viable strategy to increase rice production and productivity, especially in countries with limited cultivable land for agriculture and irrigation water, along with costlier chemical inputs. The three-line hybrid rice technology adoption rate is slowing down because of restricted heterosis per se, the availability of better combining ability in cytoplasmic male sterile lines, lower hybrid seed reproducibility, and limited market acceptability of hybrids. Two-line heterosis breeding could overcome these shortcomings. However, the wide-scale adoption and use of two-line hybrid rice technology are possible through systematic research and breeding efforts to develop temperature-sensitive genetic male sterile (TGMS) lines with low (<24 °C) critical sterility temperature point, which is discussed in this chapter. Research on the genetics, breeding, grain quality, and resistance to insect pests and diseases for TGMS line development and physiological characterization is also discussed. In addition, the identification and validation of natural sites for TGMS self-seed multiplication and hybrid rice seed production through GIS mapping and climatic data analytical tools are also tackled. The development of high-yielding two-line rice hybrids and improvement in hybrid rice seed reproducibility could help in their wide-scale adoption.

Keywords Temperature-sensitive genetic male sterile (TGMS) lines · Critical sterility temperature point (CSTP) · Physiological characterization · Genetics · Rice

J. Ali (✉) · M. Dela Paz · C. J. Robiso
Rice Breeding Platform, International Rice Research Institute, Los Baños, Philippines
e-mail: j.ali@irri.org

© The Author(s) 2021
J. Ali, S. H. Wani (eds.), *Rice Improvement*,
https://doi.org/10.1007/978-3-030-66530-2_4

1   Introduction

Global rice production in 2018 was 782 million tons from 167.1 million hectares with an average productivity of 4.68 t/ha (FAOSTAT 2020). However, production needs to keep pace with the increasing food demand in the coming decades, especially when the global human population is predicted to reach 9.73 billion by 2050 (Worldometer 2020). Increasing rice production under declining resources such as cultivable land and irrigation water and costlier agricultural inputs will become a great challenge in the coming decades. Furthermore, climate change is going to increase the pressure on stable and sustainable rice production.

Hybrid rice technology is a viable approach to increase rice production under limited resources and climate change. This technology took roots as early as 1964 in China, and around the same time, international scientific communities were discussing its prospects, especially in India, the United States, and the Philippines (Carnahan et al. 1972; Swaminathan et al. 1972; Athwal and Virmani 1972). However, it was China under Professor Yuan Longping that demonstrated hybrid rice technology on a commercial scale in 1976 with requisite cytoplasmic male sterile (CMS), maintainer, and restorer lines. This early success led China to expand hybrid rice significantly to reach 16.7 million ha, accounting for 57% of the country's rice area. Hybrid rice now accounts for more than 65% of China's total national rice production. In recent years, the average productivity of rice in China has been 6.45 t/ha: 7.50 t/ha for hybrid rice and 6.15 t/ha for conventional rice. The increased production of hybrid rice each year provides food for more than 70 million people (Yuan 2014). The International Rice Research Institute (IRRI) made a significant effort to deploy hybrid rice technology outside China by sharing the requisite hybrid rice parental lines directly to both the public and private sectors. Parental lines developed by IRRI have been used quite extensively in the release of several commercial hybrids from the private and public arenas in India, Nepal, Pakistan, Vietnam, the Philippines, Bangladesh, and Indonesia. IRRI has directly released 17 hybrids in the Philippines alone.

Despite the enormous research and extension efforts that have gone into hybrid rice from the early 1990s, especially in Asia, hybrid rice area is growing slowly. Among the major reasons for the slow growth is, first, the available level of heterosis or hybrid rice yield advantage over the best checks is from 15% to 20%. Second, hybrid rice seed reproducibility is still below 2 t/ha for most hybrids outside China, besides being cumbersome and expensive, which is not attractive to the private seed industry to adopt the technology on a wide scale. Third, hybrids do not possess the required amount of disease and insect pest resistance in the target regions. Fourth, the grain quality of hybrids does not meet market needs, and decreased head rice recovery is keeping farmers from adopting hybrid rice. In addition, the rapid rise in labor wages in India and China is causing the seed industry to look for alternative approaches to decrease the cost of hybrid rice seed and make it more efficient based on parental line improvement to entice farmers to adopt hybrid rice and benefit. In this regard, the Hybrid Rice Development Consortium (HRDC) at IRRI is considering these factors and developing market-oriented parental materials. Ongoing hybrid rice research at IRRI seeks to improve the levels of outcrossing and hybrid seed reproducibility, especially by developing newer CMS lines. The HRDC has been sharing these improved materials with both the public and private sectors in an aggressive manner since 2016. Currently, the area of hybrid rice outside China is approximately 8 million ha, and pushing hybrid rice technology is vital to overcome its shortcomings. In this context, it is crucial to revisit other alternative technologies such as two-line hybrid rice technology for efficient seed production and increased heterosis.

2 The Emergence of Two-Line Hybrid Rice Technology with a Historical Perspective

Two-line hybrid breeding began with the discovery of a photoperiod-sensitive genic male sterile (PGMS) mutant, Nongken 58S, in Hubei Province, China, which remains male sterile under long-day conditions (>13.45 h) or fertile under shorter day (<13 h) conditions (Shi 1981, 1985; Shi and Deng 1986). Likewise, the discovery of thermosensitive genic male sterility (TGMS) that renders the plant male sterile at higher mean temperatures and reverts it to fertility at lower mean temperatures allowed significant development of the technology. Several TGMS sources of spontaneous or induced origin were discovered such as Annong S-1 and Anxiang S (Tan et al. 1990; Lu et al. 1994) in China, Norin PL 12 (Maruyama et al. 1990, 1991) in Japan, IR32364 at IRRI (Virmani and Voc 1991), and SM 5, F61, and SA 2 in India (Ali 1993; Ali et al. 1995; Hussain et al. 2012; Reddy et al. 2000) (Table 1). Moreover, photo-thermosensitive genic male sterility systems were also discovered, for which researchers found the interaction of photoperiod and temperature that governs male sterility-fertility alteration. Based on these three male sterility-fertility alteration systems involving photoperiod, temperature, and photo-thermo interactions, Yuan (1987) put forward a new strategy of hybrid rice breeding that did not involve a maintainer line, and it was called the two-line method. Any fertile line with a dominant gene for this trait could be used as a pollen parent to develop rice hybrids (Lu et al. 1994). Two-line hybrid rice technology has several advantages over the three-line system, including a wider range of germplasm resources as pollen parents, thus allowing opportunities to exploit higher heterosis and simpler procedures for breeding and hybrid seed production (Ali et al. 2018; Chen et al. 2020).

In tropical conditions, day length differences are marginal, and therefore, the TGMS system is more useful than the PGMS and PTGMS systems. Consistent temperature differences are found at different altitudes and over different seasons in the same location or region, which could be exploited for two-line hybrid rice development. However, successful exploitation of this novel male sterility system relies on knowledge of the fertility behavior of TGMS lines (Chandirakala et al. 2008).

Table 1 Origin and fertility-sterility transformation behavior of photoperiod-thermosensitive and temperature-sensitive male sterile sources in rice

Source	Ecotype	Origin of gene	Place of development	Critical temp. and photoperiod for inducing sterility (h)	(°C)	CFTP (°C)	Sensitive stage (days before heading)	References
Photoperiod-thermosensitive genic male sterile (PTGMS) (interaction of h and °C)								
02428 S	–	NK58S	JAAS					Li (2009)
108 S	–	NK58S/9022	NAAS					Li (2009)
1541 S	L-*japonica*		YIAS	13.75–14.00	28.0	22.0		Lu et al. (1994); Li (2009)
1647 S	–		BAAS					Li (2009)
2177 S	*indica*		AGAI					
26 Zhai Zao	*indica*	Induced (R), China		12.00–14.00	23.0–25.0			Shen et al. (1994)
3008 S	*japonica*	NK58S/	HAC					Li (2009)
31111 S	L-*japonica*	NK58S/31111	HAU	14.00–14.75	28.0	22.0		Li (2009)
31301-1S					28.0	24.0		Zhang et al. (1994)
3502 S	L-*japonica*	7001S/Pecos	AAAS	14	22.6			Li (2009)
3516 S	L-*japonica*	N5047S/(7001S/Zhao107)	AAAS	14	23.5			Li (2009)
4008 S	L-*japonica*	7001S/Reyan 2	AAAS	14	24.0			Li (2009)
5021 S	–	SDL of MS type. S.mutant	NAU and JAAS					Li (2009)
5047 S					30.0	26.0		Zhang et al. (1994)
6334 S	L-*japonica*		HNU	13.75–14.00	24.0–30.0			Li (2009)
7001 S	E-*japonica*	NK58S/917 (HuXuan19/IR661/C57)	AAAS	13.50–14.0	30.0	22.0		Lu and Wang (1988); Lu et al. (1994); Zhang et al. (1994); Mou et al. (2003); Li (2009)
8087 S	E-*japonica*	7001S/Zhao 107	AAAS	14	23.0			Li (2009)
8801 S	*indica*		HXAU					

Source	Ecotype	Origin of gene	Place of development	Critical temp. and photoperiod for inducing sterility (h)	(°C)	CFTP (°C)	Sensitive stage (days before heading)	References
8902 S	indica		WU	13.25–13.45	27.0–30.0			Zhang et al. (1994)
8906 S	indica		WU	13.25–13.45	26.0	24.0		Zhang et al. (1994)
89-7S					30.0	24.0		Zhang et al. (1994)
8912 S	indica		WU	13.25–13.45	30.0	26.0		Zhang et al. (1994)
9044 S	japonica		HAA		32	28.0		Zhang et al. (1994)
916 S	–	NK58S	JAAS					Li (2009)
AB0195	japonica		WDAU					
C407S	japonica	Eyi MR	CAAS					Li (2009)
CIS 28-10S	indica	S. mutant, China		12.00–14.00				Huang and Zhang (1991)
Double 8-2S	L-japonica	NK58S/Double 8-2	WU	14.00–14.25				Li (2009)
EGMS	japonica	Induced (C), USA		13.00–14.00				Rutger and Schaeffer (1989)
HN5-2S	indica				24.0	24.0		Zhang et al. (1994)
HS-1	indica	HPGMR	FU					Lu et al. (1994)
HS-3	indica	NK58S	FAU	12.5	23.0c			Mou et al. (2003)
J-3S	–	NK58S	JAAS					Li (2009)
K14 S	indica		GAU					
K7 S	indica		GAU					
K9 S	indica		GAU					
Liuqianxin S	–	NK58S	JAAS					Li (2009)
M 201	japonica	Induced (C), USA		12.00–14.00				Oard and Hu (1995)
M105 S	–	60 Coγ radiating 105	WU					Li (2009)

(continued)

Table 1 (continued)

Source	Ecotype	Origin of gene	Place of development	Critical temp. and photoperiod for inducing sterility (h)	(°C)	CFTP (°C)	Sensitive stage (days before heading)	References
M901 S	indica				26.0	24.0		
MSr 54A(B)	japonica	S. mutant, China		13.00–14.00				Lu and Wang (1988)
N422 S	japonica	7001S/lun hui 422	HHRRC, CAU					Li (2009)
N5047 S	L-japonica	NK58S/5047	HAAS	14.00–14.25	26.0–30.0			Lu et al. (1994); Li (2009)
N5088 S	L-japonica	NK58S/Nonghu26	HAAS	13.50–14.00	22.0–30.0	22.0		Zhang et al. (1994); Lu et al. (1994)
N95076 S	L-japonica	5088S/7001S	HAAS		24.0			Li (2009)
N9643 S	L-japonica	NK58S/9643	HAAS	>14.00	24.0			Li (2009)
Nongken58S	L-japonica	S. mutant from NK58, China	Hubei	13.75–14.00	30.0	24.0		Shi and Deng (1986); Zhang et al. (1994)
Pei'ai64S	indica	NK58S-derived, China	HHRRC	13.00–13.30	24.0	22.0		Yang et al. (2002)
Shuanggung S	japonica		HHRRC		32.0	28.0		Zhang et al. (1994)
Shuguang612S	indica	NK58S	SAU	12.5	23.5c			Lu et al. (1994) (T), Mou et al. (2003) (P)
W6154 S	indica		HAAS	13.00–13.30	26.0	24.0		
W7415 S	indica		HAAS	13.00–13.30	26.0	24.0		
W91607S	indica				26.0	24.0		Zhang et al. (1994)
W9593 S	indica	NK58S	HAAS	13	23.5c			Mou et al. (2003)
WD 1S	L-japonica	NK58S/WD1	WU	14.00–14.50				Li (2009)
Wuxiang S (WXS)	indica							
X 88	japonica			>13.75			10–25	Lu et al. (1994)

Source	Ecotype	Origin of gene	Place of development	Critical temp. and photoperiod for inducing sterility (h)	(°C)	CFTP (°C)	Sensitive stage (days before heading)	References
Zhenong 1S	L-japonica	NK58S	ZAAS					Li (2009)
Temperature-sensitive male sterile (TGMS) (°C)								
9201	indica	560 S	FU					Lu et al. (1994)
1103 S	indica	HPGMR	WU					Lu et al. (1994)
1356 S	indica	Annong S-1	HHRR	–	24.5^c			Mou et al. (2003)
3418 S	indica	HPGMR	AAAS					Lu et al. (1994)
545 S	indica		Hunan					
5460 S	indica	Induced (R), China	Fujian		28.0–26.0			Yang et al. (1990)
6442S	indica	HPGMR	JAAS					Lu et al. (1994)
810 S	indica	AnnongS-1	AJAU,	–	24.0^c			Mou et al. (2003)
Annong S-1	indica	S. mutant	Hunan		30.2–27.0			Tan et al. (1990)
Anxiang S	indica	Annong S	HHRRC					Lu et al. (1994)
ATG-1	indica							
C815 S	indica							
Dianxin 1A	japonica	CMS	Yunan		20.0–23.0			Lu et al. (1994)
DRR 1S			DRR		30.0			Ramakrishna et al. (2006)
F 61	indica	Induced mutation (C) India	IARI		22.0–30.9		19	Ali et al. (1995)
GD 2S	indica	HPGMR	GDAAS					Lu et al. (1994)
Guangzhan63S	indica	NK58S-derived, China						
H 89-1	japonica	Induced (R), Japan			31.0–28.0			Maruyama et al. (1991)

(continued)

Table 1 (continued)

Source	Ecotype	Origin of gene	Place of development	Critical temp. and photoperiod for inducing sterility (h)	(°C)	CFTP (°C)	Sensitive stage (days before heading)	References
Hengnong S-1	indica	Cross breeding, China	Hunan		29.0–30.0			Lu et al. (1994)
ID24					29.5–25.9		10–14	Sanchez and Virmani (2005)
IR32364-20-1-3-2B	indica	Induced (R), IRRI	IRRI		32.0–24.0			Virmani and Voc (1991)
IR38949	indica	Introgression from Norin PL12	IRRI		30.0–24.0			Virmani (1992)
IR68298		Introgression from Norin PL12	IRRI		31.5–27.1		11–17	Sanchez and Virmani (2005)
IR68935		Introgression from Norin PL12	IRRI		32.4–27.7		5–14	Sanchez and Virmani (2005)
IR68945	indica	Introgression from Norin PL12	IRRI		30.0–24.0		15–21	Virmani (1992); Sanchez and Virmani (2005)
IR71018		Introgression from Norin PL12	IRRI		32.2–27.4		12–24	Sanchez and Virmani (2005)
IR73827-23S			IRRI		35.9		19	Ramakrishna et al. (2006)
IR72093		Introgression from Norin PL12	IRRI		30.4–26.3		8–16	Sanchez and Virmani (2005)
IV A	indica	Cross breeding, China			24.0–28.0			Zhang et al. (1991)
J207S	indica	S. mutant, China	IARI		31.0– >31.0			Jia et al. (2001)
JP 2	indica	S. mutant, India	IARI		23.0–33.9		19	Ali et al. (1995)
JP 24A	indica	CMS, India	IARI		23.0–33.8			Ali (1993)

Source	Ecotype	Origin of gene	Place of development	Critical temp. and photoperiod for inducing sterility (h)	(°C)	CFTP (°C)	Sensitive stage (days before heading)	References
JP 8-1A-12	indica	Breeding population, India	IARI		20.0–30.9	20.0–24.0	23	Ali et al. (1995)
KS 1S	indica	HPGMR	GAAS					Lu et al. (1994)
N8 S	indica		Hunan					
Norin PL12	japonica	Irradiation with 20 kr of gamma rays, Japan			21.4–29.4		9–16	Lopez et al. (2000); Sanchez and Virmani (2005)
R 59TS	indica	Induced (R), China	Fujian					Yang et al. (1990)
SA 2	indica	Induced mutation (C) India	IARI		20.0–31.7		17	Ali et al. (1995)
SE21S	indica	NK58S	FAU	–	23.0c			Mou et al. (2003)
SM 3	indica	S. mutant, India			22.0–32.0		22	Ali et al. (1995)
SM 5	indica	S. mutant, India			22.0–32.3		24	Ali et al. (1995)
TianfengS	indica	AnnongS-1	GZAAS,	–	24.5 c			Mou et al. (2003)
TGMS 74S	indica		TNAU		34.2	23.0		Rajesh et al. (2017)
TGMS 81S	indica		TNAU		32.9	24.2		Rajesh et al. (2017)
TGMS 82S	indica		TNAU		32.9	24.2		Rajesh et al. (2017)
TGMS 91S	indica		TNAU		34.2	22.7		Rajesh et al. (2017)
TGMS 92S	indica		TNAU		34.2	24.2		Rajesh et al. (2017)
TGMS 93S	indica		TNAU		34.2	24.2		Rajesh et al. (2017)
TGMS 94S	indica		TNAU		34.2	24.2		Rajesh et al. (2017)
TNAU 19S	indica		TNAU		20.0–30.0	24.0–26.0		Manonmani et al. (2016)
TNAU 27S	indica		TNAU		25.95	25.83	26-Jan	Sasikala et al. (2015)

(continued)

Table 1 (continued)

Source	Ecotype	Origin of gene	Place of development	Critical temp. and photoperiod for inducing sterility (h)	(°C)	CFTP (°C)	Sensitive stage (days before heading)	References
TNAU 39S	*indica*		TNAU		20.0–30.0	24.0–26.0		Manonmani et al. (2016); Kadirimangalam et al. (2017)
TNAU 45S	*indica*		TNAU		20.0–30.0	24.0–26.0		Manonmani et al. (2016); Kadirimangalam et al. (2017)
TNAU 60S	*indica*		TNAU		20.0–30.0	24.0–26.0		Manonmani et al. (2016); Kadirimangalam et al. (2017)
TNAU 95S	*indica*		TNAU		20.0–30.0	24.0–26.0		Manonmani et al. (2016); Kadirimangalam et al. (2017)
TS 09 12	*indica*		TNAU		26.45	25.78	26-Jan	Sasikala et al. (2015)
TS 09 15	*indica*		TNAU		25.80	25.45	26-Jan	Sasikala et al. (2015)
TS 09 25	*indica*		TNAU		26.73	26.58	26-Jan	Sasikala et al. (2015)
TS6	*indica*	Spontaneous mutant	TNAU		26.7	25.5		Latha et al. (2005)
TS16 (IR68945-433-4-14)	*indica*	Norin PL 12	IRRI		24.8	24.6	11-Jan	Latha et al. (2005, 2010)
TS18 (IR68949-11-5-31)	*indica*	Norin PL 12	IRRI		24.2	24.0	18-Jan	Latha et al. (2005, 2010)
TS29	*indica*	Spontaneous mutant	TNAU		25.6	25.3	11-Jan	Latha et al. (2005, 2010)
TS46 (IR68942-1-6-13-13-4)	*indica*	Norin PL 12	IRRI		25.4	25.3	26-Jan	Latha et al. (2005, 2010)

Source	Ecotype	Origin of gene	Place of development	Critical temp. and photoperiod for inducing sterility (h)	(°C)	CFTP (°C)	Sensitive stage (days before heading)	References
TS47 (IR68298-11-16-3 B)	indica	Norin PL 12	IRRI		35.3	25.2	26-Jan	Latha et al. (2005, 2010)
W6111 S	indica		Hubei					
W91607 S	indica	HPGMR	HAAS					Lu et al. (1994)
W9451 S	indica	HPGMR	HAAS					Lu et al. (1994)
Xiang125S	indica	Annong S-1	HHRRC	—	23.5c			Mou et al. (2003)
Xiangling628S	indica							
Xianquang	indica	Breeding population, China			24.0–30.0			Cheng et al. (1995)
XinanS	indica							Si et al. (2012)
Y58S	indica							
Zhu1S	indica							Yang et al. (2000)
Reverse thermo-sensitive genic male sterile (rTGMS)								
JP38	indica	S. mutant, India	IARI		24.0–30.5			Ali (1993)
Dianxin 1A	japonica	China	Yunnan		22c			Yiming (1988); Zhang et al. (1991)
IVA	indica	China	Yunnan		24c	27.0		
26 Zhaizao	india	Mutant, China			>23c			Shen et al. (1994)
J207S	india							Jia et al. (2001)
Reverse photoperiod-sensitive genic male sterile (rPGMS)								
YiD1S		B3/Hongjiang	China					Gao (1991)

(continued)

Table 1 (continued)

Source	Ecotype	Origin of gene	Place of development	Critical temp. and photoperiod for inducing sterility (h)	(°C)	CFTP (°C)	Sensitive stage (days before heading)	References
IVA	*indica*	From cross breeding	Yunnan					Zhang et al. (1991); Virmani et al. (2003)
N10S								Li et al. (1991)
N13S								Li et al. (1991)
go543S								Yang and Zhu (1996)
DianmongS-2								Jiang et al. (1997)
D38S								Joseph et al. (2011)
D52S								Joseph et al. (2011)

GAAS Guangxi Academy of Agricultural Sciences, *GDAAS* Guangdong Academy of Agricultural Sciences, *GZAAS* Gangzhou Academy of Agricultural Sciences, *HAAS* Hubei Academy of Agricultural University, *HAAS* Hubei Academy of Agricultural Sciences, *JAAS* Jiangxi Academy of Agricultural Sciences, *AAAS* Anhui Academy of Agricultural Sciences, *HHRRC* Hunan Hybrid Rice Research Center, *IARI* Indian Agricultural Research Institute, *WU* Wuhan University, *SAU* Sichuan Agricultural University, *FU* Fujian University, *FAU* Fujian Agricultural University, *GAU* Guangxi Agricultural University, *AJAU* An-Jiang Agricultural University, *HAU* Huazhong Agricultural University, *HAC* Hubei Agricultural College, *AGAI* Anhui Guangde Agricultural Institute, *YIAS* Yichang Institute of Agricultural Sciences, *ZAAS* Zheijiang Academy of Agricultural Sciences, *L* late, *E* early, *S* spontaneous, *SDL* short day length, *CDL* critical day length, *CSP* critical sterility point, *CFP* critical fertility point, *R* irradiation, *C* chemical mutagens. Several introgressed forms from Nongken 58S and Annong S-1 developed by Yang (1997) and Mou et al. (1998) not included here

In this regard, to address tropical Asian markets, IRRI is refocused on developing two-line hybrid rice technology with usable TGMS parental lines. The two-line hybrid rice approach via TGMS holds great promise as it does away with one step of outcrossing of parental line production, thus directly bringing down seed costs. Although the two-line system is well established, especially in Vietnam and the Philippines, expansion to other regions remains a challenge because of the lack of TGMS lines with a low critical sterility temperature point (CSTP) of 24 °C. Such low CSTP of TGMS lines could be a game changer in tropical Asia vis-à-vis earlier discovered TGMS lines with CSTP of >27 °C. Currently, the annual planting area of two-line hybrid rice in China has surpassed 5 million ha, while fully exploiting heterosis in rice (Chen et al. 2020). With recent research advances, TGMS-based two-line hybrid rice breeding is poised to replace three-line hybrid rice technology over the next decade (Ali et al. 2018).

3   Advantages and Disadvantages of the TGMS System in the Tropics

The TGMS-based two-line system has several advantages over the three-line system. First, hybrid seed production is less cumbersome as TGMS does not require maintainers and seed can be self-multiplied under fertility-conducive low-tempera-

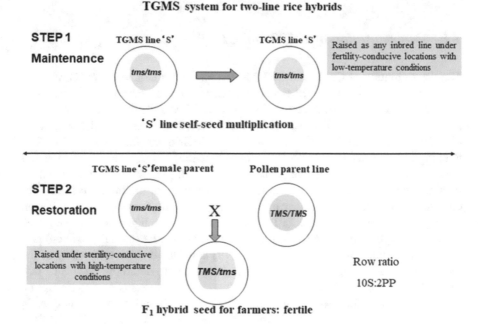

Fig. 1 TGMS system for the production of two-line rice hybrids

ture conditions (Fig. 1). Second, there is a higher probability of identifying the heterotic pool and market-oriented hybrids, as any nonTGMS parent is a potential pollen parent. Third, the current CMS three-line system is primarily based on a single source of wild abortive (WA) cytoplasm that continues to pose a constant threat because of the adverse effects associated with it. However, the two-line approach also has certain shortcomings, such as the adverse effect of low-temperature fluctuations due to sudden/unforeseen weather changes that could trigger self-seeds in hybrid seed production plots. In addition, the higher temperature fluctuations in self-seed multiplication plots could result in lower self-seed yields of the TGMS lines. Therefore, the right choice of locations based on historical agrometeorological data is essential to identify ideal places for hybrid rice seed production and self-seed multiplication.

4 Physiological Characterization of the TGMS Trait

Homozygous and true-breeding TGMS lines need to be physiologically characterized, especially for CSTP and CFTP, besides determining the temperature-sensitive stage for sterility-fertility alteration. The deployment of TGMS lines needs to match the target location requirements. Furthermore, precise information on these two indices is essential for choosing an appropriate source for the development of two-line hybrids (Ali et al. 1995).

4.1 Determination of CSTP and CFTP

The determination of CSTP and CFTP is essential for characterizing TGMS lines for their proper exploitation in target regions. CSTP pertains to the lowest mean temperature among the temperatures inducing sterility, while the highest mean temperature causing fertility is considered as the CFTP (Chandirakala et al. 2008; Latha and Thiyagarajan 2010; Sasikala et al. 2015; Kadirimangalam et al. 2017). The tracking technique (Ali et al. 1995) was used to identify the CSTP and CFTP based on the sensitive stage of a line. Using this method, the CSTP is determined by obtaining the lowest among the maximum temperatures of the three tracking dates coinciding with the sensitive stage of the three panicles that caused complete pollen sterility. At the same time, the CFTP is the temperature range in which the plants produced a higher proportion of fertile and unaborted sterile (partially stained) pollen. Further studies by Vinodhini et al. (2019) considered the lowest value of the mean maximum temperature during the sensitive stage to determine the CSTP of a TGMS line. Viraktamath and Virmani (2001) proved that the maximum temperature is what influences the expression of fertility-sterility alteration of TGMS lines in tropical countries. Moreover, Kadirimangalam et al. (2017) identified TGMS lines with a CSTP at a mean temperature of above 29 °C. It is essential to understand that a given TGMS gene varies for its CSTP and CFTP when transferred to different genetic backgrounds (Sasikala et al. 2015). The fertility of PTGMS rice lines is affected by both temperature and light duration. Usually, PTGMS rice lines tend to produce low purity of hybrid seeds because of selfing at a low temperature (23–24 °C) in seed production. The spikelets of PTGMS lines during anthesis could not normally open at high temperature (HT, $\geq$35 °C), thereby severely decreasing hybrid seed yields (Chen et al. 2020). This, along with other factors, makes PTGMS unfavorable for use in tropical conditions. However, PTGMS materials may still be useful in temperate conditions where day length is more crucial.

4.1.1 Characterization Under Controlled-Temperature Screening Conditions

Sterile single-plant selections identified in a mutation population of the M_2 generation or selections from segregating materials derived from TGMS × pollen parent (PP) crosses need to be stubbled and screened at low temperature to check for fertility reversion in the new emerging panicles. The crosses need to be bagged, and the generations correctly advanced under low-temperature facilities. At IRRI, the focus is on TGMS traits with low CSTP; thus, screening of the stable mutants and fixed materials is done under a phytotron in three mean temperature treatments (23, 24, and 25 °C) to determine their critical temperature for sterility/fertility induction (Fig. 2). This helped in identifying several TGMS lines with sterility at 24 °C and above and fertility at 23 °C. A few sterile plants were also identified in all three temperature conditions and are currently being evaluated for fertility reversion at <22 °C. Wongpatsa et al. (2014) carried out a similar study using two TGMS lines (KU-TGMS1 and KU-TGMS3) screened at the panicle initiation stage under growth chambers using day/night temperature parameters of 26/22 °C, 26/20 °C, 24/18 °C, and 22/20 °C, along with 11.5 h light/12.5 h dark periods and 75% relative humidity. Their results suggest that night temperatures of 18–22 °C induced maximum pollen viability and seed set. Furthermore, the highest seed rate was observed for KU-TGMS3 under 24/18 °C, peaking at 33.63%. In conclusion, this revealed that night temperature has a more significant effect on pollen viability than day temperature.

4.1.2 Field Screening Through Sequential Seeding

The physiological characterization of fixed TGMS lines can also be carried out through continuous seeding or sequential sowing. Sequential seeding is done in such a way that flowering is observed throughout the year at the candidate target sites to study pollen sterility and spikelet sterility (bagged and unbagged conditions). Such studies help in evaluating the stability of promising TGMS lines and determining the sterile phase window for hybrid rice seed production. Based on the tracking method (Ali et al. 1995), one can determine the CSTP and the sensitive

Fig. 2 Physiological characterization of TGMS lines in (**a**) plant growth facility bay, (**b**) cold-water facility, and (**c**) reach-in chamber

stage for sterility. A study done by Ramakrishna et al. (2006) observed six TGMS lines planted in three staggered sowing at intervals of 10 days. The lines were seen over two different seasons, postrainy 2002 (October–December) for fertility reversion with lower temperature range (25.5/16.1 °C) and prerainy 2003 (February–April) for sterility reversion with higher temperature range (35.7/23.8 °C), especially during the panicle initiation stage (Ramakrishna et al. 2006). Shuttle breeding of the selected sterile plants from segregating materials and their stubbles then transfers them to low-temperature conditions for obtaining self-seeds to advance the generations under low-temperature conditions. It would help to identify suitable TGMS lines for such environments. At IRRI, the sterile plant stubbles are sent to Lucban and Benguet in the Philippines for self-seed multiplication and generation advancement. Likewise, researchers at Tamil Nadu Agricultural University, India, evaluated TGMS lines in two sterility-inducing environments, Coimbatore and Sathiyamangalam, during rabi season starting in December 2013 and 2014. The same lines were stubble-planted and evaluated for pollen sterility in pollen fertility-inducing environments during kharif season in July 2013 and 2014 at the Hybrid Rice Evaluation Centre, Gudalur, a high altitude (1500 masl) with colder climate (Manonmani et al. 2016). Latha and Thiyagarajan (2010) also recommended a high-altitude area such as Gudalur for TGMS self-seed multiplication of lines such as

TS29, which was observed to have only 16 days of fertile phase during December in Coimbatore. In Gudalur, TS29 had more than 60% pollen fertility and seed set when the mean temperature was 22 °C (28/17 °C) and below from June to November.

4.2 Determination of the Critical Stage for Fertility-Sterility Alteration

The critical stages of panicle development sensitive to temperature could be determined from the stages exhibiting a significant correlation with pollen sterility (Chandirakala et al. 2008). The stamen pistil primordial stage, which is 15–24 days before heading, was considered as the sensitive stage (Ali et al. 1995; Salgotra et al. 2012). Furthermore, Viraktamath and Virmani (2001) found 4–8 days after panicle initiation as the most sensitive stage. For the lines that Latha and Thiyagarajan (2010) had examined, those were sensitive to temperature from stamen pistil primordial differentiation to pollen ripening except for two lines that were sensitive from the meiotic division of the pollen mother cell to pollen ripening. The sensitive stages observed to vary with the four TGMS lines (TNAU 27S, TS 09 12, TS 09 15, and TS 09 25) showed a significant amount of positive correlation between pollen sterility and maximum and mean temperatures (Sasikala et al. 2015). The period of partial sterility was considered as the phase of fertility transition (Ali et al. 1995; Latha and Thiyagarajan 2010). Sanchez and Virmani (2005) observed differentiation of secondary branch primordium and the filling stage of pollen, that is, 24 to 5 days before heading was considered a sensitive stage for temperature. The results showed that the critical stage for most of the TGMS lines occurred during panicle developmental stages and approximately 26 to 5 days before heading (Kadirimangalam et al. 2017). Based on all these studies, we can demarcate the critical stage for sterility expression from 5 to 26 days before heading that coincides with the differentiation of secondary branch primordium and the filling stage of pollen. These sensitive days before heading also varied with early-, medium-, and late-duration TGMS lines and depending on the synchronous flowering habit.

4.3 Evaluation of TGMS Lines for Sterility-Fertility Alteration in Different Environments

TGMS-based two-line breeding programs require natural sites with low temperatures in higher altitudes in the tropics that are essential for advancing generations of selected TGMS lines. However, it will be worthwhile to select sterile plants with low CSTP in the range of 23–25 °C as they are stable under high-temperature conditions (28–30 °C) for sterility. A recent discovery at IRRI of A07 with low CSTP of 24 °C is an excellent example of this type of TGMS line (Ali et al. 2018). Regular self-seed multiplication of TGMS lines is carried out for their use in hybrid rice seed production plots under high-temperature conditions. IRRI has two locations (Lucban and Benguet) for self-seed multiplication in the Philippines. Multilocation trials for two-line hybrid rice seed reproducibility trials are essential for understanding the stability of the TGMS parental lines and their outcrossing features.

4.4 Improvement of Outcrossing Traits in TGMS and Pollen Parental Lines

Outcrossing is directly correlated as a function of floral morphology and flowering behavior for the male-sterile parental line (Oka and Morishima 1967). The wider angle of lemma and palea correlated with greater exsertion and surface area of the stigma, leading to higher seed-set percentage. Visual phenotypic selection can be used efficiently to identify higher seed-set potential (Ramakrishna et al. 2006; Salgotra et al. 2012). According to the standards set by Chen et al. (2010), female parents should possess a panicle exsertion rate of >70%, along with an excellent outcrossing rate, early and short flowering span, and well-closed lodicules and lemmas after pollination. On the other hand, pollen parents should exhibit large anthers and pollen quantity, pollen vigor, and vigorous growth ability (Chen et al. 2010).

Better panicle exsertion from the sheath in male-sterile lines would help increase the number of spikelets for outcrossing than lines with incomplete panicle exsertion (Rahul Roy and Kumaresan 2019; Abeysekera et al. 2003; Virmani 1994). The lines with higher panicle exsertion percentage coupled with higher seed set and higher spikelet fertility percentage influence outcrossing ability and could be well exploited for the development of hybrid rice (Arasakesary et al. 2015).

Many of the traits for outcrossing in CMS, such as greater glume opening angle and more stigma exsertion, lead to higher seed setting (Mahalingam et al. 2013), which could be used as well for TGMS breeding. Outcrossing of relevant traits, especially the longer feathery stigma protrusion on either side of the lemma-palea and full glume opening, is highly attractive for increased pollination reception, germination, and seed set. Developing synchronous flowering habits in TGMS lines is essential for successful seed production. At IRRI, a few long feathery stigma-protruding types of TGMS lines with synchronous flowering patterns were successfully identified (Fig. 3) (Ali et al. 2018). Similarly, at TNAU, the TGMS lines developed through pedigree breeding, mutation breeding, and identification of spontaneous mutants in the breeding material were addressing the market requirements for medium duration, better agronomic characteristics, and excellent floral traits and requirements such as high stigma exsertion, wider glume opening, and acceptable grain quality characteristics such as medium slender grain type, etc. (Manonmani et al. 2016). However, the outcrossing traits translating into higher hybrid seed yields need to be verified under hybrid seed production geographies.

Fig. 3 Newly developed
TGMS line with long
feathery stigma

The floral traits of the pollen parents are also equally important to obtain higher seed setting. The pollen parents need to be highly diverse from the TGMS parental lines. At the same time, they need to possess floral traits similar to those of a restorer in the three-line system, especially in terms of plant height, profuse tillering, heavy pollen load, and pollen dehiscence. Moreover, the pollen parents should possess a staggered flowering habit to provide good pollen dehiscence during hybrid seed production. Consideration should be given to the synchrony of the timing of pollen dehiscence of pollen parents. It should match the TGMS parent's spikelet opening, and stigma receptivity is essential. In addition, pollen parents need to possess all the market-required traits such as appropriate grain shape and quality, abiotic stress tolerance, and insect pest and disease resistance.

5   Genetics of TGMS Lines

The recent discovery of new low-CSTP TGMS lines that showed complete sterility at a mean temperature of 24 °C has sparked renewed interest in two-line hybrid rice technology. The genetics of the TGMS trait is essential for the exploitation of this technology.

5.1   Identification of Genes Governing the TGMS Trait

A single recessive nuclear gene governs the TGMS trait in TGMS lines (Hussain et al. 2012). So far, 13 TGMS genes and their alleles (*tms1, tms2, tms3, tms4, tms5, tms6, tms6(t), tms7(t), tms8, tms9, tms9-1, tms10,* and *tmsX*) found in 5460S, Norin PL 12, IR32364, SA 2, Annong S-1, SoKcho-MS, 0A15-1, UPRI-95-140TGMS, F61, Zhu1S, Hengnong S-1, *japonica* cv. 9522, and Xian S, respectively, have been identified based on their allelic relationship as well as molecular marker studies. (Wang et al. 1995; Subudhi et al. 1997; Yamaguchi et al. 1997; Reddy et al. 2000;

Table 2 Molecular markers associated with EGMS genes in rice (modified from Ali et al. 2018)

Trait	Gene	Source	Chromosome	Closest flanking markers	Distance (cM)	Reference	
TGMS	tms_1	5460S	8	RZ562-RG978	6.7	Wang et al. (1995)	
	tms_2	Norin PL12	7	R643A-R1440 (D24156)	0.3	Yamaguchi et al. (1997)	
		Norin PL12	7	RM11-RM2	5.0, 16.0	Lopez et al. (2003)	
		KDML105	7	Os7g2690	15.4, 16.9	Pitnjam et al. (2008)	
	tms_3	IR32364S	6	$OPAC3_{640}$-$OPAA7_{550}$	7.7, 10.0	Subudhi et al. (1997)	
		IR32364S	6	F18F, F18RM, F18FM/F18RM	2.7	Lang et al. (1999)	
	tms_4	TGMS-VN1	2	$E5/M12_{600}$	3.3	Dong et al. (2000)	
		SA2	9	RM257, EAA/MCAG	6.2, 5.3	Reddy et al. (2000)	
	tms_5	Annong S-1	2	RM174, R394	0, 2.5	Jia et al. (2000)	
		Annong S-1	2	C365-1, G227-1	1.04, 2.08	Wang et al. (2003)	
		M105S	2	RM174	0	Nas et al. (2005)	
		Annong S-1 and Y58S	2	4039-1 and 4039-2	–	Yang et al. (2007)	
		103S	2	RM3294, RM6378, RM7575 and RM71	–	Hien and Yoshimura (2015)	
		Annong S-1	2	dCAPS-172	–	Song et al. (2016)	
		IR68301S	2	RM12676, 2gAP0050058			Khlaimongkhon et al. (2019)	New
	tms_6	Sokcho-MS	5	RM3351, E60663	0.1, 1.9	Lee et al. (2005)	
	$tms_{6(t)}$	0A15-1	3	S187-770	1.3	Wang et al. (2004)	
		UPRI 95-140TGMS	3	–	–	Li et al. (2005)	
		G20S	10	RM3152, RM4455	3.0, 1.10	Liu et al. (2010)	
	$tms_{7(t)}$	UPRI 95-140TGMS	7	–	–	Li et al. (2005)	
	tms_8	F61	11	RM21, RM224	4.3, 3.0	Hussain et al. (2012)	
	tms_9	Zhu1S	2	Indel 37, Indel 57	0.12, 0.31	Sheng et al. (2013)	
		Zhu1S	2	Indel 91, Indel 101	–	Sheng et al. (2015)	
	tms_{9-1}	HengnongS-1	9	QY-9-19, QY-9-27	0.22, 0.07	Qi et al. (2014)	
	$tms10$	*japonica* cv. 9522	2	Os02g18320			Yu et al. (2017)	New
	$tmsX$	XianS	2	RMAN81, RMX21	–	Peng et al. (2010)	

Trait	Gene	Source	Chromosome	Closest flanking markers	Distance (cM)	Reference
PGMS/PTGMS	pms_1	32001S	7	RG477-RG511, RZ272	3.5–15.0	Zhang et al. (1994)
				RG477/R277, R1807	0.1, 6.0	Liu et al. (2001)
	$pms_{1(t)}$	Pei'ai64S	7	RM21242, YF11	0.2, 0.2	Zhou et al. (2011)
	pms_2	32001S	3	RG348, RG191	10.6, 7.0	Zhang et al. (1994)
	pms_3	Nongken 58S	12	RZ261/C751, R2708	5.5, 9.0	Mei et al. (1999)
		Nongken 58S	12	LJ47 and LJ265	0	Lu et al. (2005), Ding et al. (2012)
	pms_4	Mian 9S	4	RM6659, RM1305	3.0, 3.5	Huang et al. (2008)
	p/tms_{12-1}	Pei'ai64S	12	PA301, PAIDL2	–	Zhou et al. (2012)
	$ptgms_{2-1}$	Guangzhan63S	2	S2-40, S2-44	0.08, 0.16	Xu et al. (2011)
rTGMS	$rtms_1$	J207S	10	RM239-RG257	3.6, 4.0	Jia et al. (2001)
rPGMS	$rpms_1$	YiD1S	8	RM22980, RM23617	0.9, 1.8	Peng et al. (2008)
	$rpms_2$	YiD1S	9	RM23898, YDS925	0.9, 0.9	Peng et al. (2008)
	$rpms_{3(t)}$	D52S	10	RM5271 and RM244	6.6, 4.6	Joseph et al. (2011)
	csa					Zhang et al. (2013)

Jia et al. 2000; Wang et al. 2004; Lee et al. 2005; Li et al. 2005; Peng et al. 2010; Hussain et al. 2012; Sheng et al. 2013; Qi et al. 2014; Yu et al. 2017) (Table 2).

The identified *tms* genes could be further exploited for developing TGMS pyramiding lines by using two to three *tms* genes for improving stability during the sterility phase. However, only a few studies have been attempted on the pyramiding of these alleles, studying them for improving the stability of the TGMS lines (Nas et al. 2005). So far, 13 *tms*, seven *pms,* and three *rtms* genes have been identified governing the EGMS trait that is spread across all 12 rice chromosomes.

The TGMS trait is governed by a single major gene and could have several modifier genes that exist in different backgrounds. Therefore, it is crucial to characterize the TGMS lines physiologically before their commercial exploitation. The TGMS trait is much easier to transfer to other backgrounds through the marker-assisted backcross (MABC) approach, and one has to take care of modifier genes as well that may influence trait expression. In this context, it is essential to understand the molecular function of the TGMS trait (Ding et al. 2012; Zhou et al. 2012; Wang et al. 2013; Pan et al. 2014; Kim and Zhang 2017; Mishra and Bohra 2018).

Earlier studies on TGMS focused on the physiological aspects and how the gene is phenotypically expressed in the population. However, the first genetic study to confirm the location of the TGMS gene was begun by Wang et al. (1995) using an F_2 cross from a mutant TGMS line (5460S) and Hong Wan 52. Bulk segregant analysis and QTL mapping using RAPD markers identified the first TGMS gene as *TGMS1.2*, located within chromosome 8 (Wang et al. 1995). Succeeding genetic studies are all compiled and given in Table 2 with the corresponding molecular markers.

5.2   Molecular Mechanisms of the TGMS Trait

With the advent of new technologies in the field of genomics and transcriptomics, Luo et al. (2020) confirmed the location of the *tms* gene, which was begun by Wang et al. (1995), for the identification of *tms1* on chromosome 8 using RFLP markers. This transition from RFLP to SSRs and more recently with transcriptomics in confirming the *tms1* loci led to the unraveling of the mechanism behind *tms* genes (Luo et al. 2020). Furthermore, Pan et al. (2014) showed that, in line TGMS-Co27, male sterility is based on the cosuppression of a UDP-glucose pyrophosphorylase gene (*Ugp1*), and the underlying molecular mechanisms need to be unraveled. Zhou et al. (2014) uncovered the molecular mechanism of rice *tms5,* which functions in RNase ZS1-mediated UbL40 mRNA regulation during pollen development. Under permissive (low) temperature conditions, the level of UbL40 mRNAs remains low in the *tms5* mutant plants, allowing the production of normal pollen. However, at restrictive (high) temperature, UbL40 mRNAs are not processed by RNase ZS1, which leads to their high-level accumulation, causing male sterility (Zhou et al. 2014). Wang et al. (2019) carried out a comparative quantitative proteomic analysis of the anthers of TGMS line Annong S-1 grown at permissive (low) (21 °C) and restrictive

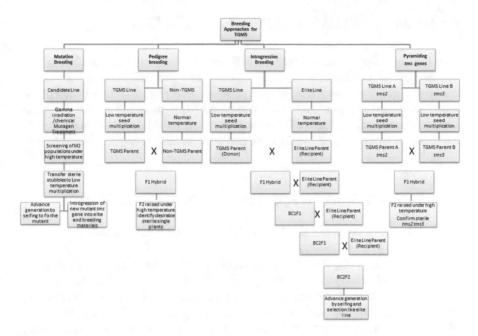

Fig. 4 Current breeding approaches for TGMS followed at IRRI

(high) temperatures (>26 °C). The restrictive high temperatures resulted in 89 differentially accumulated proteins (DAPs) in the anthers as compared to permissive low-temperature conditions. Out of the 89 DAPs, 46 had increased abundance and 43 had decreased abundance, which are distributed in most of the subcellular compartments of anther cells. Most have catalytic and binding molecular functions. Moreover, the gene ontology analysis for biological processes done by Wang et al. (2019) indicated that high-temperature induction caused the fertility-sterility conversion. This mainly adversely affects the metabolism of protein, carbohydrate, and energy and decreases the abundance of vital proteins closely related to defense and stress. This further impedes the growth and development of the pollen and weakens the overall defense and stress ability of Annong S-1.

Li et al. (2020) carried out RNA-Seq on rice TGMS lines at the microspore mother cell and meiosis stages under sterile and fertile conditions that revealed 1070 differentially expressed genes found to be enriched in protein folding, protein binding, regulation of transcription, transcription factor activity, and metabolic-related processes. They showed that hub genes (such as UbL40s) were predicted to interact with proteolysis-related genes and DNA-directed RNA polymerase subunit, and heat shock proteins (HSPs) interacted with kinases to play significant roles in regulating fertility alteration. Their study suggested that, besides UbL40s, DNA-directed RNA polymerase subunit, kinases, and HSPs might be involved in TGMS fertility alteration and could be applied for TGMS breeding (Li et al. 2020). Despite several of these in-depth studies, the TGMS trait mechanism still needs to be unraveled entirely for its immediate exploitation by breeders.

6 Breeding of TGMS and Pollen Parental Lines

Two-line breeding strategies for TGMS are currently carried out using four approaches: (a) the use of mutagenesis to induce new *tms* gene mutants from current materials, (b) conventional crossing and pedigree selection, (c) introgression of currently identified *tms* genes into elite lines, and (d) pyramiding known *tms* genes from different sources (Fig. 4). For each strategy, parental line selection remains the most crucial part to ensure hybrid vigor and address market segment requirements.

6.1 Different Available Approaches to Breed TGMS Lines

6.1.1 Mutation Breeding for the Identification of TGMS Mutants

Mutation breeding for the development of TGMS lines was first reported by Maruyama et al. (1991) for the development of Norin PL12 using gamma radiation. Furthermore, Ali et al. (1995) developed and characterized several TGMS lines using chemical and physical mutagens. Interestingly, Ali and Siddiq (1999) also identified a spontaneous mutant (JP38s) that showed a reverse TGMS trait, behaving as sterile at lower temperatures (<24 °C) and as fertile at higher temperatures (>30.5 °C). IRRI began a mutation breeding program using chemical mutagens in 2015 to discover new TGMS mutants, which are currently being characterized. The mutation populations in the M_2 generation need to be screened under high-temperature conditions to identify complete male sterility, and these are then stubbled and taken to low-temperature conditions to check for fertility reversions. Depending upon their seed settings in the stubbles, they are further generation advanced under low-temperature conditions to fix the TGMS mutants quickly. Upon fixation, these mutants are studied in different temperature regimes to characterize them physiologically (Ali et al. 1995, 2020 Unpublished).

6.1.2 Pedigree Breeding

It is also essential to breed new materials through crossing TGMS parents with elite lines and selection in the F_2 generation for male-sterile single plants under high-temperature conditions. At IRRI, conventional crosses were made with the TGMS line A07 as a pollinator and elite breeding materials as the female parents (Ali et al. 2018). After the initial cross in the F_2 generation, the selected male-sterile single plants in high-temperature regimes are then stubbled and selfed seeds are produced under low-temperature conditions. These selected single plants are verified for the presence of the *tms5* gene across succeeding generations. Using this approach, a new TGMS line with the *tms5* gene will be developed (Ali et al. 2020 Unpublished).

6.1.3   Transfer from a Known TGMS Gene Source to Elite Lines

Another strategy for integrating TGMS in two-line hybrid rice is by introgression of *tms* genes. At IRRI, the TGMS line A07 is used as a donor for introgressing the *tms5* gene into elite breeding materials by two backcrosses and selecting the progenies in BC_2F_2 onward for the *tms5* gene. By using foreground markers and high-density background SNP markers, introgression of the *tms5* gene into elite materials is possible. However, it is essential to accurately characterize these materials upon fixation for their fertility-sterility alteration behavior.

6.1.4   Pyramiding TGMS Genes for Better Stability

Despite the independent successes in characterizing and isolating different TGMS genes in rice, only a few studies have dealt with the additive effect and pyramiding of different TGMS genes (Nas et al. 2005). Two- and three-gene pyramids constructed using the three TGMS donors, Norin PL 12 (*tms2*), SA2 (*tgms*), and DQ200047-21 (*tms5*), possessing the RM11 allele of Norin PL 12, RM257 allele of SA2, and RM174 allele of DQ200047-21 were selected. As expected, all selected progenies were male-sterile in sterility-inducing conditions (Nas et al. 2005). The pyramids developed from this effort were designated as IR80775-46 (with *tms2* and *tms5*) and IR80775-21 (with *tms2*, *tgms,* and *tms5*). Pyramiding *tms* genes is useful to improve the stability of the TGMS line and to widen the sterility phase. Currently, at IRRI, efforts are ongoing to pyramid *tms2* and *tms5* genes to understand the mechanisms of the genes and to improve the stability of the TGMS trait. The current

	A07	IR 132941-B-1-9-AC 43	IR 75589-31-27-8-33-1 (TGMS)
Maturity (days)	100	120	117
Plant height (cm)	81.7	54.0-60.0	75.0
Tiller number	7-11	10-12	15-20
Panicle length (cm)	23.0	17.0-20.0	23.0
Flag-leaf length (cm)	45.8	35.0	21.5-30.0
Flag-leaf width (cm)	1.6	1.2	1.6
Value-added genes	GS3	GS3+Wxg2 and Wxg3	GS3+Wx-e6, Bph18+Bph3

Fig. 5 New IRRI stable TGMS lines with low critical sterility temperature point at 24 °C

Table 3 TGMS lines developed at IRRI (modified from Ali et al. 2018)

S. no.	TGMS line
1	A07
2	A32
3	A36
4	A37
5	IR75589-31-27-8-33-1 (TGMS)
6	IR68301-11-6-4-4-3-6-6 (TGMS)
7	IR73827-23-26-15-7 (TGMS)
8	IR73834-21-26-15-25-4 (TGMS)
9	IR75589-31-27-8-33 (TGMS)
10	IR77271-42-5-4-36 (TGMS)

Sterility-fertility alteration in newly bred TGMS lines

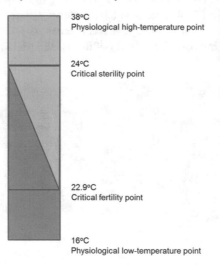

38°C
Physiological high-temperature point

24°C
Critical sterility point

22.9°C
Critical fertility point

16°C
Physiological low-temperature point

PHENOTYPIC CHARACTERISTICS
Maturity: 125 days
Plant height (cm): 96
Tiller number: 10
Yield (t/ha): 4.1 (under <23°C)
Flag leaf length (cm): 45.8
Flag leaf width (cm): 1.6
Grain shape: Medium slender
Low critical sterility-temperature point (Mean):24°C
Highly stable line and synchronous flowering

Fig. 6 New IRRI stable TGMS line with low critical sterility temperature point. (Source: Ali et al. 2018)

TGMS pyramiding studies at IRRI used line A07 as a pollinator parent (Ali et al. 2020 Unpublished).

6.2 Rapid Fixation of Segregating TGMS Lines

Conventionally, generation advancement is accomplished by growing the plants under natural low-temperature locations, for example, Lucban-Quezon (14.0805°N, 121.5427°E) and Tublay-Benguet (16.50805°N, 120.63524°E). This method remains the most popular as it is the most cost-efficient and requires the least technical work. This method, however, has its disadvantages as well. First, the environmental variables (temperature, humidity, and day length) at the location could cause genetic purities, mainly if fluctuations occurred during the plant's panicle initiation. Second, it requires labor-intensive cultural management of the field to prevent pests and diseases, especially under higher altitude locations in the tropics. Regardless of the fixation method, marker-assisted selection (MAS) is integral to the generation advancement of TGMS lines. MAS ensures the integrity of the *tms* gene and the genetic purity of the TGMS lines across generations.

Pedigree breeding and generation advancement of desirable TGMS segregants and mutants are challenging until the lines are correctly fixed. Recently, using new techniques of speed breeding under rapid generation advancement (RGA) facilities with specialized lighting, one can fix the segregating TGMS trait within 2–3 years, and this can be put to use to develop hybrid combinations. The use of the RGA method is a viable alternative to save on time and costs vis-à-vis field conditions. RGA hastens the fixation of new lines by advancing single seeds per line from a segregating population under controlled conditions (Collard et al. 2017). Instead of the usual dry and wet seasons, RGA allows several generations of advancement in a single season by growing the plants in trays instead of transplanting in the field to facilitate faster growth. Generations of TGMS breeding lines are advanced at low temperature (<22 °C) in plant growth facility (PGF) chambers. It is essential to maintain the critical temperature and humidity necessary to induce pollen fertility and self-seed setting in plants, thus requiring more labor costs, a PGF, and technical expertise.

To speed up the fixation of TGMS traits in the mutants and segregants, one can use a doubled-haploid (DH) approach. It is essential to identify the right segregants and mutants for fixation through the DH approach (Fig. 5). Many times, the DH TGMS lines, once fixed, may not be the ideal ones to match the market requirements. IRRI has previously developed four TGMS lines using DH technology: A07, A36, A32, and A37 (Ali et al. 2018) (Table 3). Among them, A07 has already been validated as highly stable and it has a low CSTP of 24 °C (Fig. 6).

Moreover, the *tms* gene present in this line (*tms5*) is the most extensively studied *tms* gene and it is used in different breeding programs as well (Wang et al. 2003; Nas et al. 2005; Yang et al. 2007; Kadirimangalam et al. 2019). Finally, DH technology offers the best potential among the three approaches. The use of DH technology ensures the fastest method of fixing recombinant genotypes, encompassing six generations of population advancement typically required to fix the population in just a single season (Yao et al. 2018). Moreover, the use of DHs eliminates the presence of deleterious alleles and background noise, which are typically observed in a natural population.

6.3   Breeding Pollen Parents

Heterotic pool-based breeding of pollen parents, more diverse and distinct from the TGMS pool, is required. These materials need to be improved within the pool and avoid contamination from materials nearer the TGMS heterotic pool. Breeders need to select for the target traits that help in pollen dehiscence, staggering flowering characteristics, and heavy tillering to provide a continuous pollen supply. Furthermore, the pollen parents need to address market segment needs so that the hybrids developed fit well. Pedigree breeding, single seed descent with genomic selection, along with RGA approaches could help to speed up the pollen parental breeding process. Specific traits such as genes with resistance against major insect pests and diseases that address market segment needs could be incorporated through a marker-assisted backcross (MABC) breeding approach.

6.4   Two-Line indica/japonica Hybrids

The two-line system is ideal for exploiting *indica/japonica* hybrids as there is no barrier for the identification of pollen parents, which could be any parent other than the TGMS parent. The TGMS gene could preferably be in the *indica* parental background, and with the use of a wide-compatibility (WC) gene in any one of the parents, one could develop *indica/japonica* hybrids. Shukla and Pandey (2008) suggested brighter prospects of combining improved *japonica* and *tropical japonica* germplasm having WC genes with *indica* TGMS lines for the exploitation of intersubspecific heterosis. Recently, with the discovery of reliable WC gene-based markers, ones such as *S5* could be highly useful for selection to combine with *tms* gene-based markers. At IRRI, the *S5* gene from different sources is backcrossed into TGMS line A07. Hybrid rice seed production of intersubspecific hybrids may be challenging due to the varying timing of spikelet opening and pollination of the two subspecies, especially in tropical environments. So, we need to carefully identify parental lines from these two subspecies closer to each other.

7   Breeding Two-Line Hybrids

Two-line rice hybrids have higher heterosis than three-line rice hybrids as any nonTGMS line could be used as a pollen parent, thus creating more extensive opportunities. Unlike the three-line system, CMS requires only restorers with restorer fertility (*Rf*) genes to restore fertility in the F_1 hybrid. Thus, it is a much narrower range within which heterosis needs to be exploited. On the other hand, TGMS-based two-line hybrids open up more opportunities to use the intersubspecific hybrids (*indica/japonica*) as the *japonica* subspecies has a low frequency of restorer genes. At IRRI, all source nurseries are genotyped, and heterotic pools are formed based on the genetic distances. Heterotic pool-based breeding is being followed to identify the best combinations for the two-line hybrids. To improve the heterotic pools, we have to make crosses within the pool. There is a need to maintain different heterotic pools carefully and to avoid contamination from other heterotic pools. To develop new heterotic hybrids, we can attempt crosses between distant pools.

Fig. 7 IR134554H, a multiple-stress-tolerant two-line rice hybrid developed at IRRI

Table 4 Performance of hybrids over the best inbreds and hybrids at IRRI-South Asia Hub, Hyderabad, in WS 2018, and at ISARC-Varanasi in kharif 2019

Designation	IRRI-South Asia Hub, Hyderabad (WS2018)									ISARC Varanasi (kharif 2019)		
								Yield advantage (%) over check			Yield advantage (%) over check	
	Days to flowering	Plant height (cm)	Panicle length (cm)	Productive tillers (no.)	Total spikelets per panicle	Spikelet fertility (%)	Total grain yield (kg/ha)	US337 (comm. hybrid)	MTU1010 (best variety)	Total grain yield (kg/ha)	Arize 6444Gold (hybrid check)	Sava127 Pro (hybrid check)
IR134554 H	101	92	21	12	158	80	7767.0	10.0	21.0	7956.7	6.4	12.9
IR81958H (Mestiso 77)	92	94	22	12	161	78	7638.0	8.0	19.0	7987.0	6.8	13.4
IR90872 H	95	88	22	18	173	79	7532.0	7.0	17.0	7606.0	1.7	8.0
IR81265H (Mestiso 61)	92	84	23	16	199	84	7164.0	2.0	11.0	7704.7	3.1	9.4
IR82391H (Mestiso 68)	95	88	21	13	163	78	6885.0	-2.0	7.0	7790.7	4.2	10.6
IR121020 H	87	78	20	13	153	79	6218.0	-12.0	-3.0	7040.7	-5.8	-0.1
IR106638 H	94	80	22	14	201	82	6153.0	-13.0	-4.0	7746.3	3.6	10.0
IR106616 H	90	85	22	12	221	83	6032.0	-14.0	-6.0	7676.0	2.7	9.0
IR81255H (Mestiso 89)	91	83	21	16	199	81	5947.0	-16.0	-8.0	7818.3	4.6	11.0
IR82386H (Mestiso 71)	93	82	22	13	168	68	5477.0	-22.0	-15.0	7927.0	6.0	12.5
DRR DHAN 44	83	81	21	13	157	76	3398.0	-	-	-	-	-
US 312	102	87	21	12	148	71	6881.0	-2.0	7.0	-	-	-
US 382	106	95	20	11	169	77	6756.0	-4.0	5.0	-	-	-
US 337	94	101	23	11	187	73	7051.0	0.0	10.0	-	-	-
MTU 1010	89	93	21	16	196	80	6436.0	-9.0	0.0	-	-	-

Designation	IRRI-South Asia Hub, Hyderabad (WS2018)									ISARC Varanasi (kharif 2019)		
	Days to flowering	Plant height (cm)	Panicle length (cm)	Productive tillers (no.)	Total spikelets per panicle	Spikelet fertility (%)	Total grain yield (kg/ha)	Yield advantage (%) over check		Total grain yield (kg/ha)	Yield advantage (%) over check	
								US337 (comm. hybrid)	MTU1010 (best variety)		Arize 6444Gold (hybrid check)	Sava127 Pro (hybrid check)
Arize 6129 Gold (hybrid check)	–	–	–	–	–	–	–	–	–	7249.3	–3.0	2.9
VNR 2355 Plus (hybrid check)	–	–	–	–	–	–	–	–	–	6636.0	–11.2	–5.8
Super Moti (hybrid check)	–	–	–	–	–	–	–	–	–	6617.3	–11.5	–6.1
BPT 5204 (inbred check)	–	–	–	–	–	–	–	–	–	6499.3	–13.1	–7.8
Sarju 52 (inbred check)	–	–	–	–	–	–	–	–	–	5559.3	–25.6	–21.1
MTU 7029 (inbred check)	–	–	–	–	–	–	–	–	–	6958.0	–6.9	–1.2
Arize 6444 Gold (hybrid check)	–	–	–	–	–	–	–	–	–	7476.7	0.0	6.1
Sava 127 pro (hybrid check)	–	–	–	–	–	–	–	–	–	7045.3	–5.8	0.0

7.1 Combining Ability Nurseries

The general combining ability (GCA) of an inbred is its average performance across a series of hybrid combinations, and it is primarily due to the additive effects of genes. The GCA effects of the parental lines help in the identification of suitable parental lines (Chandirakala et al. 2012). The most promising TGMS lines developed in high combining ability backgrounds could be used to further identify and validate their general combining ability. For this, a line × tester design could be used to identify high GCA of lines. This also allows identifying combinations with high specific combining ability that could be immediately exploited. A combining ability nursery needs to be regularly created to identify TGMS lines with high GCA and pollen parents from the breeding pipelines. Chen et al. (2010) stressed the importance of identifying PTGMS with high combining ability as this is the basis of robust heterotic hybrid rice varieties. Cao and Zhao (2014) showed that successful hybrids are directly determined by the combining ability of the sterile line, and sterile lines with high GCA have higher chances to produce heterotic combinations. In situations with poor GCA of TGMS lines, it is good to have pollen parents with high GCA to develop heterotic hybrids. Shukla and Pandey (2008), with a broad set of line × tester crosses, found that the parents with good GCA did not always produce the best hybrid combinations due to a lack of higher-order additive interaction, and they suggested evaluating the specific combinations. They found TGMS line 365-8S to be the best general combiner for all six traits: grain yield, panicle length, grain number per panicle, earliness in flowering, panicle number per plant, and 1000-grain weight.

7.2 Breeding Trials

Once hybrid combinations are identified, small-scale seed production either by hand crosses or in field conditions should be sufficient to carry out an observation yield trial (OYT). An OYT evaluation of the F_1s under best management conditions would allow the identification of good performing hybrids, and these should be forwarded to an advanced yield trial (AYT) in a larger plot size with proper replications and the best market checks. Simultaneously, the AYT is screened for resistances to insect pests and diseases. The highly performing hybrids should be identified and sent for grain quality evaluation. Based on all the data, the best hybrids need to be produced in large quantities and also evaluated for their hybrid seed reproducibility for ensuring their success when screened in multienvironment trials (METs). The best candidate hybrids tested under METs lay the foundation for the identification of the best hybrids for a given target location and market segment. IRRI conducted two demonstration trials in India to evaluate the performance of some newly developed two-line hybrids. One hybrid (IR134554H) performed exceedingly well at both Hyderabad and Varanasi (Table 4, Fig. 7).

Table 5 Market needs segment per region for hybrid rice

Countries	Product concept	Duration (days)	Potential hybrid area (m ha)	Core target geography	Yield advantage over best hybrid (%)	Yield advantage over best OPV (%)	Check variety	Plant height (cm)	Key defensive traits	Key agronomic traits	Grain quality	Producibility benchmark (yield/ha)	Key gaps in the present products
India	Mid-early segment	110–125	8	PUN, HAR, CG, CU, JH, MP, GJ, AS, UP, WB, KAR, W, C, Assam	5–8	15–20	MTU1010 and US 312 (inbred)	110–120	Blast,* neck blast,* BLB, BPH, false smut,* stem borer,* drought tolerant	High vigor, >85% spikelet fertility, non-lodging, cold tolerant (seedling), MTU 1010 & IET 4786 grain type, fine grain	HRR: 55–65% MRR: >70% AC: 18–24% LS	>3.0 t/ha, staggering <15 days	Lack of stress-tolerant materials, producibility, lodging,* false smut*
Bangladesh	MS grain segment (boro)	125–140	4	Sylhet, Dhaka, Chittagong, Rajshahi, Ranpur, Khulna, Barisal Divisions	10	20 (>9.0 t/ha)	BRRI dhan 81 (inbred) BRRI dhan 5 (hybrid)	90–100	BLB, SB, BPH	High vigor, spikelet fertility, cold tolerant, Jeerasail grain type (>85%)	MS with high amylose (>24%), milling (>65%)	>3.0 t/ha, staggering <15 days	Lack of stress-tolerant materials (salinity, drought, cold, diseases, insects
	Mid-early segment (T. aman)	115–120	2		10	20 (>7.0 t/ha)	BRRI dhan 49 (inbred) BRRI Hybrid 6 (hybrid)	100–110	*False smut,* BLB*	High vigor, spikelet fertility, early, drought tolerant, BRRI dhan49 grain type (>85%)	Slender grain with high amylose (>24%), milling outturn (>65%)	>3.0 t/ha, staggering <10 days	Lack of stress-tolerant materials (salinity, drought, cold, submergence, stagnation at 15–30-cm depths, diseases, insects)

(continued)

Table 5 (continued)

Countries	Product concept	Duration (days)	Potential hybrid area (m ha)	Core target geography	Yield advantage over best hybrid (%)	Yield advantage over best OPV (%)	Check variety	Plant height (cm)	Key defensive traits	Key agronomic traits	Grain quality	Producibility benchmark (yield/ha)	Key gaps in the present products
Philippines	Mid-early segment	95–115	2	Region 1, Region 3, Region 4A/4B	5	15–20	NSIC Rc 222 (inbred) Mestiso 3 (hybrid)	≥100	BLB, BPH, blast, stem borer,* RTV, non-lodging	Stability, more tillers, uniform grain maturity, non-lodging, non-shattering	AC 17–24%, >55% HRR, >65% MRR, LS, less chalk, slight aroma	≥2.5 t/ha	Susceptible to SB, poor standability, yield coupled with GQ, BLB, BPH, low HRR, susceptibility to fungal diseases
Indonesia	Mid-late segment	120–130	2	Bali, Central Java, East Java, Aceh, South Sulawesi, Lampung	≥10	≥20	Ciherang (inbred) Hipa 18, Hipa 20 (hybrid)	90–115	Blast, neck blast,* BLB, BPH, false smut,* stem borer,* drought tolerant, RTV, BPH, RRSV/RGSV	More productive tillers, more filled grains, lodging tol., wide adaptability, threshability	AC 18–22%, low-med GT, soft GC, >50% HRR, >65% MRR, <10% chalkiness, good palatability, transluscent, S-LS	>2.5 t/ha, staggering <10 days, high OCR (>50%)	RSV-BPH, BLB, blast, drought
Vietnam	Medium segment	> 110–130	2046.1 t/ha	North	1–5	>20	Nhi Uu 838, Thien uu, BC 15 (inbreds)	110–120	BLB, BPH	Yield	Intermediate AC, high HRR, low chalk	>2.5 t/ha	LS grain hybrid with BLB, BPH, SB traits
	Early segment	85–105	1.4 mt (early/LS)	South & Central	1–3	<20	TH3-3 (hybrid)	110–120	BLB, BPH	Very early	Intermediate AC, high HRR, low chalk	>2.5 t/ha	LS grain hybrid with BLB & BPH traits

7.3 Insect Pest and Disease Resistance

Two-line rice hybrid yield potential could be fully realized by incorporating resistance to major diseases and insect pests (bacterial leaf blight (BLB), blast (BL), false smut (FS), sheath blight (SHB), tungro, green leafhopper (GLH), brown planthopper (BPH), stem borer, leaf folder, and gall midge). Most insect pest and disease resistances are governed by major genes and could be easily introgressed into parental lines depending on market segment requirements. Two-line breeding offers better opportunities to convert the TGMS parent to acquire disease and insect pest resistance as compared to a CMS/maintainer parent, which is more cumbersome and requires more time (Ali et al. 2018). In this regard, IRRI has developed a global product concept addressing different market needs, which could be useful, require fewer resources, and result in higher impact. Researchers at Huazhong Agricultural University (HAU) introgressed *Xa7*, *Xa21*, and *Xa23* genes into C815S, a popular TGMS parental line, to develop five BLB-resistant cultivars: Hua1005S, Hua1002S, Hua 1009S, Hua 1006S, and Hua1001S (Jiang et al. 2015). Two-line hybrids with *Xa23* showed a resistance reaction to seven *Xanthomonas oryzae* pv. *oryzae* (Xoo) strains. Hua1006S was the most promising TGMS parent among them with a higher degree of resistance based on *Xa23* besides better plant type and grain quality features (Jiang et al. 2015). Currently, at IRRI, introgression of BLB and blast resistance genes into elite TGMS and pollen parental lines is carried out through marker-assisted selection.

7.4 Grain Quality Considerations Addressing Market Needs

A wider array of heterotic two-line rice hybrids opens up better options for developing customized grain quality that caters to market needs (Table 5). IRRI's two-line rice hybrid Mestiso 61 with good grain quality matched the market needs of the Philippines. It was successfully licensed to SL Agritech Company in the Philippines with limited exclusivity for a 6-year period. However, it is still available for license to the private seed industry for other countries under the Hybrid Rice Development Consortium. This hybrid gave an average yield of 6.7 t/ha during the dry season and 6.4 t/ha during the wet season across the Philippines. The yield potential of this hybrid was nearly 10 t/ha, with 55% head rice recovery and amylose content of 20.5%, ideally fitting Philippine market needs. We developed a strategy to breed and customize grain quality as per market requirements (Allahgholipour et al. 2006; Pang et al. 2016). In this approach, breeders identify good-quality lines that will cater to the varied interests of consumers across rice-consuming countries by screening the breeding materials for eating and cooking quality (ECQ) and keeping the popularly preferred good-quality varieties as controls in the study. Furthermore, work is ongoing to identify advanced rice breeding lines/cultivars with similar apparent amylose content (AAC), gelatinization temperature (GT), and rapid viscosity analysis (RVA), properties like those of the popular high-quality rice varieties, through simple cluster analysis. A two-line hybrid from China, Pei-Liang-you 1108, has relatively good ECQ, and through our study, we identified seven lines in the HC21 cluster clade with similar AAC, GT, and RVA and hence with comparable ECQ. Likewise, another two-line hybrid with good ECQ, Jin-ke-you651, allowed us to identify 11 hybrid lines within the HC18 cluster clade that had similar AAC, GT, and ECQ (Pang et al. 2016).

It is essential to develop rice hybrids with better ECQ that address market needs, paving the way for the expansion and adoption of rice hybrids in Asia and Africa. Higher hybrid rice yields have no value if they do not translate into higher percentage head rice recovery (>55%), leading to increased farmers' income.

8   Seed Production Challenges

Two-line hybrid rice technology largely depends on the identification of TGMS lines that need to be multiplied under low-temperature conditions, and hybrid rice seed production requires a minimum of 2 weeks of stable high temperature to reach the sterile phase. To achieve these two different aspects of seed production carefully, we have different approaches to identify appropriate locations based on agrometeorological data. However, this needs validation before large-scale seed production.

Key Challenges
- Addressing market requirements for different target places varies: for example, long-duration hybrids for the Indian market segment may require a longer duration of TGMS and pollen parents.
- Identification and exploitation of hybrid rice seed production and TGMS self-seed multiplication sites.
- Development of usable and stable TGMS parental lines matching market segment requirements.
- The relative heterosis of two-line rice hybrids needs to be superior to that of the existing best three-line hybrids in the market.
- Two-line hybrid rice technology should assure lower seed costs on account of better hybrid seed reproducibility rates of 3 t/ha and higher self-seed multiplication rates (>4.5 t/ha), making this seed feasible for use by farmers.

8.1   Identification of Ideal Locations for Self-Seed Multiplication of TGMS and Hybrid Rice Seed Production

TGMS-based two-line hybrid rice technology mainly depends on the identification of suitable areas for both self–seed multiplication and hybrid rice seed production (Table 6). Earlier, a systematic analysis of 50 years of agrometeorological data helped in the identification of appropriate sites in India (Siddiq and Ali 1999). Interestingly, the authors identified places located in India between 500 and 700 m above sea level from May to September for both hybrid seed and self-seed multiplication of the TGMS lines. Furthermore, through experimental validation, these places were confirmed as suitable for hybrid rice seed production, TGMS seed multiplication, and locations ideal for both operations (Siddiq and Ali 1999).

Critical considerations for the choice of place could be the hills, coastal plains, or interior plains, keeping within the physiological sterility limits of <40 °C to >16 °C. The Two-line Hybrid Rice Research Station was established under Tamil Nadu Agricultural University in the Nilgiris hills at 1200 m above sea level in a place known as Gudalur as early as 1995 in India (Soundararaj et al. 2002). Male-sterile TGMS selections at high temperatures at Trichy were made and immediately sent as stubbles to Gudalur to allow their self-seed multiplication and generation advancement. The most suitable time for matching the temperature conducive to self-fertility was from June to November. Shuttle breeding helped to identify 15 highly stable TGMS lines with better stigma exsertion of 40–66%, and many are in the pipeline. Nearly 800 ha of paddy lands are available for commercial self-seed multiplication of promising TGMS lines (Soundararaj et al. 2002). In the Philippines, Lucban, Nueva Vizcaya, and Benguet are all identified as highly suitable for self-seed multiplication of TGMS lines. In Nueva Vizcaya, the mean temperature from the beginning of October to the end of February in the next year is less than 22 °C, making it a suitable place to reproduce TGMS line seed. The mean temperature at Lucban from January to February was <23 °C, and so all the TGMS lines possessing a CFTP of <23 °C could be multiplied at Lucban. The TGMS lines should be completely male sterile to ensure the safety of hybrid seed production. Interestingly, we observed that the mean temperature at IRRI, Los Baños, was higher than 25 °C almost all year. So, the CSTP of fertility-sterility alteration of TGMS lines in the Philippines could be set at >24 °C for ensuring completely safe hybrid seed production, especially from April to June.

Pollen of A07 was partially fertile to completely sterile at Lucban as observed from 5 May to 17 June and completely sterile (with no pollen type) at Los Baños. A07 possesses a lower CFTP to turn completely fertile at <24 °C. A07 seeds produced in Nueva Vizcaya are possible where lower temperature prevails as compared to Lucban (Ali et al. 2020 Unpublished). Recently, with GIS technologies, IRRI has successfully identified a suitable choice of sites for hybrid seed production and TGMS self-seed multiplication based on 20 years of agrometeorological data. The potential GIS maps for the Philippines, identifying the places suitable for self-seed multiplication and hybrid rice seed production, are shown in Fig. 8. A map with a 0.08° spatial resolution and limited climatic data from 2010 to 2018 was used to avoid results affected by climate change trends. The following assumptions were used for locations selected based on temperature meeting a stable criterion for 28 days minimum each year, especially for hybrid seed production: (a) average daily temperature of >28 °C and ≤36 °C and (b) a minimum temperature of >24 °C. Likewise, for TGMS self-seed multiplication, a criterion of average temperature of <24 °C and Tmin >16 °C was used.

Table 6 Ideal locations for two-line hybrid seed production and TGMS self-seed multiplication (modified from Ali et al. 2018)

Seed production operation	Ideal places
Hybrid seed production	India: Aduthurai, Trichy, Killikulum, Madurai, Karnal, Delhi Philippines: Los Banos
TGMS seed production	India: Aduthurai, Gudalur, Samalkota, Karnal; Philippines: Lucban, Benguet, Nueva Vizcaya
Hybrid seed production & TGMS seed multiplication	India: Aduthurai and Samalkota

9   Wide-Scale Adoption and Use of Two-Line Hybrid Rice Technology

To achieve wide-scale adoption of two-line hybrid rice technology, we need ideal TGMS lines that should possess a higher combining ability, outcrossing rate, and market-desirable grain quality features along with insect and disease resistance (Fig. 6). During the sterile induction phase, the plants must be 100% male sterile with more than 99.5% pollen sterility and must behave stably under well-defined fertility-sterility alteration conditions. Higher seed setting above 45% in the self-seed multiplication phase is essential. Ideal TGMS lines should have lower CSTP (24 °C) and lower CFTP (22 °C). However, researchers are still attempting to lower the CSTP to 23 °C (mean temperature), which will render the TGMS lines highly stable, especially during the sterile phase, and make them highly suitable for hybrid rice seed production. The frequency of heterotic hybrids is much higher for two-line hybrids than for three-line hybrids as any nonTGMS parent could be used as a pollen parent, thereby increasing hybrid breeding efficiency. Furthermore, as there is no need for restorer genes in the male parents of two-line hybrids, this is highly ideal for developing *indica/japonica* hybrids as most *japonica* lines do not possess restorer genes. Since there is no need for a maintainer line for seed multiplication, this makes seed production much simpler and highly cost-effective. Two-line hybrids have obvious superiority over three-line hybrids for rice grain yield, quality, and insect pest and disease resistance (Chen et al. 2010). In this regard, the best two-line hybrids should address market segment requirements with a 30–35% yield advantage over market check inbreds and with higher seed reproducibility rates (>3 t/ha).

Two-line hybrid rice technology is feasible for tropical conditions for which the temperature regimes are highly suitable for its exploitation. TGMS parental lines with lower CSTP of 23 °C are highly essential for the success of this technology. At IRRI, we are trying to reach 22 °C for CSTP, which is even more stable and would ensure the wide-scale adoption of two-line hybrid rice technology. In this regard, the Two-Line Hybrid Rice Study Group involving key hybrid rice seed companies agreed to join hands in 2019 primarily to test-verify and validate potential TGMS lines, pollen parents, and F_1 hybrids in the target geographies. The study group will be able to jointly confirm the strength of two-line hybrid rice technology, especially

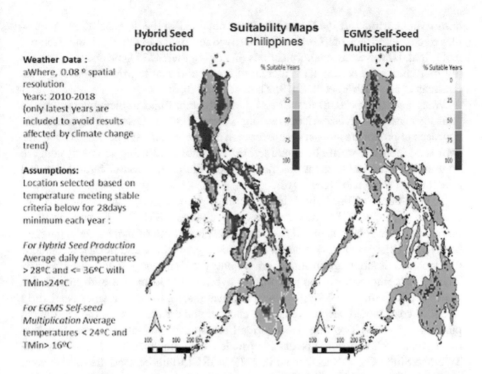

Hybrid Seed Production

Suitability Maps Philippines

EGMS Self-Seed Multiplication

Weather Data :
aWhere, 0.08 ° spatial resolution
Years: 2010-2018 (only latest years are included to avoid results affected by climate change trend)

% Suitable Years
0
25
50
75
100

% Suitable Years
0
25
50
75
100

Assumptions:
Location selected based on temperature meeting stable criteria below for 28days minimum each year :

For Hybrid Seed Production
Average daily temperatures > 28°C and <= 36°C with TMin>24°C

For EGMS Self-seed Multiplication Average temperatures < 24°C and TMin> 16°C

100 0 100 200 km

100 0 100 200 km

Fig. 8 Suitability maps for hybrid rice seed production and self-seed multiplication of TGMS lines for the Philippines developed by the IRRI GIS team

for its feasibility in South Asia. IRRI will continue to invest in this crucial technology for bringing the benefits of two-line rice hybrids to the rice farmers in South Asia. The accomplishment of this study group would ensure a lower cost of hybrid seeds, higher heterosis of two-line hybrids, and potential combinations meeting the market needs of the target regions. The success of two-line hybrid rice technology in tropical Asia would shift the attention of hybrid rice development in China toward South Asia, thus triggering widespread adoption of two-line rice hybrids.

10   Future Directions and Conclusions

The recent discovery of genome editing tools has opened up more opportunities to correct the genes of interest, including the *tms* gene, and to make them more stable and with precise expression. However, in many countries, genome editing is still under the genetically modified (GM) domain, including the Philippines. Li et al. (2019) introduced specific mutations into the *TMS5*, *Pi21*, and *Xa13* genes in Pinzhan intermediate breeding material using the CRISPR/Cas9 multiplex genome editing system. They demonstrated multiplex gene editing by finding transgene-free homozygous triple *tms5/pi21/xa13* mutants obtained in the T_1 generation that displayed characteristics of TGMS with improved resistance to rice blast and bacterial leaf blight. However, recent publications on editing the *TMS5* gene and also achieving multiplex gene editing have increased our confidence to improve TGMS lines (Barman et al. 2019; Li et al. 2019; Zhou et al. 2016).

Wang and Deng (2018) described the development and implementation of the "third-generation" hybrid rice breeding system that is based on a transgenic approach to propagate and use stable recessive nuclear male-sterile lines. Using this approach, the male-sterile lines and hybrid rice produced using such a system are nontransgenic and hold great promise to boost the production of hybrid rice and other crops (Wang and Deng 2018).

To conclude, two-line hybrid rice technology primarily concentrates on the identification of proper TGMS parental lines with a lower CSTP (23 °C) and matching market segment requirements. The hybrids developed out of these TGMS parental lines should also meet market needs by achieving consumer and farmer acceptance that includes duration, grain shape, grain quality, and insect pest and disease resistance. Furthermore, these top-performing hybrids should have a high hybrid rice seed reproducibility of >3 t/ha to allow the private sector to adopt them. Also, hybrid rice seed costs would become relatively cheaper and enable farmers to invest in the purchase of seeds. Two-line rice hybrids have several advantages over three-line rice hybrids, and they could be easily upscaled once they match market needs. The Two-line Study Group was formed in 2019 at IRRI to understand the fundamental challenges for the wide-scale adoption of two-line hybrid rice technology and validate the research efforts by IRRI to meet these challenges and make the technology feasible. The study group is in the process of testing and verifying IRRI TGMS materials in the target regions. Recent advances in the field of GIS and the precise identification of suitable locations for hybrid rice seed production and TGMS self-seed multiplication, especially in the tropical countries in Asia and Africa, have given us the confidence to scale up two-line hybrid rice technology.

References

Abeysekera SW, Abeysiriwardana DS, Dehideniyz E (2003) Characteristics associated with outcrossing rate of cytoplasmic male sterile (CMS) lines in rice under local conditions. Ann Sri Lanka Dept Agric 5:1–6

Ali J (1993) Studies on temperature sensitive genic male sterility and chemical induced sterility towards development of two line hybrids in rice (*Oryza sativa* L.). Dissertation, Indian Agricultural Research Institute, New Delhi, India

Ali J, Siddiq E (1999) Isolation and characterization of a reverse temperature sensitive genic male sterile mutant in rice. Indian J Genet 59(4):423–428

Ali J, Siddiq E, Zaman FU, Abraham MJ, Ahmed IM (1995) Identification and characterization of temperature sensitive genic male sterile sources in rice (*Oryza sativa* L.). Indian J Genet 55(3):243–259

Ali J, Dela Paz M, Marfori CM, Nicolas KL (2018) Environment sensitive genic male sterility (EGMS) based hybrid breeding in rice. In: Rao PV, Muralidharan K, Siddiq EA (eds) Modern breeding strategies for crop improvement. Proceedings of one-day dialogue on 10 July 2017, Professor Jayashankar Telangana State Agricultural University (PJTSAU), Rajendranagar, Hyderabad, India, p 250

Allahgholipour MA, Ali AJ, Alinia F, Nagamine T, Kojima Y (2006) Relationship between rice grain amylose and pasting properties for breeding better quality rice varieties. Plant Breed 125(4):357–362

Arasakesary SJ, Manonmani S, Pushpam R, Robin S (2015) New temperature sensitive genic male sterile lines with better out-crossing ability for production of two line hybrid rice. Rice Sci 22(1):49–52

Athwal DS, Virmani SS (1972) Cytoplasmic male sterility and hybrid breeding in rice. In: Rice breeding. International Rice Research Institute, Los Baños, pp 615–620

Barman HN, Sheng Z, Fiaz S, Zhong M, Wu Y, Cai Y, Wang W, Jiao G, Tang S, Wei X, Hu P (2019) Generation of a new thermo-sensitive genic male sterile rice line by targeted mutagenesis of *TMS5* gene through CRISPR/Cas9 system. BMC Plant Biol 19(1):109. https://doi.org/10.1186/s12870-019-1715-0

Cao L, Zhao X (2014) Chinese experiences in breeding three-line, two-line and super hybrid rice. In: Yan W (ed) Rice: germplasm, genetics and improvement. https://doi.org/10.5772/56821

Carnahan HL, Erikson JR, Tseng ST, Rutger JN (1972) Outlook for hybrid rice in the USA. In: Rice breeding. International Rice Research Institute, Los Baños, pp 603–607

Chandirakala R, Kandasamy G, Thiyagarajan K (2008) Determination of fertility behaviour of thermo sensitive genic male sterile lines in rice. Oryza 45(2):110–114

Chandirakala R, Kandasamy G, Thiyagarajan K (2012) Genetic variability and combining ability for quality characters in two line hybrids in rice. Electron J Plant Breed 3(3):843–847

Chen LY, Xiao YH, Lei DY (2010) Mechanism of sterility and breeding strategies for photoperiod/thermo-sensitive genic male sterile rice. Rice Sci 17(3):161–167. https://doi.org/10.1016/S1672-6308(09)60012-3

Chen J, Xu Y, Fei K, Wang R, He J, Fu L, Shao S, Li K, Zhu K, Zhang W, Wang Z, Yang J (2020) Physiological mechanism underlying the effect of high temperature during anthesis on spikelet-opening of photo-thermo-sensitive genic male sterile rice lines. Sci Rep 10(2210):24–27. https://doi.org/10.1038/s41598-020-59183-0

Cheng S, Sun Z, Si H, Zhuo L (1995) Response to photoperiod and temperature in photo-thermo period sensitive genic male sterile line Xinguang S (*Oryza sativa* L. subsp. *indica*). Chin J Rice Sci 9(2):87–91

Collard BCY, Beredo JC, Lenaerts B, Mendoza R, Santelices R, Lopena V, Verdeprado H, Raghavan C, Gregorio GB, Via L, Demon M, Biswas PS, Iftekharuddaula KM, Rahman MA, Cobb JN, Islam MR (2017) Revisiting rice breeding methods: evaluating the use of rapid generation advance (RGA) for routine rice breeding. Plant Prod Sci 20(4):337–352. https://doi.org/10.1080/1343943X.2017.1391705

Ding JH, Lu Q, Yidan OY, Mao HL, Zhang PB, Yao JL, Xu CG, Li XH, Xiao JH, Zhang QF (2012) A long noncoding RNA regulates photoperiod-sensitive male sterility, an essential component of hybrid rice. Proc Natl Acad Sci U S A 109:2654–2659

Dong NV, Subudhi PK, Luong PN, Quang VD, Quy TD, Zheng HG, Nguyen HT (2000) Molecular mapping of a rice gene conditioning thermosensitive genic male sterility using AFLP, RFLP and SSR techniques. Theor Appl Genet 100(5):727–734

FAO (Food and Agriculture Organization of the United Nations) (2020) FAOSTAT database. FAO, Rome. www.fao.org/faostat/en/#data. Accessed 13 Jul 2020

Gao YZ (1991) Discovery and preliminary study of short photoperiod sensitive male-sterile rice. J Yichun Univ (Nat Sci) 1:1–5

Hien V, Yoshimura A (2015) Identifying map location and markers linked to thermosensitive genic male sterility gene in 103S line. J Sci Dev 13(3):331–336

Huang QC, Zhang XT (1991) CIS 28-10s, a new *indica* photoperiod sensitive genic male sterile rice. Int Rice Res Newsl 16(2):8–9

Huang TY, Wang Z, Hu YG, Shi SP, Peng T, Chu XD, Liu DY (2008) Genetic analysis and primary mapping of *pms4*, a photoperiod-sensitive genic male sterility gene in rice (*Oryza sativa*). Rice Sci 15(2):153–156

Hussain A, Ali J, Siddiq E, Gupta V, Reddy U, Ranjekar P (2012) Mapping of *tms8* gene for temperature-sensitive genic male sterility (TGMS) in rice (*Oryza sativa* L.). Plant Breed 131:42–47

Jia J, Li C, Qu X, Wang Q, Deng Q, Weng M, Wang B (2000) Construction of genetic linkage map and mapping of thermo-sensitive genic male sterile gene *tms5* in rice. http://icgr.caas.net. cn/973/abstract/bwang.htm

Jia JH, Zhang DS, Li CY, Qu XP, Wang SW, Chamarerk V, Wang B (2001) Molecular mapping of the reverse thermo-sensitive genic male-sterile gene (*rtms1*) in rice. Theor Appl Genet 103(4):607–612

Jiang YM, Rong Y, Tao GX, Tang LJ (1997) Breeding of DiannongS2, a thermosensitive genic male sterile line with new cytoplasm from *japonica* rice. Southwest Chin J Agric Sci 3:21–24

Jiang J, Mou T, Yu H, Zhou F (2015) Molecular breeding of thermo-sensitive genic male sterile (TGMS) lines of rice for blast resistance using *Pi2* gene. Rice 8(1):11. https://doi.org/10.1186/s12284-015-0048-3

Joseph CA, Chen Z, Ma D, Zeng HL (2011) Analysis of short photoperiod-sensitive genic male sterility and molecular mapping of *rpms3*(t) gene in rice (*Oryza sativa* L.) using SSR markers. Genes Genomics 33:513–519. https://doi.org/10.1007/s13258-010-0074-x

Kadirimangalam SR, Kumar M, Saraswathi R, Mannonmani S, Raveendran M (2017) Study on critical sages and critical sterility point of thermo-sensitive genic male sterile lines of rice for two line hybrid production. Int J Curr Microbiol Appl Sci 6(5):2128–2135. https://doi. org/10.20546/ijcmas.2017.605.238

Kadirimangalam SR, Hifzur R, Saraswathi R, Kumar M, Raveendran M, Robin S (2019) Fine mapping and expression analysis of thermosensitive genic male sterility gene (tms) in rice (*Oryza sativa* L.). Plant Gene 19:100186. https://doi.org/10.1016/j.plgene.2019.100186

Khlaimongkhon S, Chakhonkaen S, Pitngam K, Ditthab K, Sangarwut N, Panyawut N, Wasinanon T, Mongkolsiriwatana C, Chunwongse J, Muangprom A (2019) Molecular markers and candidate genes for thermo-sensitive genic male sterile in rice. Rice Sci 26(3):147–156. https://doi. org/10.1016/j.rsci.2018.08.006

Kim YJ, Zhang D (2017) Molecular control of male fertility for crop hybrid breeding. Trends Plant Sci 23(1):53–65

Lang NT, Subudhi PK, Virmani SS, Brar DS, Khush GS, Li Z, Huang N (1999) Development of PCR-based markers for thermosensitive genetic male sterility gene *tms3* (t) in rice (*Oryza sativa* L.). Hereditas 131(2):121–127

Latha R, Thiyagarajan K (2010) Fertility alteration behaviour of thermosensitive genic male sterile lines in rice *Oryza sativa* L. Electron J Plant Breed 1(July):1118–1125

Latha R, Senthilvel S, Thiyagarajan K (2005) Critical temperature and stages of fertility alteration in thermo-sensitive genic male sterile lines of rice. In: Poster papers of the 4th international crop science congress, Brisbane, 2004

Lee DS, Chen LJ, Suh HS (2005) Genetic characterization and fine mapping of a novel thermo-sensitive genic male-sterile gene *tms6* in rice (*Oryza sativa* L.). Theor Appl Genet 111(7):1271–1277

Li CQ (2009) Accelerating the development of *japonica* hybrid rice in China. In: Xie F, Hardy B (eds) Accelerating hybrid rice development. International Rice Research Institute, Laguna, pp 267–289

Li RB, Pandey MP, Sharma P (2005) Inheritance of thermosensitive genic male sterility in rice (*Oryza sativa* L.). Curr Sci 88(11):1805–1815

Li SF, Shen L, Hu P, Liu Q, Zhu XD, Qian Q, Wang KJ, Wang YX (2019) Developing disease-resistant thermosensitive male sterile rice by multiplex gene editing. J Integr Plant Biol 61(12):1201–1205. https://doi.org/10.1111/jipb.12774

Li C, Tao RF, Li Y, Duan MH, Xu JH (2020) Transcriptome analysis of the thermosensitive genic male-sterile line provides new insights into fertility alteration in rice (*Oryza sativa*). Genomics 112(3):2119–2129. https://doi.org/10.1016/j.ygeno.2019.12.006

Liu N, Shan Y, Wang FP, Xu CG, Peng KM, Li XH, Zhang QF (2001) Identification of an 85-kb DNA fragment containing *pms*1 a locus for photoperiod-sensitive genic male sterility in rice. Mol Gen Genomics 266:271–275

Liu X, Li X, Zhang X, Wang S (2010) Genetic analysis and mapping of a thermosensitive genic male sterility gene, *tms*6(t), in rice (*Oryza sativa* L.). Genome 53(2):119–124

Lopez MT, Tojinda T, Vanavichit A, Tragoonrung S (2000) Introgression of thermosensitive male sterility gene to aromatic Thai rice. In: Posters of the 4th International Rice Genetics Symposium, International Rice Research Institute, Laguna, Philippines, 22–27 October 2000

Lopez MT, Toojinda T, Vanavichit A, Tragoonrung S (2003) Microsatellite markers flanking the gene facilitated tropical TGMS rice line development. Crop Sci 43(6):2267–2271

Lu X, Wang J (1988) Fertility transformation and genetic behavior of Hubei photoperiod-sensitive genic male sterile rice. In: Hybrid rice. International Rice Research Institute, Los Baños, pp 129–138

Lu XG, Zhang ZG, Maruyama K, Virmani SS (1994) Current status of two-line method of hybrid rice breeding. In: Virmani SS (ed) Hybrid rice technology: new developments and future prospects. International Rice Research Institute, Manila, pp 37–49

Lu Q, Li XH, Guo D, Xu CG, Zhang Q (2005) Localization of *pms*3, a gene for photoperiod-sensitive genic male sterility, to a 28.4-kb DNA fragment. Mol Gen Genomics 273(6):507–511

Luo Y, Tingchen M, Teo J, Luo Z, Li Z, Yang J, Yin Z (2020) Marker-assisted breeding of thermo-sensitive genic male sterile line 1892S for disease resistance and submergence tolerance. Rice Sci 27:89–98

Mahalingam A, Saraswathi R, Ramalingam J, Jayaraj T (2013) Genetics of floral traits in cytoplasmic male sterile (CMS) and restorer lines of hybrid rice (*Oryza sativa* L.). Pak J Bot 45(6):1897–1904

Manonmani S, Pushpam R, Robin S (2016) Stability of TGMS lines under different temperature regimes for pollen sterility. J Rice Res 9(1):17–19

Maruyama K, Araki H, Amano E (1990) Enhancement of out-crossing habits of rice plant by mutation breeding. Gamma Field Symp 29:11–25

Maruyama K, Araki H, Kato H (1991) Thermosensitive genetic male sterility induced by irradiation. In: Rice genetics II. International Rice Research Institute, Manila, pp 227–235

Mei MH, Dai XK, Xu CG, Zhang QF (1999) Mapping and genetic analysis of genes for photoperiod-sensitive genic male sterility in rice using the original mutant Nongken 58S. Crop Sci 39:1711–1715

Mishra A, Bohra A (2018) Non-coding RNAs and plant male sterility: current knowledge and future prospects. Plant Cell Rep:1–15

Mou TM, Xing-Gui L, Hoan NT, Virmani SS (2003) Two-line hybrid rice breeding in and outside China. In: Virmani SS, Mao CX, Hardy B (eds) Hybrid rice for food security, poverty alleviation, and environmental protection. Proceedings of the 4th international symposium on hybrid rice, Hanoi, 14–17 May 2002. International Rice Research Institute, Manila, pp 31–52

Nas TMS, Sanchez DL, Diaz MGQ, Mendioro MS, Virmani SS (2005) Pyramiding of thermosensitive genetic male sterility (TGMS) genes and identification of a candidate *tms*5 gene in rice. Euphytica 145:67–75

Oard JH, Hu J (1995) Inheritance and characterization of pollen fertility in photoperiodically sensitive rice mutants. Euphytica 82:17–23

Oka HT, Morishima H (1967) The ancestors of cultivated rice and their evolution. Department of Applied Genetics, National Institute of Genetics, Japan. p 145

Pan Y, Li Q, Wang Z, Wang Y, Ma R, Zhu L, He G, Chen R (2014) Genes associated with thermosensitive genic male sterility in rice identified by comparative expression profiling. BMC Genomics 15(1):1114. https://doi.org/10.1186/1471-2164-15-1114

Pang Y, Ali J, Wang X, Franje NJ, Revilleza JE, Xu J, Li Z (2016) Relationship of rice grain amylose, gelatinization temperature and pasting properties for breeding better eating and cooking quality of rice varieties. PLoS One 11(12):e0168483. https://doi.org/10.1371/journal. pone.0168483

Peng HF, Zhang ZF, Wu B, Chen XH, Zhang GQ, Zhang ZM, Lu YP (2008) Molecular mapping of two reverse photoperiod-sensitive genic male sterility genes (*rpms*1 and *rpms*2) in rice (*Oryza sativa* L.). Theor Appl Genet 118(1):77–83

Peng HF, Chen XH, Lu YP, Peng YF, Wan BH, Chen ND, Wu B, Xin SP, Zhang GQ (2010) Fine mapping of a gene for non-pollen type thermosensitive genic male sterility in rice (*Oryza sativa* L.). Theor Appl Genet 120:1013–1020

Pitnjam K, Chakhonkaen S, Toojinda T, Muangprom A (2008) Identification of a deletion in tms2 and development of gene-based markers for selection. Planta 228:813–822. https://doi. org/10.1007/s00425-008-0784-3

Qi Y, Liu Q, Zhang L, Mao B, Yan D, Jin Q, He Z (2014) Fine mapping and candidate gene analysis of the novel thermo-sensitive genic male sterility *tms*9-1 gene in rice. Theor Appl Genet 127(5):1173–1182

Rahul Roy R, Kumaresan D (2019) Genetic variability and association studies for yield and floral traits in temperature sensitive male sterile lines (TGMS) of rice (*Oryza sativa* L). Electron J Plant Breed 10(3):1200–1209. https://doi.org/10.5958/0975-928X.2019.00152.2

Rajesh T, Radhakrishnan VV, Pressana KK, Francies RM, Sreenivasan E, Ibrahim KK, Latha A (2017) Critical stages of thermo-sensitive genic male sterile lines of rice in Kerala. ORYZA Int J Rice 54(4):401–406. https://doi.org/10.5958/2249-5266.2017.00054.6

Ramakrishna S, Mallikarjuna SBP, Mishra B, Virakthamath BC, Illyas Ahmed M (2006) Characterization of thermo sensitive genetic male sterile lines for temperature sensitivity, morphology and floral biology in rice (*Oryza sativa* L.). Asian J Plant Sci 5(3):421–428. https:// doi.org/10.3923/ajps.2006.421.428

Reddy OUK, Siddiq EA, Sarma NP, Ali J, Hussain AJ, Nimmakayala P, Reddy AS (2000) Genetic analysis of temperature-sensitive male sterility in rice. Theor Appl Genet 100(5):794–801

Rutger JN, Schaeffer GW (1989) An environmentally sensitive genetic male sterile mutant in rice. In: Annual meeting of the American Society of Agronomy, Las Vegas, Nev., USA. Agron. Abstr. p 98

Salgotra RK, Gupta BB, Ahmed MI (2012) Characterization of thermo-sensitive genic male sterility (TGMS) rice genotypes: (*Oryza sativa* L.) at different altitudes. Aust J Crop Sci 6(6):957–962

Sanchez DL, Virmani SS (2005) Identification of thermosensitive genic male-sterile lines with low critical sterility point for hybrid rice breeding. Philipp J Crop Sci 30:19–28

Sasikala R, Kalaiyarasi R, Paramathama M (2015) Influence of weather factors on fertility alteration in thermosensitive genic male sterile lines of rice (*Oryza sativa* L.). Int J Trop Agric 33(2):773–779. https://doi.org/10.13140/RG.2.2.24654.41280

Shen Y, Gao M, Cai Q (1994) A novel environment induced genic male sterile (EGMS) mutant in *indica* rice. Euphytica 76:89–96

Sheng Z, Wei X, Shao G, Chen M, Song J, Tang S, Chen L (2013) Genetic analysis and fine mapping of *tms*9, a novel thermosensitive genic male-sterile gene in rice (*Oryza sativa* L.). Plant Breed 132(2):159–164

Sheng Z, Tang L, Shao G, Xie L, Jiao G, Tang S, Hu P (2015) The rice thermo-sensitive genic male sterility gene *tms*9: pollen abortion and gene isolation. Euphytica 203(1):145–152

Shi MS (1981) Report on breeding and application of two-line system in later *japonica*. Hubei Agric Sci 7:1–3

Shi MS (1985) The discovery and study of the photosensitive recessive male sterile rice (*Oryza sativa* L. subsp. *japonica*). Sci Agric Sin 2:44–48

Shi MS, Deng JY (1986) The discovery, determination and utilization of Hubei photosensitive genic male sterile rice (*Oryza sativa* L. subsp. *japonica*). Acta Genet Sin 13(2):107–112

Shukla SK, Pandey MP (2008) Combining ability and heterosis over environments for yield and yield components in two-line hybrids involving thermosensitive genic male sterile lines in rice (*Oryza sativa* L.). Plant Breed 127(1):28–32. https://doi.org/10.1111/j.1439-0523.2007.01432.x

Si HM, Fu YP, Liu WZ, Sun ZX, Hu GC (2012) Pedigree analysis of photoperiod thermo-sensitive genic male sterile rice. Acta Agron Sin 38:394–407

Siddiq EA, Ali J (1999) Innovative male sterility systems for exploitation of hybrid vigour in crop plants: environment sensitive genic male sterility system. Proc Indian Natl Sci Acad B 65:331–350

Song FS, Ni JL, Qian YL, Li L, Ni DH, Yang JB (2016) Development of SNP-based dCAPS markers for identifying male sterile gene *tms*5 in two-line hybrid rice. Genet Mol Res 15(3). https://doi.org/10.4238/gmr.15038512

Soundararaj AK, Ali AJ, Thiyagarajan P, Arumugachamy S (2002) Prospects of two-line hybrid rice breeding in Tamil Nadu, India. Int Rice Res Notes 27(1):1

Subudhi PK, Borkakati RP, Virmani SS, Huang N (1997) Identification of RAPD markers linked to rice thermosensitive genetic male sterility gene by bulk segregant analysis. Rice Genet Newsl 12:228–231

Swaminathan MS, Siddiq EA, Sharma SD (1972) Outlook for hybrid rice in India. Rice breeding. International Rice Research Institute, Laguna, pp 609–613

Tan ZC, Li YY, Chen LB, Zhou GQ (1990) Studies on ecological-adaptability of dual purpose line Annong S-1. Hybrid Rice 3:35–38

Vinodhini M, Saraswathi R, Viswanathan PL, Raveendran M (2019) Studies on sterility behaviour in thermo-sensitive genic male sterile lines of rice. Int J Chem Stud 7(6):55–61

Viraktamath BC, Virmani SS (2001) Expression of thermosensitive genic male sterility in rice under varying temperature situations. Euphytica 122(1):137–143. https://doi.org/10.1023/A:1012607608792

Virmani SS (1992) Transfer and induction of thermosensitive genic male sterile mutant in *indica* rice. In: Proceedings of the second international symposium on hybrid rice, 21–25 April 1992. International Rice Research Institute, Manila

Virmani SS (1994) Heterosis and hybrid rice breeding. Monographs on theoretical and applied genetics. Springer, Berlin, 192 p

Virmani SS, Voc PC (1991) Induction of photo- and thermo-sensitive male sterility in *indica* rice. Agron Abstr 119

Virmani SS, Sun ZX, Mou TM, Ali J, Mao CX (2003) Two-line hybrid rice breeding manual. International Rice Research Institute, Los Banos, p 88

Wang H, Deng XW (2018) Development of the "third-generation" hybrid rice in China. Genomics Proteomics Bioinformatics 16(6):393–396. https://doi.org/10.1016/j.gpb.2018.12.001

Wang B, Xu WW, Wang JZ et al (1995) Tagging and mapping the thermo-sensitive genic male-sterile gene in rice (*Oryza sativa* L.) with molecular markers. Theor Appl Genet 91:1111–1114. https://doi.org/10.1007/BF00223928

Wang YG, Xing QH, Deng QY, Liang FS, Yuan LP, Weng ML, Wang B (2003) Fine mapping of the rice thermo-sensitive genic male-sterile gene *tms*5. Theor Appl Genet 107(5):917–921

Wang C, Zhang P, Ma Z, Zhang M, Sun G, Ling D (2004) Development of a genetic marker linked to a new thermo-sensitive male sterile gene in rice (*Oryza sativa* L.). Euphytica 140(3):217–222

Wang K, Peng X, Ji Y, Yang P, Zhu Y, Li S (2013) Gene, protein, and network of male sterility in rice. Front Plant Sci 4:92. https://doi.org/10.3389/fpls.2013.00092

Wang SY, Tian QY, Zhou SQ, Mao DD, Chen LB (2019) A quantitative proteomic analysis of the molecular mechanism underlying fertility conversion in thermo-sensitive genetic male sterility line AnnongS-1. BMC Plant Biol 19:65. https://doi.org/10.1186/s12870-019-1666-5

Wongpatsa U, Kaveeta L, Sriwongchai T, Khamsuk O (2014) Effects of temperature on male sterility of two inbred lines of hybrid rice. Kasetsart J Nat Sci 48(4):525–533

Worldometer (2020). https://www.worldometers.info/world-population/#/table-forecast. Accessed 13 Jul 2020

Xu J, Wang B, Wu Y, Du P, Wang J, Wang M, Liang G (2011) Fine mapping and candidate gene analysis of *ptgms2-1*, the photoperiod-thermo-sensitive genic male sterile gene in rice (*Oryza sativa* L.). Theor Appl Genet 122(2):365–372

Yamaguchi Y, Hirasawa H, Minami M, Ujihara A (1997) Linkage analysis of thermosensitive genic male sterility gene, *tms-2* in rice (*Oryza sativa* L.). Jpn J Breed 47(4):371–373

Yang QH, Zhu J (1996) Breeding of go543S, an *indica* PTGMS rice with its sterility induced by short day-length and low temperature. Hybrid Rice 1:9–10

Yang RC, Wang NY, Mang K, Chau Q, Yang RR, Chen S (1990) Preliminary studies on application of *indica* photo-thermo genic male sterile line 5460S in hybrid rice breeding. Hybrid Rice 1:32–34

Yang YZ, Tang PL, Yang WC, Liu AM, Chen YQ, Ling WB, Shi TB (2000) Breeding and utilization of TGMS line Zhu1S in rice. Hybrid Rice 15(2):6–9

Yang ZY, Zhang GL, Zhang CH, Chen JJ, Wang HQ, Zhang JJ, Yan Z (2002) Breeding of fine quality PTGMS line Guangzhan 63S in medium *indica* rice. Hybrid Rice 17(4):4–6

Yang QK, Liang CY, Zhuang W, Jun Li J, Deng HB, Deng QY, Wang B (2007) Characterization and identification of the candidate gene of rice thermo-sensitive genic male sterile gene *tms5* by mapping. Planta 225:321–330

Yao L, Zhang Y, Liu C, Liu Y, Wang Y, Liang D, Liu J, Sahoo G, Kelliher T (2018) OsMATL mutation induces haploid seed formation in indica rice. Nat Plants 4:530–533. https://doi.org/10.1038/s41477-018-0193-y

Yiming J (1988) Studies on the effect of high temperature on fertility of the male sterile lines in Dian-type hybrid rice [J]. J Yunnan Agric Univ 2000

Yu J, Han J, Kim YJ, Song M, Yang Z, He Y, Fu R, Luo Z, Hu J, Liang W, Zhang D (2017) Two rice receptor-like kinases maintain male fertility under changing temperatures. Proc Natl Acad Sci U S A 114(46):12327–12332. https://doi.org/10.1073/pnas.1705189114

Yuan LP (1987) Strategic assumption of hybrid rice breeding. Hybrid Rice 1:1–3

Yuan LP (2014) Development of hybrid rice to ensure food security. Rice Sci 21(1):1–2. https://doi.org/10.1016/S1672-6308(13)60167-5

Zhang Z, Zeng H, Yang J, Yuan S, Zhang D (1994) Conditions inducing fertility alteration and ecological adaptation of photoperiod-sensitive genic male-sterile rice. Field Crops Res 38(2):111–120. https://doi.org/10.1016/0378-4290(94)90005-1

Zhang H, Xu CX, He Y, Zong J, Yang X, Si HM, Sun ZX, Hu JP, Liang WQ, Zhang DB (2013) Mutation in CSA creates a new photoperiod-sensitive genic male sterile line applicable for hybrid rice seed production. Proc Natl Acad Sci U S A 110:76–81

Zhou YF, Zhang XY, Xue QZ (2011) Fine mapping and candidate gene prediction of the photoperiod and thermo-sensitive genic male sterile gene *pms1*(t) in rice. J Zhejiang Univ Sci B 12(6):436–447

Zhou H, Liu Q, Li J, Jiang D, Zhou L, Wu P, Chen L (2012) Photoperiod-and thermo-sensitive genic male sterility in rice are caused by a point mutation in a novel noncoding RNA that produces a small RNA. Cell Res 22(4):649–660

Zhou H, Zhou M, Yang Y et al (2014) RNase ZS1 processes UbL40 mRNAs and controls thermosensitive genic male sterility in rice. Nat Commun 5:4884. https://doi.org/10.1038/ncomms5884

Zhou H, He M, Li J, Chen L, Huang Z, Zheng S, Zhu L, Ni E, Jiang D, Zhao B, Zhuang C (2016) Development of commercial thermo-sensitive genic male sterile rice accelerates hybrid rice breeding using the CRISPR/Cas9-mediated TMS5 editing system. Sci Rep 6:37395. https://doi.org/10.1038/srep37395

Growing Rice with Less Water: Improving Productivity by Decreasing Water Demand

Balwant Singh, Shefali Mishra, Deepak Singh Bisht, and Rohit Joshi ⓘ

Abstract Rice is a staple food for more than half of the global population. With the increasing population, the yield of rice must correspondingly increase to fulfill the requirement. Rice is cultivated worldwide in four different types of ecosystems, which are limited by the availability of irrigation water. However, water-limiting conditions negatively affect rice production; therefore, to enhance productivity under changing climatic conditions, improved cultivation practices and drought-tolerant cultivars/varieties are required. There are two basic approaches to cultivation: (1) plant based and (2) soil and irrigation based, which can be targeted for improving rice production. Crop plants primarily follow three mechanisms: drought escape, avoidance, and tolerance. Based on these mechanisms, different strategies are followed, which include cultivar selection based on yield stability under drought. Similarly, soil- and irrigation-based strategies consist of decreasing non-beneficial water depletions and water outflows, aerobic rice development, alternate wetting and drying, saturated soil culture, system of rice intensification, and sprinkler irrigation. Further strategies involve developing drought-tolerant cultivars through marker-assisted selection/pyramiding, genomic selection, QTL mapping, and other breeding and cultivation practices such as early planting to follow escape strategies and decreasing stand density to minimize competition with weeds. Similarly, the identification of drought-responsive genes and their manipulation will provide a technological solution to overcome drought stress. However, it was the Green Revolution that increased crop production. To maintain the balance, there is a need for another revolution to cope with the increasing demand.

Keywords Aerobic rice · Drought · Molecular breeding · *Oryza sativa* · Transgenic · Water deficit

B. Singh · S. Mishra · D. S. Bisht
ICAR-National Institute for Plant Biotechnology, New Delhi, India

R. Joshi (✉)
Division of Biotechnology, CSIR-Institute of Himalayan Bioresource Technology, Palampur, Himachal Pradesh, India

Academy of Scientific & Innovative Research (AcSIR), Palampur, Himachal Pradesh, India

© The Author(s) 2021
J. Ali, S. H. Wani (eds.), *Rice Improvement*,
https://doi.org/10.1007/978-3-030-66530-2_5

147

1   Introduction

More than 60% of the human population consumes rice (Joshi et al. 2018). An FAO report that considered rice production and growth of the human population in the past decade suggested the urgent need to increase rice production by 70% to fulfill upcoming demand by 2050 (FAO 2018; Schroeder et al. 2013). In view of this increased demand for rice production, an urgent need exists to study rice cultivation practices. Basically, rice cultivation is grouped into four ecosystems: irrigated (50% of the total rice grown), rainfed lowland (34%), rainfed upland (9%), and flood-prone (7%) (IRRI 2014). Cultivation under the irrigated rice ecosystem is the most productive and plays a significant role in meeting global food demand. However, irrigated rice itself is strongly affected by water availability, irrigation patterns, water quality, and the duration of water standing in the rice field during the growth period (Joshi et al. 2018). The rainfed ecosystem has a higher opportunity for yield improvement as it covers 43% of rice cultivation that still has limited yield potential. The most common factors that limit rice production in the rainfed ecosystem are irregular water supplies (i.e., severe drought, flood, and sometimes both in a single cropping season) and infertile soil due to its acidic or saline nature. Such conditions further complicate rice genetic improvement programs, which increases pressure on the irrigated ecosystem (He et al. 2020).

In recent years, changes in environmental conditions imposed multiple abiotic stresses that severely affect rice production in all ecosystems by strongly inhibiting plant growth and development (Joshi et al. 2020). According to one estimate, to produce 1 kg of rice, 2000–5000 L of water are required (Joshi et al. 2009; Caine et al. 2019). The increased competition of accelerated urbanization and industrial development further limit freshwater resources for rice production. Therefore, the need for "more rice with less water" is the need of the hour for global food security (Maneepitak et al. 2019; He et al. 2020). Thus, water availability is the key requirement for rice cultivation in each of the rice ecosystems. This forces us to develop new techniques of water management for rice cultivation that specifically improve production in different ecosystems (Carracelas et al. 2019). Further, water use can be managed either by cultivating water-stress-tolerant cultivars that can yield more under less water availability or by managing soil conditions suitable for growing rice under water-deficient conditions. Stress-tolerant cultivars can be identified from crop germplasm resources so that stress tolerance can be transferred into high-yielding cultivated varieties (Singh et al. 2015a,b; Mishra et al. 2016a, b). Aerobic rice (AR) is one of the promising rice cultivation systems for managing water and growing rice under water-limited conditions, thus decreasing water losses by 27–51% and increasing water productivity by 32–88% (Nie et al. 2007; Joshi et al. 2009). Aerobic rice varieties are usually grown in upland conditions in unpuddled soil in non-flooded conditions, that is, unsaturated (aerobic) soil with less water requirement (Bouman et al. 2006; Joshi and Kumar 2012). Under these conditions, the cultivation of high-yielding aerobic rice genotypes may help to save water. Other approaches that decrease water consumption are alternate wetting and drying of the field; saturated soil culture (SSC) that relies on forming farming beds, separated by furrows in which a shallow depth of water is maintained; mid-season drainage; delayed flooding; and sprinkler irrigation. Keeping this in mind, the Indian Council of Agricultural Research-National Rice Research Institute (ICAR-NRRI), India, has released promising rice varieties for cultivation in varying rice ecosystems (https://crri.icar.gov.in/popular_var.pdf).

Water-limiting conditions are usually designated as drought and different plant species respond in different ways to cope with drought conditions: avoidance, tolerance, and escape (Turner 2003). These adaptation strategies include physiological and metabolic adjustment by plants to minimize damage caused by drought (IPCC 2001; Singh et al. 2015c). Escape is the capability of the plant to not reach drought conditions but instead complete the life cycle before drought onset in water-sufficient conditions (Boyer 1996; Joshi et al. 2014). This is of crucial importance as it is related to early establishment of the crop, inhibition of stomatal conductance, and water management. Thus, the goal is to have early flowering and maturity along with rapid germination and seedling establishment. Avoidance is a means to avoid the stress by maintaining ample water during the stress period (Bodner et al. 2015; Urban et al. 2017). Plants achieve this by changing the shape and decreasing the number and size of leaves. Plants also roll their leaves and change their orientation to decrease absorption of radiation and prevent water loss (Caine et al. 2019). Moreover, an increase in waxiness of leaves, root density, and deep rooting enables plants to uptake water from depth for sustaining themselves during adverse conditions (Ashraf et al. 2011; Joshi and Karan 2014). Cultivars that have these traits are suitable for the rainfed cropping system (Bodner et al. 2015; Korres et al. 2016). According to Levitt (1980), drought avoidance via an efficient water uptake methodology is the best method to achieve higher yield. Additionally, areas that are prone to frequent drought conditions should be cropped with cultivars that are early maturing and have high vigor (Gouache et al. 2012). The tolerance mechanism actually allows plants to survive and grow in stress conditions. This is done by maintaining turgor through osmoregulation, producing antioxidants, and accumulating compatible solutes (Joshi and Chinnusamy 2014).

2 Current Rice Cultivars/Varieties Grown Under Water-Limiting Conditions

Availability of water for irrigation is increasingly a limiting factor in attaining the full-yield potential among many crops (Boyer 1982). Various techniques have been devised and discovered to counteract the effects of water-limiting conditions and climate changes by using acquired plant adaptations. The appropriate choice of cultivar as per its adaptation to the rice ecosystem/local environment is important as different varieties show different mechanisms to cope with drought (Turner 2003). A field experiment using two rice genotypes, Hanyou 113 (HY113) and Yangliangyou

6 (YLY6), under flooding and drought stress revealed that drought stress at the reproductive stage strongly affects physiological traits, yield, and grain quality (Yang et al. 2019). IRRI has successfully developed and released 17 high-yielding drought-tolerant rice varieties, which include Sahod Ulan and Katihan (Philippines), Hardinath and Sookha Dhan (Nepal), Sahbhagi Dhan (India), BRRI dhan (Bangladesh), Inpago LIPI Go 1/2 (Indonesia), M'ZIVA (Mozambique), and UPIA3 (Nigeria) (Kumar et al. 2014). In India, Sahbhagi Dhan was reported to produce 4 t/ha under normal conditions and 1–2 t/ha under severe drought conditions. Because of its early maturity (105 days) and low irrigation requirements, farmers can save up to USD 60 per crop (Basu et al. 2017). Similarly, in Nepal, drought-tolerant cultivar Sookha Dhan 2 showed higher yield from an altitude of 1000–1600 masl (Dhakal et al. 2020). Further, ICAR-NRRI has developed and released drought-tolerant rice cultivar DRR-Dhan 45, which is moderately resistant to major diseases and pests such as rice tungro viruses, sheath rot, and blast, with average yield of 6 t/ha (Nirmala et al. 2016).

3 Existing Rice Cultivation Practices Under Water-Deficit Conditions

Farmers practice a traditional way of cultivation and selection of cultivars as per their natural adaptations toward changes in environmental conditions (Gala Bijl and Fisher 2011). However, with rice cultivation, emphasis has been given to the development of rice cultivation techniques that result in a lot of technological options for cultivation to enhance production under water-limiting conditions (Fig. 1). A cumulative approach of water management and cultivation of high-yielding varieties that performed well under water-limiting conditions was supposed to diminish the yield penalty. Therefore, water management practices, including short-term, long-term, and anticipatory phenological adaptation measures, are required before assessing the impact of water-limiting conditions, and they usually aim at mitigating effects (Nguyen 2005; IPCC 2007). A study on phenological and water-saving adaptation strategies of crop plants showing higher yield stability under water-limiting conditions has further proved the utility of cumulative approaches (Bodner et al. 2015).

3.1 Plant-Based Strategies

3.1.1 Selection of Cultivars/Varieties

The right choice of cultivar plays a significant role in rice cultivation under less water because of specificity of the tolerance mechanism of a cultivar to drought: tolerance, avoidance, and escape. Cultivar selection basically depends on

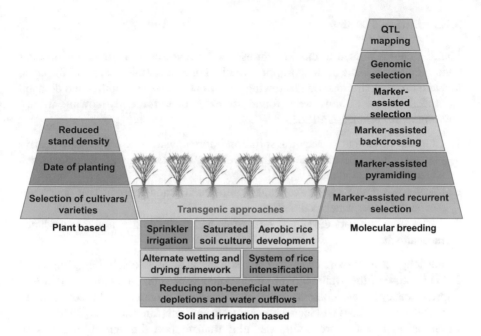

Fig. 1 Different strategies to improve rice productivity by decreasing water demand

demography and availability of water for irrigation. Following an escape strategy, plants complete their life cycle before drought onset in water-insufficient conditions (Boyer 1996). Cultivars having early flowering and maturity date and also rapid germination and establishment time completed their life cycle early and were therefore selected for cultivation in rainfed upland, lowland, and typhoon-prone coastal areas (Fukai et al. 1999). Areas that are prone to frequent drought conditions should be cropped with cultivars that are early maturing and have high vigor (Gouache et al. 2012; Bodner et al. 2015). Similarly, cultivars adapted for drought avoidance traits such as decreasing size and number of leaves, leaf rolling, and an increase in waxiness of leaves, density of root, and deep rooting are suitable for cropping in the rainfed ecosystem (Farooq et al. 2009; Bodner et al. 2015; Korres et al. 2016). Plants having a tolerance mechanism maintain turgor pressure through osmotic adjustment via generating osmolytes and osmoprotectant and producing antioxidants. In addition, the development of screening tools to identify drought stress tolerance at the seedling stage is crucial for developing rice cultivars suitable for water-limited environments. Thus, 100 *tropical japonica* rice genotypes were studied under pot conditions for their drought tolerance and a cumulative drought stress response index (CDSRI) was developed by combining individual response indices of all the varieties that were found to be important for identifying tolerant cultivars for early-season drought (Lone et al. 2019). Similarly, taking 15 rice cultivars commonly grown in Mississippi (USA), early-season drought-tolerant cultivars were selected by analyzing total drought response index (TDRI) (Singh et al. 2017).

3.1.2 Date of Planting

Planting date is related to the drought escape mechanism so as to escape drought conditions and it is the most appropriate method of escape (Ding et al. 2020). Thus, to avoid drought, optimization of sowing time as per water availability and demand is crucial. Three reasons were found to be critical for early sowing in dry environments (Bodner et al. 2015):

1. Climatic variations in evapotranspiration improve water-use efficiency of early-planted cultivars because most of the developmental stages have to face decreased water potential gradients.
2. Early sowing shifts sensitive stages (germination and reproductive stages) to periods of better water availability.
3. Early-sown cultivars develop deeper roots and facilitate the drought avoidance mechanism.

For long-term climatic changes, early sowing is a suitable solution (Ding et al. 2017) because of the availability of ample amounts of water and nutrients that will improve canopy development, yield, and biomass production. In contrast, an increase in canopy area will increase evapotranspiration (Lin et al. 2020). Therefore, variations in biomass production per unit transpiration through adjustments in planting dates will be beneficial for drought tolerance or drought escape. Although early planting could enhance spikelet sterility caused by high temperatures (Jagadish et al. 2015), by using early-maturing cultivars, both a drought and heat stress escape strategy will be a beneficial approach (Mukamuhirwa et al. 2019).

3.1.3 Decreased Stand Density

A decrease in stand density focuses on a decrease in intraspecific competition and improved water availability for a single plant and thus is a measure related to water saving. Although decreased stand density also relates to higher evapotranspiration, it is beneficial under certain conditions. An increase in soil evaporation by decreasing plant density and/or widening row spacing depends on the prevalence of rainfall and is more during intermittent drought than in a prolonged dry span. Besides evaporation, decreased radiation interception due to scattered stands might diminish growth and increase weed competition with crop plants such as wheat (Chen et al. 2008), rice (Rees and Khodabaks 1994), maize (Barbieri et al. 2012), barley (McKenzie et al. 2005), sugar beet (Ehlers and Goss 2016), and sorghum (Buah and Mwinkaara 2009). These have been investigated under different conditions and high yield has been observed with low stand density.

3.2   Soil- and Irrigation-Based Strategies

Environmental changes will influence water accessibility, especially in rice in zones where water is scarce. Expanded high-temperature environments diminish rice yield amid the dry season when prevention measures are lacking. Water system alterations or improvement of appropriate water system frameworks allow water reserves while decreasing the yield penalty (Krishnan et al. 2011). The existing water-saving technologies, for example, the alternate wetting and drying (AWD) water system, SSC, and aerobic rice system, appear to be the most appropriate advances in current rice research work (Joshi et al. 2018). With a deprived water system and poor administration, rice production is more affected by climatic vagaries, especially in tropical countries (Wassmann and Dobermann 2014). Improving technologies for increasing water-use efficiency will provide long-term economic as well as environmental benefits. This would also decrease soil salinization problems that arise from irrigation (Wang et al. 2016).

3.2.1   Alternate Wetting and Drying

As indicated by Tuong (1999), by just considering evapotranspiration, 500–2000 kg of water are required to produce 1 kg of rice, which gives 33–50% water profitability (Bouman and Tuong 2001). The AWD technique primarily relies on water management by alternately applying water in either flooded or non-flooded conditions (Maneepitak et al. 2019). This alteration in watering the field has been determined by a fixed number of non-flooded days, extent of soil potential, appearance of cracks on the soil surface, symptoms shown by plants, and a drop in water level below the soil surface (Pascual and Wang 2017; Sriphirom et al. 2019). Further, in the AWD system, water is connected to non-flooded soil for a few days after flooding recedes (Bouman et al. 2007). Soil type also influences the measure of water reserves through AWD in contrast to customary flooded rice. In loamy and sandy soils with deep groundwater tables, water input decreases by using AWD, with a 20% yield decrease, in contrast to waterlogged cropping (Singh et al. 2002). However, in soils with shallow groundwater tables, water input diminishes by 15–30%, accompanied by a noteworthy yield decrease (Carracelas et al. 2019). Grain production in AWD is usually lower than in flooded rice. However, water-use efficiency (the estimation of aggregate water used) in AWD is higher, based on decreased water inputs (62%) and decreased yield (25%) (Bouman et al. 2007; Wang et al. 2016). This shows the higher efficiency of AWD technology in comparison to persistent overflowed rice production in connection with water use per unit that results in a 24.6% increase in income from rice cultivation (Uddin and Dhar 2020). In addition to decreased water loss, AWD has been reported to decrease methane emissions from rice fields and decrease heavy metal accumulation in rice grain (Carrijo et al. 2018; Wang et al. 2019; He et al. 2020).

3.2.2 Saturated Soil Culture

In this system, soil is kept soaked as much as could reasonably be expected to bring about a diminished hydraulic head to flooding. This diminishes water loss by decreasing leakage and permeation streams (Borrell et al. 1997; Bouman et al. 2007). This shallow-water system of around 1 cm of water profoundly diminishes water consumption from 10 to 25% in comparison with continuous flooding (Bouman and Tuong 2001). This framework, in light of information from Tabbal et al. (2002), transfers the superiority of wet-seeded rice to transplanted rice with decreased rice yield under continuous flooding (i.e., 4% vs. 10%). Therefore, in both wet-seeded and transplanted rice, the water profitability under SSC was found to be higher than that in consistently flooded rice, in addition to the cost effectiveness for farmers' acceptability (Kima et al. 2014).

3.2.3 Aerobic Rice Development

Aerobic rice development is used to decrease water needs since the rice is grown as an upland harvest with optimum yield and a supplementary water system just when precipitation is inadequate (Joshi et al. 2018). In this system, rice cultivars were sown in non-puddled and unsaturated (vigorous) soils (Bouman et al. 2007). Vigorous rice cultivars have been achieved by consolidating the positive attributes of upland rice with those of high-yielding flooded rice (Atlin et al. 2006). During the mid- to late 1990s, early-maturing, oxygen-consuming, nitrogen-efficient, and high-yielding aerobic rice cultivars were released, such as Han Dao 502, Han Dao 297, and Han Dao 277 (Yang et al. 2005; Joshi et al. 2018). These new cultivars have 50–70% less water consumption than flooded rice due to their more extended roots that encourage water retention and enhance air dissemination (Mitin 2009). Under field conditions, these aerobic cultivars produce from 4.7–6.6 t/ha to 8.0–8.8 t/ha under flooded conditions (Xue et al. 2008). In addition, rice cultivars bred for the aerobic system must also be bred for competitive ability with weeds because of enhanced weed problems as soon as flooding is removed (Korres et al. 2016). Thus, the traits related to water and nutrient acquisition that affect weed-suppressive ability of the crop include root surface area, water uptake rate, root length, and root density (Korres et al. 2016).

3.2.4 Decreasing Non-beneficial Water Depletions and Water Outflows

A decrease in evaporation during different stages of development is achieved by early canopy closure via either manipulating crop density or selecting rice cultivars with good seedling vigor (Gouache et al. 2012; Bodner et al. 2015). These measures also increase the competitive ability of the crop by decreasing transpiration from weeds (Korres et al. 2016). In addition, other methods to control weeds include using herbicides, manual or mechanical weeding, timely flooding, and land leveling, which can help to diminish non-beneficial water losses that occur due to transpiration by weeds (Rodenburg et al. 2011). Soil mulching is also an effective approach to increase water productivity and decrease water inputs in rice, especially under non-saturated aerobic soil conditions (Dittert et al. 2002). Puddling in clay soils (Tuong et al. 2005) or soil compaction in sandy-loamy soils with clay content greater than 5% or shallow tillage before flooding was reported to be beneficial in decreasing water outflows (Cabangon and Tuong 2000; Tuong et al. 2005).

3.2.5   System of Rice Intensification

To increase rice productivity, a climate-shrewd agroecological method is required to increase rice yield by altering water, soil, plant, and supplement management. The SRI philosophy depends on the following four fundamental rules that are connected with each other: early, snappy, and sound plant foundation; decreased density of plants; upgraded soil conditions through augmentation with organic supplements; and controlled and decreased application of water (Uphoff 2004; SRI-Rice 2010). In light of these standards, farmers can adjust prescribed SRI practices according to their agroecological and financial conditions. Adjustments are frequently embraced to handle changing soil conditions, climate designs, water control, work accessibility, access to natural resources, and the choice to completely depend on organic farming (Uddin and Dhar 2020). Notwithstanding flooded rice, SRI standards have been connected to rainfed rice and to different harvests, for example, wheat, finger millet, sugarcane, beets, and teff, demonstrating expanded profitability over current old cropping practices. At the point when SRI standards are connected to different products, we allude to it as the system of crop intensification or SCI. The advantages of SRI included up to a 90% decrease in seed requirement, 20–100% or more expanded yield, and up to half water reserves. SRI standards and practices have been developed for rainfed rice and also for different harvests, with yield increments and related financial advantages (Duttarganvi et al. 2014).

3.2.6   Sprinkler Irrigation

The majority of cultivated rice across the globe is grown under flooded conditions, through which a huge amount of water is lost via deep percolation, seepage, surface runoff, and evapotranspiration (Vories et al. 2013; Materu et al. 2018). Among the various techniques developed for water-saving irrigation, mechanized sprinkler irrigation systems are gaining attention among farmers in several countries because of easy management of irrigation combined with improved water-use efficiency and enhanced productivity (Kahlown et al. 2007; Spanu et al. 2009; Vories et al. 2017; Kar et al. 2018; Mandal et al. 2019; Pinto et al. 2020). In comparison to 1168 mm in flooded rice, a total depth of 414 mm can be achieved by sprinkler irrigation with a 20–50% decrease in water consumption (Vories et al. 2013; Pinto et al. 2016; Kumar et al. 2018). Additionally, sprinkler irrigation enables farmers to adopt soil conservation techniques such as no-till farming and crop rotation (Pinto et al. 2020).

4 Molecular Breeding for Rice Improvement

To attain global food security, a promising approach is to cope with drought by developing drought-tolerant cultivars (Xiao et al. 2009). However, drought tolerance is a complex trait that involves changes at developmental, physiological, biochemical, and molecular levels (Joshi and Karan 2014). These changes involve alterations in photosynthesis, osmotic adjustment, guard cell regulation, root growth, and synthesis of specific proteins and antioxidants. In addition, breeders can attempt to improve yield through improved harvest indices, manipulating transpiration rate, and decreasing non-beneficial depletions (Tuong 1999; Bennett 2003). In this regard, considerable progress has been made and several QTLs (Quantitative Trait Loci) for drought-related traits that lead to improved grain yield under water-limiting conditions have been identified and transferred into suitable varieties through marker-assisted breeding (MAB). However, most of the studies were conducted on biparental or multiparental populations that use only allelic variations present within the selected parents. In addition, there is limited exploration of genetic resources in identifying novel QTLs regulating drought-related traits (Kumar et al. 2014; Pascual et al. 2016).

4.1 QTL Mapping

QTL mapping is the genetic association between the genotypic constituents of a population and the trait of interest. Therefore, to map a QTL, it is mandatory to develop a mapping population, genotype it, and make a linkage map out of it. Mapped QTLs need to be identified by their robustness and contribution toward the trait of interest by estimating an LOD score and phenotypic variation (PV). PV of more than 10% was considered as a major QTL and less than that considered as a minor QTL. Much progress has already been made toward identifying drought-related traits and associated genetic factors, that is, QTLs/genes that demarcate tolerant rice cultivars. Subsequently, identified genetic factors have been transferred into high-yielding drought-susceptible rice varieties. Using rice genetic resources, different QTLs targeting major drought-related traits, including yield under water-limiting conditions, deep rooting, osmotic- and dehydration-responsive traits, etc., have been identified and transferred. For drought tolerance, several QTLs have been identified, although only a few have a significant effect on rice under water-limited conditions (Table 1). One of the QTLs for deep rooting has been identified from japonica cultivar Kinandang Pantong (KP) (Uga et al. 2013). Multiple QTLs related to yield under water-limiting conditions have been identified in different *indica* cultivars and wild progenitors of cultivated rice *Oryza rufipogon*. Bernier et al. (2007) identified a QTL on chromosome 12 (*Qtl12.1*) that accounted for about 50% of the genetic variation and functionally reported an increased water uptake of plants under drought stress. QTL *qDTY3.1* had been identified from a cross between

Table 1 Drought tolerance QTLs mapped and used for rice breeding programs

QTL/gene	QTL/gene	Identification method	Cultivar/varietal group	References
DRO1	Deep rooting	Fine mapping	Kinandang Patong (KP)/*japonica*	Uga et al. (2011, 2013)
DRO2	Deep rooting	Fine mapping	KP/*japonica*	Uga et al. (2013)
DRO3	Deep rooting	Fine mapping	KP/*japonica*	Uga et al. (2015)
QTL2, 9, and *11*	Controlling root traits	Mapping	Azucena/*japonica*	Steele et al. (2006, 2007, 2013)
qtl12.1	Plant water uptake	Mapping	Way Rarem	Bernier et al. (2009)
qDTY2.3 and *qDTY3.2*	Grain yield under drought	Mapping	Vandana/cross between *indica* and cross product of *japonica* and *indica*	Dixit et al. (2012)
qDTY1.1	Grain yield under drought	Mapping	N22/*aus*	Vikram et al. (2011)
qDTY2.2, *qDTY4.1*, *qDTY9.1*, and *qDTY10.1*	Grain yield under drought	Mapping	Aday Sel	Swamy et al. (2013)
qDTY1.1	Grain yield under drought	Mapping	Dhagaddeshi	Ghimire et al. (2012)
qDTY2.1 and *qDTY3.1*	Grain yield in lowland drought stress	Mapping	Apo (IR55423-01)/*indica*	Venuprasad et al. (2009)
Multiple root traits	Multiple root traits		Yuefu/*japonica*	Li et al. (2005)
Rooting			Moroberekan/*japonica*	Champoux et al. (1995)
QTL (osmotic adjustment)	QTL (osmotic adjustment)		Co39/*indica*	Lilley et al. (1996)
Multiple QTLs	Multiple QTLs		Aus 276/*aus*	Sandhu et al. (2014)
Multiple QTLs	Multiple QTLs		*O. rufipogon*/wild	Zhou et al. (2009)
Polygenes/ multiple genes	Polygenes/ multiple genes		*O. rufipogon*/wild	Hu et al. (2011)
qRL6.1	Root length			Gowda et al. (2011)

tolerant variety Apo and susceptible variety Swarna showing a large effect on drought tolerance (Venuprasad et al. 2009). Different studies identified multiple DTY QTLs from different donor rice cultivars such as Dhagaddeshi, Apo, N22, Aday Sel, Way Rarem, etc., and incorporated them into rice breeding programs for improving drought tolerance in rice (Sandhu and Kumar 2017).

4.2   Marker-Assisted Selection

Marker-assisted selection (MAS) is a practice to substitute phenotypic screening by using molecular markers linked to particular loci. MAS precisely isolates the desired genotype at particular marker loci from a population without being a phenotype (Qing et al. 2019). MAS could be applied in various ways for crop improvement programs such as the marker-assisted evaluation of breeding material, early-generation selection, marker-assisted backcross breeding, gene pyramiding, and combined MAS (Collard and Mackill 2008). Kumar et al. (2018) used early-generation selection by combining both phenotyping and genotyping for the selection of drought-tolerant progenies and subsequently incorporated them into their breeding programs.

4.3   Marker-Assisted Backcrossing

Marker-assisted backcrossing (MABC) is an efficient genetic method to transfer a locus controlling a trait of interest from wild relatives, landraces, and known trait-specific genes from a genetic material into desired cultivars, called recurrent parents, without altering their essential characteristics (Dixit et al. 2017). The MABC scheme includes foreground selection, recombinant selection, and background selection. Integrating linkage map information with a QTL map helps span the markers in the target locus. Foreground selection was performed with peak markers, which assisted in the selection of a linked gene/QTL in the progenies while flanking markers of the target locus were used for recombinant selection. Foreground selection was performed in each filial generation. Recombinant selection was performed to minimize linkage drag and decrease the size of the target locus containing the gene of interest in an elite background (Collard and Mackill 2008). Background selection must be performed at later stages of breeding programs to minimize cost and labor. After that, the BC_2F_2 or BC_3 generation should be selected for background selection (Ab-Jalil et al. 2018). This method is used to validate the function of QTLs from identified genotypes by transferring them into different genetic backgrounds (Ha et al. 2016). MABC is employed for transferring QTLs for different drought stress-related traits such as qDTYs for yield under drought conditions (Kumar et al. 2014); DRO1, DRO2, and DRO3 for deep rooting (Uga et al. 2011, 2013, 2015); qRL6.1 for root length (Gowda et al. 2011); and QTL12.1 for plant water uptake (Bernier et al. 2009).

4.4   Marker-Assisted Pyramiding

Several morphological and physiological characters have been reported that contributed to drought tolerance and each of the traits can be controlled by a QTL (Sandhu et al. 2019). Moreover, individual QTLs can contribute to yield under drought stress. Several important traits controlling drought tolerance are root traits, plant morphology, and yield under drought stress, and QTLs for these have been mapped (Muthu et al. 2020). Pyramiding of QTLs/genes is a widely followed approach in disease resistance breeding. However, polygenic traits governed by more than one gene within the identified QTLs are complex to integrate. A significant amount of work has to be done for pyramiding multiple QTLs. A suitable approach for integrating multiple QTLs is equally important. Sometimes, integrated multiple QTLs may not work as they work independently. Nevertheless, the approach of gene/QTL integration depends on the number of QTLs to be integrated, the presence of QTLs in the same genetic backgrounds or different ones, the distance between the QTL and flanking marker, the filial stage, and the recovered recurrent parent genome (Shamsudin et al. 2016). Less breeding time is required if the QTLs to be integrated are present in the same genetic background in the advanced filial generation that recovered a higher proportion of recurrent parent genome. Genetic parameters such as interaction between alleles, within QTLs, and with the genetic background; pleiotropic effect of genes; and linkage drag of the introgressed loci need to be addressed (Kumar et al. 2018). For analyzing positive effects of alleles and other genetic effects, a large number of progenies need to be phenotyped and genotyped, which may correspondingly increase with the complexity of the trait (Kumar et al. 2018). Still, some success stories are present for MAP for drought, including DTY QTLs $qDTY_{2.2}$ + $qDTY_{4.1}$ (Swamy et al. 2013); $qDTY_{12.1}$ + $qDTY_{3.1}$ (Shamsudin et al. 2016); $qDTY_{3.2}$ and $qDTY_{12.1}$ (Dixit et al. 2017); $qDTY_{2.2}$, $qDTY_{3.1}$, and $qDTY_{12.1}$ (Shamsudin et al. 2016); and root QTLs for drought tolerance, qRT_{6-2}, qRT_{11-7}, qRT_{6-2}, and qRT_{19-1+7} (Selvi et al. 2015).

4.5   Marker-Assisted Recurrent Selection

Marker-assisted recurrent selection (MARS) is basically increasing the frequency of beneficial alleles with additive and small individual effects (Bankole et al. 2017). Selection cycles started with parental selection (called the C_0 cycle) and after that three to four rounds of recurrent selection. Parental selection can be carried out through genomic selection by calculating the genomic estimated breeding value (GEBV) across all the lines in the original populations (C_0). The best linear unbiased predictor helps in predicting GEBV. Lines with the highest GEBVs were planted and intercrossed. Thereafter, subsequent recombinant selection cycles (C_1 to C_3) were performed based on recombination of selected associated markers (Grenier et al. 2015; Sevanthi et al. 2019). While performing MARS, there is an increase in the frequency of favorable alleles and this minimizes genetic drift (Bankole et al. 2017). Important points to be considered while performing MARS are allelic interaction, genotype by environment interactions, functions of alleles in different genetic backgrounds, and cost and time duration of performing MARS. In rice, MARS is employed for incorporating DTY QTLs ($qDTY_{1.1}$, $qDTY_{2.1}$, $qDTY_{3.1}$, and $qDTY_{11.1}$) into a Samba Mahsuri background (Sandhu et al. 2018).

4.6 Genomic Selection

Genomic selection (GS) is a next-generation breeding strategy that ensures speedy breeding and selection of desired genotypes for cultivar improvement. GS is completed in two phases: training phase and breeding phase (Nakaya and Isobe 2012). The training population is used to predict genomic values; therefore, it is genotyped as well as phenotyped. Based on this information, a breeding model is developed to calculate the GEBVs of individuals in the testing population, which is only genotyped. Based on the GEBVs, progenies are selected, thus increasing the proportion of high-performing progenies in a population and increasing the breeding gain (Shikha et al. 2017). GS has a greater relevance in cases of drought as its phenotyping demands extensive field screening, cost, and labor (Cabrera-Bosquet et al. 2012). While performing GS, genetic heterozygosity, and genotype × genotype and genotype × environment interaction may affect the genomic prediction. There is also a need for a model-based prediction of GE and GG interactions. For complex traits such as drought, reaction norm model GEBV has greater accuracy than conventional models (Mulder 2016). Similarly, molecular marker-based predictions of crop traits are more accurate than pedigree-based predictions (Crossa 2012).

5 Transgenic Strategies

Transgenic studies have opened the door to the development of useful varieties that are superior in various traits. Because of a complex trait, different gene families are supposed to be upregulated or downregulated during drought stress. These genes belong to transcription factors, kinases, late embryogenesis abundant (LEA) proteins, osmoprotectants, and phytohormone families, and their transfer in different genotypes showed improved drought tolerance (Joshi et al. 2016). NAC family genes are responsive under drought stress (Nakashima et al. 2012) and overexpression of *OsNAC9* showed drought tolerance in transgenic plants in field trials (Redillas et al. 2012). Other transcription factors such as *AbEDT1* (Yu et al. 2016), *SNAC1*, *SNAC2*, and *OsNCED3* were upregulated in transgenic rice plants in response to drought stress.

Kinases represent a diversely fractioning gene family and enhanced drought tolerance was reported in transgenic plants overexpressing calcium-dependent protein kinase, including *OsCPK4* (Campo et al. 2014), *OsCDPK7* (Saijo et al. 2000), and *DcaCIPK9, -14,* and *-16* (Wan et al. 2019). Overexpression of *OsCIPK12* showed tolerance by increasing the accumulation of osmolytes such as proline and soluble sugars. Receptor-like kinase (RLK) *OsSIK1* aided in drought tolerance in transgenic rice plants by an increased accumulation of peroxidase, catalase, and superoxide dismutase and decreased stomatal density and decreased accumulation of H_2O_2 (Ouyang et al. 2010). Another enzyme of the RLK family, SIK2, also showed drought tolerance in transgenic rice plants (Chen et al. 2013).

LEA proteins are stress inducible and play a significant role in protection under stress conditions (Minh et al. 2019). Reports show enhanced tolerance of drought in transgenic plants overexpressing *OsLEA3-1* or *OsLEA3-2* (Xiao et al. 2007; Duan and Cai 2012). SNAC2 protein binds to the *OsOAT* promoter expressing ornithine δ-aminotransferase (You et al. 2012) and overexpression of the *OsOAT* gene enhanced the activity of δ-OAT in transgenic rice. This increased glutathione, proline, and ROS-scavenging enzyme activity, resulting in drought tolerance. Similarly, overexpression of the trehalose-6-phosphate synthase (*OsTPS1*) gene expressing an enzyme in trehalose biosynthesis showed improved drought tolerance in transgenic rice (Li et al. 2011). Using a fusion gene from *Escherichia coli* coding for trehalose-6-phosphate synthase/phosphatase under the control of an ABA-inducible promoter, we generated marker-free, high-yielding transgenic rice (in an IR64 background) that can tolerate high pH (~9.9), high EC (~10.0 dS/m), and severe drought (30–35% soil moisture content) (Joshi et al. 2020). Enhanced tolerance was observed in transgenic rice overexpressing the *OsPYL/RCA5* gene. Expression of stress-responsive genes was increased, resulting in enhanced drought tolerance (Kim et al. 2010). The isopentyltransferase gene (*IPT*), involved in cytokinin synthesis, under control of a SAPK promoter, exhibited changes in the expression of genes involved in hormone homeostasis and resource mobilization, delay in stress response, and enhanced drought tolerance (Peleg et al. 2011). Further, in the expression analysis of *OsM4* and *OsMB11* genes, we found these genes to be highly expressed under drought and salinity stress conditions (Kushwaha et al. 2016). We also reported that transgenic rice constitutively overexpressing *SaADF2* showed higher growth, relative water content, photosynthesis, and yield under drought conditions than the wild type under drought stress conditions (Sengupta et al. 2019).

6   Future Prospects

The current challenge is a sustainable increase in rice production corresponding to demand. The development of high-yielding varieties resistant to diseases and insects and tolerant of major abiotic stresses is essential for producing enough rice under irrigated habitats. However, production is limited in upland and rainfed areas and also under limited water availability. It was demonstrated earlier that farmers in drought-prone areas accept the decrease in yield variability offered by new stress-tolerant cultivars, and would be willing to pay a significant premium for these traits (Arora et al. 2019). Therefore, it is important to work concurrently on both aspects: the development of technology related to crop management and improvement of rice varieties for yield under adverse conditions. Climate changes further complicate the conditions for rice cultivation in these areas. This leads to intermittent rainfall, enhanced flash flooding, and sporadic drought, which adversely affect cultivated soil. Secondary adverse effects on crop production due to climatic fluctuation are changes in soil stature, enhanced soil salinity, and loss of genetic diversity. Therefore, to mitigate these effects, researchers seek to exploit existing genetic resources. In the context of rice, ample germplasm accessions belong to different species and have been conserved ex situ and more efforts are in progress. Improvement strategies such as marker-assisted selection, marker-assisted backcrossing, marker-assisted pyramiding, marker-assisted recurrent selection, and genomic selection rely on QTL mapping, which itself relies on pre-breeding experiments. Pre-breeding strategies for the selection of cultivars to incorporate into breeding cycles are major components for succeeding in breeding programs. Further, transgenic approaches have been applied for functional validation of genes, improving quality traits, and developing resistant and tolerant cultivars. Thus, the development of rice cultivars with improved water-use efficiency will offer significant economic and environmental benefits toward the achievement of the Sustainable Development Goals.

Acknowledgments We gratefully acknowledge the director, CSIR-Institute of Himalayan Bioresource Technology, Palampur, for providing the facilities to carry out this work. CSIR support in the form of project MLP0201 for this study is highly acknowledged. This manuscript represents CSIR-IHBT communication no. 4592.

References

Ab-Jalil M, Juraimi AS, Yusop MR, Uddin MK, Hakim MA (2018) Introgression of root trait genes for drought tolerance to a Malaysian rice variety by marker-assisted backcross breeding. Int J Agric Biol 20(1):119–126

Arora A, Bansal S, Ward PS (2019) Do farmers value rice varieties tolerant to droughts and floods? Evidence from a discrete choice experiment in Odisha, India. Water Resour Econ 25:27–41

Ashraf M, Akram NA, Al-Qurainy F, Foolad MR (2011) Drought tolerance: roles of organic osmolytes, growth regulators, and mineral nutrients. Adv Agron 111:249–296

Atlin GN, Lafitte HR, Tao D, Laza M, Amante M, Courtois B (2006) Developing rice cultivars for high-fertility upland systems in the Asian tropics. Field Crops Res 97(1):43–52

Bankole F, Menkir A, Olaoye G, Crossa J, Hearne S, Unachukwu N, Gedil M (2017) Genetic gains in yield and yield related traits under drought stress and favorable environments in a maize population improved using marker assisted recurrent selection. Front Plant Sci 8:808

Barbieri P, Echarte L, Della Maggiora A, Sadras VO, Echeverria H, Andrade FH (2012) Maize evapotranspiration and water-use efficiency in response to row spacing. Agron J 104(4):939–944

Basu S, Jongerden J, Ruivenkamp G (2017) Development of the drought tolerant variety Sahbhagi Dhan: exploring the concepts commons and community building. Int J Commons 11(1):144–170

Bennett J (2003) Status of breeding for tolerance of water deficit and prospects for using molecular techniques. In: Water productivity in agriculture: limits and opportunities for improvement. CABI Publishing, Wallingford

Bernier J, Kumar A, Venuprasad R, Spaner D, Atlin GN (2007) A large-effect QTL for grain yield under reproductive-stage drought stress in upland rice. Crop Sci 47:507–516

Bernier J, Kumar A, Venuprasad R, Spaner D, Verlukar S, Mandal NP, Sinha PK, Peeraju P, Dongre PR, Mahto RN, Atlin GN (2009) Characterization of the effect of rice drought resistance qtl12.1 over a range of environments in the Philippines and eastern India. Euphytica 166:207–217

Bodner G, Nakhforoosh A, Kaul HP (2015) Management of crop water under drought: a review. Agron Sustain Dev 35(2):401–442

Borrell A, Garside A, Fukai S (1997) Improving efficiency of water use for irrigated rice in a semi-arid tropical environment. Field Crops Res 52(3):231–248

Bouman BAM, Tuong TP (2001) Field water management to save water and increase its productivity in irrigated lowland rice. Agric Water Manag 49:11–30

Bouman BA, Yang X, Wang H, Wang Z, Zhao J, Chen B (2006) Performance of aerobic rice varieties under irrigated conditions in North China. Field Crops Res 97(1):53–65

Bouman BAM, Lampayan RM, Tuong TP (2007) Water management in irrigated rice: coping with water scarcity. International Rice Research Institute, Los Baños

Boyer JS (1982) Plant productivity and environment. Science 218:443–448

Boyer SJ (1996) Advances in drought tolerance in plants. Adv Agron 56:187–218

Buah SS, Mwinkaara S (2009) Response of sorghum to nitrogen fertilizer and plant density in the Guinea Savanna Zone. J Agron 8(4):124–130

Cabangon RJ, Tuong TP (2000) Management of cracked soils for water saving during land preparation for rice cultivation. Soil Tillage Res 56(1–2):105–116

Cabrera-Bosquet L, Crossa J, von Zitzewitz J, Serret MD, Luis Araus J (2012) High-throughput phenotyping and genomic selection: the frontiers of crop breeding converge. J Integr Plant Biol 54(5):312–320

Caine RS, Yin X, Sloan J, Harrison EL, Mohammed U, Fulton T, Biswal AK, Dionora J, Chater CC, Coe RA, Bandyopadhyay A (2019) Rice with reduced stomatal density conserves water and has improved drought tolerance under future climate conditions. New Phytol 221(1):371–384

Campo S, Baldrich P, Messeguer J, Lalanne E, Coca M, San Segundo B (2014) Overexpression of a calcium-dependent protein kinase confers salt and drought tolerance in rice by preventing membrane lipid peroxidation. Plant Physiol 165(2):688–704

Carracelas G, Hornbuckle J, Rosas J, Roel A (2019) Irrigation management strategies to increase water productivity in *Oryza sativa* (rice) in Uruguay. Agric Water Manag 222:161–172

Carrijo DR, Akbar N, Reis AF, Li C, Gaudin AC, Parikh SJ, Green PG, Linquist BA (2018) Impacts of variable soil drying in alternate wetting and drying rice systems on yields, grain arsenic concentration and soil moisture dynamics. Field Crops Res 222:101–110

Champoux MC, Wang G, Sarkarung S, Mackill DJ, O'Toole JC, Huang N, McCouch SR (1995) Locating genes associated with root morphology and drought avoidance in rice via linkage to molecular markers. Theor Appl Genet 90(7–8):969–981

Chen C, Neill K, Wichman D, Westcott M (2008) Hard red spring wheat response to row spacing, seeding rate, and nitrogen. Agron J 100(5):1296–1302

Chen LJ, Wuriyanghan H, Zhang YQ, Duan KX, Chen HW, Li QT, Lu X, He SJ, Ma B, Zhang WK, Lin Q (2013) An S-domain receptor-like kinase, *OsSIK2*, confers abiotic stress tolerance and delays dark-induced leaf senescence in rice. Plant Physiol 163(4):1752–1765

Collard BC, Mackill DJ (2008) Marker-assisted selection: an approach for precision plant breeding in the twenty-first century. Philos Trans R Soc Lond B Biol Sci 363(1491):557–572

Crossa J (2012) From genotype x environment interaction to gene x environment interaction. Curr Genomics 13(3):225–244

Dhakal S, Adhikari BB, Kandel BP (2020) Performance of drought tolerant rice varieties in different altitudes at Duradada, Lamjung, Nepal. J Agric Nat Resour 3(1):290–300

Ding Y, Wang W, Song R, Shao Q, Jiao X, Xing W (2017) Modeling spatial and temporal variability of the impact of climate change on rice irrigation water requirements in the middle and lower reaches of the Yangtze River, China. Agric Water Manag 193:89–101

Ding Y, Wang W, Zhuang Q, Luo Y (2020) Adaptation of paddy rice in China to climate change: the effects of shifting sowing date on yield and irrigation water requirement. Agric Water Manag 228:105890

Dittert K, Shan L, Kreye C, Zheng XH, Xu YC, Lu XJ, Shen QR, Fan XL, Sattelmacher B (2002) Saving water with ground cover rice production systems (GCRPS) at the price of increased greenhouse gas emissions. In: Water-wise rice production. International Rice Research Institute, Los Baños, pp 197–206

Dixit S, Swamy BM, Vikram P, Ahmed HU, Cruz MS, Amante M, Atri D, Leung H, Kumar A (2012) Fine mapping of QTLs for rice grain yield under drought reveals sub-QTLs conferring a response to variable drought severities. Theor Appl Genet 125(1):155–169

Dixit S, Singh A, Sandhu N, Bhandari A, Vikram P, Kumar A (2017) Combining drought and submergence tolerance in rice: marker-assisted breeding and QTL combination effects. Mol Breed 37(12):143

Duan J, Cai W (2012) OsLEA3-2, an abiotic stress induced gene of rice plays a key role in salt and drought tolerance. PLoS One 7(9):e45117

Duttarganvi S, Tirupataiah K, Yella RK, Sandhyarani K, Mahendra KR, Malamasuri K (2014) Yield and water productivity of rice under different cultivation practices and irrigation regimes. In: International symposium on integrated water resources management (IWRM–2014), February 2014, Kerala, India, pp 19–21

Ehlers W, Goss M (2016) Water dynamics in plant productions, 2nd edn. CABI Publishing, Wallingford, 396

FAO (2018) FAOSTAT database collections. Food and Agriculture Organization of the United Nations. Food outlook biannual report on global food markets. Rome. www.fao.org/faostat. Accessed 2018

Farooq M, Wahid A, Lee DJ, Ito O, Siddique KHM (2009) Advances in drought resistance of rice. Crit Rev Plant Sci 28:199–217

Fukai S, Pantuwan G, Jongdee B, Cooper M (1999) Screening for drought resistance in rainfed lowland rice. Field Crops Res 64(1–2):61–74

Gala Bijl C, Fisher M (2011) Crop adaptation to climate change. CSA News Magazine 5–9

Ghimire KH, Quiatchon LA, Vikram P, Swamy BM, Dixit S, Ahmed H, Hernandez JE, Borromeo TH, Kumar A (2012) Identification and mapping of a QTL (*qDTY1.1*) with a consistent effect on grain yield under drought. Field Crops Res 131:88–96

Gouache D, Le Bris X, Bogard M, Deudon O, Pagé C, Gate P (2012) Evaluating agronomic adaptation options to increasing heat stress under climate change during wheat grain filling in France. Eur J Agron 39:62–70

Gowda VR, Henry A, Yamauchi A, Shashidhar HE, Serraj R (2011) Root biology and genetic improvement for drought avoidance in rice. Field Crops Res 122(1):1–13

Grenier C, Cao TV, Ospina Y, Quintero C, Châtel MH, Tohme J, Courtois B, Ahmadi N (2015) Accuracy of genomic selection in a rice synthetic population developed for recurrent selection breeding. PLoS One 10(8):e0136594

Ha T, Khang DT, Tuyen PT, Toan TB, Huong NN, Lang NT, Buu BC, Xuan TD (2016) Development of new drought tolerant breeding lines for Vietnam using marker-assisted backcrossing. Int Lett Nat Sci 59:1–13

He G, Wang Z, Cui Z (2020) Managing irrigation water for sustainable rice production in China. J Clean Prod 245:118928. http://www.knowledgebank.irri.org/submergedsoils/index.php/rice-growing-environments/lesson-2

Hu BL, Fu XQ, Zhang T, Yong WA, Xia LI, Huang YH, Dai LF, Luo XD, Xie JK (2011) Genetic analysis on characteristics to measure drought resistance using Dongxiang wild rice (*Oryza rufipogon* Griff.) and its derived backcross inbred lines population at seedling stage. Agric Sci China 10(11):1653–1664

IPCC (2001) Impacts, adaptation and vulnerability. In: Contribution of the working group II to the third assessment report on the intergovernmental panel on climate change. Cambridge University Press, Cambridge

IPCC (2007) Climate change 2007: the physical science basis. In: Solomon S, Qin D, Manning M, Chen Z, Marquis M, Averyt KB, Tignor M, Miller HL (eds) Contribution of working group I to the fourth assessment report of the intergovernmental panel on climate change. Cambridge University Press, Cambridge

IRRI (2014) Standard evaluation system for rice, 5th edn. International Rice Research Institute, Los Baños

Jagadish SV, Murty MV, Quick WP (2015) Rice responses to rising temperatures–challenges, perspectives and future directions. Plant Cell Environ 38(9):1686–1698

Joshi R, Chinnusamy V (2014) Antioxidant enzymes: defense against high temperature stress. In: Oxidative damage to plants. Academic Press, New York, pp 369–396

Joshi R, Karan R (2014) Physiological, biochemical and molecular mechanisms of drought tolerance in plants. In: Gaur RK, Sharma P (eds) Molecular approaches in plant abiotic stress. CRC Press, Boca Raton, pp 209–231

Joshi R, Kumar P (2012) Aerobic rice: an option for growing rice under limited water availability. Indian Farming 62(2):11–14

Joshi R, Mani SC, Shukla A, Pant RC (2009) Aerobic rice: water use sustainability. Oryza 46(1):1–5

Joshi R, Ramanarao VM, Lee S, Kato N, Baisakh N (2014) Ectopic expression of ADP Ribosylation Factor1 (*SaARF1*) from smooth cordgrass (*Spartina alterniflora*) confers drought and salt tolerance in transgenic rice and *Arabidopsis*. Plant Cell Tissue Organ Cult 117:17–30

Joshi R, Wani SH, Singh B, Bohra A, Dar ZA, Lone AA, Pareek A, Singla-Pareek SL (2016) Transcription factors and plants response to drought stress: current understanding and future directions. Front Plant Sci 7:1029

Joshi R, Singh B, Shukla A (2018) Evaluation of elite rice genotypes for physiological and yield attributes under aerobic and irrigated conditions in tarai areas of western Himalayan region. Curr Plant Biol 13:45–52

Joshi R, Sahoo KK, Singh AK, Anwar K, Pundir P, Gautam RK, Krishnamurthy SL, Sopory SK, Pareek A, Singla-Pareek SL (2020) Enhancing trehalose biosynthesis improves yield potential in marker-free transgenic rice under drought, saline, and sodic conditions. J Exp Bot 71(2):653–668

Kahlown MA, Raoof A, Zubair M, Kemper WD (2007) Water use efficiency and economic feasibility of growing rice and wheat with sprinkler irrigation in the Indus Basin of Pakistan. Agric Water Manag 87(3):292–298

Kar I, Mishra A, Behera B, Khanda C, Kumar V, Kumar A (2018) Productivity trade-off with different water regimes and genotypes of rice under non-puddled conditions in Eastern India. Field Crops Res 222:218–229

Kim TH, Böhmer M, Hu H, Nishimura N, Schroeder JI (2010) Guard cell signal transduction network: advances in understanding abscisic acid, CO_2, and Ca^{2+} signaling. Annu Rev Plant Biol 61:561–591

Kima AS, Chung WG, Wang YM (2014) Improving irrigated lowland rice water use efficiency under saturated soil culture for adoption in tropical climate conditions. Water 6(9):2830–2846

Korres NE, Norsworthy JK, Tehranchian P, Gitsopoulos TK, Loka DA, Oosterhuis DM, Gealy DR, Moss SR, Burgos NR, Miller MR, Palhano M (2016) Cultivars to face climate change effects on crops and weeds: a review. Agron Sustain Dev 36(1):12

Krishnan P, Ramakrishnan B, Reddy KR, Reddy VR (2011) High-temperature effects on rice growth, yield, and grain quality. Adv Agron 111:87–206

Kumar A, Dixit S, Ram T, Yadaw RB, Mishra KK, Mandal NP (2014) Breeding high-yielding drought-tolerant rice: genetic variations and conventional and molecular approaches. J Exp Bot 65(21):6265–6278

Kumar GS, Ramesh T, Subrahmaniyan K, Ravi V (2018) Effect of sprinkler irrigation levels on the performance of rice genotypes under aerobic condition. Int J Curr Microbiol Appl Sci 3:1848–1852

Kushwaha HR, Joshi R, Pareek A, Singla-Pareek SL (2016) MATH-domain family shows response toward abiotic stress in *Arabidopsis* and rice. Front Plant Sci 7:923

Levitt E (1980) Chapter 4. Drought avoidance. In: Levitt E (ed) Responses of plants to environmental stresses, vol. 2 edn. Academic Press, New York, pp 93–103

Li Z, Mu P, Li C, Zhang H, Li Z, Gao Y, Wang X (2005) QTL mapping of root traits in a doubled haploid population from a cross between upland and lowland japonica rice in three environments. Theor Appl Genet 110(7):1244–1252

Li HW, Zang BS, Deng XW, Wang XP (2011) Overexpression of the *trehalose-6-phosphate syn-thase* gene *OsTPS1* enhances abiotic stress tolerance in rice. Planta 234(5):1007–1018

Lilley JM, Ludlow MM, McCouch SR, O'Toole JC (1996) Locating QTL for osmotic adjustment and dehydration tolerance in rice. J Exp Bot 47(9):1427–1436

Lin BS, Lei H, Hu MC, Visessri S, Hsieh CI (2020) Canopy resistance and estimation of evapotranspiration above a humid cypress forest. Adv Meteorol 2020:4232138

Lone AA, Jumaa SH, Wijewardana C, Taduri S, Redoña ED, Reddy KR (2019) Drought stress tolerance screening of elite American breeding rice genotypes using low-cost pre-fabricated mini-hoop modules. Agronomy 9(4):199

Mandal KG, Thakur AK, Ambast SK (2019) Current rice farming, water resources and micro-irrigation. Curr Sci 4:568–576

Maneepitak S, Ullah H, Paothong K, Kachenchart B, Datta A, Shrestha RP (2019) Effect of water and rice straw management practices on yield and water productivity of irrigated lowland rice in the Central Plain of Thailand. Agric Water Manag 211:89–97

Materu ST, Shukla S, Sishodia RP, Tarimo A, Tumbo SD (2018) Water use and rice productivity for irrigation management alternatives in Tanzania. Water 10(8):1018

McKenzie RH, Middleton AB, Bremer E (2005) Fertilization, seeding date, and seeding rate for malting barley yield and quality in southern Alberta. Can J Plant Sci 85(3):603–614

Minh BM, Linh NT, Hanh HH, Hien LT, Thang NX, Hai NV, Hue HT (2019) A LEA gene from a Vietnamese maize landrace can enhance the drought tolerance of transgenic maize and tobacco. Agronomy 9(2):62

Mishra S, Singh B, Panda K, Singh BP, Singh N, Misra P, Rai V, Singh NK (2016a) Association of SNP haplotypes of HKT family genes with salt tolerance in Indian wild rice germplasm. Rice 9(1):1–3

Mishra S, Singh B, Misra P, Rai V, Singh NK (2016b) Haplotype distribution and association of candidate genes with salt tolerance in Indian wild rice germplasm. Plant Cell Rep 35(11):2295–2308

Mitin A (2009) Documentation of selected adaptation strategies to climate change in rice cultivation. East Asia Rice Working Group, Quezon City, Philippines

Mukamuhirwa A, Persson Hovmalm H, Bolinsson H, Ortiz R, Nyamangyoku O, Johansson E (2019) Concurrent drought and temperature stress in rice—a possible result of the predicted climate change: effects on yield attributes, eating characteristics, and health promoting compounds. Int J Environ Res Public Health 16(6):1043

Mulder HA (2016) Genomic selection improves response to selection in resilience by exploiting genotype by environment interactions. Front Genet 7:178

Muthu V, Abbai R, Nallathambi J, Rahman H, Ramasamy S, Kambale R, Thulasinathan T, Ayyenar B, Muthurajan R (2020) Pyramiding QTLs controlling tolerance against drought, salinity, and submergence in rice through marker assisted breeding. PLoS One 15(1):e0227421

Nakashima K, Takasaki H, Mizoi J, Shinozaki K, Yamaguchi-Shinozaki K (2012) NAC transcription factors in plant abiotic stress responses. Biochim Biophys Acta 1819(2):97–103

Nakaya A, Isobe SN (2012) Will genomic selection be a practical method for plant breeding? Ann Bot 110(6):1303–1316

Nguyen NV (2005) Global climate changes and rice food security. FAO, Rome, pp 24–30

Nie L, Peng S, Bouman BAM, Huang J, Cui K, Visperas RM, Park HK (2007) Alleviation of soil sickness caused by aerobic monocropping: growth response of aerobic rice to soil oven heating. Plant Soil 300:185–195

Nirmala B, Babu VR, Neeraja CN, Waris A, Muthuraman P, Rao DS (2016) Linking agriculture and nutrition: an ex-ante analysis of zinc biofortification of rice in India. Agric Econ Res Rev 29:171–177

Ouyang SQ, Liu YF, Liu P, Lei G, He SJ, Ma B, Zhang WK, Zhang JS, Chen SY (2010) Receptor-like kinase OsSIK1 improves drought and salt stress tolerance in rice (*Oryza sativa*) plants. Plant J 62(2):316–329

Pascual VJ, Wang YM (2017) Utilizing rainfall and alternate wetting and drying irrigation for high water productivity in irrigated lowland paddy rice in southern Taiwan. Plant Prod Sci 20(1):24–35

Pascual L, Albert E, Sauvage C, Duangjit J, Bitton BF, Desplat N, Brunel D, Le Paslier M, Ranc N, Bruguier L, Chauchard B, Verschave P, Causse M (2016) Dissecting quantitative trait variation in the resequencing era: complementarity of bi-parental, multi-parental and association panels. Plant Sci 242:120–130

Peleg Z, Reguera M, Tumimbang E, Walia H, Blumwald E (2011) Cytokinin-mediated source/sink modifications improve drought tolerance and increase grain yield in rice under water-stress. Plant Biotechnol J 9(7):747–758

Pinto MA, Parfitt JM, Timm LC, Faria LC, Scivittaro WB (2016) Produtividade de arroz irrigado por aspersão em terras baixas em função da disponibilidade de água e de atributos do solo. Pesq Agropecu Bras 51(9):1584–1593

Pinto MA, Parfitt JM, Timm LC, Faria LC, Concenço G, Stumpf L, Nörenberg BG (2020) Sprinkler irrigation in lowland rice: crop yield and its components as a function of water availability in different phenological phases. Field Crops Res 248:107714

Qing D, Dai G, Zhou W, Huang S, Liang H, Gao L, Gao J, Huang J, Zhou M, Chen R, Chen W (2019) Development of molecular marker and introgression of *Bph3* into elite rice cultivars by marker-assisted selection. Breed Sci 69(1):40–46

Redillas MC, Jeong JS, Kim YS, Jung H, Bang SW, Choi YD, Ha SH, Reuzeau C, Kim JK (2012) The overexpression of *OsNAC9* alters the root architecture of rice plants enhancing drought resistance and grain yield under field conditions. Plant Biotechnol J 10(7):792–805

Rees DJ, Khodabaks MR (1994) The effects of seed source and crop density on rice grown on red rice-infected land in Nickerie, Suriname

Rodenburg J, Meinke H, Johnson DE (2011) Challenges for weed management in African rice systems in a changing climate. J Agric Sci 149(4):427–435

Saijo Y, Hata S, Kyozuka J, Shimamoto K, Izui K (2000) Over-expression of a single Ca^{2+}-dependent protein kinase confers both cold and salt/drought tolerance on rice plants. Plant J 23(3):319–327

Sandhu N, Kumar A (2017) Bridging the rice yield gaps under drought: QTLs, genes, and their use in breeding programs. Agronomy 7(2):27

Sandhu N, Singh A, Dixit S, Cruz MT, Maturan PC, Jain RK, Kumar A (2014) Identification and mapping of stable QTL with main and epistasis effect on rice grain yield under upland drought stress. BMC Genet 15(1):63

Sandhu N, Dixit S, Swamy BM, Vikram P, Venkateshwarlu C, Catolos M, Kumar A (2018) Positive interactions of major-effect QTLs with genetic background that enhances rice yield under drought. Sci Rep 8(1):1626

Sandhu N, Dixit S, Swamy BP, Raman A, Kumar S, Singh SP, Yadaw RB, Singh ON, Reddy JN, Anandan A, Yadav S (2019) Marker assisted breeding to develop multiple stress tolerant varieties for flood and drought prone areas. Rice 12(1):8

Schroeder JI, Delhaize E, Frommer WB, Guerinot ML, Harrison MJ, Herrera-Estrella L, Horie T, Kochian LV, Munns R, Nishizawa NK, Tsay YF (2013) Using membrane transporters to improve crops for sustainable food production. Nature 497(7447):60–66

Selvi GS, Hittalmani S, Uday G (2015) Root QTL pyramiding through marker assisted selection for enhanced grain yield under low moisture stress in rice (*Oryza sativa* L). Rice Res Open Access 3:157

Sengupta S, Mangu V, Sanchez L, Bedre R, Joshi R, Rajasekaran K, Baisakh N (2019) An actin-depolymerizing factor from the halophyte smooth cordgrass, *Spartina alterniflora* (*Sa ADF* 2), is superior to its rice homolog (*OsADF 2*) in conferring drought and salt tolerance when constitutively overexpressed in rice. Plant Biotechnol J 17(1):188–205

Sevanthi AM, Prakash C, Shanmugavadivel PS (2019) Recent progress in rice varietal development for abiotic stress tolerance. In: Advances in rice research for abiotic stress tolerance. Woodhead Publishing, Sawston, pp 47–68

Shamsudin NA, Swamy BM, Ratnam W, Cruz MT, Raman A, Kumar A (2016) Marker assisted pyramiding of drought yield QTLs into a popular Malaysian rice cultivar, MR219. BMC Genet 17(1):30

Shikha M, Kanika A, Rao AR, Mallikarjuna MG, Gupta HS, Nepolean T (2017) Genomic selection for drought tolerance using genome-wide SNPs in maize. Front Plant Sci 8:550

Singh AK, Choudhury BU, Bouman BAM (2002) Effects of rice establishment methods on crop performance, water use, and mineral nitrogen. In: Bouman BAM, Hengsdijk H, Hardy B, Bindraban PS, Tuong TP, Ladha JK (eds) Water-wise rice production. International Rice Research Institute, Los Baños, pp 237–246

Singh BP, Jayaswal PK, Singh B, Singh PK, Kumar V, Mishra S, Singh N, Panda K, Singh NK (2015a) Natural allelic diversity in *OsDREB1F* gene in the Indian wild rice germplasm led to ascertain its association with drought tolerance. Plant Cell Rep 34:993–1004

Singh N, Jayaswal PK, Panda K, Mandal P, Kumar V, Singh B, Mishra S, Singh Y, Singh R, Rai V, Gupta A, Singh NK (2015b) Single-copy gene based 50 K SNP chip for genetic studies and molecular breeding in rice. Sci Rep 5:11600

Singh B, Bohra A, Mishra S, Joshi R, Pandey S (2015c) Embracing new-generation 'omics' tools to improve drought tolerance in cereal and food-legume crops. Biol Plant 59(3):413–428

Singh B, Reddy KR, Redoña ED, Walker T (2017) Screening of rice cultivars for morpho-physiological responses to early-season soil moisture stress. Rice Sci 24(6):322–335

Spanu A, Murtas A, Ballone F (2009) Water use and crop coefficients in sprinkler irrigated rice. Ital J Agron 4(2):47–58

Sriphirom P, Chidthaisong A, Towprayoon S (2019) Effect of alternate wetting and drying water management on rice cultivation with low emissions and low water used during wet and dry season. J Clean Prod 223:980–988

SRI-Rice (2010) SRI International Network and Resources Center, College of Agriculture and Life Sciences. Cornell University, New York

Steele KA, Price AH, Shashidhar HE, Witcombe JR (2006) Marker-assisted selection to introgress rice QTLs controlling root traits into an Indian upland rice variety. Theor Appl Genet 112(2):208–221

Steele KA, Virk DS, Kumar R, Prasad SC, Witcombe JR (2007) Field evaluation of upland rice lines selected for QTLs controlling root traits. Field Crops Res 101(2):180–186

Steele KA, Price AH, Witcombe JR, Shrestha R, Singh BN, Gibbons JM, Virk DS (2013) QTLs associated with root traits increase yield in upland rice when transferred through marker-assisted selection. Theor Appl Genet 126(1):101–108

Swamy MBP, Ahmed HU, Henry A, Mauleon R, Dixit S, Vikram P, Tilatto R, Verulkar SB, Perraju P, Mandal NP, Variar M (2013) Genetic, physiological, and gene expression analyses reveal that multiple QTL enhance yield of rice mega-variety IR64 under drought. PLoS One 8(5):e62795

Tabbal DF, Bouman BA, Bhuiyan SI, Sibayan EB, Sattar MA (2002) On-farm strategies for reducing water input in irrigated rice; case studies in the Philippines. Agric Water Manag 56(2):93–112

Tuong TP (1999) Productive water use in rice production: opportunities and limitations. J Crop Prod 2:241–264

Tuong TP, Bouman BAM, Mortimer M (2005) More rice, less water. Plant Prod Sci 8(3):231–241

Turner NC (2003) Drought resistance: a comparison of two research frameworks. In: Saxena NP (ed) Management of agricultural drought: agronomic and genetic options. Science Publishers Inc., Enfield, pp 89–102

Uddin MT, Dhar AR (2020) Assessing the impact of water-saving technologies on boro rice farming in Bangladesh: economic and environmental perspective. Irrig Sci 14:1–4

Uga Y, Okuno K, Yano M (2011) *Dro1*, a major QTL involved in deep rooting of rice under upland field conditions. J Exp Bot 62(8):2485–2494

Uga Y, Sugimoto K, Ogawa S, Rane J, Ishitani M, Hara N, Kitomi Y, Inukai Y, Ono K, Kanno N, Inoue H (2013) Control of root system architecture by DEEPER ROOTING 1 increases rice yield under drought conditions. Nat Genet 45(9):1097

Uga Y, Kitomi Y, Yamamoto E, Kanno N, Kawai S, Mizubayashi T, Fukuoka S (2015) A QTL for root growth angle on rice chromosome 7 is involved in the genetic pathway of *DEEPER ROOTING 1*. Rice 8(1):1–8

Uphoff N (2004) System of rice intensification responds to 21st century needs. Rice Today 42:42–43

Urban J, Ingwers MW, McGuire MA, Teskey RO (2017) Increase in leaf temperature opens stomata and decouples net photosynthesis from stomatal conductance in *Pinus taeda* and *Populus deltoides x nigra*. J Exp Bot 68(7):1757–1767

Venuprasad R, Dalid CO, Del Valle M, Zhao D, Espiritu M, Cruz MS, Amante M, Kumar A, Atlin GN (2009) Identification and characterization of large-effect quantitative trait loci for grain yield under lowland drought stress in rice using bulk-segregant analysis. Theor Appl Genet 120(1):177–190

Vikram P, Swamy BM, Dixit S, Ahmed HU, Cruz MT, Singh AK, Kumar A (2011) *qDTY_{1.1}*, a major QTL for rice grain yield under reproductive-stage drought stress with a consistent effect in multiple elite genetic backgrounds. BMC Genet 12(1):89

Vories ED, Stevens WE, Tacker PL, Griffin TW, Counce PA (2013) Rice production with center pivot irrigation. Appl Eng Agric 29(1):51–60

Vories E, Rhine M, Straatmann Z (2017) Investigating irrigation scheduling for rice using variable rate irrigation. Agric Water Manag 179:314–323

Wan X, Zou LH, Zheng BQ, Wang Y (2019) Circadian regulation of alternative splicing of drought-associated CIPK genes in *Dendrobium catenatum* (Orchidaceae). Int J Mol Sci 20(3):688

Wang Z, Zhang W, Beebout SS, Zhang H, Liu L, Yang J, Zhang J (2016) Grain yield, water and nitrogen use efficiencies of rice as influenced by irrigation regimes and their interaction with nitrogen rates. Field Crops Res 193:54–69

Wang Y, Wang L, Zhou J, Hu S, Chen H, Xiang J, Zhang Y, Zeng Y, Shi Q, Zhu D, Zhang Y (2019) Research progress on heat stress of rice at flowering stage. Rice Sci 26(1):1–10

Wassmann R, Dobermann A (2014) Climate change adaptation through rice production in regions with high poverty levels. J Semi-Arid Trop Agric Res 4(1):1–24

Xiao B, Huang Y, Tang N, Xiong L (2007) Over-expression of a *LEA* gene in rice improves drought resistance under the field conditions. Theor Appl Genet 115(1):35–46

Xiao J, Zhuang Q, Liang E, Shao X, McGuire AD, Moody A, Kicklighter DW, Melillo JM (2009) Twentieth-century droughts and their impacts on terrestrial carbon cycling in China. Earth Interact 13(10):1–31

Xue C, Yang X, Bouman BA, Deng W, Zhang Q, Yan W, Zhang T, Rouzi A, Wang H (2008) Optimizing yield, water requirements, and water productivity of aerobic rice for the North China Plain. Irrig Sci 26(6):459–474

Yang XG, Bouman BAM, Wang HQ, Wang ZM, Zhao JF, Chen B (2005) Performance of temperate aerobic rice under different water regimes in North China. Agric Water Manag 74:107–122

Yang X, Wang B, Chen L, Li P, Cao C (2019) The different influences of drought stress at the flowering stage on rice physiological traits, grain yield, and quality. Sci Rep 9(1):1–2

You J, Hu H, Xiong L (2012) An ornithine δ-aminotransferase gene OsOAT confers drought and oxidative stress tolerance in rice. Plant Sci 197:59–69

Yu LH, Wu SJ, Peng YS, Liu RN, Chen X, Zhao P, Xu P, Zhu JB, Jiao GL, Pei Y, Xiang CB (2016) *Arabidopsis* EDT 1/HDG 11 improves drought and salt tolerance in cotton and poplar and increases cotton yield in the field. Plant Biotechnol J 14(1):72–84

Zhou L, Chen L, Jiang L, Zhang W, Liu L, Liu X, Zhao Z, Liu S, Zhang L, Wang J, Wan J (2009) Fine mapping of the grain chalkiness QTL *qPGWC-7* in rice (*Oryza sativa* L.). Theor Appl Genet 118(3):581–590

Crop Establishment in Direct-Seeded Rice: Traits, Physiology, and Genetics

Fergie Ann Quilloy, Benedick Labaco, Carlos Casal Jr, and Shalabh Dixit

Abstract The changing climate and water availability strongly affect the current state of agricultural production. While the global temperature rises, the occurrence of extreme climatic conditions becomes erratic. This current scenario has driven the development of rice varieties and cultivation practices that require less water and favor mechanization. Although puddled transplanted rice has been more widely used in the past, direct seeding has been gaining popularity in recent years, especially due to its water- and labor-saving features. This technique allows full crop establishment from seeds that were directly sown in the field, thus avoiding puddling, transplanting, and maintaining standing water. Consequently, it offers promising positive environmental effects including decreasing the release of greenhouse gases and increasing water-use efficiency. Historically, rice varieties bred for transplanting are also used in direct seeding, which limits the maximum yield potential of field trials. The success of direct seeding relies strongly on the development of rice varieties with robust crop establishment. Anaerobic germination, seed longevity, and early seedling vigor are the key traits required to achieve this. This chapter expounds on the physiology, molecular mechanisms, genetics, and relevance of the enumerated traits for direct seeding. A brief discussion of breeding for rice varieties with improved germination under direct seeding is also provided.

Keywords Direct seeded rice · Anaerobic germination · Seed longevity · Early seedling vigor

1 Introduction

Rice is grown on close to 160 million ha of land across the world, producing 700 million t of paddy. Notably, more than 90% of the worldwide rice supply is cultivated and consumed in Asia (Fig. 1). A significant volume of this production is

---

The original version of this chapter was revised. The author sequence has now been updated. The correction to this chapter is available at https://doi.org/10.1007/978-3-030-66530-2_15

F. A. Quilloy · B. Labaco · C. Casal Jr · S. Dixit (✉)
Rice Breeding Platform, International Rice Research Institute, Metro Manila, Philippines
e-mail: f.quilloy@irri.org; b.labaco@irri.org; s.dixit@irri.org; c.casal@irri.org

© The Author(s) 2021, Corrected Publication 2021
J. Ali, S. H. Wani (eds.), *Rice Improvement*,
https://doi.org/10.1007/978-3-030-66530-2_6

171

Fig. 1 2017 worldwide (**a**) milled rice production (tons) and (**b**) total rice consumption. (Data retrieved from http://ricestat.irri.org:8080/wrs2/, June 2018)

contributed by small and marginal farm holdings that principally grow rice with limited inputs under an unpredictable water supply. Rice can be categorized into different types based on the topographical location where it is grown and the prevailing ecosystem. Topographically, rice is grouped into five main classes: upland, shallow lowland, mid-lowland, semi-deep, and deepwater ecosystems. Further, these ecosystems are classified as irrigated or rainfed based on the availability of irrigation water. In cases where water is maintained throughout most of the crop growth period (80%), the area is classified as irrigated. However, rainfed areas are those where the sole water source is rainfall and ponded water availability is uncertain.

The start of the Green Revolution led to a marked change in plant architecture and cultivation practices that primarily suited highly productive irrigated environments. Several varieties of rice with promising yield potential and input responsiveness were developed and adapted across irrigated and rainfed rice-growing areas. Although this was advantageous in increasing grain yield, the consequences of the Green Revolution came with an environmental cost. As noted by Pingali (2012), the policy underlying the Green Revolution prompted misguided overuse of fertilizer and crop intensification in unbefitting environments. On a sizable scale, the environment is subjected to unintended consequences such as soil degradation, water-table diminution, and chemical runoff, which ultimately contribute to the escalating climate change.

This review aims to summarize crop establishment practices in rice in relation to climate change and water scarcity. We begin by describing the current state of climate change and the need for new crop management approaches to mitigate the worsening conditions. The review then summarizes major crop establishment methods in rice and demonstrates their suitability to varying growth scenarios. Further, the review discusses crop establishment in direct-seeded rice in detail covering relevant traits with their underlying physiology and genetics. Finally, we briefly describe the breeding approach to systematically include new traits in programs on direct-seeded rice breeding.

2 Climate Change and Water Scarcity

The state of the environment strongly affects the world's agricultural capacity (FAO 2016). Climate change and water scarcity are two developing stories that confound the downside in agricultural production. Annually, a total of 3853 km^3 of groundwater is withdrawn, 69% of which (i.e., 2769 km^3) is used for agricultural irrigation while 19% and 12% are allocated for industry and municipalities, respectively (FAO 2016). The repercussions of climate change are believed to drastically affect the water and agricultural status of the world. Predictions suggest that climate change can elicit an increase in temperature, a shift in the patterns of precipitation, the occurrence of more extreme climatic events, and further water deficit (Barker 2009).

Climate is described as the environmental condition over a location in a span of 30 years. Temperature and precipitation are two of the most pressing factors that regulate Earth's climate. Remarkably, over the past 100 years, an average increase of 0.74 °C has been recorded in the global surface temperature (IPCC 2007) as a result of the increase in atmospheric methane, nitrous oxide, and other greenhouse gases (GHGs). Since rainfall has a direct relationship with temperature, the amount of annual rainfall scales up with temperature. In the past 100 years, there has been an increase of 2% in overall precipitation; however, because this has large regional discrepancies, the benefits have been limited (IPCC 2007). Moreover, the occurrence of extreme climatic events is increasing and this leads to a decline in rice production. Long-term experiments on rice yields coupled with crop simulation models have reported that the yield decline can partly be attributed to the fluctuating climatic conditions (Aggarwal 2008). Figure 2 illustrates the average yearly precipitation across different countries in 2011. Notably, some of the major rice-producing countries such as Vietnam, Myanmar, Thailand, the Philippines, Bangladesh, Nepal, India, Burundi, and Nigeria also have the highest precipitation rates (Kreft et al. 2016). In India, for example, it was estimated that the variability in temperature and rainfall would result in a 10–40% loss in agricultural production by 2080–2100 (Aggarwal et al. 2010). Projection models suggest that direct climate impacts on maize, soybean, wheat, and rice production can account for losses of 400–1400 Pcal (8–24% of the current total) (Elliott et al. 2014). As mentioned by Barker et al. (1998), the world must develop and advocate for policies, investments, and infrastructure that can adapt to the ever-changing climatic parameters.

Water scarcity is a state when water is insufficient to meet the demand of all sectors of a particular place or demography. It has diverse origins, including decreasing water tables, declining water quality, faulty irrigation systems, and growing water competition from industrial and urban sectors (Bouman et al. 2007). It can be categorized into two major types: physical and economic water scarcity.

Physical water scarcity, in its simplest form, occurs when water resources are depleted; hence, the water supply is not able to meet human and environmental demand (Molden 2007; OCHA 2010). As further discussed by Molden (2007), physical scarcity is commonly associated with, but not limited to, arid areas. It also occurs in areas with an abundant water supply yet the resources (i.e., irrigation) are allocated only to certain sectors.

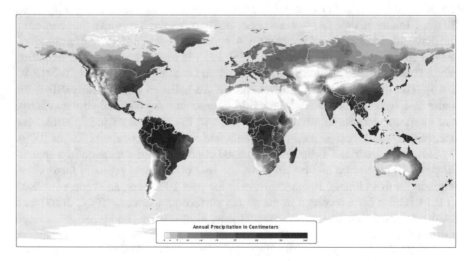

Fig. 2 Average yearly precipitation (cm) across different countries. (Image retrieved from https://nelson.wisc.edu, July 2018)

Economic water scarcity is due to the lack of capability to withdraw water given that there is a sufficient amount of reserve (Molden 2007; OCHA 2010). This relates to the lack of infrastructure to enable the withdrawal of water. However, this is also linked to the inequitable allocation of infrastructure; hence, the water supply is limited to certain groups. Signs indicative of economic water scarcity are inadequate infrastructure development, high cost of irrigation, fluctuating water availability, flooding, and long-term drought. Therefore, it is important to address the proper management of water resources to maximize water utility and prevent water contamination.

Physical and economic water scarcities are widespread across different countries. As of 2016, it was revealed that about four billion people suffer from water scarcity for at least 60 days in a year (Mekonnen and Hoekstra 2016). In Asia alone, it is expected that per capita water availability will decline across different countries by 2050 (Fig. 3). This decline could range from 1911 to 63,135 m^3 for different countries (Kumar and Ladha 2011). Several reasons are reported to be causing this decline, such as booming population, decreasing water table, deteriorating water quality, inept irrigation, and rising competition with nonagricultural sectors (Kaur and Singh 2017).

It is estimated that approximately four billion people worldwide are constrained by extreme water scarcity for blue water (fresh surface water and/or groundwater that is extracted from the earth and not returned due to evaporation into the atmosphere or incorporation into a product), of which more than half are from China and India (Mekonnen and Hoekstra 2016), two of the highest rice-producing Asian

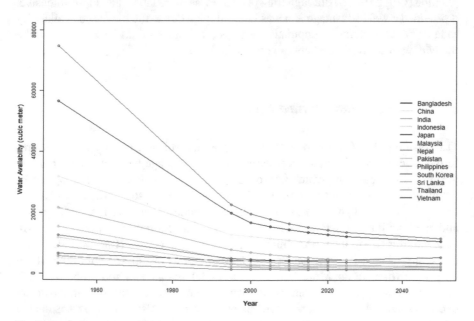

Fig. 3 Per capita water availability (in cubic meters) based on population growth rate among different rice-producing countries across 100 years (1950–2050). (Source: Kumar and Ladha 2011)

countries. Therefore, it can be expected that physical and economic water scarcities are going to induce a sharp deterioration in agriculture's portion of freshwater allocation in Asia.

Water is also lost in the field throughout the cropping season because of factors such as surface runoff, seepage, percolation, evaporation, and transpiration (Bouman et al. 2007). However, only water loss through transpiration is considered to be productive since this is useful for plant growth and development. Although 1432 L of water are needed to produce 1 kg of rough rice (Rice Knowledge Bank 2017), only about 10% of the total absorbed water is used by the plant for its development (Chavarria and dos Santos 2012). To this end, a large portion of water input becomes unused, leading to a loss in farmers' investments. In the present-day scenario of water scarcity, the puddled transplanted system of rice is becoming increasingly unsustainable. With more than 90% of global rice being cultivated and expended in Asia and the large water requirements of conventional rice cultivation systems, it is evident that water scarcity is going to severely affect rice production on the continent. The high water requirement of puddled transplanted rice makes it more vulnerable to water scarcity and demands more drastic changes in cultivation practices than in other crops.

3   Rice Establishment Methods

Historically, in Asia, direct seeding was a preferred cultivation practice over transplanting (Grigg 1974). Although transplanting was more labor- and input-intensive, it provided a yield advantage and was later adopted due to the increased availability of labor because of drastic population expansion. In the sections below, we describe the two main cultivation practices popular in Asia.

3.1   Puddled Transplanted Rice

Transplanting 20–30-day-old seedlings in puddled soil is currently the most preferred practice in Asia. Puddling decreases water seepage by compacting the soil during land preparation, which improves weed suppression, eases seedling establishment, and creates an anaerobic condition that increases soil nutrient availability (e.g., zinc, iron, and phosphorus) (Fanish 2016). Puddling ensures high crop growth and yield but it is not the most water- and labor-efficient option (Dixit et al. 2016). Rice requires two to three times more water than other cereals (Barker et al. 1998), especially when it is grown under puddled conditions lasting to about 80% of the crop's growth duration. Moreover, repeated puddling damages soil properties such that the soil aggregates are dismantled, the permeability of the soil subsurface layers decreases, and, at shallow depth, hard-pans are formed. Also, puddled transplanted rice (PTR) requires longer crop duration due to the delay in plant maturity caused by transplanting shock and nursery time. Although transplanting is favored by low wages and adequate water supply, the decreasing labor and water availability in most rice-growing regions in Asia requires a shift in cultivation practices to make rice farming a sustainable and profitable enterprise.

3.2 Direct-Seeded Rice

The current climate and social scenarios in Asia have led to increased efforts in developing rice varieties and cultivation practices requiring less water and favoring mechanization. These include operations such as direct seeding and alternate wetting and drying. Among all the water-saving technologies, dry direct seeding is the most water-efficient and it favors mechanization, which reduces labor requirements.

Direct seeding is the crop establishment practice from seeds that were directly sown in the field instead of transplanting seedlings grown from the nursery. It comes mainly in three forms: dry direct-seeded rice (dry-DSR), wet direct-seeded rice (wet-DSR), and water seeding. Kumar and Ladha (2011) described and differentiated these methods based on field conditions. Dry-DSR involves broadcasting, dibbling, or drilling of dry seeds on unpuddled soil, which was either conventionally tilled or not tilled at all. Meanwhile, wet-DSR is done by sowing pregerminated rice seeds onto or into puddled soil and is known as aerobic and anaerobic wet-DSR, respectively. Lastly, water seeding uses pregerminated seeds that are broadcast onto the field with standing water. In this case, the field may be either puddled (wet-water seeding) or unpuddled (dry-water seeding).

DSR is considered as an opportunity to advance rice production practices in water-scarce areas into a high-water-use-efficient cultivation practice (Fanish 2016). Through dry direct seeding, three basic field operations are avoided: puddling, transplanting, and maintaining 4–5 cm of standing water throughout the season. Further, PTR fields are one of the biggest sources of greenhouse gases, particularly methane. As reported by Pathak et al. (2013), in the districts of Punjab, DSR decreased total global warming potential by about 33%, from 2.0 to 4.6 t CO_2 eq./ha to 1.3–2.9 t CO_2 eq./ha. In DSR, the production of both methane and carbon dioxide was less than that in PTR. This shows the promising positive effects of DSR on the environment.

DSR is also known for its water- and labor-saving attributes. It is reported to save around 30% of water (Fanish 2016) and 11.2% of labor costs (Akhgari and Kaviani 2011). Apart from this, the labor requirement for DSR is spread out through the season, promoting mechanization and the use of family labor instead of depending on hired labor. This makes rice farming more profitable for farmers and also allows continued operations throughout the season despite labor unavailability. Furthermore, DSR saves the plants from transplanting injury; hence, the plants reach physiological maturity in fewer days and this leads to early crop maturity.

In general, PTR varieties are also used in direct seeding. The unavailability of proper varieties developed for direct seeding is a major constraint to exploiting its maximum yield potential. Historically, previous approaches to the improvement of crop establishment in DSR involved refining cultural practices rather than rice varietal improvement. Ultimately, the success of DSR relies heavily on breeding for varieties with anaerobic germination (AG) tolerance, seed longevity, early seedling vigor, and the ability to germinate from deep soil.

4 Traits, Physiology, Genetics, and Breeding

4.1 Anaerobic Germination

Poor crop establishment remains a concern in areas that experience flooding after sowing and where fields are not properly leveled (Ismail et al. 2009). Flooding negatively affects germination and survival in most rice genotypes (Ismail et al. 2012). Conversely, it suppresses the growth of weeds and shrinks the cost of physical weeding and/or herbicide application. It is projected that approximately 30% of the total cultivation cost is saved when weed emergence is suppressed by flood.

The ability of seeds to sprout, grow, and survive under low oxygen (hypoxia) or very little to no oxygen (anoxia) is known as AG. This trait is a must-have for all direct-seeded rice systems. The trait ensures risk mitigation at the early stages of the crop and can be used as a weed control mechanism. The majority of the modern high-yielding rice varieties do not show any germination underwater. However, landraces that can maintain germination under flooded conditions have evolved in various parts of rice-growing areas. Some examples of such tolerant landraces are Khao Hlan On, Ma Zhan (red), Kalarata, Nanhi, and Khaiyan. Results of donor identification studies have shown tolerant landraces spread across all major groups of rice. However, *indica* and *aus* landraces have been used in genetic studies and breeding programs more than the other groups so far. Khao Hlan On and Ma Zhan (red) are currently used extensively as donors of the trait in breeding and marker-assisted selection (MAS) programs at the International Rice Research Institute (IRRI).

4.1.1 Physiology and Molecular Mechanisms of AG

Rice is the only cereal that can withstand water submergence; hence, it grows even under hypoxic conditions. In extreme conditions, submerged plants can experience anoxia when subjected to prolonged flooding. Under anoxia, the plant shifts to an alcoholic fermentation (AF) pathway rather than respiration for energy, in the form of adenosine triphosphate (ATP). During AF, only two ATPs are being produced vis-à-vis 38 ATPs are produced in aerobic respiration (Magneschi and Perata 2009). Therefore, the plant is 19 times less efficient in ATP production when exposed to anaerobic conditions. Notably, three common physiological responses allow rice seedlings to survive under anaerobic conditions: (1) longer coleoptile, (2) greater water imbibition, and (3) higher starch reserves.

According to Pradet and Bomsel, in 1978 (as cited by Kennedy et al. 1980), the only plant organ that can grow under anoxia is the rice coleoptile. Through coleoptile elongation, the plant can gain access to aeration, which enables the germination of the developing embryo (Ismail et al. 2012). Because of this, AG tolerance can be indirectly measured using coleoptile length. Hsu and Tung (2015) referred to this trait as the "anaerobic response index." Alpi and Beevers (1983) demonstrated that seedlings develop coleoptiles at a higher speed when subjected to low environmental O_2 concentration. However, they further observed that, despite its greater length, the coleoptile is thin and fragile with less fresh weight. This phenomenon can be referred to as the snorkel effect, whereby anoxia induces the development of a hollow coleoptile to access a better-aerated environment (Kordan 1974). This provides oxygen supply for the plant's root and endosperm, which supports complete and vigorous seedling establishment. Adachi et al. (2015) evaluated the germination of tolerant (IR06F459) and intolerant (IR42) rice lines and revealed that the coleoptiles of IR06F459 had significantly longer coleoptiles than those of IR42. IR06F459 is an AG-tolerant line developed from backcrossing IR64 to Khao Hlan On. In addition, Ismail et al. (2009) revealed that water stress during germination decreased shoot and root length of intolerant genotypes by 81% and 68%, respectively (Fig. 4). This was incomparable to the decrease of 61% in shoot length and 7% in root length of the tolerant genotypes.

During flooding, seeds experience oxygen-limiting conditions. Under these conditions, the plant requires above-normal tolerance of anoxia to exhibit strong seed germination and growth (Ismail et al. 2009). In 1996, Yamauchi revealed that the use of genotypes that can withstand anaerobic conditions could help improve seedling establishment, weed control, and lodging resistance in DSR. Hence, the utility of genotypes with high germination rate (i.e., coleoptile and mesocotyl elongation) and seedling vigor under anaerobic conditions can curtail problems in crop establishment and even weed competition (Azhiri-Sigari et al. 2005). As revealed by Ismail et al. (2009) in their evaluation of tolerant (Cody, Khaiyan, Nanhi, and Khan Hlan On) and intolerant (FR13A, IR22, IR28, IR42, and IR64) genotypes, the

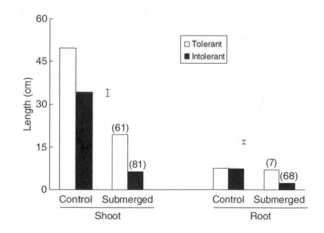

Fig. 4 Mean shoot and root lengths of five flooding-tolerant and five flooding-sensitive genotypes of rice measured 21 days after sowing under control and flooded (10 cm) conditions. Values in parentheses are percentage decreases in length relative to the length in the control. Vertical bars indicate l.s.d. at $p = 0.05$. (Source: Ismail et al. 2009)

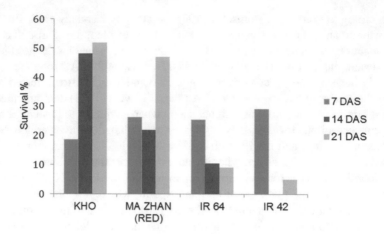

Fig. 5 Percentage survival of two tolerant (Khao Hlan On and Ma Zhan (red)) and two intolerant (IR64 and IR42) varieties after 7, 14, and 21 days of submergence (DAS) in 8 cm of water

survival of both groups decreased but there was a markedly larger decrease in the survival of the intolerant genotypes (Fig. 5). Among these tolerant genotypes, Khaiyan and Khao Hlan On displayed the highest survival rates, whereas the intolerant genotypes recorded a survival percentage of only about 5%.

The rapid germination of AG lines was credited by El-Hendawy et al. (2011) to rapid water imbibition under submerged conditions. They evaluated 58 contrasting genotypes, which showed a strong negative correlation ($r = -0.46$, $p < 0.001$) between water uptake and germination time. Further, the authors performed cluster analysis of the lines based on water content over different times of seed imbibition and revealed that tolerant genotypes clustered together within the group, which was characterized by rapid water uptake. Together, this suggests that high water uptake can induce rapid seed germination under anoxia. A proposed mechanism of action is that water uptake promotes sugar mobilization from the endosperm to the developing embryo (El-Hendawy et al. 2011), which provides energy for plant growth.

Apart from efficient water imbibition, carbohydrate content has the potential to enable plant growth under water stress. Ella and Setter (1999) suggested that the adverse effect of decreased ATP supply during AF is relieved when more carbohydrate is available for breakdown. As such, seeds with high starch content are deemed more tolerant of AG than seeds with high fat content (Magneschi and Perata 2009). Under anoxia, plant survival was found to be strongly correlated to starch content ($r = 0.73$–0.88) (Ella and Setter 1999). However, starch cannot be used as it is; hence, some also hypothesize the role of α-amylases, which are responsible for the breakdown of starch into soluble sugar (El-Hendawy et al. 2011). Rice, among the cereal crops under anoxia, is the only crop that expresses α-amylase mRNA (Perata et al. 1992, 1993). Also, a rice seed grown under anoxia expresses the whole set of enzymes needed for starch breakdown into its soluble forms (Ismail et al. 2009). These amylases regulate starch degradation to glucose, which is needed in the plants' fermentative metabolism that is activated for ATP production under anoxia

(Septiningsih et al. 2013). Adachi et al. (2015) reported in their study that the α-amylase activity in their tolerant line (IR06F459) was significantly higher than in the intolerant line (IR42). This was also in agreement with the findings of Illangakoon et al. (2016), for which the survival rates of their genotypes were positively correlated with α-amylase activity ($r = 0.79$) and soluble sugar content ($r = 0.74$). Ismail et al. (2009) reported that the soluble sugar concentration remained low for both tolerant and intolerant genotypes until the tolerant genotypes had higher soluble sugar at 4–8 days after sowing. These findings suggest that tolerant genotypes can break down carbohydrate reserves in the seed and mobilize the resulting monosaccharide, which can enable germination and growth under hypoxia. In 2015, Kretzschmar et al. found that *OsTPP7* was expressed in the germinating tissues of a NIL-AG1 (tolerant) line while its expression was absent in IR64 (susceptible). Given that *OsTPP* genes (i.e., *OsTPP1* and *OsTPP2*) in rice convert trehalose-6-phosphate (T6P) to trehalose, this mechanism can also be true for *OsTPP7*. Further analysis revealed that trehalose and sucrose were 2.3-fold and 2.0-fold higher in NIL-AG1 than in IR64. This suggests that *OsTPP7* affects the conversion of T6P to trehalose, which signals low sugar availability; consequently, the plant enables starch mobilization from the endosperm reserve (source) to the coleoptile (sink).

Expression analysis, through RNA sequence (RNA seq), revealed that RAmy3D was highly expressed in tolerant genotypes under anoxia (Ismail et al. 2009). RAmy3D is a member of the Amy3 gene subfamily, which was found to be upregulated in rice embryos under anoxia (Hwang et al. 1999). RAmy3D, unlike RAmy1A from the Amy1 gene subfamily, lacks the gibberellic acid (GA)-responsive element in its promoter region (Lu et al. 1998) and is important in oligosaccharide degradation (Terashima et al. 1997). Under starvation, a glucose and sucrose receptor, RAmy3D, is activated by protein kinase SnRK1A (Lu et al. 2007). Meanwhile, SnRK1A is stimulated by calcineurin B-like protein kinase (CIPK15) under oxygen-limiting conditions (Kudahettige et al. 2010). Additionally, the interaction of calcineurin B-like proteins, such as CBL4 and CBL5, with CIPK15 under anoxia is viewed as a regulating mechanism in plant response during anoxia (Ho et al. 2017; Sadiq et al. 2011).

Although many other rice enzymes are downregulated during anoxia to inactivate energy-conserving steps, pyruvate decarboxylase (PDC), alcohol dehydrogenase (ADH), and aldehyde dehydrogenase (ALDH) are active in genotypes tolerant of anoxia (i.e., Khaiyan), as reported by Ismail et al. (2009). The upregulation of these enzymes proposes enhanced alcohol fermentation during anoxia, which uses nicotinamide adenine dinucleotide (NAD) for glycolysis and substrate-level phosphorylation (Saika et al. 2006; Shingaki-Wells et al. 2011). As a result, enough energy and carbon are supplied to the developing coleoptile under anoxia. Moreover, rice coleoptiles under anoxic conditions have been reported to exhibit a decrease in pH, which also suggests alcoholic fermentation pathway activation. In 2009, Ismail et al. showed that the concentration of ethylene increased significantly after only 3 days of imbibition. They also reported that the ethylene produced affects starch hydrolysis through the reduction of abscisic acid (ABA) synthesis while upregulating the synthesis of and sensitivity to gibberellic acid (GA) of the internode tissue. This enhances starch catabolite enzyme activity under stress, which enables seedling growth and survival under AG. Further, ethylene promotes coleoptile growth through cell expansion by regulating enzymes such as peroxidases, which are responsible for cell wall rigidity. Peroxidase activity decreases the plant cell wall's lignin content and protein assembly (Waffenschmidt et al. 1993), which in turn decreases its rigidity and permits cellular elongation. Figure 6 shows the proposed mechanism of coleoptile elongation under anoxia as interpreted by Ismail et al. (2009).

Other factors that are deemed responsible for cell expansion are a group of non-enzymatic proteins called expansins, which can affect cell wall loosening. These proteins attach at the cell wall interface between cellulose microfibrils and matrix polysaccharides, which results in a disturbance of non-covalent bonds, enabling cell wall loosening (McQueen-Mason and Cosgrove 1995). Thus far, EXPA2, EXPA4, EXPA7, EXPB12, EXPA1, EXPB11, and EXPB17 were reported to be induced during submergence stress (Juang et al. 2000; Lasanthi-Kudahettige et al. 2007; Takahashi et al. 2011).

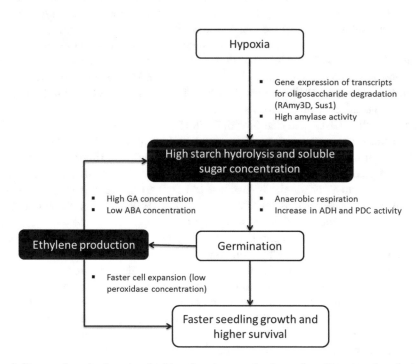

Fig. 6 Proposed mechanisms involved in enhancing germination and seedling growth under low-oxygen stress in rice. (Source: Ismail et al. 2009)

4.1.2 Genetic Factors Underlying the AG Trait

Before evaluating the genetics behind AG, it was necessary to identify lines that can germinate under very limited oxygen supply. The largest AG screening conducted has been that of Angaji et al. in 2010. They screened 8000 rice accessions from the IRRI Genetic Resources Center, which contains collections from gene bank accessions, elite breeding lines, IR64 mutants, and introgression lines. However, only 0.23% (19 lines) showed greater than or equal to 70% survival, a subset of which is present in Table 1. Nevertheless, Hsu and Tung (2017) evaluated gene expression in six genotypes and discovered 3597 genes affected by water stress irrespective of the plant's genetic background, 5100 genes differentially affected across genotypes, and 471 genes affected in a genotype-dependent manner. Hence, the different genotypes are expected to display different degrees of tolerant response to anoxia.

Both quantitative trait loci (QTL) mapping and genome-wide association studies (GWAS) are useful in identifying genomic regions underlying different phenotypic traits. QTL mapping uses markers for a trait of interest, which can categorize the sample population into genotypic groups (Sehgal et al. 2016), while GWAS link common single nucleotide polymorphisms (SNPs) to phenotypic traits, which can provide information on the association between particular genetic variants and phenotype (Pal et al. 2015). Table 2 summarizes the results of some of the previous QTL mapping studies for AG tolerance using different rice landraces as sources.

In 2004, Jiang et al. evaluated 81 recombinant inbred lines (RILs) from a *japonica/indica* cross, Kinmaze/DV85. Kinmaze germinates well under anoxia; hence, it was used as the donor parent. These RILs were screened using 137 restriction fragment length polymorphism (RFLP) markers and five QTLs were detected on chromosomes 1, 2, 5, and 7. These QTLs explained phenotypic variation ranging from

Table 1 Rice accessions tolerant of anaerobic germination together with three sensitive checks, IR42, IR64, and FR13A. Source: Angaji et al. (2010)

Cultivar	Origin	Survival (%)
Khaiyan	Bangladesh	90
Ma-Zhan (red)	China	90
Kalongchi	Bangladesh	90
Sossoka	Guinea	85
Kaolack	Guinea	85
Dholamon 64-3	Bangladesh	80
Nanhi	India	80
Khao Hlan On	Myanmar	75
IR68552-100-1-2-2-2	NPT-IRRI	75
Cody	USA	70
Liu-Tiao-Nuo	China	70
IR64 (sensitive)		20
FR13A (sensitive)		20
IR42 (sensitive)		5

Table 2 Summary of QTLs detected for anaerobic germination identified from different studies

Locus	Chr	LOD	R^2	Donor	Population	Source
qAG-1	1	3.66	11.0	Khaiyan	BILs	Angaji (2008)
qAG-2	2	4.44	14.5	IR64		
qAG-1-2	1	2.73	2.83	Khao Hlan On	BILs	Angaji et al. (2010)
qAG-3-1	3	2.55	2.78	Khao Hlan On		
qAG-7-2	7	7.12	9.58	Khao Hlan On		
qAG-9-1	9	3.88	3.26	Khao Hlan On		
qAG-9-2	9	15.32	20.59	Khao Hlan On		
qAG2.2	2	2.43	9.79	Nanhi	F_2	Baltazar et al. (2014)
qAG3	3	2.42	18.21	IR64		
qAG7	7	13.93	14.06	Nanhi		
qAG-1	1	3.25	12.1	Kinmaze	RILs	Jiang et al. (2004)
qAG-2	2	4.09	15.6	Kinmaze		
qAG-5a	5	2.9	19.6	DV85		
qAG-5b	5	2.61	16.6	DV85		
qAG-7	7	2.74	10.5	Kinmaze		
qAG-5	5	3.78	15.51	USSR5	F_2	Jiang et al. (2006)
qAG-11	11	2.97	10.99	USSR5		
qAG-2	2	3.7	9.3	Ma Zhan (red)	$F_{2:3}$	Septiningsih et al. (2013)
qAG-5	5	5.1	11.8	Ma Zhan (red)		
qAG-6	6	4.5	1.5	Ma Zhan (red)		
qAG-7-1	7	16.5	26.7	Ma Zhan (red)		
qAG-9	9	3.1	1.7	Ma Zhan (red)		
qAG-12	12	3.5	5.3	Ma Zhan (red)		

10.5% to 19.6% of the shoot lengths. According to the calculated additive effects, the positive effect of three QTLs (qAG-1, qAG-2, and qAG-7) was contributed by Kinmaze. Meanwhile, AG tolerance conferred by qAG-5a and qAG-5b was improved due to the recurrent parent, DV85.

Another set of rice lines was evaluated by Jiang et al. in 2006 using USSR5 and N22 as parents. USSR5 is a *japonica* subspecies from the former Soviet Union while N22 is an *indica* type. These were crossed and the generated F_2 plants were evaluated for survival under AG and then genotyped using 121 simple sequence repeat (SSR) markers. Mapping revealed two QTLs on chromosomes 5 (qAG-5) and 11 (qAG-11) contributing to AG tolerance. Both QTLs were contributed by the USSR5 allele, which accounted for 11.0–15.5% of the coleoptile length variation.

Four QTLs were identified by Angaji in 2008 from a cross using Khaiyan, an *aus*-type tolerant variety from Bangladesh, as a donor. Khaiyan was crossed with semi-dwarf nonaromatic *indica* lowland modern rice variety IR64, and the F_1s were backcrossed to IR64 to produce 150 BC_2F_2 lines. The identified QTLs are qAG-1, qAG-2-2, qAG-11, and qAG-12, which explained 12.0–29.2% of the variation in survival of the lines, while together these QTLs accounted for a total of 51.4% of the phenotypic variation. However, based on the additive values, only the QTL *qAG-1* had the allele from Khaiyan, which increased AG tolerance.

Another promising donor for AG tolerance is Khao Hlan On, which is a tall aromatic *japonica* upland landrace from Myanmar used by Angaji et al. in their study in 2010. They crossed Khao Hlan On with IR64 and the F_1 progenies were backcrossed and selfed to produce 423 BC_2F_2 lines. QTL mapping was done using 135 SSRs and one InDel marker. Five QTLs accounting for 17.9–33.5% of the variation in percentage survival were detected from chromosomes 1 (*qAG-1-2*), 3 (*qAG-3-1*), 7 (*qAG-7-2*), and 9 (*qAG-9-1* and *qAG-9-2*). Further, from the values of the additive effects of the QTLs, it was inferred that AG tolerance was contributed by alleles from Khao Hlan On.

More recently, in 2013, Septiningsih et al. used 118 SSR markers to assess AG tolerance alleles from Ma-Zhan (red), a highly AG-tolerant landrace from China with 90% survival (Angaji et al. 2010). Ma Zhan (red) was crossed with IR42 and then backcrossed to produce BC_2F_3 families. The survival rates and genotype data of 118 SSR markers in 175 families revealed six QTLs related to AG on chromosomes 2, 5, 6, and 7, accounting for 5.6–30.3% of the phenotypic variation. These QTLs are *qAG-2*, *qAG-5*, *qAG-6*, *qAG-7-1*, *qAG-7-2*, and *qAG-7-3*. It must also be noted that all five of these QTLs were donated by Ma Zhan (red). QTL *qAG-7-2* explained the highest amount of phenotypic variation, amounting to 30.3% in interval mapping (IM) and 31.7% in composite interval mapping (CIM), with LOD scores of 13.7 and 14.5, respectively. This QTL was further evaluated to confirm its presence and effect. The authors developed introgression lines bearing the homozygous alleles of *qAG-7-2* from Ma Zhan (red) and revealed that these introgression lines have significantly higher survival (47.58 ± 4.72%) than the introgression lines bearing homozygous IR42 alleles (16.67 ± 2.99%). Moreover, the results confirmed the effect of the QTL, which rendered an additional 30% survival in line with the homozygous allele of Ma Zhan (red). Given that Ma Zhan (red) still had the highest survival rate (73.7 ± 14.19%) among all the lines analyzed, this suggests the effect of other QTLs distributed widely in the rice genome to improve AG tolerance.

Hsu and Tung (2015) used both GWAS and QTL mapping to detect genomic regions related to AG tolerance. A total of 144 RILs from a Nipponbare and IR64 cross were used in QTL mapping with 35,501 SNPs. The authors identified a peak on chromosome 1 using both single marker analysis (SMA) and interval mapping based on the linkage map of 355 SNPs filtered after Bonferonni correction with $\alpha = 0.05$. For GWAS, 153 accessions and 36,901 SNPs were used. This revealed three genomic regions on chromosome 7 (25.12, 26.92, and 27.72 Mb); however, the major QTL on chromosome 1, which was detected in QTL mapping, was not identified using GWAS. Since this QTL on chromosome 1 cannot be tagged as a rare variant based on haplotype analysis, it was proposed that the haplotype and varietal relationships can account for the inconsistency in findings. As mentioned by Kang et al. (2008), genetic association studies are confounded by population structure and relatedness, which can produce several false positives. However, Brachi et al. (2011) noted that overcorrection for these factors can lead to false-negative results.

Remarkably, some of the identified QTLs overlap across different studies. Some of these are (1) *qAG-2* detected by Jiang et al. (2004) and Septiningsih et al. (2013), (2) *qAG-7-2* by Angaji et al. (2010) and *qAG-7-3* from Septiningsih et al. (2013), and (3) *qAG-1-2* from Angaji et al. (2010) and chromosome 1 QTL from Hsu and Tung (2015). Further assessment should be done to detect whether these QTLs are controlled by the same gene(s).

At present, two QTLs have been considered of prime importance for AG tolerance, AG1 (*qAG-9-2*) and AG2 (*qAG-7-1*) identified from Khao Hlan On and Ma Zhan (red), respectively (Miro and Ismail 2013). In 2015, Kretzschmar et al. fine-mapped QTL *qAG-9-2* to a 50-kb region on chromosome 9. Based on Nipponbare as the reference genome, this region encompasses four genes and one transposable element. Contrasting this region to a de novo assembled IR64 sequence revealed a 20.9-kb deletion of AG1 spanning across LOC_Os09g20390/Os09g0369400 (OsTPP7) and a deletion of neighboring loci. Meanwhile, the QTL did not show any structural variation against Nipponbare apart from a 4-kb deletion that is also present in IR64.

4.2 Seed Longevity

The life cycle of seeds involves fertilization, dormancy, and germination. Bewley (1997) defined seed dormancy as "the failure of an intact viable seed to completely germinate under favorable conditions." Associated with dormancy is seed longevity, which is the ability of the seed to remain viable after harvesting, and this is usually evaluated through seed germination ratio or percentage.

Seed longevity varies among different rice varieties as it is governed by both genetic and nongenetic factors. Some notable nongenetic factors are ambient environmental conditions during seed development, maturity at seed harvest, mechanical damage, presence of pathogens, moisture content, and storage conditions after harvest (Copeland and McDonald 1995). Seed longevity and viability usually decrease when exposed to high temperature and moisture (Roberts 1961; McDonald 1999). Yamauchi and Winn (1996) reported that low seed longevity or seed aging is a major issue in the establishment of seeds in anaerobic environments where direct seeding is practiced. If seed longevity fails, crop establishment decreases and weed pressure proliferates. Both scenarios can dramatically decrease crop yield; hence, the importance of varieties tolerant of seed aging.

4.2.1 Physiology and Molecular Mechanisms Affecting Seed Longevity

Seeds battle aging through protection and repair. The seed protection mechanism involves the development of a glassy cytoplasm, which limits cellular metabolic activities and antioxidant production during seed storage to inhibit the buildup of oxidized macromolecules that can diminish seed viability. Conversely, the seed repair system exploits DNA glycosylases and methionine sulfoxide reductases, which prevent damage accumulation in the DNA, RNA, and/or proteins during seed imbibition (Sano et al. 2016).

Seed aging is oftentimes linked with the oxidation of nucleic acids, proteins, and lipids (Bailly 2004), causing oxidative stress. This can be combated through passive mechanisms, such as nonenzymatic reactive oxygen species (ROS)-scavenging systems, and active mechanisms, such as enzymatic ROS detoxification (Sano et al. 2016). In enzymatic responses, seeds rely on antioxidant enzymes such as superoxide dismutase, catalase, glutathione, ascorbate peroxidase, monodehydroascorbate, dehydroascorbate, and glutathione reductase (Bailly 2004; Kumar et al. 2015). Petla et al. (2016) also reported the role of the repair enzyme L-isoaspartyl O-methyltransferase, overexpressed in transgenic rice, by restricting ROS accumulation and by repairing homeostasis-disruptive isopartlyl residues in proteins.

Among the nonenzymatic responses, the most important antioxidants are tocopherols or vitamin E, which prevents nonenzymatic lipid peroxidation. Vitamin E-mediated free radical scavenging in rice decreases lipid membrane peroxidation, which results in greater seed longevity (Hameed et al. 2014).

Goufo and Trindade (2013) enumerated other biochemical factors affecting the longevity of rice seeds that include phenolic acid, flavonoids, coumarins, tannins, and anthocyanins. For example, antioxidant polyphenols such as flavonoids in the seed coat, embryo, and endosperm of *Arabidopsis* have been shown to increase seed longevity (Debeaujon et al. 2000). It is worth noting that AG donor lines such as Khao Hlan On have red pericarps, which may have putative roles in both seed longevity and AG tolerance.

Interest has increased in the epigenetic basis of the seed life cycle from fertilization to dormancy until germination. van Zanten et al. (2013) reported that seeds require epigenetic factors such as chromatin condensation to enable the transition from seed dormancy to germination in *Arabidopsis thaliana*. These epigenetic modifications include condensation and methylation (Exner and Hennig 2008; Xiao et al. 2006), which affect chromatin accessibility under stress. Xiao et al. (2006) stated that mutated genes for DNA methylation result in nonviable seedlings and improperly developed embryo in *A. thaliana*. Interestingly, using a drought-tolerant barley variety, it was found that HvDME gene was highly expressed after 10 days of drought stress (Kapazoglou et al. 2013). This gene codes for DNA glycolases, which are responsible for DNA methylation affecting chromatin accessibility.

Recent work on the environmental effect on seed longevity likewise showed the influence of the pretreatments of seed drying at 45–60 °C (Whitehouse et al. 2017). It was revealed that this improves seed longevity in rice germplasm prior to storage. Furthermore, environmental transgenerational changes such as warmer parental growth environment affect the subsequent generation's seed longevity in alpine trees (*Silene vulgaris*) by providing transcripts of SvHSP17.4 and SvNRPD12 mRNAs to withstand heat stress (Mondoni et al. 2014). Moreover, the maternal environment of rice might have a putative role in seed longevity as was shown in barley (Nagel et al. 2015).

4.2.2   Genetic Factors Affecting Seed Longevity

Genome-wide association mapping in barley (*Hordeum vulgare*) showed that seed longevity is inherently affected by genetics, maternal environment, and seed deterioration while negating the effects of lipid-soluble tocochromanols, oil, and proteins (Nagel et al. 2015). Studies of different rice genotypes have shown that *indica*-type varieties have higher seed longevity than *japonica*-type (Chang 1991; Ellis et al. 1992). Table 3 summarizes some of the QTLs identified from different rice populations related to seed longevity.

In rice, three QTLs for seed longevity (*qLG-2*, *qLG-4*, and *qLG-9*) were detected on chromosomes 2, 4, and 9, respectively, using backcross inbred lines from a Nipponbare/Kasalath cross (Miura et al. 2002). QTL *qLG-9* accounted for 59.5% of the phenotypic variation, which has a larger effect than the phenotypic variation effect of *qLG-2* with 13.4% and *qLG-4* with 11.6%. Further investigation by Sasaki et al. (2005) detected 12 germination QTLs as indices of seed longevity. In the latter study, *qLG-9* was also detected and identified as *RC9-2*, which affected phenotypic variation by 12.5–12.8%.

Sasaki et al. (2015) fine-mapped the putative gene located in *qLG-9* using advanced backcross progenies from the cross between *japonica*-type Nipponbare and *indica*-type Kasalath. They found that the candidate genes in the region encode for trehalose-6-phosphatase (Os09g039369400), TPP, and an unknown protein (Os09g039369500). Notably, Kretzschmar et al. (2015) mapped a candidate gene in the same region, which enhances anaerobic germination. This further affirms that the gene for seed longevity can be used for marker-assisted selection (MAS) for the breeding of aging tolerant rice for direct-seeded conditions to enable seedling germination and establishment even in anaerobic conditions.

4.3   Early Seedling Vigor

Seedling vigor reflects the plant's potential to quickly emerge from the soil or water, which is a relevant trait for crop establishment, especially for direct-seeded rice. It is associated with germination and seedling growth, which enable good crop establishment and weed competition (Diwan et al. 2013). It is a complex trait that is governed by several genetic factors that affect the physiological, morphological, and biochemical processes of the plant.

4.3.1   Physiology and Molecular Mechanisms of Early Seedling Vigor

Some of the physiological traits that affect early seedling vigor are shoot length, shoot fresh weight, number of tillers per plant, mesocotyl length, root fresh and dry weight, germination rate, germination index, amylase activity, reducing sugar content, root activity, and chlorophyll content. Teng et al. (1992) evaluated the seedling

Table 3 Summary of major QTLs for seed longevity-related traits identified from different rice populations

Trait	QTL	Chr	LOD	R^2	Donor	Population	Reference
GECAP	qMT-SGC7.1	7	4.8	11.6	Milyang 23	RILs	Jiang et al. (2011)
	qMT-SGC1.1	1	6.19	10.7	Tong 88-8		
	qMT-SGC5.1	5	5.96	12.5	Milyang 23		
	qMT-SGC7.2	7	6.76	12.5	Milyang 23		
	qMT-SGC9.1	9	5.97	11.8	Milyang 23		
	qDT-SGC2.1	2	7.5	12.4	Dasanbyeo	RILs	
	qDT-SGC3.1	3	5.79	15.6	Dasanbyeo		
	qDT-SGC9.1	9	10.29	20.5	Dasanbyeo		
GEC	qRC9-2	9	5.3	12.5	Milyang 23	RILs	Sasaki et al. (2005)
	qRC9-1	9	4.4	11.3	Milyang 23		
	qRC9-2	9	5.4	12.8	Milyang 23		
	qSSnj-2-1	2	4.1	12.3	Nanjing35	BILs	Lin et al. (2015)
GEP	qSSn-5	5	5.3	15.0	N22		
	qSSn-6	6	4.1	13.6	N22		
	qSSn-9	9	4.3	13.1	N22		
	qSSn-1	1	4.9	11.3	N22		
	qSSn-9	9	6.8	13.7	N22		
	qSSn-1-1	1	4.2	9.9	N22		
	qLG-9	9	13.883	59.5	Kasalath	BILs	Miura et al. (2002)
	qSD-3	3	5.11	14.5	ZYQ8	DH	Guo et al. (2004)
GER	qSS4-1	4	4.99	18.68	Shennong 265	RILs	Dong et al. (2017)
	qSS12-2	12	10.0	44.3	Lijiangxing-tuanheigui		
	qSS9-2	2	5.22	29.09	Shennong 265		
	qSS-6	6	6.74	14.45	Kasalath	BILs	Li et al. (2012)
	qSS-9	9	5.36	10.63	Kasalath		
SI	qRGR-1	1	6.95	28.63	IR24	RILs	Xue et al. (2008)

DH doubled haploids, *RILs* recombinant inbred lines, *BILs* backcross-derived inbred lines, *GECAP* germination capability, *GEC* germination count, *GEP* germination percentage, *GER* germination rate, *SI* storage index growth of 38 rice varieties with differences in grain length, width, and weight. They revealed that the most vigorous seedlings were those with larger embryos and heavier endosperms.

Within the endosperm, the stored starch granules are composed of both amylose and amylopectin. These starch granules were distinguished based on their glucose residues (Huang et al. 2017). Amylose is made up of α-1,4-linked glucose residues; hence, it can be broken down into maltose by amylases. Amylopectin, however, cannot be split due to the presence of both α-1,4 and α-1,6 linkages. Starch hydrolysis is an important step for ATP production and for the generation of carbon skeletons for the development of new cellular components (Mitsui et al. 1996). Prior to photoautotrophy, seedlings gain the energy required for plant development from the breakdown of starch granules (i.e., amylose) in the endosperm (Cui et al. 2002b). This is regulated by starch-hydrolyzing enzymes such as α-amylases, which facilitate the cleavage of α-1,4-glucan bonds of starch (Hakata et al. 2012). Notably, this is considered as the first committed step in starch degradation.

α-Amylases are known to be initially expressed in the scutellar epithelial tissue and in the aleurone layer; they are then subsequently secreted into the starchy endosperm (Mitsui et al. 1996) to initiate starch breakdown. Therefore, at the seedling stage, plant vigor can be accounted for by higher amylose content and amylase activity, which provide the growing embryo with enough energy for development. Huang et al. (2017) used two cultivars, Yuxiangyouzhan and Huanghuazhan, to assess the morphophysiological traits governing seedling vigor. Yuxiangyouzhan (96%) showed a higher germination percentage than Huanghuazhan (87%). It was further revealed that Yuxiangyouzhan had higher seed amylose content and amylase activity. Additionally, Huang et al. (2017) cited the utility of using cultivars with thinner hulls, hence requiring less mechanical strength for coleoptile emergence.

4.3.2 Genetic Factors Affecting Early Seedling Vigor

Diwan et al. (2013) highlighted the importance of producing genotypes with early seedling vigor due to economic and environmental concerns. To address this, it is important to determine the genetics underlying seedling vigor. Several studies have been conducted to identify QTLs related to early seedling vigor. Table 4 summarizes some of the major QTLs related to early vigor that have been identified from different studies. Anumalla et al. (2015) surmised that RM259 and RM84 of chromosome 1; RM282, RM148, and RM85 of chromosome 3; RM26 of chromosome 5; and RM11 of chromosome 7 were the most promising markers for several seedling vigor traits. Nagavarapu et al. (2017) validated eight of the previously identified QTLs from different studies: *qGR-1*, *qGP*, and *qGI-11* (Wang et al. 2010); *RZ448* and *RZ395* (Redoña and Mackill 1996); *qFV-3-2*, *qFV-5-2*, and *qFV-10* (Zhou et al. 2007). The authors used 47 genotypes suitable for diverse ecosystems and estimated seedling vigor traits, followed by QTL analysis. Among the eight QTLs used for validation, six (*qGR-1*, *qFV-10*, *qFV-3-2*, *qFV-5-2*, *qGP-6*, and *RZ395*) were found to be associated with some of the selected genotypes. QTL *qGR-1* for germination

Table 4 Summary of major QTLs for early seedling vigor-related traits identified from different rice populations

Trait	QTL	Chr	LOD	R^2	Donor	Population	Reference
EUE	qEMM1.1	1	–	43.0	–	BILs	Dixit et al. (2015)
	qEUE3.1	3	11.3	18.6	Swarna	BILs	Singh et al. (2017)
	qEUE3.2	3	6.49	11.2	Swarna		
	qEUE4.1	4	7.1	12.2	Swarna		
	qEUE5.1	5	11.0	18.3	Swarna		
	qEUE5.2	5	10.7	12.6	Swarna		
	qEUE6.1	6	7.4	12.5	Swarna		
EV	qEVV9.1	9	–	5.8	AUS276/3	BILs	Sandhu et al. (2014)
	qEVV9.1	9	–	6.4	AUS276/3		
	qEV3.1	3	4.13	7.2	Swarna	BILs	Singh et al. (2017)
	qEV3.2	3	7.7	13.1	Swarna		
	qEV4.1	4	7.6	12.1	Swarna		
	qEV5.1	5	4.2	7.4	Swarna		
	qEV5.2	5	4.2	7.5	Swarna		
	qEV6.1	6	6.0	10.6	Swarna		
GI	qGI-7	7	4.3	13.9	Daguandao	RILs	Wang et al. (2010)
	qGI-11	11	5.1	54.9	Daguandao		
GEP	qGP-4	4	5.7	12.7	IR28		
	qGP-6	6	5.3	24.0	Daguandao		
	qGP-8	8	4.3	12.2	Daguandao		
GER	qGR3-2	3	5.2	5.8	Zhenshan97	RILs	Cui et al. (2002a)
	qGR3-3	3	4.3	4.9	Minghui63		
	qGR5-1	5	15.9	17.4	Zhenshan97		
	qGR5-2	5	6.4	8.3	Minghui63		
	qGR6-1	6	5.1	5.6	Zhenshan97		
	qGR6-2	6	4.9	6.3	Zhenshan97		
	qGR3	3	9.69	31.3	Arroz da Terra	BILs	Fukuda et al. (2014)
	qSV-3-2	3	5.4	14.7	Teqing	RILs	Zhang et al. (2005)
GRR	qGR2	2	4.33	10.7	M-203	RILs	Cordero-Lara et al. (2016)
	qMel-1	1	5.4	15.9	Kasalath	BILs	Lee et al. (2012)
	qMel-3	3	4.2	11.5	Kasalath		
RAL	qSV-1	1	11.9	22.6	Minghui63	RILs	Xie et al. (2014)
	qSV-5c	5	14.8	30.5	Zhenshan97		
	qSV-6a	6	5.33	12.6	Zhenshan97		
	qSV-6b	6	5.63	13.3	Zhenshan97		
RODW	qRDW5-1	5	19.4	24.1	Zhenshan97	RILs	Cui et al. (2002a)
	qRDW5-2	5	10.4	13.2	Zhenshan97		
	qRDW10-1	10	5.6	7.3	Zhenshan97		
	qRDW3.1	3	4.8	8.4	Moroberekan	BILs	Singh et al. (2017)

(continued)

Table 4 (continued)

Trait	QTL	Chr	LOD	R^2	Donor	Population	Reference
ROHD	qRHD8.1	8	–	6.6	AUS276/3	BILs	Sandhu et al. (2014)
	qRHD5.1	5	–	6.3	AUS276/3		
	qRHD1.1	1	–	6.3	AUS276/3		
ROHL	qRHL8.1	8	–	25.3	AUS276/3		
	qRHL1.1	1	–	4.1	AUS276/3		
ROL	qMRL1-1	1	9.8	11.4	Minghui63	RILs	Cui et al. (2002a)
	qMRL1-2	1	4.1	5.1	Zhenshan97		
	qMRL5-1	5	11.0	13.1	Zhenshan97		
	qMRL6-1	6	12.4	15.0	Zhenshan97		
	QSV-5a	5	5.5	10.9	Teqing	RILs	Zhang et al. (2005)
	qSV-8-2	8	5.3	10.1	Teqing		
	qSV-6	6	4.9	13.2	Lemont		
SERUE	qSRUE4.4	4	4.15	14.6	IR26	RILs	Cheng et al. (2013)
SEDW	qFV-5-1	5	5.58	11.6	Teqing	RILs	Zhou et al. (2007)
	qFV-10b	10	4.99	8.0	Teqing		
SHE	qCL11	11	4.2	14.8	Arroz da Terra	BILs	Fukuda et al. (2014)
	qSV-5a	5	6.78	12.7	Zhenshan97	RILs	Xie et al. (2014)
	qSV-5b	5	6.53	14.9	Minghui63		
	qSV-8	8	6.03	13.4	Zhenshan97		
	qSV-11-1	11	13.0	28.8	Minghui63		
	qSV-11-2	11	5.27	13.4	Minghui63		
	qHES8	8	5.64	12.3	M-206	RILs	Cordero-Lara et al. (2016)
	qHES9	9	4.37	9.38	M-203		
	qHLS10	10	4.99	11.0	M-203		
	qHLS12	12	4.37	9.54	M-206		
	qPHS1	1	5.5	7.2	Dunghan Shali	RILs	Abe et al. (2012)
	qPHS3-1	3	5.3	6.6	Dunghan Shali		
	qPHS3-2	3	17.5	26.2	Dunghan Shali		
	qPHS4	4	4.4	5.2	Dunghan Shali		
	qCSH2	2	5.2	16.6	Jileng	$F_{2:3}$	Han et al. (2007)
	qCSH12	12	4.68	17.9	Jileng		
	qFV-5-2	5	8.76	14.0	Teqing	RILs	Zhou et al. (2007)
	qFV-10a	10	6.91	9.8	Teqing		
	qSHL5.2	5	5.0	8.8	Swarna	BILs	Singh et al. (2017)
	QSV-5b	5	4.2	6.3	Teqing	RILs	Zhang et al. (2005)
	qSV-7	7	5.9	10.7	Lemont		

(continued)

Table 4 (continued)

Trait	QTL	Chr	LOD	R^2	Donor	Population	Reference
SHDW	qSDW4.2	4	7.06	16.8	IR26	RILs	Cheng et al. (2013)
	qSDW8.2	8	4.43	14.4	IR26		
	qSDW1-1	1	8.4	9.7	Zhenshan97	RILs	Cui et al. (2002a)
	qSDW3-1	3	7.2	8.2	Minghui63		
	qSDW5-1	5	7.9	8.6	Zhenshan97		
	qSDW5-2	5	4.7	6.6	Zhenshan97		
	qSDW6-2	6	8.7	10.3	Zhenshan97		
	qSDW9-1	9	5.7	6.2	Minghui63		
	qSV-3-1	3	4.4	6.3	Lemont	RILs	Zhang et al. (2005)
	qSV-5	5	7.2	12.1	Teqing		
	qSDW3	3	8.24	32.2	Arroz da Terra	BILs	Fukuda et al. (2014)
SHFW	qFW1	1	18.4	34.6	M-203	RILs	Cordero-Lara et al. (2016)
	qFW9	9	4.28	7.2	M-203		
STL	qSL3.1	3	4.0	7.0	Moroberekan	BILs	Singh et al. (2017)
	qSL5.2	5	4.9	8.5	Moroberekan		

DH doubled haploids, *RILs* recombinant inbred lines, *BILs* backcross-derived inbred lines, *EUE* early uniform emergence, *EV* early vigor, *GI* germination index, *GEP* germination percentage, *GER* germination rate, *GRR* growth rate, *RAL* radicle length, *RODW* root dry weight, *ROHD* root hair density, *ROHL* root hair length, *ROL* root length, *SERUE* seed reserve use efficiency, *SEDW* seedling dry weight, *SHE* seedling height, *SHDW* shoot dry weight, *SHFW* shoot fresh weight, *STL* stem length

rate was detected in 12 genotypes; *qFV-10* for seedling height under flooding in eight genotypes; *qFV-3-2* for coleoptile emergence under flooding, *qFV-5-2* for seedling height upon water draining, and *RZ395* for mesocotyl elongation in three genotypes; and *qGP-6* for germination percentage in two genotypes. Notably, some of the genotypes also showed multiple QTLs such as Dinesh with four and Pooja Sabita and Vivekadhan 62 each with two. Furthermore, Abe et al. (2012) identified *OsGA20ox1* as a candidate gene for seedling vigor in rice. Using RILs from a cross between Kekashi and Dinghan Shali, the authors identified a putative QTL on chromosome 3, *qPHS3-2*, which affects 26.2% of the phenotypic variation in seedling height. This QTL region was then introduced to NILs with a background of an elite cultivar, Iwatekko. The region was fine-mapped into an 81-kb interval bearing the *OsGA20ox1* gene that is related to gibberellin biosynthesis. Their further analysis revealed that both Dunghan Shali and the NILs had higher expression of *OsGA20ox1* relative to Iwatekko, which suggests the utility of this gene for seedling vigor.

4.4 Breeding Rice with Improved Germination

Genetic studies on traits such as AG, seed longevity, and early seedling vigor have shown high genetic variation between modern rice varieties, breeding lines, and traditionally evolved cultivars. Systematic extraction of genes that confer stronger germination from traditional varieties is required to develop high-yielding varieties with good crop stand and improved weed competitiveness. With DSR covering a larger proportion of rice area and quickly becoming an ask of the future, it is clear that rice varieties have to evolve from their current state and reach much higher tolerance of adverse conditions during seed germination. Although a large number of DSR breeding programs target selection for early vigor, a thorough understanding of this trait is most important to make genetic gains. For example, it is believed that increased seed size allows better germination capacity. However, in the case of rice, grain shape is an important seed quality and has to be kept in line with the product requirements of specific environments. This limits the amount of improvement for the trait as the best lines have to be selected within grain shape classes. This may also lead to higher popularity of varieties with larger grains under DSR. Similarly, grain pericarp color has been associated with higher AG tolerance and seed longevity and this needs to be addressed through systematic genetic studies to be able to develop white rice with better germination and seed longevity. Trait development and breeding methods must be combined systematically to be able to leverage the advantage of alleles with major and minor effects.

Figure 7 shows a set of pipelines that allow a systematic flow of required traits into a breeding program. Since most modern rice varieties are deficient in these traits, the work begins in the trait development pipeline with the development of a

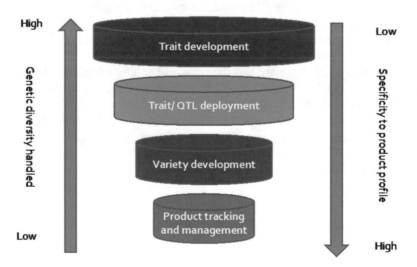

Fig. 7 Breeding program structure for systematically combining exotic traits in a breeding program with minimum linkage drag

phenotyping protocol and donor identification. Once this is complete, adequate mapping populations are to be developed that can support the identification of robust QTLs. It is also to be kept in mind during this process that the recipient be chosen correctly. Ideally, these recipients should have elite genetic backgrounds that can be used in a breeding program in the future or those DSR-adapted breeding lines or varieties that are deficient in these traits. Upon the identification of the QTLs, screening a part of the mapping population in a few key locations can test their effect in the target environments. This allows developing confidence in the effect of the QTL and investment in further fine-mapping can be carried out. Fine-mapped QTL regions and associated markers are delivered to the trait deployment pipeline, which rapidly develops high-yielding NILs of a wide range of recipient parents. Ideally, these should be the latest elite breeding lines from the variety development pipeline. These NILs are entered into the pipeline and checked for yield potential and trait performance across key testing sites. High-performing NILs may be advanced for release as well as recycled into the crossing process to develop the next generation of breeding lines. This cyclic process allows the rapid recycling of high-performing lines as parents and allows building on the achieved genetic gain to advance it further. Once released, the varieties developed through this pipeline are managed through a seed systems team for adequate placement and deployment in collaboration with national seed systems. This team also collects feedback and provides it to the above pipeline so that any possible problems can be addressed in the next wave of varieties. The improvement of breeding lines for seed germination-related traits is important for both inbred programs and hybrid breeding programs. Hybrids provide great potential for achieving high yields in direct-seeded environments. However, for hybrids to be successful in DSR, germination and seedling survival must be high, and good crop stand be achieved with low seed rates to keep seed costs down.

All three major groups of traits discussed previously should be targeted to be able to improve germination under direct-seeded conditions. Although anaerobic germination is one of the most important traits, it cannot provide the needed robustness without other traits. Early seedling vigor helps to boost the effect of anaerobic germination and leads to better tolerance. Vigorous and rapidly growing coleoptiles can emerge from deeper soil layers efficiently and allow more uniform germination. This is specifically helpful in cases where seeding is mechanized and field preparations are not adequate. Lastly, higher seed longevity helps in maintaining germination for relatively older seeds. In cases where farmers use their own seeds, the age of seeds normally varies from 6 to 8 months. It is observed that although this duration does not affect germination to a large extent, it may have an effect on early vigor and anaerobic germination. In some cases where seeds are not adequately stored, germination may also become affected. It is observed, however, that genetic variation exists for the trait and it can be harnessed to improve modern varieties similar to the previous group of traits. In general, multiple donor lines for the three groups of traits can be crossed to two to three elite recipient lines, each belonging to a different grain shape group. This will allow the development of nested association mapping populations with a common recipient parent within the grain shape group.

Trait development activities can then be undertaken to identify and fine-map robust QTLs for each of the grain shape groups. Finally, the identified QTLs can be introgressed singly or pyramided into the recipient lines and can be further deployed in the breeding program as donor lines.

5 Conclusions

Globally, DSR systems are of high importance to sustain profitable rice production under the current agricultural scenario. Declining water and labor resources are driving rice away from the traditional transplanted system. DSR systems are profitable and more amenable to mechanization and are able to decrease water and labor requirements considerably vis-à-vis transplanted systems. However, rice varieties require a specific set of traits at various growth stages to be able to accumulate biomass and produce yield. In the absence of such varieties, the full potential of direct-seeded rice systems cannot be exploited. Further, the use of unsuitable rice varieties makes these systems risky and vulnerable to crop failure. These result in a low adoption of the technology despite its obvious benefits. Rice varieties containing DSR adaptation traits are thus critical to success in these systems.

An important phase in DSR systems is crop establishment. Although the use of high-yielding varieties is important, this will be futile if crop establishment fails. This becomes even more important in the case of hybrid rice, for which adoption depends on the minimization of seed rate and good crop establishment. Anaerobic germination, seed longevity, and early seedling vigor are some traits relevant to crop establishment under direct seeding. Anaerobic germination enables the embryo to sprout and the seedling to survive during anoxia, which can occur upon seeding when the field is intentionally or accidentally irrigated with water. Seed longevity renders aging tolerance to seeds, such that the seeds remain viable under temperature or moisture changes for a longer period. Lastly, early seedling vigor enables the growing seedlings to quickly emerge from the soil and/or water and gives them strong early crop establishment and weed competition.

The physiology and genetics of the aforementioned traits have been studied in the past; however, a lot remains to be unraveled to fully exploit these traits. Further, there remains a need to use these traits in breeding programs to develop DSR-specific varieties. This requires well-designed trait development, pre-breeding, and breeding pipelines with set criteria to advance only the most robust of products. The success of rice in DSR systems largely depends on the robustness of rice varieties at early stages to avoid the risk of crop failure and weed competition. Developing rice varieties with stronger germination capacity under varying conditions is one of the most important factors that will determine success in this area.

References

Abe A, Takagi H, Fujibe T, Aya K, Kojima M, Sakakibara H et al (2012) *OsGA20ox1*, a candidate gene for a major QTL controlling seedling vigor in rice. Theor Appl Genet 125:647–657

Adachi Y, Sugiyama M, Sakagami JI, Fukuda A, Ohe M, Watanabe H (2015) Seed germination and coleoptile growth of new rice lines adapted to hypoxic conditions. Plant Prod Sci 18:471–475

Aggarwal PK (2008) Global climate change and Indian agriculture: impacts, adaptation and mitigation. Indian J Agric Sci 78(11):911

Aggarwal PK, Kumar SN, Pathak H (2010) Impacts of climate change on growth and yield of rice and wheat in the Upper Ganga Basin. WWF report. WWF, New Delhi, pp 1–44

Akhgari H, Kaviani B (2011) Assessment of direct seeded and transplanting methods of rice cultivars in the northern part of Iran. Afr J Agric Res 6(31):6492–6498

Alpi A, Beevers H (1983) Effects of O_2 concentration on rice seedlings. Plant Physiol 71(1):30–34

Angaji S (2008) Mapping QTLs for submergence tolerance during germination in rice. Afr J Biotechnol 7(15):2551–2558

Angaji S, Septiningsih EM, Mackill D, Ismail AM (2010) QTLs associated with tolerance of flooding during germination in rice (*Oryza sativa* L.). Euphytica 172:159–168

Anumalla M, Anandan A, Pradhan S (2015) Early seedling vigour, an imperative trait for direct-seeded rice: an overview on physio-morphological parameters and molecular markers. Planta 241(5):1027–1050

Azhiri-Sigari T, Gines H, Sebastian LS, Wade L (2005) Seedling vigor of rice cultivars in response to seedling depth and soil moisture. Philipp J Crop Sci 30:53–58

Bailly C (2004) Active oxygen species and antioxidants in seed biology. Seed Sci Res 14:93–107

Baltazar M, Ignacio J, Thompson M, Ismail A, Mendioro M, Septiningsih E (2014) QTL mapping for tolerance to anaerobic germination from IR64 and the aus landrace Nanhi using SNP genotyping. Euphytica 197(2):251–260

Barker T (2009) Climate change and EU labor markets. Employment in Europe 2009. EU, Brussels

Barker R, Dawe D, Tuong T, Bhuiyan S, Guerra L (1998) The outlook for water resources in the year 2020: challenges for research on water management in rice production. In: Assessment and Orientation Towards the 21st Century. Proceedings of 19th Session of the International Rice Commission, Cairo, Egypt, 7–9 September 1998. FAO, Rome, pp 96–109

Bewley JD (1997) Seed dormancy and germination. Plant Cell 9:1055–1066

Bouman B, Lampayan R, Tuong T (2007) Water management in irrigated rice: coping with water scarcity. International Rice Research Institute, Los Baños

Brachi B, Morris GP, Borevitz JO (2011) Genome-wide association studies in plants: the missing heritability is in the field. Genome Biol 12:232

Chang TT (1991) Findings from a 28-year seed viability experiment. Int Rice Res Newsl 16:5–6

Chavarria G, dos Santos HP (2012) Plant water relations: absorption, transport and control mechanisms. In: Monanaro G (ed) Advances in selected plant physiology aspects. InTech, Rijeka. ISBN: 978-953-51-0557-2, https://www.intechopen.com/books/advances-in-selected-plantphysiology-aspects/plant-water-relations-absorption-transport-and-control-mechanisms

Cheng X, Cheng J, Huang X, Lai Y, Wang L, Du W (2013) Dynamic quantitative trait loci analysis of seed reserve utilization during three germination stages in rice. PLoS One 8:e80002

Copeland LO, McDonald MB (1995) Seed science and technology, 3rd edn. Chapman and Hall, London, pp 181–220

Cordero-Lara KI, Kim H, Tai TH (2016) Identification of seedling vigor-associated quantitative trait loci in temperate japonica rice. Plant Breed Biotechnol 4:426–440

Cui K, Peng S, Xing Y, Xu C, Yu S, Zhang Q (2002a) Molecular dissection of seedling-vigor and associated physiological traits in rice. Theor Appl Genet 105:745–753

Cui K, Peng S, Xing Y, Yu S, Xu C (2002b) Molecular dissection of relationship between seedling characteristics and seed size in rice. Acta Bot Sin 44(6):702–707

Debeaujon I, Léon-Kloosterziel KM, Koornneef M (2000) Influence of the testa on seed dormancy, germination, and longevity in *Arabidopsis*. Plant Physiol 122(2):403–414

Diwan J, Channbyregowda M, Shenoy V, Salimath P, Bhat R (2013) Molecular mapping of early vigour related QTLs in rice. Res J Biol 1:24–30

Dixit S, Grondin A, Lee CR, Henry A, Olds TM, Kumar A (2015) Understanding rice adaptation to varying agro-ecosystems: trait interactions and quantitative trait loci. BMC Genet 16:86

Dixit S, Kumar A, Woldring H (2016) Water scarcity in rice cultivation: current scenario, possible solutions, and likely impact. In: Regional: development and dissemination of climate-resilient rice varieties for water-short areas of South Asia and Southeast Asia. International Rice Research Institute, Los Baños, pp 3–26

Dong XY, Fan SX, Liu J, Wang Q, Li MR, Jiang X, Liu ZY, Yin YC, Wang JY (2017) Identification of QTLs for seed storability in rice under natural aging conditions using two RILs with the same parent Shennong 265. J Integr Agric 16(5):1084–1092

El-Hendawy S, Sone C, Ito O, Sakagami J (2011) Evaluation of germination ability in rice seeds under anaerobic conditions by cluster analysis. Res J Seed Sci 4(2):82–93

Ella ES, Setter TL (1999) Importance of seed carbohydrates in rice seedling establishment under anoxia. Acta Hortic 504:209–216

Elliott J, Deryng D, Müller C, Frieler K, Konzmann M, Gerten D et al (2014) Constraints and potentials of future irrigation water availability on agricultural production under climate change. Proc Natl Acad Sci U S A 111(9):3239–3244

Ellis RH, Hong TD, Roberts EH (1992) The low-moisture-content limit to the negative logarithmic relation between seed longevity and moisture content in three subspecies of rice. Ann Bot 69(1):53–58

Exner V, Hennig L (2008) Chromatin rearrangements in development. Curr Opin Plant Biol 11(1):64–69

Fanish SA (2016) Enhancing resource use efficiency (RUE) under direct seeded rice (DSR) system: a review. Am Eurasian J Agric Environ Sci 16(9):1534–1544

FAO (2016) AQUASTAT main database. Food and Agriculture Organization of the United Nations (FAO), Rome. http://www.fao.org/nr/water/aquastat/data/query/index.html?lang=eng

Fukuda A, Kataoka K, Shiratsuchi H, Fukushima A, Yamaguchi H, Mochida H et al (2014) QTLs for seedling growth of direct seeded rice under submerged and low temperature conditions. Plant Prod Sci 17(1):41–46

Goufo P, Trindade H (2013) Rice antioxidants: phenolic acids, flavonoids, anthocyanins, proanthocyanidins, tocopherols, tocotrienols, c-oryzanol, and phytic acid. Food Sci Nutr 2:75–104

Grigg DE (1974) The agricultural systems of the world: an evolutionary approach. Cambridge University Press, Cambridge

Guo L, Zhu L, Xu Y, Zeng D, Wu P, Qian Q (2004) QTL analysis of seed dormancy in rice (*Oryza sativa* L.). Euphytica 140:155–162

Hakata M, Kuroda M, Miyashita T, Yamaguchi T, Kojima M, Sakakibara H et al (2012) Suppression of alpha-amylase genes improves quality of rice grain ripened under high temperature. Plant Biotechnol J 10(9):1110–1117

Hameed A, Rasheed A, Gul B, Khan MA (2014) Salinity inhibits seed germination of perennial halophytes *Limonium stocksii* and *Suaeda fruticosa* by reducing water uptake and ascorbate dependent antioxidant system. Environ Exp Bot 107:32–38

Han L, Qiao Y, Zhang S, Zhang Y, Cao G, Kim J, Lee K, Koh H (2007) Identification of quantitative trail loci for cold response of seedling vigor traits in rice. J Genet Genom 34(3):239–246

Ho V, Tran A, Cardarelli F, Perata P, Pucciariello C (2017) A calcineurin B-like protein participates in low oxygen signalling in rice. Funct Plant Biol 44(9):917–928

Hsu SK, Tung CW (2015) Genetic mapping of anaerobic germination-associated QTLs controlling coleoptile elongation in rice. Rice (N Y) 8(1):1–12

Hsu SK, Tung CW (2017) RNA-Seq analysis of diverse rice genotypes to identify the genes controlling coleoptile growth during submerged germination. Front Plant Sci 8:1–15

Huang M, Zhang R, Chen J, Cao F, Jiang L, Zou Y (2017) Morphological and physiological traits of seeds and seedlings in two rice cultivars with contrasting early vigor. Plant Prod Sci 20(1):95–101

Hwang YS, Thomas B, Rodriguez R (1999) Differential expression of rice α-amylase genes during seedling development under anoxia. Plant Mol Biol 40:911–920

Illangakoon T, Ella E, Ismail A, Marambe B, Keerthisena R, Bentota A et al (2016) Impact of variety and seed priming on anaerobic germination-tolerance of rice (Oryza sativa L.) varieties in Sri Lanka. Trop Agric Res 28(1):26–37

IPCC (2007) Climate change 2007: mitigation. In: Metz B, Davidson OR, Bosch PR, Dave R, Meyer LA (eds) Working Group III contribution to the Fourth Assessment Report of the Intergovernmental Panel on Climate Change (IPCC). Technical summary. IPCC, Geneva. Chapters 3 (Issues related to mitigation in the long term context) and 11 (Mitigation from a cross sectoral perspective)

Ismail AM, Ella ES, Vergara GV, Mackill DJ (2009) Mechanisms associated with tolerance to flooding during germination and early seedling growth in rice (Oryza sativa). Ann Bot 103(2):197–209

Ismail AM, Johnson DE, Ella ES, Vergara GV, Baltazar AM (2012) Adaptation to flooding during emergence and seedling growth in rice and weeds, and implications for crop establishment. AoB Plants 2012:1–18

Jiang L, Hou MY, Wang CM, Wan JM (2004) Quantitative trail loci and epistatic analysis of seed anoxia germinability in rice (Oryza sativa). Rice Sci 11(5–6):238–244

Jiang L, Liu S, Hou M, Tang J, Chen L, Zhai H et al (2006) Analysis of QTLs for seed low temperature germinability and anoxia germinability in rice (Oryza sativa L.). Field Crop Res 98(1):68–75

Jiang W, Lee J, Jin YM, Qiao Y, Piao R, Jang SM, Woo MO, Kwon SW, Liu X, Pan HY, Du X, Koh HJ (2011) Identification of QTLs for seed germination capability after various storage periods using two RIL populations in rice. Mol Cell 31:385–392

Juang J, Takano T, Akita S (2000) Expression of α-expansin genes in young seedlings of rice (Oryza sativa L.). Planta 211(4):467–473

Kang HM, Zaitlen NA, Wade CM, Kirby A, Heckerman D, Daly MJ, Eskin E (2008) Efficient control of population structure in model organism association mapping. Genetics 178:1709–1723

Kapazoglou A, Drosou V, Argiriou A, Tsaftaris AS (2013) The study of a barley epigenetic regulator, HvDME, in seed development and under drought. BMC Plant Biol 13:172

Kaur J, Singh A (2017) Direct seeded rice: prospects, problems/constraints and researchable issues in India. Curr Agric Res J 5(1):13–32

Kennedy RA, Barrett SC, Vander Zee D, Rumpho ME (1980) Germination and seedling growth under anaerobic conditions in Echinochloa crus-galli (barnyard grass). Plant Cell Environ 3:243–248

Kordan HA (1974) Patterns of shoot and root growth in rice seedlings germinating under water. J Appl Ecol 1(2):685–690

Kreft S, Eckstein D, Melchior I (2016) Global climate risk index 2017: who suffers most from extreme weather events? Weather-related loss events in 2015 and 1996 to 2015. Germanwatch Nord-Süd Initiative eV, Bonn

Kretzschmar T, Pelayo MF, Trijatmiko KR, Gabunada LM, Alam R, Jimenez R et al (2015) A trehalose-6-phosphate phosphatase enhances anaerobic germination tolerance in rice. Nat Plants 1(9):1–5

Kudahettige N, Pucciariello C, Parlanti S, Alpi A, Perata P (2010) Regulatory interplay of the Sub1A and CIPK15 pathways in the regulation of α-amylase production in flooded rice plants. Plant Biol 13(4):611–619

Kumar V, Ladha J (2011) Direct seeding of rice: recent developments and future research needs. Adv Agron 111:297–413

Kumar SJ, Prasad SR, Banerjee R, Thammineni C (2015) Seed birth to death: dual functions of reactive oxygen species in seed physiology. Ann Bot 116:663–668

Lasanthi-Kudahettige R, Magneschi L, Loreti E, Gonzali S, Licausi F, Novi G et al (2007) Transcript profiling of the anoxic rice coleoptile. Plant Physiol 144(1):218–231

Lee HS, Sasaki K, Atsushi H, Ahn SN, Sato T (2012) Mapping and characterization of quantitative trait loci for mesocotyl elongation in rice (Oryza sativa L.). Rice 5(1):13

Li L, Lin Q, Liu S, Liu X, Wang W, Hang NT, Liu F, Zhao Z, Jiang L, Wan J (2012) Identification of quantitative trait loci for seed storability in rice (Oryza sativa L.). Plant Breed 131:739–743

Lin Q, Wang W, Ren Y, Jiang Y, Sun A, Qian Y, Zhang Y, He N, Hang NT, Liu Z, Li L, Liu L, Jiang L, Wan J (2015) Genetic dissection of seed storability using two different populations with a same parent rice cultivar N22. Breed Sci 65:411–419

Lu CA, Lim EK, Yu SM (1998) Sugar response sequence in the promoter of a rice α-amylase gene serves as a transcriptional enhancer. J Biol Chem 273:10120–10131

Lu CA, Lin CC, Lee KW, Chen JL, Huang LF, Ho SL et al (2007) The SnRK1A protein kinase plays a key role in sugar signaling during germination and seedling growth of rice. Plant Cell 19(8):2484–2499

Magneschi L, Perata P (2009) Rice germination and seedling growth in the absence of oxygen. Ann Bot 103(2):181–196

McDonald MB (1999) Seed deterioration: physiology, repair and assessment. Seed Sci Technol 27:177–237

McQueen-Mason SJ, Cosgrove DJ (1995) Expansin mode of action on cell walls. Plant Physiol 107:87–100

Mekonnen MM, Hoekstra AY (2016) Four billion people facing severe water scarcity. Sci Adv 2(2):e1500323

Miro B, Ismail AM (2013) Tolerance of anaerobic conditions caused by flooding during germination and early growth in rice (Oryza sativa L.). Front Plant Sci 4:1–18

Mitsui T, Yamaguchi J, Akazawa T (1996) Physicochemical and serological characterization of alpha-amylase isoforms and identification of their corresponding genes. Plant Physiol 110(4):1395–1404

Miura K, Lin SY, Yano M, Nagamine T (2002) Mapping quantitative trait loci controlling seed longevity in rice (Oryza sativa L.). Theor Appl Genet 104:981–986

Molden D (2007) Water for food, water for life: a comprehensive assessment of water management in agriculture. International Water Management Institute, London

Mondoni A, Orsenigo S, Donà M, Balestrazzi A, Probert RJ, Hay FR, Abeli T (2014) Environmentally induced transgenerational changes in seed longevity: maternal and genetic influence. Ann Bot 113(7):1257–1263

Nagavarapu T, Keshavulu K, Subba Rao LV, Sengguttuvel P, Maganti SM (2017) Validation of seedling vigour QTLs in rice (Oryza sativa L.). Curr Trends Biotechnol Pharm 11(1):24–33

Nagel M, Kranner I, Neumann K, Rolletschek H, Seal CE, Colville L, Fernández-Marín B, Börner A (2015) Genome-wide association mapping and biochemical markers reveal that seed ageing and longevity are intricately affected by genetic background and developmental and environmental conditions in barley. Plant Cell Environ 38(6):1011–1022

OCHA (2010) Water scarcity and humanitarian action: key emerging trends and challenges. In: Policy Development and Studies Branch. Occasional policy briefing series No. 4. OCHA, New York, NY. 15 p

Pal LR, Yu CH, Mount S, Moult J (2015) Insights from GWAS: emerging landscape of mechanisms underlying complex trait disease. BMC Genomics 16(8):1–15

Pathak H, Sankhyan S, Dubey D, Bhatia A, Jain N (2013) Dry direct-seeding of rice for mitigating greenhouse gas emission: field experimentation and simulation. Paddy Water Environ 11:593–601

Perata P, Pozueta-Romero J, Akazawa T, Yamaguchi J (1992) Effects of anoxia on starch breakdown in rice and wheat seeds. Planta 188:611–618

Perata P, Geshi N, Yamaguchi J, Akazawa T (1993) Effect of anoxia on the induction of α-amylase in cereal seeds. Planta 191:402–408

Petla BP, Kamble NU, Kumar M, Verma P, Ghosh S, Singh A, Rao V, Salvi P, Kaur H, Saxena SC, Majee M (2016) Rice protein L-isoaspartyl O-methyltransferase isoforms differentially accumulate during seed maturation to restrict deleterious isoAsp and reactive oxygen species accumulation and are implicated in seed vigor and longevity. New Phytol 211(2):627–645

Pingali PL (2012) Green revolution: impacts, limits, and the path ahead. Proc Natl Acad Sci U S A 109(31):12302–12308

Redoña E, Mackill D (1996) Mapping quantitative trait loci for seedling vigor in rice using RFLPs. Theor Appl Genet 92:395–402

Rice Knowledge Bank (2017) How to manage water. http://www.knowledgebank.irri.org/step-by-step-production/growth/water-management

Roberts EH (1961) The viability of rice seed in relation to temperature, moisture content, and gaseous environment. Ann Bot 25(3):381–390

Sadiq I, Fanucchi F, Paparelli E, Alpi E, Bachi A, Alpi A et al (2011) Proteomic indentification of differentially expressed proteins in the anoxic rice coleoptile. J Plant Physiol 168:2234–2243

Saika H, Matsumura H, Takano T, Tsutsumi N, Nakazono M (2006) A point mutation of *Adh1* gene is involved in the repression of coleoptile elongation under submergence in rice. Breed Sci 56:69–74

Sandhu N, Torres RO, Sta Cruz MT, Maturan PC, Jain R, Kumar A, Henry A (2014) Traits and QTLs for development of dry direct-seeded rainfed rice varieties. Exp Bot 66(1):225–244

Sano N, Rajjou L, North HM, Debeaujon I, Marion-Poll A, Seo M (2016) Staying alive: molecular aspects of seed longevity. Plant Cell Physiol 57(4):660–674

Sasaki K, Fukuta Y, Sato T (2005) Mapping of quantitative trait loci controlling seed longevity of rice (*Oryza sativa* L.) after various periods of seed storage. Plant Breed 124:361–366

Sasaki K, Takeuchi Y, Miura K, Yamaguchi T, Ando T, Ebitani A, Higashitani T, Yamaya A, Yano M, Sato T (2015) Fine mapping of a major quantitative trait locus, qLG-9, that controls seed longevity in rice (*Oryza sativa* L.). Theor Appl Genet 128(4):769–778

Sehgal D, Singh R, Rajpal V (2016) Quantitative trait loci mapping in plants: concepts and approaches. In: Rajpal VR, Rao SR, Raina SN (eds) Molecular breeding for sustainable crop improvement. Springer International Publishing, Cham, pp 31–59

Septiningsih EM, Ignacio JI, Sendon PM, Sanchez DL, Ismail AM, Mackill DJ (2013) QTL mapping and confirmation for tolerance of anaerobic conditions during germination derived from the rice landrace Ma-Zhan Red. Theor Appl Genet 126:1357–1366

Shingaki-Wells RN, Huang S, Taylor NL, Carroll AJ, Zhou W, Millar A (2011) Differential molecular responses of rice and wheat coleoptiles to anoxia reveal novel metabolic adaptations in amino acid metabolism for tissue tolerance. Plant Physiol 156:1706–1724

Singh UM, Yadav S, Dixit S, Ramayya P, Devi M, Raman K et al (2017) QTL hotspots for early vigor and related traits under dry direct-seeded system in rice (*Oryza sativa* L.). Front Plant Sci 8:286

Takahashi H, Saika H, Matsumura H, Nagamura Y, Tsutsumi N, Nishizawa NK et al (2011) Cell division and cell elongation in the coleoptile of rice alcohol dehydrogenase 1-deficient mutant are reduced under complete submergence. Ann Bot 108(2):253–261

Teng S, Vergara BS, Alejar AA (1992) Relationship of grain length, width, and weight to seedling vigor in rice (*Oryza sativa* L.). Philipp J Crop Sci 17(1):17–20

Terashima M, Hosono M, Katoh S (1997) Functional roles of protein domains on rice α-amylase activity. Appl Microbiol Biotechnol 47(4):364–367

Waffenschmidt S, Woessner JP, Beer K, Goodenough UW (1993) Isodityrosine cross-linking mediates insolubilization of cell walls in *Chlamydomonas*. Plant Cell 5(7):809–820

Wang Z, Wang J, Bao Y, Wang F, Zhang H (2010) Quantitative trait loci analysis for rice seed vigor during the germination stage. J Zhejiang Univ Sci B (Biomed and Biotechnol) 11:958–964

Whitehouse KJ, Hay FR, Ellis RH (2017) High-temperature stress during drying improves subsequent rice (*Oryza sativa* L.) seed longevity. Seed Sci Res 27(4):281–291

Xiao W, Custard KD, Brown RC, Lemmon BE, Harada JJ, Goldberg RB, Fischer RL (2006) DNA methylation is critical for *Arabidopsis* embryogenesis and seed viability. Plant Cell 18(4):805–814

Xie L, Tan Z, Zhou Y, Xu R, Feng L, Xing Y et al (2014) Identification and fine mapping of quantitative trait loci for seed vigor in germination and seedling establishment in rice. J Integr Plant Biol 56(8):749–759

Xue Y, Zhang SQ, Yao QH, Peng RH, Xiong AS, Li X, Zhu WM, Zhu YY, Zha DS (2008) Identification of quantitative trait loci for seed storability in rice (*Oryza sativa* L.). Euphytica 164:739–744

Yamauchi M (1996) Development of anaerobic direct-seeding technology for rice in the tropics. Crop research in Asia: achievements and perspectives. In: Proceedings of the 2nd Asian Crop Science Conference, Fukui, Japan, pp 198–203

Yamauchi M, Winn T (1996) Rice seed vigor and seedling establishment in anaerobic soil. Crop Sci 36(3):680–686

van Zanten M, Liu Y, Soppe WJ (2013) Epigenetic signalling during the life of seeds. In: Epigenetic memory and control in plants. Springer, Berlin, pp 127–153

Zhang Z, Su L, Li W, Chen W, Zhu Y (2005) A major QTL conferring cold tolerance at the early seedling stage using recombinant inbred lines of rice (*Oryza sativa* L.). Plant Sci 168:527–534

Zhou L, Wang J-K, Yi Q, Wang Y-Z, Zhu Y-G, Zhang Z-H (2007) Quantitative trait loci for seedling vigor in rice under field conditions. Field Crop Res 100(2–3):294–301

Genetics and Breeding of Heat Tolerance in Rice

Changrong Ye, Xiaolin Li, Edilberto Redoña, Tsutomu Ishimaru, and Krishna Jagadish

Abstract Extreme weather events, especially heat waves, have become more frequent with global warming. High temperature significantly affects world food security by decreasing crop yield. Rice is intensively planted in tropical and subtropical areas in Asia, where high temperature has become a major factor affecting rice production. Rice is sensitive to high temperature, especially at booting and flowering stages. Rice varieties tolerant of high temperature are rare, and only a few heat-tolerant rice varieties have been identified. High temperature at booting and flowering stages causes sterile pollen, decreased pollen shedding, and poor pollen germination, which finally lead to a yield decrease. Heat-tolerant QTLs have been identified in different studies, but new breeding lines with considerable heat tolerance have not been bred using identified heat-tolerance donors and QTLs. Research on heat-tolerant donor identification, QTL mapping, gene cloning, and large-scale phenotyping technology is important for developing heat-tolerant rice varieties.

Keywords Rice · Heat tolerance · Global warming · High temperature

C. Ye (✉)
Center for Molecular Breeding Technology Invention and Development, Huazhi Biotechnology Co. Ltd., Changsha, Hunan, China
e-mail: changrong.ye@higentec.com

X. Li
Institute of Food Crops, Yunnan Academy of Agricultural Sciences, Kunming, Yunnan, China

E. Redoña
Delta Research and Extension Center, Mississippi State University, Stoneville, MS, USA
e-mail: edredona@drec.msstate.edu

T. Ishimaru
Hokuriku Research Station, Central Region Agricultural Research Center, National Agriculture and Food Research Organization, Niigata, Japan
e-mail: cropman@affrc.go.jp

K. Jagadish
Department of Agronomy, Kansas State University, Manhattan, KS, USA
e-mail: kjagadish@ksu.edu

© The Author(s) 2021
J. Ali, S. H. Wani (eds.), *Rice Improvement*,
https://doi.org/10.1007/978-3-030-66530-2_7

203

Global temperature has been increasing rapidly since the last century, and high temperature has become a major factor affecting agricultural production. Rice is one of the crops most affected by high temperature. Heat tolerance of rice has been studied since the 1970s, but the progress attained has been quite slow. Evaluation of rice heat tolerance at different growth stages revealed that flowering stage is the most sensitive stage, and booting stage is the second most sensitive stage (Yang and Heilman 1993; Yoshida et al. 1981). Thus, recent findings and future prospects on heat tolerance of rice at these reproductive stages are the main focus in this review.

1　Climate Change and Global Warming

Climate change is the long-term change in weather patterns. Climate change is caused by factors such as biotic processes, variations in solar radiation received by Earth, plate tectonics, and volcanic eruptions. Certain human activities have been identified as primary causes of ongoing climate change, often referred to as global warming (Pachauri and Reisinger 2007).

Global warming is the long-term rise in the average temperature of Earth's climate system. Climate proxies show that the temperature had been relatively stable over 1000 or 2000 years before 1850. From 1880 to 2012, the global average (land and ocean) surface temperature increased by 0.85 (0.65–1.06) °C. From 1906 to 2005, Earth's average surface temperature rose by 0.74 ± 0.18 °C (Solomon et al. 2007). Since the early twentieth century, Earth's mean surface temperature has increased by 0.8 °C, with 0.6 °C of this hike occurring since 1980 (Jansen et al. 2007). Human influence has been the dominant cause of the observed warming since the mid-twentieth century (IPCC 2013). The largest human influence has been the emission of greenhouse gases (GHG) such as carbon dioxide, methane, and nitrous oxide. The greenhouse effect is the process by which absorption and emission of infrared radiation by gases in a planet's atmosphere warm its lower atmosphere and surface. Global mean surface temperatures for 2081–2100, relative to 1986–2005, are likely to increase by 0.3–1.7 °C for the lowest and by 2.6–4.8 °C for the highest GHG emission scenarios (IPCC 2013).

Global warming caused by human activities has become a major constraint for agricultural development and crop production. Studies have shown that the annual mean maximum and minimum temperatures increased by 0.35 and 1.13 °C for the period 1979–2003 at the International Rice Research Institute, Los Baños, Philippines (Peng et al. 2004). By 2080, most cropping areas in the world are likely to be exposed to record average air temperature (Battisti and Naylor 2009). Relatively small changes in mean temperature can result in disproportionately large changes in the frequency of extreme events (Rosenzweig et al. 2001). These extreme temperature events are likely to become more frequent with global warming (Tabaldi et al. 2006). Of the 13 warmest years since 1880, 11 occurred from 2001 to 2011 (i.e., every year starting with 2001), while 2011 was the warmest La Niña year in the period from 1950 to 2011 (NOAA 2011).

Future climate change and associated impacts will differ from region to region. Anticipated effects include increasing global temperature, rising sea level, changing precipitation, and expansion of deserts in the subtropics (IPCC 2014). Other likely changes involve more frequent extreme weather events such as heat waves, droughts, heavy rainfall with floods, heavy snowfall, ocean acidification, and species extinctions due to shifting temperature regimes. Effects significant to human beings are the threat to food security from decreasing crop yields (Battisti and Naylor 2009).

2 Rice Production and Heat Damage

Rice is the most widely produced and consumed staple food for a large part of the world's human population, providing more than 20% of the calories. Rice is widely planted around the world, with about 90% of the total production from Asia (southern Asia 31.44%, East Asia 30.68%, Southeast Asia 27.75%, and West and Central Asia 0.27%), 4.86% from the Americas, 4.39% from Africa, 0.57% from Europe, and 0.04% from Oceania (data for 2016 from FAO statistics).

The world's top rice producers are mainly in Asia. Many of these Asian countries are located in the tropics and subtropics, where the temperature is high during the rice crop season, including part of China, India, Indonesia, Bangladesh, Vietnam, Myanmar, Thailand, the Philippines, Cambodia, Lao PDR, and Sri Lanka.

Temperature stress is a complex interaction of temperature intensity, duration, rapidity, and plant growth stage. Damage from extreme high temperature is particularly severe when it occurs at critical crop developmental stages, particularly the reproductive period. Optimum temperature is 20–30 °C during the reproductive stage, but temperatures surpassing 35 °C have critical negative effects on rice growth. High daytime temperatures in some of the major tropical rice-growing regions are already close to the threshold beyond which yield begins to decline (Prasad et al. 2006; Wassmann et al. 2009a).

Rice has been cultivated in a wide range of climatic environments. High temperature has more effects in the tropics and low-altitude valleys in some of the temperate regions. Above 32 °C, spikelet sterility becomes a major factor affecting rice yield, even if sufficient growth occurs in other yield components (Matthews et al. 1995). Traditionally, farmers grow rice at optimal seasonal temperatures to maximize grain yield. However, with the increase in the frequency of extreme temperature events, climate change may increase the probability of overlapping peaks of temperature and the flowering period, which diminishes final yield (Teixeira et al. 2011).

Although there has been no systematic monitoring and evaluation of temperature stress-induced yield losses worldwide, heat-vulnerable regions were geographically mapped based on the critical temperatures at flowering stage (Jagadish et al. 2014; Laborte et al. 2012; Wassmann et al. 2009a). Regional high-temperature damage was observed in many tropical and subtropical countries, such as Pakistan, India, Bangladesh, China, Thailand, Laos, Japan, Sudan, Australia, and the United States

(Hasegawa et al. 2009; Ishimaru et al. 2016; Matsushima et al. 1982; Osada et al. 1973; Tian et al. 2009). Analysis of temperature and rice yield during 1992–2003 at the International Rice Research Institute (IRRI) showed that rice grain yield declined by 10% for each 1 °C increase in growing-season minimum temperature (Peng et al. 2004). Tian et al. (2009) reported that at least six severe heat events damaged the rice crop in the past 50 years in China. Studies on the Yangtze River basin in China showed that an estimated 3 million ha of rice were damaged and 5.18 million tons of paddy rice were lost in 2003 because of a heat wave with the temperature above 38 °C lasting for more than 20 days (Li et al. 2004; Xia and Qi 2004; Yang et al. 2004). In Laos and southern India, the combined stress of heat and intense solar radiation during daytime increases the spikelet sterility of local popular cultivars when heading coincides with high temperatures (Ishimaru et al. 2016). In the record hot summer of 2007, the percentage of spikelet sterility rose to 25% when the maximum daily temperature was around 38 °C in the temperate regions of Japan (Hasegawa et al. 2011). High temperature after heading significantly decreased rice grain quality in many rice-growing regions of Japan in 2010 (Morita et al. 2016). Thus, heat stress at flowering is a real threat to sustained rice production not only in tropical and subtropical regions but also in temperate regions.

The effect of extreme temperature events on crop production is likely to become more frequent in the near future. Significant yield losses have also been predicted by using different crop models. Short-term predictions indicated that, by 2030, rice production in South Asia could decrease by up to 10% (Lobell et al. 2008). Medium- to long-term predictions, that is, by 2080, estimated rice yields in developing countries to decrease by 10–25%, while yields in India could drop by 30–40% (Cline 2008). By 2100, rice and maize yields in the tropics are expected to decrease by 20–40% because of higher temperatures, without accounting for the decrease in yields as a result of drought enhanced by temperature increases (Battisti and Naylor 2009). Spatial model simulation indicated that yield of boro rice in Bangladesh could decrease by 20% and 50% by 2050 and 2070, respectively (Basak et al. 2010), and, on average, rice yields could decline by up to 33% by 2081–2100 (Karim et al. 2012).

Besides high day temperatures, night temperatures greater than 29 °C can decrease spikelet fertility in rice with a subsequent decrease in seed set and grain yield (Satake and Yoshida 1978; Ziska et al. 1996). The increase in night temperature from 27 to 32 °C decreased grain length and grain width, thereby decreasing grain yield (Counce et al. 2005; Mohammed and Tarpley 2011; Morita et al. 2005).

3 Heat Tolerance of Rice

Heat tolerance is defined as the ability of the plant to grow and produce economic yield at high temperature (Wahid et al. 2007). The developmental stage at which the plant is exposed to heat stress determines the severity of the possible damage to the crop (Wahid et al. 2007). High temperature causes injury to the rice plant at different growth stages, such as poor germination, retarded seedling growth, leaf yellowing, inhibited rooting and tillering, inhibited panicle initiation and development, spikelet degeneration, disturbed pollen formation, poor panicle exsertion, inhibited anther dehiscence and pollination, and poor grain filling and development (Yang and Heilman 1993; Yoshida et al. 1981). Thus, heat tolerance is actually the responses of different traits to high temperature at different growth stages, for example, seedling growth or survival at the seedling stage, spikelet fertility at the reproductive stage, and chalkiness of grains at the grain-filling stage (Ishimaru et al. 2009; Lanning et al. 2011). At the grain-filling stage, high temperature affects cellular and developmental processes, leading to decreased fertility and grain quality (Barnabas et al. 2008). Decreased grain weight, decreased grain filling, and higher percentage of chalky grains are common effects of exposure to high temperature during the ripening stage in rice (Osada et al. 1973; Yoshida et al. 1981). In addition, increased temperature causes a serious decrease in grain size and amylase content (Yamakawa et al. 2007; Zhu et al. 2005).

Rice is relatively tolerant of high temperatures during the vegetative phase (Prasad et al. 2006; Yoshida et al. 1981), but is highly susceptible during the reproductive phase, particularly at flowering stage (Jagadish et al. 2008; Matsui et al. 2001b). High temperature surpassing 35 °C during flowering stage increases pollen and spikelet sterility, which leads to significant yield losses, low grain quality, and low harvest index (Matsui et al. 1997a, b; Matsushima et al. 1982; Osada et al. 1973; Prasad et al. 2006; Zhong et al. 2005). The response of spikelet fertility is a major factor determining rice production under high-temperature conditions. Thus, spikelet fertility at high temperature has been widely used as a screening index for heat tolerance at the reproductive stage (Prasad et al. 2006).

Wide genetic variation exists in tolerance of heat stress (Matsui and Omasa 2002). Large cultivar variation exists in spikelet sensitivity to high-temperature damage, and the primary cause of this cultivar variation in heat tolerance at flowering is the number of viable pollen grains shed on the stigma, which is positively correlated with basal anther dehiscence (Matsui 2009).

It has been suggested that *indica* varieties are more tolerant of higher temperatures than *japonica* cultivars (Matsui et al. 2000; Satake and Yoshida 1978), although heat-tolerant genotypes have been found in both subspecies (Matsui et al. 2001b; Prasad et al. 2006).

Humidity also plays an important role in rice yield, as higher relative humidity (RH) at the flowering stage at increased temperature affects spikelet fertility negatively (Yan et al. 2010). Field observations in some high-yielding rice areas with a drier climate and high temperatures (e.g., New South Wales and southern Iran) suggested no significant increase in spikelet sterility even at temperatures above 40 °C (Wassmann et al. 2009b). The fertility of spikelets at high air temperatures decreased further with increased humidity (Matsui et al. 1997a; Nishiyama and Satake 1981). An RH of 85–90% at the heading stage induced almost complete grain sterility in rice at a day/night temperature of 35/30 °C (Abeysiriwardena et al. 2002). Increasing both air temperature and RH significantly increased spikelet sterility, while high temperature-induced sterility decreased significantly with decreasing RH. A reduction in sterility with decreased RH was more due to decreased spikelet temperature than to air temperature. Thus, both air temperature and humidity are equally important in determining pollen viability, splitting of anthers, pollen shedding, and spikelet sterility in rice. The impact of RH should be considered when interpreting the effect of high temperature on grain sterility.

Heat tolerance at flowering is often tested at 37.5–38.0 °C (with relative humidity of 60–70%) to have a great contrast in spikelet fertility between susceptible and tolerant genotypes (Mackill et al. 1982; Satake and Yoshida 1978; Ye et al. 2015b, 2012). At the booting stage, 38 °C is a threshold for most rice varieties. Similarly, we confirmed that 37 °C is a threshold at flowering stage for most varieties. Above this limit, pollen development will fail and spikelet fertility will decline significantly (Ye et al. 2015b).

4 Heat-Tolerant Rice Genetic Resources

The response of rice to high temperatures differs according to the developmental stage. High-temperature tolerance at one developmental stage may or may not necessarily lead to tolerance during other stages (Wassmann et al. 2009a). Hence, the effect of high temperature during different developmental stages has to be partitioned and evaluated separately for assessing, identifying, and characterizing for genetic manipulation of tolerance mechanisms (Wahid et al. 2007). Since flowering stage is the most sensitive stage to high temperature, most screenings for heat-tolerant germplasm were done at flowering stage.

Rice genetic resources tolerant of high temperature have been identified in both *indica* (Matsui et al. 1997b) and *japonica* subspecies (Matsui et al. 2001b). N22 and Dular are excellent sources of genes for heat tolerance (Manigbas et al. 2014). Among the heat-tolerant varieties, N22, an Indian *aus*-type landrace, was identified as one of the most heat-tolerant genotypes in both chamber and open-field experiments (Jagadish et al. 2010a; Mackill et al. 1982; Manigbas et al. 2014; Poli et al. 2013; Prasad et al. 2006; Ye et al. 2012; Yoshida et al. 1981). N22 has been used as a check variety for many studies on heat tolerance. Akitakomachi is the most tolerant genotype found among *japonica* rice (Matsui et al. 2001b).

A recent investigation using a representative set of popular cultivars grown across highly vulnerable rice-growing regions of South and Southeast Asia, Latin America, and West Africa concluded that most of the popular rice cultivars were susceptible to heat stress at reproductive stages (Shi et al. 2015). More than 80% of hybrid rice combinations in China are heat-susceptible, but some heat-tolerant combinations were found in hybrid rice (Hu et al. 2012; Zhou et al. 2009). The hybrid rice Guodao 6 was considered as a heat-tolerant variety (Tao et al. 2008). An accession of wild rice, *Oryza meridionalis* Ng, was also identified as a heat-tolerant species (Scafaro et al. 2010).

In *japonica* cultivars, Akitakomachi, Nipponbare, Hitomebore, and Todorokiwase (Maruyama et al. 2013; Matsui et al. 2001b; Tenorio et al. 2013) are classified as heat-tolerant genotypes. In *indica* cultivars, IR24, IR36, Ciherang, ADT36, BG90-2, Dular, Huanghuazhan, AUS17, M9962, Sonalee, and AUS16 (Cao et al. 2008; Cheabu et al. 2018; Maruyama et al. 2013; Shi et al. 2015; Tenorio et al. 2013) are known as heat-tolerant genotypes. It is notable that Giza178, an Egyptian cultivar developed from a *japonica-indica* cross, has considerable heat tolerance at booting stage as well as flowering stage (Tenorio et al. 2013).

Varieties such as Agbede, Carreon, Dular, N22, OS4, P1215936, and Sintiane Diofor have high spikelet fertility even at high temperatures (Yoshida et al. 1981). The local Iranian landraces Anbori and Hoveaze (probably the same as Hoveyzeh) are tolerant of high temperatures (Gilani et al. 2009). Cultivars KRN, Citanduy, Belle patna, and BPB were tolerant of high-temperature treatment at the ripening stage (Zakaria et al. 2002). An *indica* cultivar (HT54) from China was tolerant of high temperature at both seedling and grain-filling stages. HT54 seedlings could tolerate high temperature up to 48 °C for 79 h (Wei et al. 2013). New Rice for Africa line 44 (NERICA-L-44) was also identified as heat tolerant at both vegetative and reproductive stages (Bahuguna et al. 2015).

As part of IRRI's initiative to develop improved breeding lines tolerant of high temperature, studies were conducted to identify genetic donors of the heat-tolerance trait from the IRRI Genebank. A series of trials were conducted using a set of 455 IRRI Genebank accessions coming from "hot" countries (Pakistan, India, Afghanistan, Iran, and Iraq). However, few varieties (about 5%) showed some degree of heat tolerance. Twenty-three accessions were selected as potential donors for heat tolerance. Dular and Todorokiwase are tolerant at the booting stage, while Milyang23 and IR2006-P12-12-2-2 are tolerant at the flowering stage. Giza178 is tolerant at both booting and flowering stages. Darbari Roodbar, Larome, Mulai, Giza178, IR2006-P-12-12-2-2, Milyang23, and Todorokiwase were tolerant of high temperature in the field and growth chambers. Other potential donors identified based on at least one trait were IR22, IR2307-247-2-2-3, IR6, IR8, MRC603-383, Ganjay, Todorokiwase, Giza 178, Giza 159, and Toor Thulla (Tenorio et al. 2013). The accessions with heat tolerance at booting and flowering stage are useful genotypes for a breeding program to improve heat resilience in terms of spikelet sterility.

5 Physiology of Heat Tolerance in Rice

Flowering stage is the most susceptible to high temperatures, followed by booting stage. High temperature is more injurious if it occurs just before or during anthesis (Satake and Yoshida 1978). Exposure to 41 °C for 4 h at flowering caused irreversible damage and plants became completely sterile, whereas this high temperature (41 °C) had no effect on spikelet fertility at 1 day before or after flowering (Yoshida et al. 1981). The same study also found that pollination of heat-stressed stigmas with unstressed pollen as well as self-pollination at 1 h before heat stress application did not affect spikelet fertility. These analyses indicated that the heat-sensitive stage is about 1 h before and after flowering. High temperature affects anther dehiscence, pollination, and pollen germination, which then leads to spikelet sterility and yield loss (Yoshida et al. 1981). Exposure at anthesis even for just 1–2 h of high temperature may result in high spikelet sterility (Jagadish et al. 2007).

High temperature at booting stage mainly decreases the fertility of pollen grains, while at flowering stage, it mainly decreases the number of pollens shedding on the stigma and the germination of pollen grains. The decreased production of pollens at elevated temperatures may be attributable to impaired cell division of the microspore mother cells (Takeoka et al. 1992). The major causes of high temperature-induced sterility were decreased pollen shedding and decreased viability of pollen grains, resulting in a lower number of germinated pollen grains on the stigma (Mackill et al. 1982; Satake and Yoshida 1978).

Among physiological processes occurring at anthesis, anther dehiscence is perceived to be the most critical stage affected by high temperature (Matsui et al. 1997a, b, 2000, 2001a). Spikelet opening triggers rapid pollen swelling, leading to anther dehiscence and pollen shedding from the anthers' apical and basal pores (Matsui et al. 2000). Increased basal pore length in a dehisced anther was found to contribute significantly to successful pollination (Matsui and Omasa 2002). The anthers of heat-tolerant cultivars dehisce more easily than those of susceptible cultivars under high-temperature conditions (Mackill et al. 1982; Matsui et al. 1997a, b, 2001b; Satake and Yoshida 1978). This is because of the tight closure of the locule by the cell layers, which delays locule opening and decreases spikelet fertility at high temperature (Matsui and Omasa 2002).

In heat-tolerant cultivar N22, the dehiscence of anthers begins soon after the glumes open and is completed when the anthers are still situated inside the glumes on short filaments; thus, pollen grains of N22 could be easily shed onto stigma at that time (Satake and Yoshida 1978). The heat-tolerant cultivar Nipponbare had well-developed cavities in anthers and thick locule walls, which enabled easy rupture of the septa in response to swelling of pollen (Matsui et al. 2001b).

Rice plants, when exposed to high temperatures during critical stages, can avoid heat by maintaining their microclimate temperature below critical levels by efficient transpiration cooling (Wassmann et al. 2009a). Lower relative humidity of 60% at 38 °C leads to a higher vapor pressure deficit, facilitating the plant in exploiting its transpiration cooling ability (Jagadish et al. 2007). On the basis of the interaction between high temperature and relative humidity, rice cultivation regions in the tropics and subtropics can be classified into hot/dry and hot/humid regions. It can be assumed that rice cultivation in hot/dry regions where temperatures may exceed 40 °C (e.g., Pakistan, Iran, and India) has been facilitated through unintentional selection for efficient transpiration cooling under sufficient supply of water. An exceptionally high temperature difference of 6.8 °C between crop canopy and ambient air temperature (34.5 °C) was recorded in the Riverina region of New South Wales, Australia (Matsui et al. 2007). Rice pollen is extremely sensitive to temperature and relative humidity and loses its viability within 10 min of shedding (Matsui et al. 1997a). Tolerant cultivar Shanyou63 showed a significantly slower decrease in pollen activity, pollen germination, and rate of floret fertility vis-à-vis susceptible cultivar Teyou559 at 39 °C (Tang et al. 2008).

The temperature inside the spikelet decreases with a reduction in relative humidity, possibly because of the enhancement of transpiration at low relative humidity (Weerakoon et al. 2008). This decrease in temperature inside the spikelet increases the viability of pollen grains. Viable pollen grains absorb moisture and swell at moderate to high relative humidity and create the required pressure for the rupture of the septum, which helps in the deposition of pollen on stigma and thus produces a fertilized spikelet (Weerakoon et al. 2008). The panicle temperature of Chinese hybrid rice exceeded the ambient air temperature by 4 °C under humid and low wind conditions and also caused a severe decrease in spikelet fertility (Tian et al. 2010).

Cultivar NL-44 has high heat tolerance at both vegetative and reproductive stages. NL-44 under extreme heat stress retained the ability to maintain higher chlorophyll (relative greenness) and photosynthesis, a feature that could sustain its survival under severe heat stress during both the vegetative and reproductive stages. NL-44 and the heat-tolerant check N22 consistently displayed lower membrane damage and higher antioxidant enzyme activity across leaves and spikelets (Bahuguna et al. 2015).

6   Genetics of Heat Tolerance in Rice

Heat tolerance is controlled by not only one major gene but also several genes. The identification of quantitative trait loci (QTLs) is a promising approach to dissect the genetic basis of heat tolerance. By using genetic resources with heat tolerance at flowering, QTL mapping studies for heat tolerance (spikelet fertility) have been conducted on various rice populations at booting (Zhao et al. 2006) and flowering stages (Cao et al. 2003; Chen et al. 2008; Cheng et al. 2012; Jagadish et al. 2010a; Xiao et al. 2011b; Ye et al. 2015b, 2012; Zhang et al. 2009, 2008). About 60 QTLs associated with heat tolerance at flowering stage have been identified so far. For example, two major QTLs were identified on chromosome 1 (*qHTSF1.1*) and chromosome 4 (*qHTSF4.1*) in an IR64/N22 population. These two major QTLs could explain 12.6% (*qHTSF1.1*) and 17.6% (*qHTSF4.1*) of the variation in spikelet fertility at high temperature (Ye et al. 2012). Four QTLs were identified in an IR64/Giza178 population, two other QTLs were identified in a Milyang23/Giza178 population, and five QTLs were identified in the three-way cross population IR64//Milyang23/Giza178. Three of these QTLs were identified in both biparental and three-way populations (Ye et al. 2015b). Recently, two QTLs with high genetic effect (*qSTIPSS9.1* and *qSTIY5.1/qSSIY5.2*) were mapped in less than 400 kbp genomic regions (Shanmugavadivel et al. 2017). QTLs for other heat tolerance-related traits such as anther length, apical dehiscence length, basal dehiscence length, and percentage of longitudinal dehiscence (Tanveer et al. 2015) and pollen fertility (Xiao et al. 2011a) were also detected.

Among the several identified QTLs, the most promising QTLs for heat tolerance across different genetic backgrounds and locations have been identified on chromosomes 1 and 4 (Jagadish et al. 2010a; Xiao et al. 2011b; Ye et al. 2015b, 2012). The heat-tolerant QTL (*qHTSF4.1*) on chromosome 4 was identified in different populations of heat-tolerant rice varieties 996, N22, Milyang23, and Giza178 (Raddatz et al. 2001; Xiao et al. 2011b; Ye et al. 2012). The QTL interval was fine-mapped to 1.2 Mb. The heat tolerance (spikelet fertility) of the near-isogenic line (NIL) carrying *qHTSF4.1* increased consistently in all of the backcross populations. In BC_3F_3, the spikelet fertility of plants with *qHTSF4.1* (34.7 ± 14.2%) was significantly higher than in those without the QTL (22.5 ± 7.9%) and in the recurrent parent IR64 (15.1 ± 6.3%), whereas, in BC_5F_2, the spikelet fertility of plants with *qHTSF4.1* (44.6 ± 13.1%) was significantly higher than in plants without the QTL (27.1 ± 9.6%) and in the recurrent parent IR64 (19.4 ± 8.4%) (Ye et al. 2015a).

Recently, a thermotolerance gene (*TT1*) in African rice (*O. glaberrima*) variety CG14 was identified and cloned (Li et al. 2015). Gene *TT1* encodes an α2 subunit of the 26S proteasome involved in the degradation of ubiquitinated proteins. Ubiquitylome analysis indicated that *OgTT1* protects cells from heat stress through more efficient elimination of cytotoxic denatured proteins and more effective maintenance of heat-response processes than achieved with *OsTT1*. Overexpression of *OgTT1* was associated with markedly enhanced thermotolerance in rice at seedling, flowering, and grain-filling stages (Li et al. 2015). A gene for heat tolerance at seedling stage (*OsHTAS*) was also cloned and characterized. *OsHTAS* encodes a ubiquitin ligase localized in the nucleus and cytoplasm. *OsHTAS* was responsive to multiple stresses and was strongly induced by exogenous ABA. *OsHTAS* modulated hydrogen peroxide accumulation in shoots, altered the stomatal aperture status of rice leaves, and promoted ABA biosynthesis. The RING finger ubiquitin E3 ligase *OsHTAS* functions in leaf blades to enhance heat tolerance through modulation of hydrogen peroxide-induced stomatal closure and is involved in both ABA-dependent and drought- and salt-tolerance-mediated pathways (Liu et al. 2016).

Reverse genetic approaches were also employed to identify the genes for heat tolerance at anthesis. Expression analyses revealed that at least 13 genes were designated as high temperature-repressed genes in the anther. These genes were expressed specifically in the immature anther, mainly in the tapetum at the microspore stage, and downregulated after 1 day of high temperature. High temperatures may disrupt some of the tapetum functions required for pollen adhesion and germination on the stigma (Endo et al. 2009).

A proteomic analysis compared proteins expressed in heat-stressed anthers from three rice varieties with different temperature tolerances. The temperature-tolerant rice genotype (N22) showed a higher accumulation of small heat shock proteins (sHSP) than the temperature-sensitive rice genotype (Moroberekan). The moderately tolerant rice genotype (IR64) showed intermediate sHSP accumulation. The accumulation of sHSP may confer greater heat tolerance in N22 rice (Jagadish et al. 2010b).

7 Breeding of Heat Tolerance in Rice

Breeding heat-tolerant rice is one of the most important strategies used to mitigate the effects of climate change, particularly in the hot Asian countries where most rice is grown. However, breeding rice varieties tolerant of high temperature has so far received little attention as compared to other abiotic stresses such as drought and salinity. After one comprehensive study in the early 1980s (Mackill 1981; Mackill and Coffman 1983; Mackill et al. 1982), high-temperature tolerance of rice has been treated only within region-specific breeding programs, with limited success. Using identified genetic resources and QTLs to improve heat tolerance in rice varieties has not been achieved.

QTLs for rice heat tolerance at flowering have been mapped on all chromosomes by using various rice populations (Cao et al. 2003; Chen et al. 2008; Cheng et al. 2012; Jagadish et al. 2010a; Xiao et al. 2011b; Zhang et al. 2009, 2008). However, the additive effect of each QTL is low. Introducing one or a few QTLs into a variety may not sufficiently increase its heat tolerance. Therefore, it is necessary to validate and characterize more QTLs and design SNP chips with QTL-linked markers to accelerate selection and incorporation of multiple QTLs to improve the efficiency of heat-tolerance breeding.

To mitigate heat-induced spikelet sterility, two strategies have been proposed. One is to develop cultivars that shed larger numbers of pollen grains or produce pollen grains able to germinate at high temperatures. Another strategy is to breed cultivars that escape heat at flowering because of their early-morning flowering (EMF) trait (Satake and Yoshida 1978). The EMF trait could be beneficial for decreasing yield loss from rising temperatures. The use of germplasm with the EMF trait could help to diminish anticipated yield losses caused by spikelet sterility at anthesis as a result of expected global warming (Ishimaru et al. 2010).

Spikelets are highly susceptible to heat stress at flowering; however, they remain fertile when flowering occurs 1 h prior to heat stress, because fertilization is completed within 1 h after flowering (Satake and Yoshida 1978). Shifting the time of anthesis to early hours of the cooler morning will help plants to escape high-temperature stress during processes of pollen shed, pollination, and fertilization and can thus minimize sterility caused by high temperatures. It has been suggested that there is a potential for genetic improvement to advance flowering to an earlier time of day in current high-yielding cultivars (Nishiyama and Blanco 1980). The EMF strategy has been used to produce introgression lines with the EMF trait transferred from wild rice *O. officinalis*. EMF NILs carrying *qEMF3* had earlier flower opening time by 1.5–2.0 h than recurrent parents, which decreased heat-induced sterility at flowering at elevated temperature. It was demonstrated that the shift in flower opening time to early morning is effective for escaping from heat stress at flower opening (Hirabayashi et al. 2015; Ishimaru et al. 2010). Pyramiding lines with QTLs for heat tolerance (*qHTSF4.1*) and EMF (*qEMF3*) effectively improved heat tolerance at flowering in both controlled and field conditions.

The development of new heat-tolerant rice varieties is among the best approaches to address changing climatic conditions in affected farming communities. Breeding heat-tolerant rice began in 2010 in the Philippines to develop new rice genotypes that could adapt to changing climatic and local farming conditions. By combining a heat-tolerant donor parent, such as N22, with high-yielding and better cultivars, selecting new genotypes with better adaptation to emerging climatic conditions is possible (Manigbas et al. 2014).

To increase the heat tolerance of a rice variety named Improved White Ponni (IWP), heat-tolerance QTLs *qHTSF1.1* and *qHTSF4.1* (Ye et al. 2012) were introgressed from Nagina 22 into IWP through marker-assisted breeding. The progenies harboring both *qHTSF1.1* and *qHTSF4.1* showed higher fertility percentages under high-temperature stress at the flowering stage. The results confirmed that these QTLs were responsible for maintaining membrane integrity and yield under elevated-temperature conditions (Vivitha et al. 2017).

Moreover, recent studies showed that heat tolerance at flowering stage in rice is controlled by recessive genes (Fu et al. 2015; Ye et al. 2015b, 2012). Thus, both parents should possess high-temperature tolerance to develop heat-tolerant F_1 combinations. Male parents play a more important role in heat-tolerant combinations than female parents. The heat susceptibility of hybrid rice in China is mainly due to the wide application of heat-susceptible restorer lines with high yield in three-line hybrid rice breeding (Fu et al. 2015). Therefore, it is important to improve the heat tolerance of both parents of hybrid rice combinations.

8 Future Prospects

Booting and flowering are the stages most sensitive to high temperature, which may sometimes lead to significant sterility. Great variation exists among rice germplasm in response to temperature stress. Flowering at cooler times of day, more pollen viability, larger anthers, longer basal dehiscence, and the presence of long basal pores are some of the phenotypic markers for high-temperature tolerance. Replacement of heat-sensitive cultivars with heat-tolerant ones, adjustment of sowing time, choice of varieties with a growth duration allowing avoidance of peak stress periods, and exogenous application of plant hormones are some of the adaptive measures that will help to mitigate the forecast yield decrease due to global warming (Shah et al. 2011). Staggered planting dates and short-duration varieties are advocated as some of the options to escape from high-temperature stress. Synchronizing critical growth stages with most favorable weather is another practice for avoiding extreme temperature. However, cultural practices alone are not adequate and yield loss can be minimized further by combining such methods with genetic improvement.

There is a continuous need to integrate disciplines, such as structural genomics, transcriptomics, proteomics, and metabolomics, with plant physiology and plant breeding (Varshney et al. 2005). By using the wide diversity of rice germplasm, we will be able to explore the novel QTLs and alleles that are expected to have different effects from the identified QTLs. However, conventional breeding still offers an opportunity for significant and predictable incremental improvements in high-temperature tolerance of new rice cultivars. Among the QTLs identified for rice heat tolerance at flowering stage, even QTLs with a large effect can explain only approximately 20% of the variation, and the additive effect of each QTL is low. Introducing one or a few QTLs into a genetic background may not be sufficient to significantly increase its heat tolerance. More heat-tolerance donors and QTLs need to be identified and used in our breeding programs.

Heat-induced spikelet sterility at flowering and early-morning flowering is difficult traits for precise phenotyping in large mapping populations. Future research activities should be aimed at identifying and breeding heat-tolerant germplasm accessions that exploit the variation in both genotypic and morphological characters. Several approaches should be actively exploited to improve heat tolerance in current cultivars, including discovery and exploitation of new genes and alleles, improved breeding efficiency, marker-assisted selection, and genetic modification (Shah et al. 2011). Marker-assisted gene pyramiding and marker-assisted recurrent selection can be used to improve breeding efficiency for heat tolerance. The cloning of causal genes will unveil the complex genetic control of each trait under heat stress. Further genetic efforts are required for the development of heat-resilient rice varieties to cope with the challenges of climate change.

References

Abeysiriwardena DS, Ohba K, Maruyama A (2002) Influence of temperature and relative humidity on grain sterility in rice. J Natl Sci Found Sri Lanka 30:33–41

Bahuguna RN, Jha J, Madan P, Shah D, Lawas ML, Khetarpal S, Jagadish S (2015) Physiological and biochemical characterization of NERICA-L 44: a novel source of heat tolerance at the vegetative and reproductive stages in rice. Physiol Plant 154:543–559

Barnabas B, Jager K, Feher A (2008) The effect of drought and heat stress on reproductive processes in cereals. Plant Cell Environ 31:11–38

Basak J, Ali M, Islam MN, Rashid M (2010) Assessment of the effect of climate change on boro rice production in Bangladesh using DSSAT model. J Civ Eng 38:95–108

Battisti DS, Naylor R (2009) Historical warning of future food insecurity with unprecedented seasonal heat. Science 323:240–244

Cao L, Zhao J, Zhan X, Li D, He L, Cheng S (2003) Mapping QTLs for heat tolerance and correlation between heat tolerance and photosynthetic rate in rice. Chin J Rice Sci 17:223–227

Cao YY, Duan H, Yang LN, Wang ZQ, Zhou SC, Yang J (2008) Effect of heat stress during meiosis on grain yield of rice cultivars differing in heat tolerance and its physiological mechanism. Acta Agron Sin 34:2134–2142

Cheabu S, Moung-Ngam P, Arikit S, Vanavichit A, Malumpong C (2018) Effects of heat stress at vegetative and reproductive stages on spikelet fertility. Rice Sci 25:218–226

Chen Q, Yu S, Li C, Mou T (2008) Identification of QTLs for heat tolerance at flowering stage in rice. Sci Agric Sin 41:315–321

Cheng L, Wang J, Uzokwe V, Meng L, Wang Y, Sun Y, Zhu L, Xu J, Li Z (2012) Genetic analysis of cold tolerance at seedling stage and heat tolerance at anthesis in rice. J Integr Agric 11:359–367

Cline W (2008) Global warming and agriculture. International Monetary Fund, Washington, DC. http://www.imf.org/external/pubs/ft/fandd/2008/03/pdf/cline.pdf. Accessed 5 Aug 2018

Counce PA, Bryant RJ, Bergman CJ, Bautista RC, Wang YJ, Siebenmorgen TJ, Modenhauer KAK, Meullenet J (2005) Rice milling quality, grain dimensions, and starch branching as affected by high night temperatures. Cereal Chem 82:645–648

Endo M, Tsuchiya T, Hamada K, Kawamura S, Yano K, Ohshima M, Higashitani A, Watanabe M, Kawagishi-Kobayashi M (2009) High temperatures cause male sterility in rice plants with transcriptional alterations during pollen development. Plant Cell Physiol 50:1911–1922

Fu G, Zhang C, Yang Y, Xiong J, Yang X, Zhang X, Jin Q, Tao L (2015) Male parent plays a more important role in heat tolerance in three-line hybrid rice. Rice Sci 22:116–122

Gilani AA, Siadat SA, Alami Saeed K, Bakhshandeh AM, Moradi F, Seidnejad M (2009) Effect of heat stress on grain yield stability, chlorophyll content and cell membrane stability of flag leaf in commercial rice cultivars in Khuzestan. Iranian J Crop Sci 11(1):82–100

Hasegawa T, Yoshimoto M, Kuwagata T, Ishigooka Y, Kondo M, Ishimaru T (2009) The impact of global warming on rice production: lessons from spikelet sterility observed under the record hot summer of 2007. NIAES annual report 2008. NIAES, Tsukuba, pp 23–25

Hasegawa T, Ishimaru T, Kondo M, Kuwagata T, Yoshimoto M, Fukuoka M (2011) Spikelet sterility of rice observed in the record hot summer of 2007 and the factors associated with its variation. J Agric Meteorol 67:225–232

Hirabayashi H, Sasaki K, Kambe T, Gannaban RB, Miras MA, Mendioro MS, Simon EV, Lumanglas PD, Fujita D, Takemoto-Kuno Y, Takeuchi Y, Kaji R, Kondo M, Kobayashi N, Ogawa T, Ando I, Jagadish KS, Ishimaru T (2015) *qEMF3*, a novel QTL for the early-morning flowering trait from wild rice, *Oryza officinalis*, to mitigate heat stress damage at flowering in rice, *O. sativa*. J Exp Bot 66:1227–1236

Hu SB, Zhang YP, Zhu DF, Lin XQ, Xiang J (2012) Evaluation of heat resistance in hybrid rice. Chin J Rice Sci 26:751–756

IPCC (2013) Climate change 2013: the physical science basis. Cambridge University Press, Cambridge; New York, NY

IPCC (2014) Climate change 2014: impacts, adaptation and vulnerability: Part A: Global and sectoral aspects: Working Group II contribution to the IPCC Fifth Assessment report. Cambridge University Press, Cambridge

Ishimaru T, Horigan AK, Ida M, Iwasawa N, San-Oh YA, Nakazono M, Nishizawa NK, Masumura T, Kondo M, Yoshida M (2009) Formation of grain chalkiness and changes in water distribution in developing rice caryopses grown under high-temperature stress. J Cereal Sci 50:166–174

Ishimaru T, Hirabayashi H, Ida M, Takai T, San-Oh YA, Yoshinaga S, Ando I, Ogawa T, Kondo M (2010) A genetic resource for early-morning flowering trait of wild rice *Oryza officinalis* to mitigate high temperature-induced spikelet sterility at anthesis. Ann Bot 106:515–520

Ishimaru T, Seefong X, Nallathambi J, Rajendran S, Yoshimoto M, Phoudalay L, Benjamin S, Hasegawa T, Hayashi K, Gurusamy A, Muthurajan R, Jagadish S (2016) Quantifying rice spikelet sterility in potential heat-vulnerable regions: field surveys in southern Laos and southern India. Field Crop Res 190:3–9

Jagadish SVK, Craufurd PQ, Wheeler T (2007) High temperature stress and spikelet fertility in rice. J Exp Bot 58:1627–1635

Jagadish SVK, Craufurd PQ, Wheeler T (2008) Phenotyping rice mapping population parents for heat tolerance during anthesis. Crop Sci 48:1140–1146

Jagadish SVK, Cairns J, Lafitte R, Wheeler TR, Price AH, Craufurd PQ (2010a) Genetic analysis of heat tolerance at anthesis in rice. Crop Sci 50:1633–1641

Jagadish SVK, Muthurajan R, Oane R, Wheeler TR, Heuer S, Bennett J, Craufurd PQ (2010b) Physiological and proteomic approaches to address heat tolerance during anthesis in rice. J Exp Bot 61:143–156

Jagadish SVK, Murty MVR, Quick W (2014) Rice responses to raising temperatures: challenges, perspectives and future directions. Plant Cell Environ 38:1686–1698

Jansen E, Overpeck J, Briffa KR, Duplessy JC, Joos F, Masson-Delmotte V, Olago D, Otto-Bliesner B, Peltier WR, Rahmstorf S, Ramesh R, Raynaud D, Rind D, Solomina O, Villalba R, Zhang D (2007) Palaeoclimate. In: Solomon S, Qin D, Manning M, Chen Z, Marquis M, Averyt KB, Tignor M, Miller HL (eds) Climate change 2007: the physical science basis. Contribution of Working Group I to the Fourth Assessment report of the intergovernmental panel on climate change. Cambridge University Press, Cambridge; New York, NY

Karim MR, Ishikawa M, Ikeda M, Islam M (2012) Climate change model predicts 33% rice yield decrease in 2100 in Bangladesh. Agron Sustain Dev 32:821–830

Laborte A, Nelson A, Jagadish K, Aunario J, Sparks A, Ye C, Redoña E (2012) Rice feels the heat. Rice Today 3:30–31

Lanning SB, Siebenmorgen TJ, Counce PA, Ambardekar AA, Mauromoustakos A (2011) Extreme night time air temperatures in 2010 impact rice chalkiness and milling quality. Field Crop Res 124:132–136

Li C, Peng C, Zhao Q, Xie P, Chen W (2004) Characteristic analysis of the abnormal high temperature in 2003 midsummer in Wuhan City. J Centr China Normal Univ 38:379–381

Li X, Chao D, Wu Y, Huang X, Chen K, Cui L, Su L, Ye W, Chen H, Chen H, Dong N, Guo T, Shi M, Feng Q, Zhang P, Han B, Shan J, Gao J, Lin H (2015) Natural alleles of a proteasome $\alpha 2$ subunit gene contribute to thermotolerance and adaptation of African rice. Nat Genet 47:827–833

Liu J, Zhang C, Wei C, Liu X, Wang M, Yu F, Xie Q, Tu J (2016) The RING finger ubiquitin E3 ligase OsHTAS enhances heat tolerance by promoting H_2O_2-induced stomatal closure in rice. Plant Physiol 170:429–443

Lobell DB, Burke MB, Tebaldi C, Mastrandrea MD, Falcon WP, Naylor RL (2008) Prioritizing climate change adaptation needs for food security in 2030. Science 319:607–610

Mackill DJ (1981) Studies on the mechanism and genetics of high temperature tolerance in rice. University of California-Davis, Davis, CA

Mackill DJ, Coffman W (1983) Inheritance of high temperature tolerance and pollen shedding in a rice cross. Z Pflanzen 91:61–69

Mackill DJ, Coffman WR, Rutger J (1982) Pollen shedding and combining ability for high temperature tolerance in rice. Crop Sci 22:730–733

Manigbas NL, Lambio LAF, Madrid LB, Cardenas C (2014) Germplasm innovation of heat tolerance in rice for irrigated lowland conditions in the Philippines. Rice Sci 21:162–169

Maruyama A, Weerakoon W, Wakiyama Y, Ohba K (2013) Effects of increasing temperatures on spikelet fertility in different rice cultivars based on temperature gradient chamber experiments. J Agron Crop Sci 199:416–423

Matsui T (2009) Floret sterility induced by high temperatures at the flowering stage in rice. Jpn J Crop Sci 78:303–311

Matsui T, Omasa K (2002) Rice cultivars tolerant to high temperature at flowering: anther characteristics. Ann Bot 89:683–687

Matsui T, Omasa K, Horie T (1997a) High temperature induced spikelet sterility of japonica rice at flowering in relation to air humidity and wind velocity conditions. Jpn J Crop Sci 66:449–455

Matsui T, Namuco OS, Ziska LH, Horie T (1997b) Effects of high temperature and CO_2 concentration on spikelet sterility in indica rice. Field Crop Res 51:213–219

Matsui T, Omasa K, Horie T (2000) High temperature at flowering inhibits swelling of pollen grains, a driving force for thecae dehiscence in rice. Plant Prod Sci 3:430–434

Matsui T, Omasa K, Horie T (2001a) Comparison between anthers of two rice cultivars with tolerance to high temperature at flowering or susceptibility. Plant Prod Sci 4:36–40

Matsui T, Omasa K, Horie T (2001b) The differences in sterility due to high temperature during the flowering period among japonica rice varieties. Plant Prod Sci 4:90–93

Matsui T, Kobayasi K, Yoshimoto M, Hasegawa T (2007) Stability of rice pollination in the field under hot and dry conditions in the Riverina Region of New South Wales, Australia. Plant Prod Sci 10:57–63

Matsushima S, Ikewada H, Maeda A, Honda S, Niki H (1982) Studies on rice cultivation in the tropics. I. Yielding and ripening responses of the rice plant to the extremely hot and dry climate in Sudan. Jpn J Trop Agric 26:19–25

Matthews RB, Kropff MJ, Bachelet D, Laar HV (1995) Modelling the impact of climate change on rice production in Asia. Springer, International Rice Research Institute, Berlin, Los Baños

Mohammed RA, Tarpley L (2011) Effects of night temperature, spikelet position and salicylic acid on yield and yield-related parameters of rice plants. J Agron Crop Sci 197:40–49

Morita S, Yonemaru J, Takanashi J (2005) Grain growth and endosperm cell size under high night temperatures in rice. Ann Bot 95:695–701

Morita S, Wada H, Matsue Y (2016) Countermeasures for heat damage in rice grain quality under climate change. Plant Prod Sci 19:1–11

Nishiyama I, Blanco L (1980) Avoidance of high temperature sterility by flower opening in the early morning. Jpn Agric Res Q 14:116–117

Nishiyama I, Satake T (1981) High temperature damage in the rice plant. Jpn J Trop Agric 25:14–19

NOAA (2011) State of the climate: global analysis for annual 2011. NOAA, Washington, DC. http://www.ncdc.noaa.gov/sotc/global/2011/13. Accessed 6 Aug 2018

Osada A, Sasiprapa V, Rahong M, Dhammanuvong S, Chakrabandho H (1973) Abnormal occurrence of empty grains of indica rice plants in the dry hot season in Thailand. Proc Crop Sci Soc Jpn 42:103–109

Pachauri RK, Reisinger A (2007) Climate change 2007: synthesis report. IPCC, Geneva

Peng S, Huang J, Sheehy JE, Laza RC, Visperas RM, Zhong X, Centeno GS, Khush GS, Cassman KG (2004) Rice yields decline with higher night temperature from global warming. Proc Natl Acad Sci U S A 101:9971–9975

Poli Y, Basava RK, Panigrahy M, Vinukonda VP, Dokula NR, Voleti SR, Desiraju S, Neelamraju S (2013) Characterization of a Nagina22 rice mutant for heat tolerance and mapping of yield traits. Rice 6:36

Prasad PVV, Boote KJ, Allen LH, Sheehy JE, Thomas J (2006) Species, ecotype and cultivar differences in spikelet fertility and harvest index of rice in response to high temperature stress. Field Crop Res 95:398–411

Raddatz G, Dehio M, Meyer TF, Dehio C (2001) PrimeArray: genome-scale primer design for DNA-microarray construction. Bioinformatics 17:98–99

Rosenzweig C, Iglesias A, Yang XB, Epstein PR, Chivian E (2001) Climate change and extreme weather events. Glob Change Hum Health 2:90–104

Satake T, Yoshida S (1978) High temperature induced sterility in Indica rice at flowering. Jpn J Crop Sci 47:6–17

Scafaro AP, Haynes PA, Atwell B (2010) Physiological and molecular changes in *Oryza meridionalis* Ng., a heat-tolerant species of wild rice. J Exp Bot 61:191–202

Shah F, Huang J, Cui K, Nie L, Shah T, Chen C, Wang K (2011) Impact of high-temperature stress on rice plant and its traits related to tolerance. J Agric Sci 149:545–556

Shanmugavadivel PS, Mithra SVA, Prakash C, Ramkumar MK, Tiwari R, Mohapatra T, Singh N (2017) High resolution mapping of QTLs for heat tolerance in rice using a 5K SNP array. Rice 10:28

Shi W, Ishimaru T, Gannaban RB, Oane W, Jagadish S (2015) Popular rice cultivars show contrasting responses to heat stress at gametogenesis and anthesis. Crop Sci 55:589–596

Solomon S, Qin D, Manning M, Chen Z, Marquis M, Averyt K, Tigno M, Miller H (2007) IPCC AR4 WG1. Climate change 2007: the physical science basis. Cambridge University Press, Cambridge

Tabaldi C, Hayhoe K, Arblaster J, Meehl G (2006) Going to the extremes. Climate Change 79:185–211

Takeoka Y, Mamun A, Wada T, Kaufman P (1992) Primary features of the effect of environmental stress on rice spikelet morphogenesis. Dev Crop Sci 22:113–141

Tang RS, Zheng JC, Jin ZQ, Zhang DD, Huang YH, Chen L (2008) Possible correlation between high temperature-induced floret sterility and endogenous levels of IAA, GAs and ABA in rice. Plant Growth Regul 54:37–43

Tanveer T, Kobayashi Y, Koyama H, Matsui T (2015) QTL analyses for anther length and dehiscence at flowering as traits for the tolerance of extreme temperatures in rice. Euphytica 203:629–642

Tao L, Tan H, Wang X, Cao L, Song J, Chen S (2008) Effects of high-temperature stress on flowering and grain-setting characteristics of Guodao 6. Acta Agron Sin 34:609–614

Teixeira E, Fischer G, Velthuizen H, Walter C, Ewert F (2011) Global hot-spots of heat stress on agricultural crops due to climate change. Agric For Meteorol 170:206–215

Tenorio FA, Ye C, Redoña E, Sierra S, Laza M, Argayoso MA (2013) Screening rice genetic resources for heat tolerance. SABRAO J Breed Genet 45:341–351

Tian X, Luo H, Zhou H, Wu C (2009) Research on heat stress of rice in China: progress and prospect. Chin Agric Sci Bull 25:166–168

Tian X, Matsui T, Li S, Yoshimoto M, Kobayasi K, Hasegawa T (2010) Heat-induced floret sterility of hybrid rice (Oryza sativa L.) cultivars under humid and low wind conditions in the field of Jianghan basin, China. Plant Prod Sci 13:243–251

Varshney RK, Graner A, Sorrells M (2005) Genomics-assisted breeding for crop improvement. Trends Plant Sci 10:621–630

Vivitha P, Raveendran M, Vijayalakshmi D (2017) Introgression of QTLs controlling spikelet fertility maintains membrane integrity and grain yield in Improved White Ponni derived progenies exposed to heat stress. Rice Sci 24:32–40

Wahid A, Gelani S, Ashraf M, Foolad M (2007) Heat tolerance in plants: an overview. Environ Exp Bot 61:199–223

Wassmann R, Jagadish SVK, Heuer S, Ismail A, Redoña E, Serraj R, Singh RK, Howell G, Pathak H, Sumfleth K (2009a) Climate change affecting rice production: the physiological and agronomic basis for possible adaptation strategies. Adv Agron 101:59–122

Wassmann R, Jagadish SVK, Sumfleth K, Pathak H, Howell G, Ismail A, Serraj R, Redoña E, Singh RK, Heuer S (2009b) Regional vulnerability of climate change impacts on Asian rice production and scope for adaptation. Adv Agron 102:91–133

Weerakoon W, Maruyama A, Ohba K (2008) Impact of humidity on temperature-induced grain sterility in rice. J Agron Crop Sci 194:135–140

Wei H, Liu J, Wang Y, Huang N, Zhang X, Wang L, Zhang J, Tu J, Zhong X (2013) A dominant major locus in chromosome 9 of rice (Oryza sativa L.) confers tolerance to 48°C high temperature at seedling stage. J Hered 104:287–294

Xia M, Qi H (2004) Effects of high temperature on the seed setting percent of hybrid rice breed with four male sterile lines. Hubei Agric Sci 2:21–22

Xiao Y, Pan Y, Luo L, Deng H, Zhang G, Tang W, Chen L (2011a) Quantitative trait loci associated with pollen fertility under high temperature stress at flowering stage in rice. Rice Sci 18:204–209

Xiao Y, Pan Y, Luo L, Zhang G, Deng H, Dai L, Liu X, Tang W, Chen L, Wang G (2011b) Quantitative trait loci associated with seed set under high temperature stress at the flowering stage in rice. Euphytica 178:331–338

Yamakawa H, Hirose T, Kuroda M, Yamaguchi T (2007) Comprehensive expression profiling of rice grain filling-related genes under high temperature using DNA microarray. Plant Physiol 144:258–277

Yan C, Ding Y, Wang Q, Liu Z, Li G, Muhammad I, Wang S (2010) The impact of relative humidity, genotypes and fertilizer application rates on panicle, leaf temperature, fertility and seed setting of rice. J Agric Sci 148:329–339

Yang C, Heilman J (1993) Responses of rice to short term high temperature: growth development and yield. J Agric Res China 42:1–11

Yang H, Huang ZQ, Jiang ZY, Wang X (2004) High temperature damage and its protective technologies of early and middle season rice in Anhui province. J Anhui Agric Sci 32:3–4

Ye C, Argayoso M, Redoña E, Sierra S, Laza M, Dilla C, Mo Y, Thomson M, Chin J, Delaviña C, Diaz G, Hernandez J (2012) Mapping QTL for heat tolerance at flowering stage in rice using SNP markers. Plant Breed 131:33–41

Ye C, Tenorio F, Redoña E, Morales-Cortezano P, Cabrega G, Jagadish K, Gregorio G (2015a) Fine-mapping and validating *qHTSF4.1* to increase spikelet fertility under heat stress at flowering in rice. Theor Appl Genet 128:1507–1517

Ye C, Tenorio F, Argayoso M, Laza M, Koh H, Redoña E, Jagadish K, Gregorio G (2015b) Identifying and confirming quantitative trait loci associated with heat tolerance at flowering stage in different rice populations. BMC Genet 16:41

Yoshida S, Satake T, Mackill D (1981) High temperature stress in rice. IRRI Res Pap Ser 67:1–15

Zakaria S, Matsuda T, Tajima S, Nitta Y (2002) Effect of high temperature at ripening stage on the reserve accumulation in seed in some rice cultivars. Plant Prod Sci 5:160–168

Zhang T, Yang L, Jiang K, Huang M, Sun Q, Chen W, Zheng J (2008) QTL mapping for heat tolerance of the tassel period of rice. Mol Plant Breed 6:867–873

Zhang G, Chen L, Xiao G, Xiao Y, Chen X, Zhang S (2009) Bulked segregant analysis to detect QTL related to heat tolerance in rice using SSR markers. Agric Sci China 8:482–487

Zhao Z, Zhang L, Xiao Y, Zhang W, Zhai H, Wan J (2006) Identification of QTLs for heat tolerance at the booting stage in rice. Acta Agron Sin 32:640–644

Zhong L, Cheng F, Wen X, Sun X, Zhang G (2005) The deterioration of eating and cooking quality caused by high temperature during grain filling in early-season indica rice cultivars. J Agron Crop Sci 191:218–225

Zhou Y, Gong H, Li C, Hu C, Lin T, Sheng S (2009) Influence of thermal damage on seed-setting rate of 67 indica hybrid rice combinations. Acta Agric Jiangxi 21:23–26

Zhu C, Xiao Y, Wang C, Jiang L, Zhai H, Wan J (2005) Mapping QTLs for heat tolerance during grain filling in rice. Chin J Rice Sci 19:117–121

Ziska LH, Manalo PA, Ordonez R (1996) Intraspecific variation in the response of rice (*Oryza sativa* L.) to increased CO_2 and temperature: growth and yield response of 17 cultivars. J Exp Bot 47:1353–1359

Genetics and Breeding of Low-Temperature Stress Tolerance in Rice

Sofi Najeeb, Anumalla Mahender, Annamalai Anandan, Waseem Hussain, Zhikang Li, and Jauhar Ali

Abstract Low-temperature stress (LTS) is one of the major abiotic stresses that affect crop growth and ultimately decrease grain yield. The development of rice varieties with low-temperature stress tolerance has been a severe challenge for rice breeders for a long time. The lack of consistency of the quantitative trait loci (QTLs) governing LTS tolerance for any given growth stage over different genetic backgrounds of mapping populations under different low-temperature stress conditions remains a crucial barrier for adopting marker-assisted selection (MAS). In this review, we discuss the ideal location and phenotyping for agromorphological and physiological parameters as indicators for LTS tolerance and also the traits associated with QTLs that were identified from biparental mapping populations and diverse rice accessions. We highlight the progress made in the fields of genome editing, genetic transformation, transcriptomics, and metabolomics to elucidate the molecular mechanisms of cold tolerance in rice. The stage-specific QTLs and candidate genes for LTS tolerance brought out valuable information toward identifying and improving LTS tolerance in rice varieties. We showed 578 QTLs and 38 functionally characterized genes involved in LTS tolerance. Among these, 29 QTLs

Authors Sofi Najeeb, Anumalla Mahender, and Annamalai Anandan contributed equally to this work.

S. Najeeb
Rice Breeding Innovation Platform, International Rice Research Institute (IRRI),
Los Baños, Laguna, Philippines

Mountain Research Centre for Field Crops, Khudwani, Sher-e-Kashmir University of
Agricultural Science and Technology (SKAUST), Kashmir, India

A. Mahender · W. Hussain · J. Ali (✉)
Rice Breeding Innovation Platform, International Rice Research Institute (IRRI),
Los Baños, Laguna, Philippines
e-mail: J.Ali@irri.org

A. Anandan
Plant Breeding and Genetics Division, ICAR-National Rice Research Institute (NRRI),
Cuttack, Odisha, India

Z. Li
Institute of Crop Sciences/National Key Facility for Crop Gene Resources and Genetic
Improvement, Chinese Academy of Agricultural Sciences (CAAS), Beijing, P. R. China

© The Author(s) 2021
J. Ali, S. H. Wani (eds.), *Rice Improvement*,
https://doi.org/10.1007/978-3-030-66530-2_8

were found to be colocalized at different growth stages of rice. The combination of stage-specific QTLs and genes from biparental mapping populations and genome-wide association studies provide potential information for developing LTS-tolerant rice varieties. The identified colocalized stage-specific LTS-tolerance QTLs will be useful for MAS and QTL pyramiding and for accelerating mapping and cloning of the possible candidate genes, revealing the underlying LTS-tolerance mechanisms in rice.

Keywords Low-temperature stress · Physiological indicators · Stage-specific QTLs and genes · Breeding strategies · Genetic transformation

1 Introduction

Rice (*Oryza sativa* L.) is an important cereal crop, being the staple food for more than half of the world's population, providing 21% of global human per capita energy (Nalley et al. 2017). Approximately one tenth of Earth's arable land is planted to rice, which is the primary source of food. The demand for this staple crop has put more pressure on rice breeders and biotechnologists to intensify rice production systems to enhance yield productivity under drastic changes in global climatic variations (GCVs). Based on the projection of global population growth, rice production must increase its annual yield by 1.2–1.5% in the coming decades to ensure global food security (Seck et al. 2012).

Rice is grown globally in diverse ecosystems, ranging from a few meters below sea level to as high as 2700 m above mean sea level (amsl). Despite rice originating in the swampy areas of the tropics, it is susceptible to a wide range of abiotic stresses (Ranawake and Nakamura 2011). Changes in GCVs have shifted the distribution of temperature variability across the globe. These remarkable shifts have resulted in more frequent low-temperature stress/cold stress events (chilling stress and freezing stress) during the rice-growing season, especially in subtropical and temperate regions, with consequent adverse effects on rice production. Low-temperature stress (LTS) is one of the major abiotic stresses that significantly decrease rice grain yield and is experienced by 10% of the total 130 million ha of rice (Mohanty et al. 2012). For instance, rice farmers have suffered significant declines in grain yield ranging from 0.5 to 2.5 t/ha in Australia, with an average yield income loss of USD 23.2 million/year because of LTS (Farrell et al. 2001).

Low temperature affects the rice industry in Africa, Asia, Australia, Europe, and South and North America. In the mountainous regions of South Korea, extremely low temperatures severely damaged rice crops in 1980 and 1993, with grain yield dropping by 26.0% and 9.2%, respectively, compared with the national average yield during those years (Schiller et al. 2001). Also, severe grain yield losses due to LTS conditions were reported in Italy, the United States (Board et al. 1980), and Chile. In India, LTS occurs in about 60% of the rice area in the northeastern and western hill states of the Himalayas, with cold stress caused by the cold irrigation water from melted snow and low ambient air temperature. LTS also directly affects crop duration, which increases relatively with cold temperature, thereby limiting to a large extent the possibility of double cropping in areas where water control is possible (Matlon et al. 1998).

Rice cultivars vary prominently in their tolerance of LTS, with subspecies *indica* more sensitive to LTS, while *japonica* cultivars are known to tolerate cold stress (Kim and Tai 2013). The rice crop is relatively sensitive to temperatures below 15 °C, which causes varying effects across different crop growth stages such as germination, seedling, vegetative, reproductive, and grain maturity (Andaya and Mackill 2003a, b). Low temperatures directly affect the crop by causing slow growth and decreased seedling vigor (Ali et al. 2006) as well as a delayed and lower percentage of germination (da Cruz and Milach 2000). At the seedling stage, manifestations of cold stress include low numbers of seedlings, decreased tillering, increased plant mortality, and induced nonuniform crop maturity (Zhang et al. 2014b). At the vegetative stage, LTS increases the growth period as exhibited by leaf discoloration or yellowing, leaf rolling or wilting, slowed growth, poor germination and seedling establishment, and the presence of rotten and dead seedlings (Lone et al. 2018). During flowering, the most sensitive stage, low temperature brings anomalies at anthesis, resulting in the cessation of anther development, nonripening of pollen, nonemergence of anthers from spikelets, improper anther dehiscence, pollen grains remaining in anther loculi, poor pollen shedding, and failure of pollen to germinate after reaching the stigmas (Suh et al. 2010; Shakiba et al. 2017).

LTS in both temperate and high-altitude rice-growing areas in the tropics and subtropics causes damaging effects throughout the growth dynamics of the crop (Ranawake and Nakamura 2015). The effect of LTS on different plant growth stages (germination, seedling, and reproductive) is crucial. In addition, there is a need for identifying an ideal location for phenotypic screening under LTS conditions, especially for agronomic, physiological, and biochemical traits to help in the development of LTS-tolerant cultivars. The establishment of genetic and genomic resources for LTS tolerance is a vital step toward the development of LTS-tolerant varieties. Over the years, the genetic and physiological perspectives of cold tolerance have been extensively studied, giving way to the development of a diverse set of criteria for evaluating the cold-tolerance phenomenon in rice at different growth stages. The rapid development of molecular markers and next-generation sequencing technology tools such as bisulfite sequencing and whole-transcriptome shotgun sequencing have been accelerated in many crop plants (Fig. 1). Several genomic regions have been studied for LTS in rice using biparental mapping populations and association mapping procedures. In this review, we have tried to organize and discuss the stage-specific QTLs and candidate genes for LTS tolerance, which could be used in LTS-tolerance rice varietal improvement programs. We also provide here the phenotypic characterization of LTS-tolerance traits at different growth stages of rice and associated genomic regions from the literature along with the traits. We also cover genome editing, genetic transformation, transcriptomics, and metabolomics tools for elucidating the molecular mechanisms of LTS tolerance in rice.

2 Phenological, Physiological, and Biochemical Indicators of LTS Tolerance at Different Developmental Stages

Screening for LTS in rice can be done at various growth stages (Table 1). In controlled conditions, LTS screening can be achieved timely with precision; however, it restricts the population in both sample size and number of samples. Thus, to screen large-sized populations, LTS breeding programs have resorted to evaluating many populations using cold water under field conditions (Snell et al. 2008). Such cold-water screening under field conditions has been established in research stations in Japan (Nagano 1998) and Korea (Lee 2001). The air temperature thresholds at the reproductive stage for cold-sensitive and cold-tolerant varieties are 20 °C and 15 °C, respectively (Satake 1976). Hence, high-elevation areas with low air and water temperatures, especially in subtropical regions in Kunming, People's Republic of China (subsequently "China"), and in regions of Kashmir and Himachal Pradesh, India, are ideal spots for screening for cold tolerance (Jiang et al. 2012). Natural cold-screening hotspots that represent the target population of environments are vital for the systematic screening of germplasm and segregating breeding materials. The selection of such hotspots is crucial for the success of breeding and molecular genetic studies.

Rice is quite sensitive to LTS, mainly in tropical and subtropical regions at different growth stages. The critical temperature of the germination and reproductive stage at 15 and 17 °C has shown a significant impact on growth stage and yield decrease. However, the optimum temperature required for rice cultivation ranges from 25 to 35 °C (Yoshida 1981). The selection of LTS-tolerant rice varieties with a short duration is the key requirement for decreasing LTS damage. The effects of LTS in different growth stages, such as germination stage (GS), seedling stage (SS), and reproductive stage (RS), have significant impacts on agromorphological changes and yield component losses, especially in tropical zones. As compared to *indica* or *indica* × *japonica* backgrounds, *japonica* rice varieties have shown a wide range of LTS tolerance (da Cruz et al. 2013). The list of some LTS-tolerant rice varieties provided in Table 2 spans different countries, and most of these varieties are *japonica* type. However, some *indica* rice varieties also showed considerable LTS tolerance at the GS or SS (Biswas et al. 2017).

A few varieties have been proven to have a better performance for LTS in stage-specific growth conditions: for instance, Jinheung, Nipponbare, RNR 18805, and Italica Livorno for the GS (Miura et al. 2001; Fujino et al. 2004); M202, Lemont, and AAV002863 for the SS (Andaya and Mackill 2003b; Lou et al. 2007); and Norin PL8, Kirara397, RNR 17813, Akshaydhan, Taramati, WGL 44, Bhadrakali, JGL 3844, and WGL 44 for the RS (Saito et al. 2001; Kuroki et al. 2009). However, four rice varieties, B55, Banjiemang, Lijiangheig, and HSC55 from China and the United States, showed a consistent tolerance in three different growth stages (GS, SS, and RS) in rice (Basuchaudhuri 2014). For a further selection of LTS-tolerant rice varieties, several screening methods have been proposed, along with their pros and cons, for LTS-tolerant genotypes (Almeida et al. 2016). The selection of

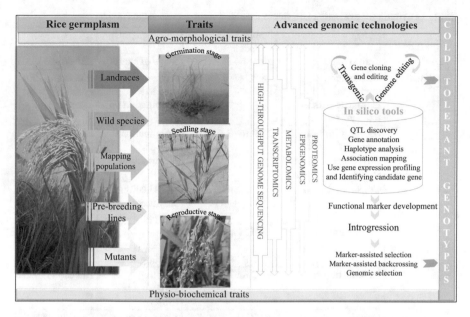

Fig. 1 Integration of high-throughput molecular approaches and phenotypic techniques to develop stage-specific desirable cold-tolerant rice genotypes

promising rice genotypes under natural LTS might favor negative results because of unpredictable climatic alterations in terms of stress intensity and duration of LTS. However, using high-throughput screening techniques such as image analysis, yield trait score, and robotics in controlled conditions of temperature, water, and air might help to detect tolerant genotypes and could also elucidate the traits related to morphological, biochemical, and yield-attributed traits during the plant growth period (Yang et al. 2014). Earlier studies of Snell et al. (2008), Suh et al. (2010), and Khatun et al. (2016) mentioned having developed reliable and straightforward screening methodologies for the selection of LTS-tolerant rice genotypes by preparing specific tanks for imposing cold-water irrigation and using a phytotron cabinet and low temperature in the glasshouse at different growth stages, which can provide the critical component traits. The primary focus traits for the GS related to germination rate, germination index, coefficient of germination, coleoptile length, and radicle length and also associated with early seedling vigor could be important traits for the selection of LTS tolerance at the GS (Li et al. 2018). In the SS, leaf discoloration, seedling survivability, leaf chlorophyll content, and estimation of the concentration of osmoprotectants (spermine and glycine betaine) and trehalose accumulation could be useful indicators to detect LTS at the SS (Han et al. 2004; Lou et al. 2007; Suh et al. 2012). Similarly, seed-setting rate, pollen growth development, incomplete panicle exsertion, days to flowering, spikelet fertility, and grain yield are the key traits for selection criteria at the RS (Ye et al. 2009; Jena et al. 2012; da Cruz et al. 2013). However, the natural incidence of LTS is significantly influenced to alter tolerance trait expression during phenotypic evaluation. Therefore, a

Table 1 Criteria for evaluating LTS tolerance at different growth stages in rice

Stage	Trait studied	Applied temperature/ duration	Stage of study	References
Germination	Germination vigor = number of germinated grains/ total grains	14 °C (7–17 days)	Incubation of seed to germination up to 17 days	Han et al. (2006)
	Seedling survival rate = (number of surviving seedlings/ sprouted seeds) × 100	2 °C for 3 days	Germination to very early seedling stage	Zhou et al. (2012)
	Coleoptile length	15 °C for 10 days	Germination to very early seedling stage	Hou et al. (2003)
	Germination rate	15 °C for 10 days	Germination to very early seedling stage	Chen et al. (2006b)
	Germination rate	5 °C for 10 days	Seedling survival rate (SSR)	Pan et al. (2015)
	Germination percentage	10 °C for 30 days	Seedling stage	Schläppi et al. (2017)
	Germination percentage	12 °C for 35 days	Dark, cold incubator set for 35 days	Shakiba et al. (2017)
Vegetative/ seedling	Changes in fresh weight after cold treatment	10 °C for 1–48 h	13 days after germination	Bonnecarrère et al. (2011)
	Number of surviving plants/total number	4 °C for 6 days	–	Zhang et al. (2011)
	Survival rate after 10 days of recovery	10 °C for 3, 6, and 9 days	–	Bertin et al. (1996)
	Survival rate after 14 days of recovery	4 °C for 6 days in dark	–	Koseki et al. (2010)
	Seedling growth (visual scale: 1–9)	9 °C for 8–18 days	3-leaf stage	Andaya and Mackill (2003a, b), Kim and Tai (2011)
	Seedling growth (visual scale: 1–9)	8 °C for 3 days	3-leaf stage	Wang et al. (2016)

(continued)

Table 1 (continued)

Stage	Trait studied	Applied temperature/ duration	Stage of study	References
Reproductive stage	Fertility/spikelet fertility percentage	12 °C for 6 days and then in greenhouse conditions up to maturity	Flowering/ booting stage	Sato et al. (2011)
	Fertility/spikelet fertility percentage	18–19 °C/cold deep irrigation water	2 months from panicle initiation to full heading stage	Shirasawa et al. (2012)
	Fertility/spikelet fertility percentage	17 °C water/air temperature for 10 days; irrigation water at 17 °C	20 DAT from tillering to grain maturity	Suh et al. (2010), Jena et al. (2012)
	Fertility/spikelet fertility percentage	17 °C for 7 days at anthesis stage	Flowering/ booting stage	da Cruz et al. (2006a)
	Fertility/spikelet fertility percentage	15.3–21.4 °C of air temperature at booting stage	Booting to milking stage	Zhu et al. (2015)
	Percent panicle sterility, number of panicles per plant, and seed weight per plant	Night-time temperature of 12 °C and daytime temperature of 27.3 °C	Dark, cold incubator set for 35 days	Shakiba et al. (2017)
	Relative seed-setting rate	15–19 °C	Booting stage	Pan et al. (2015)

combination of advanced molecular marker technology and high-throughput screening technologies provides the best method for prospecting for LTS-tolerant genotypes.

Several protocols exist to screen for cold tolerance/sensitivity in rice using different physiological and biochemical indicators. Two good indices of cold tolerance are seedling survival percentage (SSP) after subjecting seedlings to different low-temperature regimes (Morsy et al. 2007) and seedling chlorosis (Nagamine 1991). Many researchers have used SSP to analyze the resistance of transgenic plants to low temperatures (Chen et al. 2012; Huang et al. 2012). Nevertheless, the drawback of information obtained from SSP is that it is neither reproducible under natural conditions nor feasible for QTL studies. On the other hand, seedling chlorosis or the decrease in chlorophyll and leaf yellowing induced by cold stress could be captured by Soil Plant Analysis Development (SPAD) values to provide a more accurate measurement of cold stress at the seedling stage over a visual score. This indicator gives the direct association of the photosynthetic activity of the leaves, with low-temperature intensity and duration, as one of the yardsticks to screen rice germplasm and populations against cold stress (Hussain et al. 2018).

Table 2 List of popular LTS-tolerant rice varieties released in several countries

Rice varieties	Country	Remarks	References
K39, K78 (Barkat), K332, Kohsar, Jhelum, Shalimar Rice 1, Shalimar Rice 2, Shalimar Rice 3, a few varieties of VL Dhan series, Himalaya 1, Kanchan, Himali, and Bhrigu Dhan	India	Popularly grown in Kashmir valley and high hills of Himachal Pradesh, possessing a good degree of cold tolerance	Gupta et al. (2009)
Yunlu 29, B55, Lijianghegu, and Banjiemang	China	Considerable cold tolerance and remarkable recovery from cold damage	Sivapalan (2013)
Viet	Vietnam		
Jyoudeki and Tachiminori	Japan	Significant tolerance at booting/flowering stage, whereas HSC55 shows considerable tolerance at all stages	Ye et al. (2009)
M103 and M104	U.S.		
HSC55	Hungary		
Quest	Australia		
Ambar-INIA, Quila 242002, and Quila 241304	Chile	Show considerable cold tolerance at seedling stage	Donoso et al. (2015)
Doongara, Illabong, and Langi	Australia	Possess significant cold tolerance	Sivapalan (2013)
Jinbubyeo, Junganbyeo, and SR30084-F8-156	Korea	Show strong cold tolerance at booting stage	Wang et al. (2013b)
PR27137-CR153, Khazar, Hasani, and Gil2	Iran	Possess cold tolerance at germination stage	Pouramir Dashtmian et al. (2013)
L2825CA	Uruguay	Germination-stage cold-tolerant *japonica* line	Bonnecarrere et al. (2015)
Avangard and Mustaqillik	Uzbekistan	Possess flowering-stage cold tolerance	Suh et al. (2010), Jena et al. (2012)
Jinbu and Jungan	Korea		
Giza 177	Egypt		

The accumulation of more dry matter and the functionality of photosystem-II (PSII) provide quantitative information on plant performance under cold-stress conditions (Gururani et al. 2015). An increase in the efficiency of PSII photochemistry gives information on the structural and functional changes in the photosystem of different plant types or transgenics, especially when the seedlings are exposed to low temperatures (Bonnecarrère et al. 2011). A sudden drop in chlorophyll integrity parameter and chlorophyll fluorescence (Fv/Fm) indicates a gas exchange decrease caused by alterations in the photosynthetic system. Therefore, combined information on gas exchange analysis and chlorophyll fluorescence is necessary to study the photosynthetic process (Saad et al. 2012) under cold stress. The expression of the *AISAP* gene of *Aeluropus littoralis* in rice confers broad tolerance of several abiotic stresses through the maintenance of photosynthetic apparatus integrity (Saad et al. 2012), particularly for PSII. The *AISAP* gene has become the tool to precisely evaluate for cold tolerance as it is related to final photosynthetic activity (da Cruz et al. 2013).

Biochemical parameters such as electrolyte leakage (EL), proline (Pro), and ascorbic acid (AA) were reported to be higher in sensitive variety IR50 than in resistant cultivar M202 (Kim and Tai 2011). Lee et al. (1993) showed that the exogenous application of abscisic acid (ABA) biosynthetic inhibitors resulted in low accumulation and low survival of seedlings under cold stress. Breeding varieties that accumulate higher concentrations of osmoprotectants (spermine and glycine betaine) was seen to be a strategy to overcome stresses (Yang et al. 1996), which has been proven through the development of transgenic rice that accumulates higher glycine betaine and shows resistance to LTS (Sakamoto and Murata 2002). Furthermore, the significant induction in the expression of antioxidative enzymes such as catalase (CAT), superoxide dismutase (SOD), and ascorbate peroxidase (APEX) under cold stress (Kuk et al. 2003) explained the rate of cold tolerance (Morsy et al. 2007) by RNA interference (RNAi) (Song et al. 2011) and in transgenic rice encoding Cu/Zn superoxide dismutase (sodC1) (Lee et al. 2009a, b).

The high tolerance of rice of cold stress could also be attributed to trehalose accumulation (Ge et al. 2008). Song et al. (2011) also found that accumulated amounts of trehalose can be used as an index for low-temperature tolerance/sensitivity. Increased amounts of trehalose through the overexpression of the *OsNAC5* gene in transgenic rice plants were found to result in improved PSII function under abiotic stress conditions as it restricted damage due to photooxidation and exhibited soluble carbohydrates 20% higher than in nontransgenic plants (Garg et al. 2002). Similarly, at the reproductive stage, no sugar (sucrose and hexoses) accumulation has been found in anthers of low temperature-tolerant lines, resulting in no pollen grain sterility (Oliver et al. 2007). The overexpression of the gene *OsAPXa* (ascorbate peroxidase) in transgenic rice lines resulted in increased fertility under cold stress (Sato et al. 2011). It is also reported that unsaturated fatty acid content is related to plasma membrane stability at cold temperatures during the vegetative stage. Tolerant genotypes exhibited an increase in the amount of linolenic acid and a decrease in palmitic acid (da Cruz et al. 2010). Therefore, lipid peroxidation (Zhang et al. 2012a), along with EL (Huang et al. 2012), can be used to evaluate membrane lipid damage, which is an indirect assessment of cold tolerance.

Below the soil surface, roots play a crucial role under chilling stress, and root hydraulic conductivity (Lpr) is found to be profoundly affected when the plants are exposed to cold stress (Yamori et al. 2010). Murai-Hatano et al. (2008) found that Lpr decreased when susceptible rice genotypes were exposed to a temperature of 15 °C and the decrease was linked to transmembrane proteins, such as the aquaporins. These physiological and biochemical methods used for evaluating stress in rice genotypes and transgenic rice plants have played a significant role in understanding the crop's mechanism of response against cold. However, most of these procedures are destructive, time-consuming, and stage-specific and are also inadequate and inappropriate for breeding programs involving the evaluation of many lines with large sample sizes. Therefore, to better understand cold-tolerance mechanisms, it is indispensable to study the phenomenon at the molecular level.

To improve the tolerance of rice of LTS, it is imperative to understand it at the molecular and physiological levels. At changing temperatures, rice plants modify their biological pathways, and molecular alterations occur within a different growth stage (Xiao et al. 2018; Ding et al. 2019). At different growth stages of rice plants, the initial effects of LTS are a decline in plasma membrane fluidity and transportation mechanism and alterations in physiological and metabolic activities, leading to disturbance of signaling processes (Ding et al. 2019). The cascades of the signaling process were followed by adjusting their cellular metabolism by activating the plasma membrane transporters and altering the metabolic responses (Fig. 2). These changes occurred in the intracellular levels by increasing abscisic acid concentrations via changes in growth hormones such as auxin and gibberellins and cross talk between the ethylene and salicylic acid signaling mechanism (Ghosh et al. 2016; Moraes De Freitas et al. 2016). These mechanisms have occurred through an alteration of membrane fluidity and the rearrangement of the cytoskeleton by the influx of calcium, which can trigger a downstream response to LTS tolerance by C-repeat binding factor: CBF-dependent (C-repeat/drought-responsive element-binding factor-dependent) and CBF-independent transcriptional pathways (Chinnusamy et al. 2010; Ma et al. 2015). Different growth stage-specific LTS-tolerance genes can be classified into three major groups as transcription factors, protein kinase genes, and functional genes, which may be involved in signal transduction pathways. Mainly, the CBF transcription factor regulates cold-responsive gene (COR) expression by binding to the CRT/DRE element. The promoter sequence of the CBF region is activated by the bHLH transcriptional activator of the inducer of CBF expression (ICE), which can also induce the expression of CBF genes toward LTS tolerance (Ito et al. 2006; Su et al. 2010). In addition to CBF pathway-related transcription factors, two genes, *FRO1* (*FROSTBITE 1*) and *OsFAD2*, encode ferric reduction oxidase 1 and fatty acid desaturase 2, which are involved in LTS-tolerance mechanisms by maintaining membrane fluidity (Bevilacqua et al. 2015). The influx of calcium signals has also been associated with nitric oxide, reactive oxygen species, and mitogen-activated protein kinases, which can trigger the cascades of signaling pathways leading to LTS tolerance (Yuan et al. 2018). LTS tolerance at the germination stage is an important component trait for rapid seedling growth and uniform crop establishment, especially in the direct-seeding production system. The overexpression of the zeta class of glutathione S-transferases (*OsGSTZ1*) significantly improved germination rate and seedling growth under LTS (Takesawa et al. 2002). Similarly, Jin et al. (2018) identified a novel zinc finger transcription factor (*OsCTZFP8*) and it plays a key role in LTS tolerance at the reproductive stage in pollen fertility and seed setting along with yield per plant. Therefore, studying LTS-tolerance mechanisms at specific growth stages is crucial and may provide a better understating of key gene functions and their role in developing LTS-tolerant rice varieties in future breeding programs.

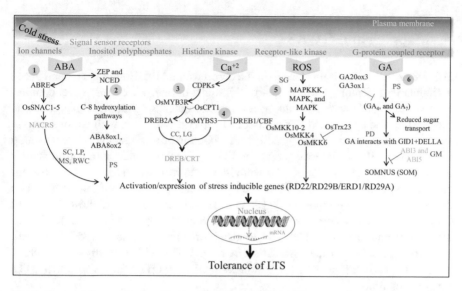

Fig. 2 Sequential steps involved in the triggering of the signaling cascades for low-temperature stress (LTS) tolerance. Schematic representations of the LTS signal mostly processed by various biological processes such as stress perceptions and physiological and molecular responses. (**1**) LTS signaling initiated by ABA accumulation, and this is transduced to ABRE-containing NAC genes, which regulate the expression of NACRS genes for tolerance of LTS (Hu et al. 2008). (**2**) The higher concentration of ABA-induced pollen sterility occurs by increasing the expression of ABA biosynthetic genes *OsZFP1* and *OSNCED3* that convert zeaxanthin to xanthoxin. The LTS-tolerant plants were followed by ABA catabolism with a higher expression of two ABA-8-hydroxylase genes and further reduced to ABA concentration in anthers via C-8 hydroxylation pathways (Ji et al. 2011; Sharma and Nayyar 2016). (**3**) Increasing the influx of Ca^{2+} signals mediated by the DREB-CRT/DRE pathway under LTS, which is transduced by calcium-dependent protein kinases (CDPKs), and MYB family transcription factors induce the stress-responsive genes. *OsMYB3R-2* regulates the LTS-tolerance mechanism at the seedling stage. *OsMYB3R-2* may regulate through OsCPT1, which is involved in the DREB/CBF pathway in rice (Su et al. 2010). (**4**) MYBS3 is a single DNA-binding repeat MYB transcription factor, which mediates sugar signaling and also tolerance of the LTS signaling pathway. Interestingly, MYBS3 has a distinct tolerance mechanism with short- and long-term adaptation of LTS tolerance by repressing the DREB1/CBF pathway and late and slow response to LTS tolerance. (**5**) The cascades of mitogen-activated protein kinase consist of three components (MAPKKK, MAPKK, and MAPK) activated by an excess of reactive oxygen species under LTS. Kumar et al. (2008) found that MAP kinases 4 and 6 are strongly regulated by LTS and salt stress at the seedling stage. Cytosolic thioredoxin (*OsTrx23*) has a potential negative regulator for MAPKs' activity. (**6**) LTS can also decrease the endogenous levels of bioactive gibberellic acid (GA) by the transcriptional repression of two bioactive GA synthesis genes (*GA20ox3* and *GA3ox1*) (Sharma and Nayyar 2016). GAs had cross-talk with other hormones to regulate the stress-response mechanism. The signal cascades of GA interact with the receptor GID1 (GA INSENSITIVE DWARF1) and GRAS family protein DELLA involved in pollen development. The two TFs (ABI3 and ABI5) bind with DELLA complex proteins, which can promote the expression of SOMNUS involved as a negative regulator of seed germination (Serrano-Mislata et al. 2017; Li et al. 2018). The cross talk with auxin and jasmonic acid-biosynthetic genes plays a major role in favoring germination under LTS. The LTS signaling and regulations of the expression of TFs and gene responses are indicated by arrows. Each pathway relates to different traits under LTS. These traits are *SC* stomata closure, *LP* lipid peroxidation, *MS* membrane stability, *RWC* relative water content, *PS* pollen sterility, *CC* cell cycle, *LG* leaf growth, *SG* seedling growth, *GM* germination, *PD* pollen development

3   Genes/QTLs Underlying LTS in Rice Detected by Linkage Mapping and GWAS

For tolerance of LTS, information on the chromosomal location of QTLs and genes is limited in different growth stage-specific traits in rice. We carried out a comprehensive literature survey, including a Gramene database (http://archive.gramene. org) search, and aggregated 578 cold-specific QTLs associated with various growth stages, including germination, seedling, and booting or reproductive stages. Among these QTLs, 239 (41.3%) were mapped through genome-wide association studies (GWAS), while 339 (58.7%) QTLs were identified from different types of biparental mapping populations, and detailed information is provided in Table 3. Based on the distribution of the reported QTLs on the chromosomes, the highest number of QTLs was noticed on chromosome 1 (65), followed by chromosome 7 (60), whereas the lowest number of QTLs was noticed on chromosome 8 (35) (Fig. 3a). Furthermore, based on the association of these QTLs with growth, stage-specific traits were classified into 214 QTLs related to GS, 249 QTLs for SS, and 115 QTLs for RS (Fig. 3b).

The physical positions of these stage-specific QTLs are depicted in Fig. 4. The QTLs were classified as main-effect QTLs (M-QTLs), based on the phenotypic variance explained by each QTL, which was ≥30%. Notably, five M-QTLs for GS-related traits (including germination rate, germination percentage, and germination index) were found on chromosomes 2, 5, and 7 (Xu et al. 2008; Cui et al. 2018); 15 M-QTLs for SS related to shoot and root growth traits on chromosomes 4, 6, 7, 8, 11, and 12 (Andaya and Mackill 2003b; Wang et al. 2009; Ranawake et al. 2014; Yu et al. 2018); and 13 M-QTLs for RS related to heading time, panicle weight, spikelet fertility, and culm length on chromosomes 1, 2, 4, 7, and 10 (Dai et al. 2004; Kuroki et al. 2009; Wainaina et al. 2018) were identified. A total of 29 QTLs were colocalized on all 12 chromosomes except on chromosomes 5 and 12 (Fig. 4 and Table 4). However, the colocalized stage-specific QTLs range from two to seven. Interestingly, the highest numbers of GS- and SS-specific QTLs were colocalized in the 22.5 Mb genomic region on chromosome 7. Three combinations of stage-specific QTLs (GS, SS, and RS) were identified on chromosome 10 in the 19.1 Mb region (Liu et al. 2013; Pan et al. 2015; Schläppi et al. 2017).

3.1   Germination Stage

Seed germination is of paramount interest for breeding varieties suitable for temperate regions and high-elevation areas, but it is given a lower priority than traits such as high yield and grain quality. To date, more than 200 QTLs (98 QTLs were identified from biparental mapping populations and 116 QTLs from GWAS) have been mapped on the 12 chromosomes (Fig. 3). The phenotypic variance of these QTLs ranges from 3.58% to 42.29%. The M-QTLs (≥30% PVE) for the GS were

Table 3 LTS-tolerance QTLs at different growth stages in rice

QTL mapping studies	Number of QTLs	PVE ranges	Chromosomes	Stage	Number of lines (accessions/parents)	References
RILs	9	10.5–16.8	1, 2, 3, 5, 6, 7, 9, and 12	RS	191 lines (M-202/IR50)	Andaya and Mackill (2003a)
RILs	15	8.7–41.7	1, 3, 4, 6, 8, 10, 11, and 12	SS	191 lines (M-202/IR50)	Andaya and Mackill (2003b)
$F_{2:3}$	6	7.2–14.9	6, 8, 11, and 12	SS	151 lines (BR1/Hbj. BVI)	Biswas et al. (2017)
ILs	3	6.5–9.5	1, 7, and 11	SS	240 lines (Xiushui 09/IR2061)	Cheng et al. (2012)
F_2	9	5.0–37.8	1, 3, 4, 6, 7, 10, and 12	RS	250 lines (Kunmingxiaobaigu/Towada)	Dai et al. (2004)
RILs	12	9.1–37.1	4, 6, and 9	SS	227 lines (Daguan dao/IR28)	Wang et al. (2009)
BC_1F_5	2	5.5–19.3	3 and 4	GS	122 lines (Livorno/Hayamasari)	Fujino et al. (2004)
$F_{2:3}$	12	5.6–42.9	1, 2, 3, 5, 7, 9, 11, and 12	SS	200 lines (Milyang 23/Jileng 1)	Han et al. (2004)
DHs	5	11.8–21.5	1, 2, 4, 10, and 11	SS	120 lines (TN1/Chunjiang 06)	Ji et al. (2010b)
RILs	9	4.8–33.5	2, 5, 7, 8, 11, and 12	GS	81 lines (Kinmaze/DV85)	Jiang et al. (2008)
RILs	3	9.1–24.1	1, 5, and 6	SS	81 lines (Kinmaze/DV85)	Jiang et al. (2008)
RILs	6	6.3–23.3	1, 4, 8, and 11	GS	124 lines (Changhui 891/02428)	Jiang et al. (2017)
RILs	6	6.1–16.5	1, 2, 4, 10, and 11	SS	123 lines (Jinbu/BR29)	Kim et al. (2014)
RILs	12	10.5–47.3	1, 2, and 10	RS	114 lines (Kirara397/Hatsushizuku)	Kuroki et al. (2009)
F_2	1	26.6	8	RS	288 lines (Hokkai-PL9/Hokkai287)	Kuroki et al. (2007)
CSSLs	4	24.3	2, 5, 6, and 10	GS	143 lines (Nipponbare/Zhenshan97)	Li et al. (2011b)
RILs	5	6.8–12.1	7, 9, and 12	GS	181 lines (USSR5/N22)	Li et al. (2013)
BC_1F_1	2	–	1	RS	161 lines (3037/02428//3037)	Li et al. (1997)
F_2	2	16.9–19.4	12	RS	121 lines (3037/02428)	Li et al. (1997)
DHs	2	–	3 and 10	GS	193 lines (Zhenshan 97B/AAV002863)	Lou et al. (2007)

(continued)

Table 3 (continued)

QTL mapping studies	Number of QTLs	PVE ranges	Chromosomes	Stage	Number of lines (accessions/parents)	References
ILs	7	8.0–20.0	1, 2, 5, 6, 7, and 10	SS	112 lines (IL112/Guichao2)	Liu et al. (2013)
DHs	6	6.4–27.4	1, 2, and 8	SS	193 lines (AAV002863/Zhenshan 97B)	Lou et al. (2007)
BC_1F_9	5	10.1–14.9	2, 4, 5, and 11	GS	98 lines (Nipponbare/Kasalath)	Miura et al. (2001)
RILs	2	7.5–16.0	1 and 4	SS	80 lines (Milyang23/Hapcheonaengmi3)	Park et al. (2013)
RILs	4	5.8–9.3	7, 8, and 11	GS	162 lines (HGKN/Hokuriku-142)	Ranawake et al. (2014)
RILs	9	5.8–35.6	2, 5, 6, 7, 8, and 11	SS	162 lines (HGKN/Hokuriku-142)	Ranawake et al. (2014)
BC_1F_5	2	–	4	RS	117 lines (Kirara397/Norin-PL8//Kirara397)	Saito et al. (1995)
F_2	6	10.4–23.0	1 and 3	GS	120 lines (Akitakomachi/Maratteli)	Satoh et al. (2016)
DHs	2	11.1–12.6	4 and 9	GS	127 lines (ZYQ8/JX17)	Teng et al. (2001)
BC_1F_5	3	7.9–19.2	7, 8, and 12	RS	77 lines (Suisei/Eikei88223)	Shinada et al. (2013)
RILs	6	5.8–10.9	3, 7, and 9	RS	153 lines (IR66160-121-4-4-2/Geumobyeo)	Suh et al. (2010)
BC_1F_5	11	3.9–8.3	7	RS	264 lines (Lijing2/Towada)	Sun et al. (2019)
F_2	16	3.1–71	1, 3, 4, 6, 7, 8, 10, and 11	RS	108 lines (Hananomai/WAB56-104)	Wainaina et al. (2018)
RILs	2	5.9–8.5	11	GS	227 lines (Daguan dao/IR28)	Wang et al. (2011)
RILs	5	5.5–22.4	3, 8, 11, and 12	SS	227 lines (Daguan dao/IR28)	Wang et al. (2011)
F_2	23	4.1–32.7	3, 4, 5, 7, 9, 10, and 11	GS	517 lines (Kunmingxiaobaigu/Towada)	Xu et al. (2008)
F_2	10	2.9–14.8	1, 4, 5, and 10	RS	517 lines (Kunmingxiaobaigu/Towada)	Xu et al. (2008)
F_3	7	–	1, 2, 5, 8, and 10	SS	10,800 lines (LPBG/Nipponbare)	Z Yang et al. (2013b)
RILs	27	4.6–42.0	2, 6, 7, 9, 11, and 12	GS	190 lines (Dongnong422/Kongyu131)	Yang et al. (2018)

(continued)

Table 3 (continued)

QTL mapping studies	Number of QTLs	PVE ranges	Chromosomes	Stage	Number of lines (accessions/parents)	References
RILs	36	4.5–35.4	2, 3, 6, 7, 9, 10, and 11	SS	190 lines (Dongnong422/Kongyu131)	Yang et al. (2018)
BILs	5	8.8–60.9	4, 8, and 12	SS	202 lines (XB//XB/DWR)	Yu et al. (2018)
RILs	15	5.0–23.1	1, 6, 7, 8, 9, 11, and 12	SS	204 lines (LTH/SHZ-2)	Zhang et al. (2014c)
RILs	5	5.5–29.8	3, 7, and 11	SS	269 lines (Lemont/Teqing)	Zhi-Hong et al. (2005)
GWAS	17	5.2–59.2	2, 3, 4, 5, 6, 8, 9, 10, 11, and 12	GS	174 Chinese rice accessions	Pan et al. (2015)
GWAS	33	5.2–59.2	1, 2, 3, 4, 5, 6, 7, 8, 10, and 12	RS	174 Chinese rice accessions	Pan et al. (2015)
GWAS	67	3.8–8.2	1, 2, 3, 4, 5, 6, 7, 8, 9, 10, and 11	SS	295 Rice diversity panel	Wang et al. (2016)
GWAS	45	3.5–11.9	1, 2, 3, 4, 5, 6, 7, 8, 9, 10, 11, and 12	GS	202 Rice mini-core collections	Schläppi et al. (2017)
GWAS	54	–	1, 2, 4, 5, 6, 7, 8, 9, 10, 11, and 12	GS	400 Rice diversity panel	Shakiba et al. (2017)
GWAS	23	3.1–13.2	1, 2, 3, 4, 5, 6, 10, 11, and 12	SS	249 Chinese rice accessions	Zhang et al. (2018)

identified on chromosome 2 (*qLTG-2-1*), chromosome 5 (*qLTG-5-2.1* and *qLTG-5-2.2*), and chromosome 7 (*qGV7-1.1* and *qGI7-1.2*) (Xu et al. 2008; Yang et al. 2018). With a comprehensive analysis of GS-QTLs, nine genetic regions on eight chromosomes (Ch3: 17.2–17.8 Mb, Ch5: 21.5–21.6 Mb, Ch6: 5.4–6.2 Mb, Ch7: 1.7–2.7 and 20.13–22.6 Mb, Ch9: 21.9–24.6 Mb, Ch10: 11.6–14.2 Mb, Ch11: 23.0–24.2 Mb, and Ch12: 7.0–7.1 Mb) had more than four GS-QTLs that were colocalized. Recently, Yang et al. (2018) identified 12 and 23 QTLs for low-temperature germinability (LTG) and cold tolerance at the seedling stage by using recombinant inbred lines (RILs) that were derived from a backcross population of Dongnong422 and Kongyu131. Interestingly, seven QTLs on chromosome 12 in the 7.1 Mb region and four QTLs on chromosome 7 in the 22.55 Mb region were colocalized, and they were associated with several GS traits such as germination time and rate, mean length of incubation time, coefficient of germination, germination value, mean daily germination, and germination index. Cloning and characterization of the major QTL *qLTG3-1*, conferring more than 30% of the variation (Fujino et al. 2004), revealed that this gene encodes for a protein of unknown function. At the same time, a microarray analysis indicated that a complex metabolic and signal

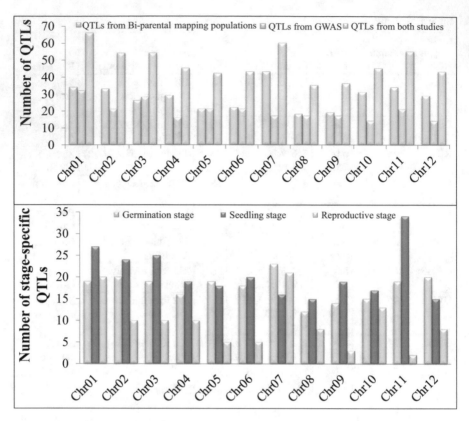

Fig. 3 Summary of QTL distributions by chromosome

pathway was involved (Fujino and Matsuda 2010). Genome-wide expression analysis suggested that genes involved in defense responses were upregulated by *qLTG3-1* and played a more general role in germination (Fujino et al. 2008), whereas correlation with proteomics indicated its involvement in rice growth and adaptability (Fujino and Sekiguchi 2011). With genome-wide association mapping studies, Pan et al. (2015) identified significant 17 QTLs from the 174 mini-core collections of Chinese rice accessions, 45 QTLs from the 202 Rice Mini-Core Collections (Schläppi et al. 2017), and 54 QTLs from a global collection of 400 Rice Diversity Panel 1 (Shakiba et al. 2017) that were detected under low-temperature germinability in rice. Given these challenges and potential in the form of effective QTLs (E-QTLs) with colocalization of QTLs, the development of molecular markers for selection for LTG would significantly contribute to identifying and developing LTS-tolerant varieties.

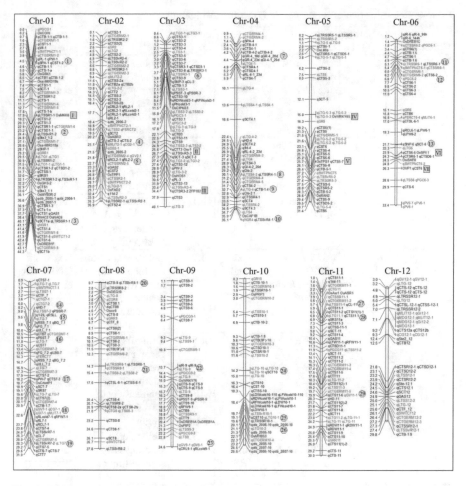

Fig. 4 Diagram of LTS-tolerance QTLs by comprehensive literature survey and Gramene database (http://archive.gramene.org) in rice. On the right side, the numerical values in chromosome bars indicate the position of QTLs and genes in Mb, and on the left side, those indicate the QTLs for stage-specific and LTS-tolerance genes. A colored font represents the stage-specific QTLs (green: germination; blue: seedling; and red: reproductive/booting stage) and genes for LTS tolerance (pink). The octagonal shapes and numerical values represent the two different stage-specific QTLs that were colocalized in the same genetic region, and the square shape indicates the QTLs and reported LTS-tolerance genes aligned together in the same genetic regions of chromosomes

3.2 Seedling Stage

Low-temperature stress severely affects the SS, causes slow growth, yellowing symptoms on leaves, drying of leaves, and decreased early seedling vigor, and ultimately leads to seedling death (Wang et al. 2011; Lone et al. 2018). Tolerance of LTS at the SS is one of the key stages to ensure stable early seedling growth in temperate and high-altitude regions. For the genetics of the SS, several researchers

Table 4 Colocalization of stage-specific QTLs for LTS tolerance

Chromosome	Number of QTLs	QTLs	Stage	Position (Mb)	Traits	References
1	4	*qPL-1*, *qPW-1*, *qSPA-1*, and *qCST1-2*	SS and RS	4.5	Panicle length, panicle weight, chlorophyll content, and seedling cold cold tolerance	Liu et al. (2013), Park et al. (2013), Wainaina et al. (2018)
	2	*qCTSR1-1* and *qCTGERM1-6*	GS and SS	22.9	Seedling survival rate and cold tolerance at germination	Shakiba et al. (2017), Zhang et al. (2018)
	2	*qSCT1a* and *qLTRSSR1-1*	SS and RS	40.1	18 days after seedling cold tolerance and seed-setting rate	Kim et al. (2014), Pan et al. (2015)
2	4	*qCTS-2*, *qMLIT2-1*, *qCG2-1.1*, and *qCG2-1*	GS and SS	21.1	Cold tolerance at seedling stage, mean length of incubation time, and coefficient of germination	Lou et al. (2007), Yang et al. (2018)
	4	*qCTGERM2-4*, *qGV2-1*, *qRCL2-1*, and *qRL2-2*	GS and SS	21.8	Cold tolerance at germination, germination value, relative conductivity of leaves, and root length	Shakiba et al. (2017), Yang et al. (2018)
3	3	*qCTS3-8*, *qLTRSSR3-1*, and *qLTSSR3-1*	SS and RS	3.7	Cold tolerance at seedling stage and seed-setting rate	Pan et al. (2015), Wang et al. (2016)

(continued)

Table 4 (continued)

Chromosome	Number of QTLs	QTLs	Stage	Position (Mb)	Traits	References
4	5	qGR-4_18d, qGR-4_20d, qGR-4_23d, qGI-4-1_20d, and qLTG4-1	GS	4.4	Germination rate after 18, 20, and 23 days of cold stress, germination index, and low-temperature germination	Miura et al. (2001), Wang et al. (2009)
	2	qLTSS4-2 and qLTSSR4-1	GS and RS	30.1	Low-temperature seedling survivability and seed-setting rate	Pan et al. (2015), Schläppi et al. (2017)
	2	qLTG-4 and qCTB-1.4	GS and RS	32.7	Germination rate at 12 days and cold tolerance at booting stage	Dai et al. (2004), Xu et al. (2008)
	2	qNGR4 and qLTSSvR4-1	GS and SS	35.1	Germination rate after 8 days of cold stress and seedling survival rate	Pan et al. (2015), Jiang et al. (2017)
6	2	qLTRSSR6-1 and qLTSSR6-1	RS	5.1	Seed-setting rate	Pan et al. (2015)
	2	qCTGERM6-2 and qCTS6-2	GS and SS	6.2	Cold tolerance at germination and cold tolerance at seedling stage	Wang et al. (2016), Shakiba et al. (2017)
	2	qSNP-6 and qSCT-6	SS and RS	21.7	Spikelet number and cold tolerance at seedling stage	Jiang et al. (2008), Wainaina et al. (2018)

(continued)

Table 4 (continued)

Chromosome	Number of QTLs	QTLs	Stage	Position (Mb)	Traits	References
7	2	*qLTSS7-2* and *qPSSR-7*	GS and RS	9.0	Low-temperature seedling survivability and percentage of seed set reduction ratio in cold-water-treated plot	Ji et al. (2010a, b), Schläppi et al. (2017)
	2	*q1-2IL*, *qRSLL*, and *qLTG-7*	GS and RS	9.5	First and second internode length, reciprocal secondary leaf length, and low-temperature germinability	Ji et al. (2008), Sun et al. (2019)
	2	*qCTB7* and *qCTGERM7-1*	GS and RS	10.5	Percentage of undeveloped spikelets and cold tolerance at germination	Andaya and Mackill (2003b), Shakiba et al. (2017)
	2	*qPW-7* and *qCTS7-3*	SS and RS	18.2	Panicle weight and cold tolerance at seedling stage	Wang et al. (2016), Wainaina et al. (2018)
	7	*qGV7-1*, *qCG7-1*, *qGI7-1*, *qMLIT7-1*, *qRLcold7-1*, *qRFW7-1*, and *qRL7-1*	GS and SS	22.5	Germination value, coefficient of germination, germination index, root fresh weight, and root length	Yang et al. (2018)
	2	*qCTGERM7-5* and *qLTSSvR7-2*	GS and SS	28.3	Cold tolerance at germination and low-temperature seedling survivability rate	Schläppi et al. (2017), Shakiba et al. (2017)
8	2	*qCTS-8* and *qLTSSvR8-1*	SS	0.7	Cold tolerance at seedling stage and seedling survivability rate	Pan et al. (2015), Wang et al. (2016)
	3	*qLTRSSR8-1*, *qLTSSR8-1*, and *qCTSSR8-1*	RS	14.3	Relative seed-setting rate and seed-setting rate under cold water	Pan et al. (2015)

(continued)

Table 4 (continued)

Chromosome	Number of QTLs	QTLs	Stage	Position (Mb)	Traits	References
9	2	qIR-9, qIR-9, and qLTG-9	GS and SS	13.6	Imbibition rate at 48 h 4 days after imbibition and 14th day of germination	Wang et al. (2009), Li et al. (2013)
	2	qGV9-1, qGV9-1.2, and qCRL9-1	GS and SS	24.6	Germination value and crimp ratio of leaves	Yang et al. (2018)
10	4	qLTG-10, qLTG-10, qLTG-10, and qSCT10	GS	14.2	Low-temperature germinability and 12th and 13th day of germination	Chen et al. (2006a), Xu et al. (2008)
	3	qCST10, qLTG10-1, and qLTRSSR10-1	GS, SS, and RS	19.1	Cold seedling tolerance, low-temperature germination, and relative seed-setting rate	Pan et al. (2015), Schläppi et al. (2017)
	2	qLTG10-2 and qctb_2005-10	GS and RS	21.3	Low-temperature germination and cold tolerance at booting stage	Kuroki et al. (2009), Schläppi et al. (2017)
11	2	qLTSS11-1 and qCL-11	GS and RS	5.7	Low-temperature seedling survivability and culm length	Schläppi et al. (2017), Wainaina et al. (2018)
	2	qLTG11.1 and qCTS11.1	GS and SS	6.8	Low-temperature germinability and cold tolerance at seedling stage	Wang et al. (2011)
	2	qCTS11-6 and qGV11-1	GS and SS	19.9	Cold tolerance at seedling stage and germination value	Wang et al. (2016), Yang et al. (2018)

GS germination stage, *SS* seedling stage, *BS/RS* booting/reproductive stage

have identified SS-specific QTLs using different genetic backgrounds of mapping populations such as RILs, near-isogenic lines (NILs), introgression lines, doubled haploids, and segregating F_2 and F_3 families (Biswas et al. 2017; Liang et al. 2018; Sun et al. 2019). However, the genetic backgrounds of *japonica* donors as LTS-tolerant cultivars have revealed that several QTLs and genes are controlling LTS tolerance during the SS in rice. So far, more than 249 QTLs (both major and minor) have been mapped on all of the rice chromosomes responsible for LTS tolerance at the seedling stage (Table 3). Among the total QTLs, 159 were identified from biparental mapping populations and 90 from GWAS. The phenotypic variance of these QTLs ranges from 4.51% to 60.96%. Fifteen M-QTLs (≥30% PVE) were identified on six different chromosomes: 4, 6, 7, 8, 11, and 12. However, only a few of them were M-QTLs that were colocalized with different stage-specific QTLs. For instance, *qSCT4.2* (57.62% PVE) on chromosome 4 (22.3–26.91 Mb) was shared with four QTLs (*qRL-4-2_23d*, *q9d-4*, *qLTG-4-2*, and *qHD-4*) for root length, percentage of plumule growth, germination rate, and heading date under LTS conditions (Miura et al. 2001; Wainaina et al. 2018; Yu et al. 2018). Similarly, *qSCT8* (60.96% PVE) on chromosome 8 (24.6–27.8 Mb) was shared with cold-tolerance seedling-stage QTL *qCTS8-1* (Andaya and Mackill 2003b), seed weight per plant QTL *qSWTCT8-4* (Shakiba et al. 2017), and seedling survival rate QTL *qLTSSvR8-2* (Pan et al. 2015). Importantly, two M-QTLs (*qCTS12a* and *qCTS12b*) for cold-induced wilting tolerance and cold-induced necrosis tolerance on chromosome 12 had PVE of 40.6% and 41.7% (Andaya and Mackill 2003b), respectively. The same genetic region of 7.1–9.3 Mb overlapped with ten QTLs for GS- and RS-related traits such as germination index, incubation time, coefficient of germination, and seed-setting rate (Pan et al. 2015; Yang et al. 2018). Other two M-QTLs (*qGAS12* and *qSCT12.1*) had PVE of 42.9% and 53.09%, respectively. These QTLs were associated with the growth ability of seedlings at low temperatures (Han et al. 2004) and seedling cold tolerance (Yu et al. 2018) on chromosome 12. The genetic region of 24.2–25.0 Mb was shared with GS- and SS-specific QTL *qCTS12.1* (Wang et al. 2011), *qLTSS12-2* (Schläppi et al. 2017), and *qSWTCT12* (Shakiba et al. 2017).

The seedling-stage QTLs (seven genetic regions on seven chromosomes) were found to overlap with more than five SS-specific QTLs. The highest number of QTLs was found to overlap in the 16.3–18.4 Mb region on chromosome 10 (12 QTLs), followed by 11 QTLs on chromosomes 2 and 11 (at 14.3–17.7 and 6.8–11.3 Mb), six QTLs on chromosome 3 (2.7–3.7 Mb) and chromosome 9 (1.1–4.8 Mb), and five QTLs on chromosome 7 (22.5–23.8 Mb) and chromosome 11 (24.9–25.4 Mb), respectively. However, two chromosomes (7 and 11) have significant M-QTLs that are also colocalized with more than five QTLs. In a deeper understanding of these two chromosomes, the M-QTL *qCTS7(2)* on chromosome 7 (20.0–20.9 Mb) had PVE of 35.3%, and it overlapped with two other QTLs (*qCT-GERM7-4* and *qCTB-1.7*) related to GS and RS (Dai et al. 2004; Ranawake et al. 2014; Shakiba et al. 2017). Another M-QTL, *qCTS11(1)-2* on chromosome 11 (20.5–26.0 Mb), had PVE of 35.6% (Ranawake et al. 2014), the same region was associated with a seed germination recovery rate QTL (*qGRR11*) (Jiang et al. 2017), and two SS-specific QTLs (*qCTS11-9* and *qCTS11-10*) were identified from the GWAS analysis (Wang et al. 2016). Therefore, the combination of stage-specific QTLs with GS, SS, and RS in the same genetic region could be a promising site for identifying potential candidate genes for improving LTS tolerance in the seedling stage.

3.3 Booting/Flowering Stage

Unlike the GS and SS, the RS is highly sensitive to LTS. Many traits, such as microspore abortion, no anther dehiscence, high spikelet sterility, incomplete panicle exsertion, delayed heading, and failure to produce pollen grains, are affected by LTS in the RS (da Cruz et al. 2013; Liang et al. 2018). Several studies have detected and mapped many major- and minor-effect QTLs responsible for LTS tolerance at the booting/flowering stage using different genetic backgrounds of mapping populations (Table 3). So far, more than 100 QTLs have been mapped on all of the rice chromosomes. The QTLs for tolerance at the RS showed that 81 QTLs from biparental mapping populations and 33 QTLs from GWAS were identified from the comprehensive literature survey. The phenotypic variance of these QTLs ranges from 2.94% to 71.0%. A total of 14 M-QTLs ($\geq$30% PVE) were located on six different chromosomes (1, 2, 4, 7, 8, and 10). Saito et al. (1995) reported two QTLs on chromosomes 3 and 4 responsible for cold tolerance at the booting stage from Norin PL8. Saito et al. (2001) also reported two OTLs (*Ctb-1* and *Ctb-2*) on chromosome 4 governing spikelet fertility under cold stress by using a set of NILs derived from a cross between cold-tolerant rice variety Norin PL8 and cold-sensitive commercial variety Kirara397 from northern Japan. By using cool-water irrigation (19 °C), Kuroki et al. (2009) identified five QTLs for cold tolerance at the booting stage, four QTLs for days to heading, and three QTLs for culm length on chromosomes 1, 2, and 10. One of the major QTLs (*qCTB.1*) flanked by RM1003 and RM3482 on chromosome 1 associated with cold tolerance at the booting stage was discovered after 3 years of field trials at the National Agricultural Research Centre for Hokkaido Region, Sapporo, Japan. Similarly, eight QTLs for the booting stage were identified on chromosomes 1, 4, 5, 10, and 11 by using a set of NILs from a cold-tolerant *japonica* landrace (Kunmingxiaobaigu) and cold-sensitive *japonica* cultivar (Towada) (Xu et al. 2008). However, four QTLs (*qCTB-1-1*, *qCTB-4-1*, *qCTB-5-1*, and *qCTB-5-2*) were detected in two different environments.

A total of nine genetic regions on chromosome 1 (28.6 and 36.6 Mb), chromosome 2 (21.5 Mb), chromosome 4 (24.5 Mb), chromosome 7 (8.1 and 18.1 Mb), chromosome 8 (11.5 Mb), and chromosome 10 (9.2–9.6 and 20.0–25.2 Mb) were associated with tolerance of RS traits such as heading time, booting stage, culm length, spikelet fertility, spikelet number, and panicle weight (Dai et al. 2004; Kuroki et al. 2009; Wainaina et al. 2018). The M-QTL (*qCTB_1*) on chromosome 1 for cold tolerance at the booting stage had PVE of 47.3% (Kuroki et al. 2009), and it was close to the genetic region associated with SS-specific QTL *qCTSR1-3* (36.9 Mb) (Zhang et al. 2018) and MYB transcription factor *OsMYB3R-2* (36.1 Mb), which is bound to the mitotic-specific activator during LTS tolerance (Ma et al. 2009). The overexpression of this gene significantly enhances the many transcripts for G2/M phase-specific genes in response to cold stress. Another M-QTL (*qSNP_1*; 28.6 Mb) in the genetic region on the same chromosome overlapped with *Osa-MIR319a*. In rice, the miR319 gene family comprises two genes, *Osa-MIR319a* and

Osa-MIR319b. Both of them are significantly involved in increasing leaf blade width under cold tolerance (Yang et al. 2013a). The colocalization of stage-specific QTLs on chromosome 2 is associated with nine QTLs (21.1–21.8 Mb) for GS- and SS-specific QTLs (Lou et al. 2007; Shakiba et al. 2017; Yang et al. 2018). Similarly, the M-QTL *qPW-7* (PVE of 41%) on chromosome 7 overlapped with cold-tolerance seedling-stage QTL *qCTS7-3* (Wang et al. 2016). Six M-QTLs were associated with tolerance of RS traits on chromosome 10. In the interval regions of M-QTLs, two QTLs (*qLTG10-2*, 21.1 Mb; *qCTGERM10-4*, 22.3 Mb) and one DNA-binding repeat MYB transcription factor (*OsMYBS3*, 22.1 Mb) are responsible for cold tolerance at the GS and SS (Su et al. 2010; Schläppi et al. 2017; Shakiba et al. 2017). The overexpression of MYBS3 confers chilling tolerance, and it has also been associated with MYBS3-mediated cold signaling pathways (Su et al. 2010). Taken together, the stage-specific QTLs of different combinations of GS, SS, and RS and associated candidate gene results could prove to be useful in breeding programs for low temperature-tolerant rice lines. These M-QTLs along with colocalized stage-specific QTLs have great potential for use in the future as their application through marker-assisted selection will hasten the process of developing cold-tolerant rice varieties for temperate and high-altitude ecosystems.

The LTS-tolerance QTLs mentioned above included a comparison of stage-specific QTL positions across different genetic backgrounds of mapping population studies. Such comparisons of stage-specific QTL positions are more informative. These QTLs may harbor potential candidate genes related to LTS, thus providing valuable information to develop LTS-tolerant rice cultivars. For this, the functionally characterized LTS-tolerance genes were collected from the OGRO database (Overview of Functionally Characterized Genes in Rice Online database) on the Q-TARO website (http://qtaro. abr.affrc.go.jp/ogro). A total of 38 candidate genes were involved in LTS tolerance in different parts of the rice plant. Specific gene functions for LTS tolerance and stages are mentioned in Table 5 and are also mapped on the genetic map (Fig. 4). Among these candidate genes, the majority of them were associated with the SS (25 genes), followed by two genes for the RS. The remaining eight genes were for the SS and RS, two genes for the GS and SS, and a single gene for the GS and RS. Functionally, the candidate genes were involved in altering the various metabolic and physiological pathways in different growth stages of LTS-tolerance mechanisms. Seven genetic regions on four chromosomes (17.9 Mb on chromosome 1; 25.6 and 34.4 Mb on chromosome 3; 15 and 25.7 Mb on chromosome 5; and 23.9 and 26.3 Mb on chromosome 6) are colocalized with candidate genes and QTLs. For example, *OsSPX1* is colocalized with *qCTS6-5* at 23.9 Mb on chromosome 6. A previous study revealed that *OsSPX1* plays a key role in the cross talk between cold tolerance, phosphate homeostasis, and oxidative stress tolerance in the SS (Wang et al. 2013a), and the cold tolerance QTL (*qCTS6-5*) was mapped in the SS by using GWAS (Wang et al. 2016). Based on the fine-tuning of the interval regions of the M-QTLs, several researchers have identified many candidate genes for LTS tolerance. The M-QTLs and colocalized QTLs within identified genetic regions, especially on chromosomes 1, 3, 5, 7, and 10, may be potential genomic regions to introgress into existing moderately LTS-tolerant genotypes or mega-varieties to improve their rate of tolerance in marker-assisted breeding programs.

4   Molecular Mechanisms of LTS Tolerance

Several genes and transcription factors are involved in regulating the molecular pathways related to alteration of physiological and metabolic compounds and, further, reprogramming of their gene expression patterns against cold-stress tolerance mechanisms (Fig. 2). During LTS response, multiple sensors and signaling elements on the plasma membrane trigger the expression of COR (cold-responsive) genes via increasing cytosolic Ca^{2+} levels. This increase in Ca^{2+} is mediated by the ligand-activated Ca^{2+} channels. Further higher levels of Ca^{2+} in the cytosol lead to signal amplification through phospholipids (Williams et al. 2005; Hashimoto and Komatsu 2007; Chinnusamy et al. 2010), which are sensed by calcium-binding proteins and other transcription factors regulating the expression of LTS-tolerance genes, which can ultimately lead to adaptation and survival during cold-stress conditions (Shinozaki and Yamaguchi-Shinozaki 2000). However, the changes in various gene expression patterns are governed by a signal cascade mechanism, which also triggers the formation of plant hormones (abscisic acid, salicylic acid, and ethylene) that may be involved in integrating various stress signal pathways and controlling downstream stress responses. CBF (C-repeat/DREB [drought-responsive element-binding factor]) regulon is a highly conserved cold-response pathway (Chinnusamy et al. 2010). The ICE1–CBF transcriptional cascade plays a crucial role in cold acclimation (Zhang et al. 2004). Constitutively expressed ICE1 (inducer of CBF Expression 1) binds to the CBF promoter to activate cold-resistance genes, and overexpressing ICE1 has significantly enhanced cold tolerance in *Arabidopsis* (Chinnusamy et al. 2003). Similarly, ICE2 (*At1g12860*, a homolog of ICE1) overexpression demonstrated enhanced freezing tolerance in *Arabidopsis* after cold acclimation (Fursova et al. 2009). The report published showed that it interacts with the alpha subunit of the sole heterotrimeric G protein, leading to a cytosolic Ca^{2+} signal or itself behaving as a cold-sensing calcium channel. A Ca^{2+} signal may be mediated by CPKs and CBL-CIPKs, which in turn activate MAP kinases (Yang et al. 2010). Phosphorylation of transcription factors such as that of CAMTAs and ICE1/2 is supposed to be caused by activated MPKs, which in turn activate COR genes (Zhu 2016).

Most COR genes carry C-repeats or DREBs (CCGAC *cis*-element) in their promoters, which bind to CBFs to activate their expression (Chinnusamy et al. 2007). The induction of CBF1, CBF2, and CBF3 precedes the COR genes in response to cold stress. Constitutive overexpression studies in *Arabidopsis* revealed redundant functional activities of CBF1, CBF2, and CBF3 and showed different functions in cold acclimation (Gilmour et al. 2004). This was observed in the cbf2 T-DNA insertion mutant with enhanced tolerance of freezing (with or without cold acclimation), dehydration, and salt stress through increased expression of CBF1 and CBF3. Furthermore, the cold-induced expression of CBF1 and CBF3 precedes that of CBF2, revealing a temporal difference in CBF expression. These results indicate that CBF1 negatively regulates CBF3 and CBF2 to optimize the expression of downstream target genes (Doherty et al. 2009). Transgenic analysis of CBF1 and

Table 5 Functionally characterized LTS-tolerance genes and stages in rice

S. No.	Chr	Position (Mb)	Gene length	Gene	Cold treatment	Stage	Expression analysis	Function	References
1	chr01	0.21	4.41 kb	OsCOIN	4 °C for 60, 72, and 84 h	SS and RS	Young root, stem, lamina, leaf sheath, young panicle, mature panicle, stem primordia, pedestal, and stipital leaf	Cold-inducible zinc finger protein for tolerance of cold, salt, and drought	Liu et al. (2007)
2	chr01	5.78	2.63 kb	OsGSK1	4 °C in Yoshida nutrient solution	SS and RS	Lamina joint in collar region, vascular bundles of coleoptile, and young panicle	Glycogen synthase kinase3-like gene for stress signal-transduction pathways and in floral developmental processes	Koh et al. (2007)
3	chr01	6.68	199 bp	Osa-MIR319b	Initially, 12 °C for 7 days and then at 4 °C for 4 days	SS	Flag leaf	Increased leaf blade width	Yang et al. (2013a)
4	chr01	17.91	5.21 kb	OsMKK6	4 °C for 0, 1, 3, 6, and 12 h	SS	Leaves	*Oryza sativa* MAPK kinase 6 for tolerance of cold and salt	Kumar et al. (2008)
5	chr01	28.58	193 bp	Osa-MIR319a	Initially, 12 °C for 7 days and then at 4 °C for 4 days	SS	Flag leaf	Increased leaf blade width	Yang et al. (2013b)
6	chr01	32.22	3.75 kb	OsGH3-2	4 °C with 14 h light/10 h dark for 5 days	SS and RS	Shoot, root, leaf size, calli, and low levels in panicles and stems	Modulation of abscisic acid and auxin levels in response to stress-tolerance mechanisms of cold and drought	Greco et al. (2012)
7	chr01	36.13	5.66 kb	OsMYB3R-2	2 °C at 0, 24, 48, 60, 72, and 84 h	SS and RS	Root, internode, leaf, lamina joint, leaf sheath, flower, and immature seed	Encodes active transcription factor involved in higher transcript levels in G2/M phase	Ma et al. (2009)

S. No.	Chr	Position (Mb)	Gene length	Gene	Cold treatment	Stage	Expression analysis	Function	References
8	chr01	38.40	2.49 kb	SNAC2	3–8 °C for 48 h	SS	Root, stem, internodes, leaf sheath, and ligule	Stress-responsive NAC gene, specifically induced in guard cells in response to cold and salt	Hu et al. (2008)
9	chr01	38.40	2.49 kb	OsNAC6	4 °C for 0–24 h	SS	2-week-old rice leaves	NAM-ATAF-CUC family 6 transcription factor, enhances tolerance of drought, cold, salt, and also blast disease	Nakashima et al. (2007)
10	chr01	42.73	838 bp	OsDREB1F	4 °C for 0.5, 1, 6, and 24 h	SS	Young roots, young leaves, mature roots, mature leaves, spike, and callus	Dehydration-responsive element-binding transcription factor 1F	Wang et al. (2008)
11	chr02	20.17	988 bp	ASR3	4 °C for 5 h	SS	Leaves and roots	ABA-dependent stress-responsive protein; induces drought- and cold-tolerance mechanism regulated by hormone and sugar signals	Joo et al. (2013)
12	chr02	26.77	4.02 kb	OsTPP1	6–8 °C for 0 and 20 min, 1, 3, 6, 12, 24, and 72 h	GS and SS	2-week-old seedlings	Trehalose-6-phosphate phosphatase1 serves as sugar storage and enhances tolerance of drought, cold, and salt stress, without alteration of growth	Ge et al. (2008)
13	chr02	29.73	5.03 kb	OsFAD2	4 °C for 4 days	GS and RS	Root, seed, stem, and leaf	Fatty acid desaturase 2 is a key enzyme responsible for increasing germination rate and grain yield under LTS	Shi et al. (2012)

(continued)

Table 5 (continued)

S. No.	Chr	Position (Mb)	Gene length	Gene	Cold treatment	Stage	Expression analysis	Function	References
14	chr03	25.63	4.02 kb	Osv1	20 °C for 22 days	SS	Pre-emerged immature leaves	Chloroplast-localized protein NUS1, actively involved in the regulation of chloroplast RNA metabolism and establishing the plastid genetic system for cold conditions	Zhang et al. (2011)
15	chr03	30.22	8.19 kb	OsHOS1	10 °C for 0 h (28 °C), 2 h (10 °C), 5 h (10 °C), and 24 h (10 °C)	SS	2-week-old seedlings	E3-ubiquitin ligase *OsHOS1* gene involved in proteasome-mediated stress response to cold stress	Lourenço et al. (2013)
16	chr03	34.42	853 bp	ZFP182	4 °C for 4 days	GS and SS	2-week-old seedlings	TFIIIA-type zinc finger protein182 promotes accumulation of various osmolytes, which involves multiple abiotic stress tolerance	Huang et al. (2012)
17	chr03	11.53	5.7 kb	OsCIPK03	4 °C for 0, 3, 6, 12, and 24 h	SS	2-week-old seedlings	Calcineurin B-like protein significantly increased the amount of proline and soluble sugar accumulation in drought- and cold-stress conditions	Xiang et al. (2007)
18	chr04	34.98	1.22 kb	OsCAF1B	4 °C for 21 days	SS and RS	3-week-old seedlings of root, shoot, leaf, sheath, leaf base, leaf tip, panicle axis, and spikelet	Rice carbon catabolite repressor 4 triggers the deadenylation mechanism in the plant-P-body and is linked with microtubules	Chou et al. (2014)
19	chr05	2.22	1.04 kb	OsLti6b	12 °C for 4 days, before heading of 5–10 days	RS	Vascular tissues of filaments, anthers, ovaries, stamens, leaves, and spikelets	Encodes for hydrophobic protein, expressed in ovaries and stamens of cold-treated flowers	Kim et al. (2007)

S. No.	Chr	Position (Mb)	Gene length	Gene	Cold treatment	Stage	Expression analysis	Function	References
20	chr05	14.99	2.1 kb	OsWRKY45	4 °C for 3–24 h.	SS	Leaves	OsWRKY45 plays a major role in ABA signaling and as a cross-talk mechanism in biotic and abiotic stresses	Tao et al. (2011)
21	chr05	25.72	7.21 kb	OsTPS1	4 °C for 0, 1, 2, 4, 6, 12, and 24 h	SS	2-week-old seedlings	Overexpression of TPS1 results in increasing trehalose and proline concentration and regulates stress-responsive genes for cold and salt	M Li et al. (2011b)
22	chr05	28.62	2.25 kb	OsRAN2	4 °C for 72 h	SS	2-week-old seedlings	Ran is a nuclear GTPase involved in GTP hydrolysis mechanism and mediates nuclear transport of RNA and proteins in cell cycle and in regulating cold tolerance	Chen et al. (2011)
23	chr06	1.44	766 bp	OsDREB1C	4 °C for 24 h	SS	17-day-old seedlings	Overexpression of the dehydration-responsive element-binding protein 1C significantly improves tolerance of drought-, cold-, and salt-stress conditions	Ito et al. (2006)
24	chr06	23.88	4.53 kb	OsSPX1	4 °C for 24 h	SS	1-week-old seedlings	SPX domain proteins are involved in phosphate (Pi) signal transduction pathways and cross talk between the oxidative pathway and cold-stress mechanism	Wang et al. (2013a)

(continued)

Table 5 (continued)

S. No.	Chr	Position (Mb)	Gene length	Gene	Cold treatment	Stage	Expression analysis	Function	References
25	chr06	24.49	1.92 kb	OsiSAP8	4 °C for 0, 2, 3, 4, 6, 12, 24, and 48 h	SS, and RS	1-week-old seedlings	SAP gene family protein transcript was detected at higher levels in root and prepollination-stage panicle and is significantly expressed in multiple abiotic stresses	Kanneganti and Gupta (2008)
26	chr06	26.28	5.77 kb	OVP1	4 °C for 12 h	SS	10-day-old seedlings	Vacuolar HD-translocating inorganic pyrophosphatase 1 involved in decreased malondialdehyde content and accumulation of more proline for tolerance of cold	Zhang et al. (2011)
27	chr06	27.30	2.17 kb	OsbZIP52/RISBZ5	4 °C for 0, 0.5, 1, 2, 4, 6, 12, 24, and 48 h	SS, and RS	Roots, leaves from 2-week-old seedlings, and stems, flag leaves, flowers, and developing seeds at 2 days after flowering and at milk grain stage	Overexpression of basic leucine zipper 52 serves as negative regulator of drought and cold stress	Liu et al. (2012)
28	chr07	18.72	8.48 kb	CRTintP1	5 °C for 3 days	SS	Leaf sheaths	Accumulation of these calreticulin-interacting proteins involved in signal transduction mechanism in cold stress	Komatsu et al. (2007)
29	chr08	2.18	3.5 kb	OsDEG10	4 °C for 0, 24, 48, and 72 h.	SS	17-day-old seedlings	Encodes small RNA-binding protein and plays a major role in response to cold, salt, anoxia, and photooxidative stress	Park et al. (2009)

S. No.	Chr	Position (Mb)	Gene length	Gene	Cold treatment	Stage	Expression analysis	Function	References
30	chr08	3.95	2.1 kb	Oscrr6	20–35 °C for 0, 24, 48, and 72 h	SS	3-week-old seedlings	Encodes an NDH-dependent cyclic electron flow and plays a key role in physiological pathways during photosynthesis and growth development at low temperature	Yamori et al. (2011)
31	chr09	14.98	1.53 kb	OsWRKY76	4 °C for 72 h	SS	2-week-old seedlings	Plays a dual role in promoting blast disease resistance and cold tolerance	Yokotani et al. (2013)
32	chr09	20.40	922 bp	OsDREB1A	4 °C for 24 h	SS	17-day-old seedlings	Overexpression of dehydration-responsive element-binding protein 1C significantly improves tolerance of drought-, cold-, and salt-stress conditions	Ito et al. (2006)
33	chr09	21.31	1.35 kb	OsPIP2	25 °C for 2 h	SS	Leaves and roots	Represents plasma membrane intrinsic proteins that are involved in water transport and maintenance of the water balance in cells under cold stress	Li et al. (2008)
34	chr10	2.90	1.37 kb	OsPRP3	4 °C for 10 h	RS	Leaves and flowers	Flower-specific proline-rich protein 3 enhances expression during cold stress	Gothandam et al. (2010)
35	chr10	22.13	8.31 kb	MYBS3	4 °C for 72 h	SS	1-week-old seedlings	Is a DNA-binding repeat MYB transcription factor and mediates cold-signaling pathways	Su et al. (2010)

(continued)

Table 5 (continued)

S. No.	Chr	Position (Mb)	Gene length	Gene	Cold treatment	Stage	Expression analysis	Function	References
36	chr11	3.28	971 bp	OsAsr1	12 °C for 4 days, before heading	SS and RS	Leaf, palea and lemma, and anther	Highly expressed with C-repeat/dehydration responsive element-binding factor 1; involved in cold tolerance at vegetative and reproductive stages	Kim et al. (2009)
37	chr11	3.28	971 bp	ASR1	4 °C for 5 h	SS	Leaves and roots	ABA-dependent stress-responsive protein; induces drought- and cold-tolerance mechanism regulated by hormone and sugar signals	Joo et al. (2013)

CBF3 RNAi lines revealed that both CBF1 and CBF3 are required for the full set of CBF regulon expression and freezing tolerance (Novillo et al. 2004). While responding to cold stress, ICE1 and calmodulin-binding transcription activators (CAMTAs) bind to CBF3 and CBF2 promoters, respectively, to respond to their expression (Doherty et al. 2009). Furthermore, in many cellular signaling pathways, particularly in response to cold stress, protein phosphorylation is considered crucial, and it predicts the involvement of one or more protein kinases to phosphorylate ICE1 to help in CBF expression (Chinnusamy et al. 2007; Yang et al. 2010).

4.1 Signaling Pathways Leading to LTS Tolerance from the Cloned Genes

Transgenic and gene expression analysis has helped to understand the physiological mechanisms responsible for tolerance against various abiotic stresses, including LTS, in plants (Gao et al. 2008; Moraes De Freitas et al. 2016). Using the OGRO database on the Q-TARO website, we collected 38 candidate genes that have been functionally characterized for LTS tolerance in different stages of the rice plant (Table 5 and Fig. 4). Among the total number of genes, eight were involved in the two stage-specific tolerance mechanisms of LTS. Four genes on chromosome 1 (*OsCOIN*, 0.2 Mb; *OsGSK1*, 5.7 Mb; *OsGH3-2*, 32.2 Mb; and *OsMYB3R-2*, 36.1 Mb), two genes on chromosome 6 (*OsiSAP8*, 24.4 Mb; and *OsbZIP52*, 27.3 Mb), and a single gene on chromosome 4 (*OsCAF1B*, 34.9 Mb) and chromosome 11 (*OsAsr1*, 3.2 Mb) were associated with tolerance at the SS and RS in rice.

The promising genetic regions of *OsGH3-2* (Greco et al. 2012) and *OsMYB3R-2* (Ma et al. 2009) showed clear evidence of seedling survival rate and seed-setting rate under cold stress. The overexpression of *OsGH3-2* significantly modulates abscisic acid (ABA) and endogenous indole-2-acetic acid (IAA) homeostasis, resulting in increased cold tolerance. Furthermore, two genes (*OsTPP1* and *OsFAD2*) on chromosome 2 and a single gene (*OsZFP182*) on chromosome 3 were associated with the GS, SS, and RS. Expression analysis of *OsTPP1* confers a tolerance mechanism for salt and cold by activating the transcriptional regulation pathways (Ge et al. 2008). The expression pattern of *OsFAD2* under LTS in different tissues in young seeds, stems, roots, and leaves plays a significant role in membrane lipid desaturation and maintenance of the lipid balance in different photosynthetic tissue (Shi et al. 2012). Meanwhile, overexpression of *OsZFP182* in transgenic lines showed an increasing accumulation of various osmolytes, which resulted in an increase in tolerance of drought, cold, and salt (Huang et al. 2012).

On chromosomes 5 and 10, two genes (*OsLti6b* and *OsPRP3*) are associated with the RS. *OsLti6a* and *OsLti6b* encode membrane proteins that contribute greatly to membrane stability (Morsy et al. 2005; Kim et al. 2007). *OsPRP3* is a novel flower-specific prorich protein (PRP) that is significantly overexpressed in the RS under LTS, mainly in flower development (Gothandam et al. 2010). The remaining

25 candidate genes are involved in LTS tolerance in the SS. The important upregulated or overexpressed genes/TFs concerning their expression and function in different LTS stages and in cold stress conditions are described briefly in Table 5. The overexpression of several TFs and protein kinases, such as *OsISAP8*, *OsbHLH1*, *OsDREB1/CBF*, *ROS-bZIP*, *SNAC2*, *OsCIPK12*, *OsNAC6*, *OsCOIN*, *OsMAPK5*, *OsMYB4*, and *OsISAP1*, confers LTS tolerance in the SS in rice (Mukhopadhyay et al. 2004; Nakashima et al. 2007; Xiang et al. 2007; Kanneganti and Gupta 2008).

For tolerance in the RS, two cell wall acid invertase genes (*OsINV1* and *OsINV4*) and one vacuolar acid invertase gene (*OsINV2*) were associated with low temperature at the pollen developmental stage. Among these genes, *OsINV4* is anther-specific and is downregulated by cold treatment, consequently causing a disturbance in hexose production and starch formation in the pollen grains in the tapetum cells. However, no decrease in expression of *OsINV4* vis-à-vis any sucrose accumulation in the anthers and pollen grains in a cold-tolerant cultivar (R31) was observed after cold treatment (Oliver et al. 2005). The *OsMAPK5* gene codes for a protein involved in kinase activity usually induced by ABA and various biotic and abiotic stresses. OX lines for the *OsMAPK5* gene exhibited increased tolerance of cold and other stresses (Xiong and Yang 2003). Thus, LTS tolerance at specific growth stages involved essential stress-responsive genes and TFs, which may be potential targets in genetic improvement for LTS tolerance in rice. However, the constitutive overexpression of these genes has led to metabolic instability and yield penalty and, as observed in so many experiments, has retarded growth under normal conditions, as shown by transgenic plants (Gilmour et al. 2000; Ito et al. 2006; Nakashima et al. 2007). Using stress-inducible promoters such as the *rd29A* promoter instead of constitutive promoters minimizes these side effects on plant growth (Kasuga et al. 2004). However, although the complex nature of the cold-tolerance phenomenon has been explained by transgenic technology, as many genes/TFs have been exploited and manipulated, its field utility is yet to be explored and assessed.

The important *COLD1* gene of Nipponbare in the background of 93-11 exhibited tolerance in the rice SS by encoding a GTPase-accelerating binding factor that regulates G-protein signaling by sensing cold to trigger Ca^{2+} signaling for cold tolerance (Ma et al. 2015). Five QTLs (on chromosomes 1, 2, 4, 6, and 8) were reported from the cross between chilling-tolerant Nipponbare (*japonica*) and chilling-sensitive 93-11 (*indica*) cultivars at 4 °C of cold treatment. Among these QTLs, three (*COLD2*, *COLD4*, and *COLD5*) were found genetically interacting with each other, and together, they contributed PV of 16.8%, while *COLD1* alone exhibited PV of 7.23%. Fine-mapping of *COLD1* leads to the identification of sequential alterations at the first exon (SNP1) and fourth exon (SNP2) and five substitutions in introns (SNP3) in the 4.78-kb region of the *COLD1^j* gene (*LOC_Os04 g51180*). The transgenic approach proved that the SNP2 allele (*COLD1^jap*) had a significant overexpression compared to WT plants and suggested that *COLD1* modulates chilling tolerance in rice. Therefore, a good combination of stage-specific QTLs and cold-tolerance genes could be helpful for developing LTS tolerance in rice. Despite the detection of cold-tolerance genes and QTLs, to date, none of the LTS-tolerant varieties were ever developed through marker-assisted backcrossing (MAB). This raises two questions: first, are the discovered LTS tolerance-related QTLs and genes veri-fiably useful for MAB? Second, is there a lack of training of LTS-tolerance rice breeders to exploit the advances in molecular genetics of LTS tolerance? However, a more reliable and reproducible QTL must be first identified to improve LTS toler-ance through more breeder-friendly MAB approaches.

4.2   Genome-Wide Association Studies for LTS Tolerance

The recent development in high-throughput genome sequencing platforms, GWAS, has become a powerful tool to exploit linkage disequilibrium to dissect traits and identify the genomic regions associated with a trait of interest. GWAS have been used in various research efforts such as drought, salinity, and deficiency and toxicity tolerance to understand the trait associated with the whole-genome sequence of genotypes using a diverse set of rice germplasm accessions. The sequencing of rice genotypes is commonly classified into SNP array genotypes and resequenced SNP genotypes. Zhang et al. (2018) identified high-quality filtered reads with a call rate of 95% for 3867 SNP markers by genotyping 249 *indica* rice varieties using a 5K SNP rice array for cold tolerance at the bud burst stage. GWAS for severity of dam-age (SD) and seed survival rate (SR) revealed 47 SNP loci significantly associated with SD and SR in cold treatment at 5 °C for 5 days. Among these SNPs, the major QTL *qCTSR1-2* on chromosome 1 overlapped with *qCTSD1-2*, which explains 13.2% of the total phenotypic variation. GWAS for germination and reproductive stages that Shakiba et al. (2017) conducted with the Rice Diversity Panel 1 (RDP1), which consisted of 400 *O. sativa* accessions belonging to five major subpopula-tions, resulted in the identification of 42 loci associated with cold tolerance, and several QTLs were colocalized with previously reported LTS-tolerance QTLs. Recently, Xiao et al. (2018) identified a potential candidate locus (*LOC_Os10g34840*) on chromosome 10, which is responsible for cold tolerance at the seedling stage, by assessing the total diversity panel of 1033 rice accessions with 289,231 SNP markers. The loci at 18.58–18.65 Mb overlap with previously reported cold-tolerance QTLs (Xiao et al. 2015), and, furthermore, they have been fine-mapped and validated by quantitative expression analysis. Similarly, using specific locus amplified fragment sequencing (SLAF-seq) technology, Song et al. (2018) conducted GWAS with 150 accessions of rice landraces by using high-density SNPs A total of 26 significant SNPs were associated with cold tolerance at the seedling stage. These SNPs had PVE ranging from 26% to 33%, and among them, three QTLs were colocalized with previously cloned genes such as *OsFAD2*, *OsMYB2*, and *OsCIPK03* related to LTS tolerance at the rice seedling stage (Yang et al. 2012). Interestingly, Song et al. (2018) noticed a strong signal of trait-marker association peaks on chromosome 1, with PVE of 27%. The expression profiling and bioinfor-matics analyses reveal that a novel candidate gene (*Os01g0620100*) showed a sig-nificant difference between the cultivars tolerant and sensitive to LTS because of the polymorphism in the WD40 domain. Thus, *Os01g0620100* is an important source for developing LTS tolerance by using marker-assisted selection. A comprehensive literature survey of GWAS for LTS tolerance resulted in a total of 239 QTLs distributed on all 12 chromosomes (Fig. 3). Among these, 116 QTLs for GS, 90 QTLs for SS, and 33 QTLs for RS were reported by several researchers (Pan et al. 2015; Wang et al. 2016; Sales et al. 2017; Schläppi et al. 2017; Shakiba et al. 2017; Singh et al. 2017; Zhang et al. 2018). The highest number of QTLs was detected on chromosome 2 (42), whereas the lowest number was detected on two chromosomes, 10 and 12 (14 QTLs). The physical position of each stage-specific QTL from GWAS revealed that three major genetic regions on chromosome 1 (41.39–41.86 Mb), chromosome 3 (3.03–3.76 Mb), and chromosome 6 (6.01–6.80 Mb) were colocalized with more than four QTLs. Two QTLs (*qCTGERM1-8* and *qSWTCT1-2*) for GS and two other QTLs (*qCTS1-4* and *qCTS1-5*) for SS stage-specific QTLs overlapped on chromosome 1 (Wang et al. 2016; Shakiba et al. 2017). Similarly, eight QTLs (*qCTS3-6*, *qCTS3-7*, *qCTSR3-1*, *qCTSD3-1*, *qCTS3-8*, *qLTRSSR3-1*, *qLTSSR3-1*, and *qCTS3-9*) detected on chromosome 3 for GS, SS, and RS were colocalized (Pan et al. 2015; Wang et al. 2016; Zhang et al. 2018). Four QTLs associated with three traits related to germination rate, cold tolerance at the seedling stage, and plumule recovery growth after cold exposure were colocalized on chromosome 6 for the GS and SS (Wang et al. 2016; Schläppi et al. 2017; Shakiba et al. 2017). Interestingly, five chromosomal regions at 30.1 Mb (chromosome 4), 5.1 Mb (chromosome 6), 28.3 Mb (chromosome 7), 14.2 Mb (chromosome 8), and 27.3 Mb (chromosome 12) were associated with the GS, and some RS-specific QTLs were also aligned together in the same genetic regions. With a large number of QTLs for stage-specific traits from GWAS information from the genome sequencing data, high-throughput phenotyping and various statistical methods could provide beneficial information for MAB programs and the discovery of potentially useful chilling-tolerance genes/alleles. Furthermore, the combination of gene expression profiling and omics technologies such as proteomics, metabolomics, epigenetics, and genome editing tools will facilitate the confirmation of more candidate gene functions in rice.

4.3 Transcriptomics Related to LTS Tolerance

Transcriptome sequencing has increased the accessibility of genomic resources in various crops, including rice. Transcriptome analysis using microarray technologies is one of the most powerful techniques that link sequence information directly to functional genomics (Sinha et al. 2018) and immensely contributes to understanding the specific tissue- or stress response-related genes in the molecular mechanisms of biotic and abiotic stress tolerance. Comparative transcriptome analysis provides a way to distinguish different genes that are regulated in stress tolerance in comparison to expression patterns between the homologous genes in various crops (Lee et al. 2019). In response to LTS tolerance, Yang and Poovaiah (2003) observed that, at low temperatures, commonly upregulated genes were associated with Ca^{2+}

signal transduction. Several genes such as Ca^{2+}-dependent protein kinases, calmodulin, mitogen-activated protein kinase 1, Ca-transporting ATPases, protein phosphatase 2C family proteins, and serine/threonine-protein kinases related to signal transduction pathways were identified in the endoplasmic reticulum-phase of chilling-stress tolerance (Yang et al. 2012; Zhang et al. 2012b). A key initial event that occurs in cold-stress response is the induction of the AP2/EREBP TF family, which includes CBF/DREB TFs, which are commonly induced within 30 min of cold treatment in plants (Zhang et al. 2004; Ito et al. 2006). CBF/DREB genes have also been shown to be gated by a circadian clock and to display cyclic behavior during cold stress (Gilmour et al. 2004). Previously, Bai et al. (2015), working on the anther transcriptome of photo-thermosensitive genic male sterile rice lines Y58S and P64S under cold stress, identified some differentially expressed genes (DEGs) involved in signal transduction, metabolism, transport, and transcriptional regulation. Among these DEGs, more differentially expressed MYB (myeloblastosis) and three zinc finger family TFs and signal transduction components such as calmodulin-/calcium-dependent protein kinases were observed in the Y58S comparison group. *LOC_Os01g62410* (*OsMYB3R-2*), identified as an upregulated gene in Y58S, encodes for an MYB domain protein activation TF that regulates the CBF pathway and cell cycle progression during cold stress, resulting in increased cold tolerance (Ma et al. 2009). In a similar study, Su et al. (2010) reported the role of MYBS3 (*LOC_Os10g41200*) in regulating signaling pathways at low temperature, suggesting that MYB family members are good candidates for improving LTS tolerance in rice. Furthermore, molecular evidence indicates that CBF responds early, and MYBS late, to chilling stress, suggesting distinct pathways that function sequentially and complementarily to promote short- and long-term cold-stress adaptation in rice. Gene profiling on chilling-tolerant *japonica* rice incubated for 24 h at 10 °C revealed that an "early response" regulatory network including ROS-bZIP1 plays a crucial role in short-term adaptive responses (Yang et al. 2012). Moreover, several regulatory clusters, including bZIP factors acting on as1/ocs/TGA-like element-enriched clusters, R2R3-MYB factors acting on MYB2-like element-enriched clusters, and ERF factors acting on GCC-box/JAre-like element-enriched clusters, are involved in early chilling response, and oxidative signaling by H_2O_2 is at the center of the regulatory network (Yun et al. 2010).

Furthermore, genes involved in gibberellic acid (GA), indole-3-acetic acid (IAA), and cytokinin biosynthesis responded to cold temperature in such a way that their expression profiles were either downregulated or upregulated in cold-susceptible and cold-tolerant rice varieties (Park et al. 2010). For example, the IAA biosynthesis genes *YUCCA1* and *TAA 1:1* showed variety-specific regulation. Among the genes involved in cytokinin biosynthesis and signaling, the expression of *LOG*, *HK1*, and *HK3* was significantly downregulated only in the cold-susceptible variety. Similarly, among the genes involved in ABA biosynthesis, neoxanthin synthase (*NSY*), and ABA-aldehyde oxidase 3 (*AAO3*) were downregulated only in the cold-tolerant variety. It is presumed that the levels of these bioactive hormones are maintained relatively high at cold temperatures in cold-tolerant varieties, which can help minimize the cold stress imposed on developing reproductive organs.

In a comparative transcriptome analysis of the shoots and roots of cold-tolerant variety TNG67 and cold-sensitive variety TCN1, the expression of OsRR4 type-A response regulators in roots of TNG67 was upregulated. The TFs *OsIAA23*, *SNAC2*, *OsWRKY1v2*, 24, 53, 71, *HMGB*, *OsbHLH*, and *OsMyb* were expressed in the roots or shoots of TNG67, and *AP2/ERF* in the shoots and roots of both varieties during cold stress, making them good candidate genes for cold-stress tolerance in rice. Also, phytohormone-related genes for ABA, polyamine, auxin, and jasmonic acid were preferentially upregulated in the shoots and roots of the cold-tolerant genotype. Functional clustering of the majority of DEGs involved in early chilling response showed their role in a complicated chilling-responsive regulatory network such as phytohormone signaling, photosynthesis pathway, ribosome translation machinery, and phenylpropanoid biosynthesis. The localization of the majority of DEGs in chloroplasts suggests a link between chilling-stress tolerance in rice and photosynthesis (Wang et al. 2016). This was observed in a comparative transcriptome profiling of the common wild rice GXWR (China)-derived chilling-tolerant chromosome segment substitution line.

4.4 Proteomics Related to LTS Tolerance

Proteins are the key components in the majority of cellular events; hence, investigating their structure, function, abundance, and interactions at a given time is advantageous to "omics" studies. Protein translational and posttranslational regulations, particularly of stressor-specific protein classes altered due to stress conditions, can also be detected by proteomics, thereby rendering a complex phenomenon such as the tolerance mechanisms for cold and other stresses well understood and addressed. Two-dimensional gel electrophoresis (2-DE) or liquid chromatography coupled with tandem mass spectrometry (LC-MS/MS) is used in protein extraction followed by protein separation and identification (Wittmann-Liebold et al. 2006; Fournier et al. 2007; Hashimoto and Komatsu 2007; Wang et al. 2013c).

Proteomic studies were undertaken pertaining to cold tolerance in rice seedlings and anthers (Cui et al. 2005; Yan et al. 2006; Hashimoto and Komatsu 2007; Zhang et al. 2014a; Lee et al. 2015). Comprehensive transcriptomic and proteomic analyses in rice have illustrated that many genes and functional proteins are involved in the crop's chilling response (Hashimoto and Komatsu 2007; Nakashima et al. 2007; Kanneganti and Gupta 2008; Oh et al. 2009; Lee et al. 2015). Furthermore, many proteins, including *otsA* and *otsB* (trehalose synthesis), choline monooxygenase (glycine betaine synthesis), and *WFT1* and *WFT2* (fructan synthesis) (Garg et al. 2002; Shirasawa et al. 2006), were found to be involved in the regulation of low-temperature tolerance in rice. Using matrix-assisted laser desorption ionization-time of flight mass spectrometry (MALDI-TOF MS), Cui et al. (2005) observed 60 protein spots progressively upregulated in response to LTS. These cold-responsive proteins include four factors of protein biosynthesis, four molecular chaperones, two proteases, eight enzymes involved in the biosynthesis of cell wall components, and seven antioxidative enzymes and proteins linked to energy pathways as in signal transduction, besides two proteins of unknown function. In addition to these proteins, chloroplast proteome is also subject to cold stress because of the localization of identified proteins in the chloroplast. Other cold-responsive proteins are sucrose synthase (Maraña et al. 1990), phenylalanine ammonia-lyase (Sanchez-Ballesta et al. 2000), and ferritin (Kawamura and Uemura 2003). Proteins involved in energy metabolism in the leaf blade were upregulated, while defense-related proteins were downregulated or even disappeared when rice seedlings were exposed to 5 °C for 48 h (Hashimoto and Komatsu 2007). Hashimoto and Komatsu (2007) used the 2D PAGE-based proteomics method for rice root plasma membrane and identified 12 cold-responsive proteins, including cold shock protein-1, which decreased significantly under cold-stress conditions. Most of the cold-responsive proteins associated with energy production, signal transduction, protein synthesis, and defense were revealed by their functional characterization.

Proteins involved in phytohormone biosynthesis play a role in stress tolerance. Asakura et al. (2004) found that GA-related proteins have increased expression during rice seed germination. The elevated expression of GA receptor GID1L2 observed in the resistant strain suggests the role played by GA in mediating the response to cold temperature in the germinating embryo. The low-temperature germination of the resistant rice line was associated with proteins involved in GA signaling, protein trafficking, and ABA-mediated stress response compared with the susceptible strain. Colebrook et al. (2014) also noticed that increased GA biosynthesis and GA signaling were linked to stress tolerance. The elevated expression of TF HBP-1b involved in the general mechanism of protein trafficking through a secretory pathway and two proteins of unknown function, UPF0041 domain-containing protein (*LOC_Os07g26700.1*) and the expressed protein (*LOC_Osg09910.1*), suggested their possible role in seed germination or cold stress (Lee et al. 2015). Phosphorylation of cellular proteins or protein kinase activation during the initial stage of cold acclimation has also been observed (Garbarino et al. 1991), and a fragment of RuBisCO large subunit protein was phosphorylated to a greater extent than others in cold-tolerant rice varieties using 2-DE analysis (Komatsu et al. 1999).

Gel-based protein separation has been used in proteomic studies, which has resulted in the identification of only high-abundant proteins, leaving TFs, kinases, and transport proteins belonging to low-abundant protein classes undetected. Although advanced LC-based separation techniques have resulted in the detectability of low-abundant proteins, phosphorylation, glycosylation, and oxidation caused by posttranslational changes that are likely to be induced by stressors are yet to be comprehensively explained with the use of these approaches. Also, metabolic processes mediated by plant hormones as well as subcellular protein translocation and protein-protein interactions are yet to be found associated with the cold-stress phenomenon and tolerance mechanisms. To move forward with our knowledge on a time-dependent response, new proteomic strategies, such as hydrogen-deuterium exchange and surface plasmon resonance-MS, together with integrated cell biology approaches such as immune precipitation and live imaging analysis, will be required.

4.5 Metabolomics Related to LTS Tolerance

Metabolites are the final response of a biological system to environmental changes so that an aberrant metabolism is linked to the most predictive of phenotypes. The successful application of metabolomics can provide a deeper insight into a plant's phenotypic response to abiotic stresses and can determine the pattern related to stress tolerance. For example, in a targeted metabolite analysis of two rice genotypes under LTS (13/12 °C), Morsy et al. (2007) observed that the chilling-tolerant genotype exhibited accumulated galactose and raffinose, while the same sugars were found to be decreasing in the chilling-sensitive variety. Also, a higher endogenous content of oxidative products and the presence of a more efficient reactive oxygen species (ROS)-scavenging metabolism were found in the chilling-tolerant genotype during chilling stress. Similarly, Arbona et al. (2013) studied photosynthetic dysfunction and effectors of osmotic readjustment at primary metabolite levels (sugars, amino acids, and Krebs cycle intermediates) and secondary levels (antioxidants) and observed a relative accumulation of some primary or secondary metabolites.

Most sugars have earlier been reported to function as osmoprotectants, nutrients, and signaling molecules in rice (Guy et al. 2008; Ma et al. 2009). Similar studies have been carried out in other crops (Urano et al. 2009; Bowne et al. 2012; Araújo et al. 2013). In a comparative metabolomics study between the varieties LTH (cold-tolerant *japonica*) and IR29 (cold-sensitive *indica*) under no-stress and chilling-stress conditions (4 °C for 2, 8, 24, and 48 h, and recovery of 24 h), it was observed that 82 of 106 metabolites exhibited significant differences and described 18.1% of the total PV (Zhao et al. 2013). Of the total of 120 stress vs. control comparisons, 85 (71%) cases had significantly increased amino acid levels, whereas only 7 (6%) cases had significantly decreased amino acid levels, which involved aspartic acid, cysteine, glutamic acid, and glycine. Compared to IR29, LTH recorded more amino acids that significantly increased at all times of the stress as well as considerably higher levels of cysteine, isoleucine, phenylalanine, proline, serine, threonine, and valine. This strongly suggests that differential amino acid accumulation is a general feature of the variety in response to chilling stress at the seedling stage. Regarding organic acids, 81 of 248 (33%) showed significantly decreased levels of LTH compared to IR29, whereas only 24 (10%) cases had significantly increased levels. Consistent decreased levels were obtained for four organic acids (oleic, quinic, eicosanoic, and sinapic) across all times of the stress in both genotypes, implying that energy production is remarkably inhibited in rice during chilling stress. Furthermore, 34 of 304 (11%) cases showed significantly increased levels of some sugars in LTH at later times of the stress, signifying that these late-accumulating sugars may be associated with the cold tolerance of LTH. Zhang et al. (2016) performed a comparative metabolomics study between *japonica* Nipponbare and *indica* 93-11 at six times during chilling treatment and found that amino acid accumulation occurred on a large scale and was consistent with the appearance of chilling injury. The accumulation of antioxidation-related compounds appeared earlier in Nipponbare than in 93-11 at the mid-treatment stage, whereas, during recovery, a higher level of ROS was observed in Nipponbare. Furthermore, metabolites related to stress tolerance and senescence were found to be induced/accumulated in Nipponbare and 93-11, respectively.

The combinational approaches of genome and metabolome are more interesting for phenotype prediction and are quite useful in breeding for stress tolerance. Selection based solely on genetic markers is highly biased because the environment profoundly influences most of the economic traits. The development of a metabolite quantitative trait locus (mQTL) and metabolome-wide association studies (MWAS) could be helpful in crop improvement by overcoming the problems emerging from differing environmental conditions during selection. This field will take advantage of the new plant genomes issued recently and of the modern and more powerful metabolite profiling tools (Dumas 2012; Wei et al. 2018). Some useful results at the metabolome level and the involvement of different metabolites in plant responses, as discussed above, have provided more insight into the complexity of cold-stress response. Even though this has extended our understanding of the molecular mechanism of plant response to stresses, their reconfiguration under stresses is still quite complex because of the involvement of multiple molecular pathways (Guy et al. 2008).

5   Breeding Approaches for LTS Tolerance in Rice

Developing breeding strategies for LTS-tolerant rice varieties is still a challenge to plant breeders because of the complex nature and lack of suitable rice varieties for high-latitude regions, mainly in China, Japan, Australia, and Korea. Therefore, identifying LTS-tolerance traits and breeding tolerant rice varieties are necessary for these regions. Several researchers have proposed different strategies such as adjusting sowing time, selecting suitable rice varieties with growth duration that avoids the peak LTS periods, and replacing LTS-susceptible rice varieties (Ye et al. 2009). However, with rapid changes in GCVs and the expected population increases, breeding for LTS tolerance and improving LTS-tolerance mechanisms are the critical factors to meet future global food demand.

5.1 Improving LTS Tolerance by Conventional Breeding Approaches

Cold tolerance is the ability of rice plants to sustain yield in the presence of low-temperature stress (Shakiba et al. 2017). Genetic breeding has been used as the approach to cope with low-temperature sensitivity. The *indica* subspecies are better adapted to tropical environments such as India, China, and Indonesia, while *japonica* cultivars have more adaptation under temperate climates such as those in Japan, Korea, and Java, Indonesia (Takahashi 1984). It has been observed that *japonica* genotypes are relatively better in tolerating a higher degree of cold stress at the germination stage (Lee 2001; da Cruz and Milach 2004) as well as at the vegetative and reproductive stages (Saito et al. 2004; da Cruz et al. 2006a; Zhang et al. 2017; Xiao et al. 2018). However, some *indica* genotypes from high-latitude regions may have moderate cold tolerance (Gautam et al. 2018). There are reports of some *javanica*, an ecotype of *japonica*, being tolerant of cold (Sweeney and McCouch 2007). Cold-tolerance genes from *javanica* cultivars such as Silewah, Lambayeque 1, and Padi Labou Alumbis were introduced into several temperate *japonica* breeding lines in Japan (Saito et al. 2004).

Selection under field conditions for cold tolerance in rice is unpredictable; hence, for effective selection, robust screening protocols using strong selective agents such as low temperature and the use of controlled air or low water temperature are highly important (da Cruz et al. 2006b). The major limitation to evaluating large plant populations in controlled-temperature environments is to provide enough space for the material. Growth chambers and phytotron facilities lead to quicker and more precise results; however, small-sample testing leads to a loss in effective population size. To deal with these limitations, some rice breeding programs use cold water under field conditions as a selection criterion, allowing the evaluation of many different populations and thousands of plants per population (Snell et al. 2008). Several experimental stations in Japan (Okamoto et al. 1986; Horisue et al. 1988; Nagano 1998; Shinada et al. 2013), Bangladesh (Khatun et al. 2016), and Korea (Jeong et al. 1998) have successfully used cold water to screen rice breeding material for cold tolerance. Different traits depend on the developmental stage, which is used as an indicator to identify cold-tolerant/susceptible lines. Some correlation studies revealed that varieties having germination and seedling tolerance under low-temperature conditions might also be tolerant at the booting and flowering stages (Ye et al. 2009). Inheritance and heritability of LTS tolerance at the germination stage showed involvement of both additive and nonadditive gene actions, with the latter component relatively more important for coleoptile length and coleoptile growth decrease. In a similar kind of study, epistatic interaction (a nonadditive effect) was found to be important for rice germination capacity at low temperature. Some tolerance genes may be more important, whereas others are stage-specific. No such genes have been demonstrated, which are responsible for cold tolerance over all the stages.

Selection for cold-tolerant genotypes in rice has shown a greater success rate due to high heritability estimates for low-temperature germinability (Sthapit and Witcombe 1998; da Cruz et al. 2006b). The nature and magnitude of gene action involved in different traits such as root and shoot length and plant stature are used as indicators under cold stress, which helps in devising a breeding program for developing cold-tolerant genotypes (Datta and Siddiq 1983; Kaw and Khush 1986; Acharya 1987). Involvement of two major genes, *Cts1* and *Cts2*, has been reported at the vegetative stage responsible for cold tolerance and estimated through leaf yellowing (Kwak 1984) and withering (Nagamine 1991). On the other hand, studies of

Andaya and Tai (2006) reported cold tolerance at the vegetative stage as a complex trait involving multiple genes and, similarly, several major genes are involved in cold tolerance at the reproductive stage. Studies on correlating cold tolerance with morphological traits in temperate *japonica* varieties show that tolerance is governed by four or more loci, which were shown to be linked to morphological marker genes (Futuhara and Toriyama 1966). It has been reported that several cold-tolerant cultivars have already been released in different countries despite the complex genetic basis and difficulties and limitations of selection for cold tolerance (da Cruz et al. 2013). These varieties may serve as a useful genetic repository rather than gene donors to develop cold tolerance in new rice varieties, which are lacking the trait. Much work on trait improvement resulted in 57% of the rice varieties in Korea having tolerance, and they are highly tolerant of low temperatures (Lee 2001). Each country developed its own strategy for breeding for cold tolerance. However, the main advances have been obtained within *japonica* cultivars. Therefore, the challenge remains to develop *indica*-type cultivars with adequate cold tolerance for high-latitude regions (Bierlen et al. 1997). An apparently simple solution could be to cross *indica* genotypes with *japonica* ones to transfer genes for cold tolerance from *japonica*. However, the differences between these two rice groups make it difficult to maintain desirable *indica* characteristics, such as the cooking quality needed for consumer acceptance. Also, gene introgression from *indica* to *japonica* rice has shown problems of high sterility and poor plant type with some linkage drag in the progenies (Khush 2005). Furthermore, overcoming the major constraints associated with conventional breeding for cold tolerance in rice has been of limited success with gene transfer from wild relatives (Flowers and Yeo 1995). Another difficulty in the selection of cold-tolerant varieties in field conditions is the lack of appropriate selection pressure, which provides critical stress-environment control (Blum 1988). Conventional breeding is an extremely slow process for generating varieties with improved tolerance of stress conditions. In addition, incompatibility in wider crosses along with limited germplasm resources for stress tolerance is a major limitation encountered in conventional breeding. Therefore, integrating traditional breeding programs with modern compatible molecular methods and elucidating the genetic mechanisms of this complex phenomenon will improve precision and speed in the development of genotypes with low-temperature tolerance.

5.2 Improving LTS Tolerance by Selective Introgression

The selective introgression breeding (SIB) strategy is a powerful tool for the simultaneous improvement of tolerance and also to dissect complex traits. This is based on two approaches, such as developing a larger number of trait-specific introgression lines via early backcross breeding and using a marker-facilitated approach to track the gene flow from donors to recipients. In addition to that, the selected early backcross breeding population can be used for identifying the genomic regions for the association of target traits of interest through QTL mapping (Pang et al. 2017;

Feng et al. 2018; Jewel et al. 2018). In the Green Super Rice (GSR) breeding program, this SIB strategy has laid the foundation for a better molecular and genetic understanding of complex traits in rice. This methodology provides more advantages than classical QTL mapping methods, such as decreasing the cost in genotyping and phenotyping with selected lines, high statistical power to detect QTLs for targeted traits, and the selected lines are expected to carry the beneficial alleles of QTLs (Ali et al. 2017; Liang et al. 2018). By dissecting complex traits and developing trait-specific introgression lines, Liang et al. (2018) successfully identified cold-tolerance QTLs on chromosomes 1, 2, 3, 4, 6, 9, 11, and 12. A total of 17 QTLs for cold tolerance at the reproductive stage were detected in 84 cold-tolerant introgression lines (ILs) selected from five BC_2F_4 populations in a Chaoyou genetic background using a consensus linkage map. In addition, 310 random ILs from the same BC populations were used for dissecting the genetic networks underlying cold tolerance by detecting QTLs and functional genetic units (FGUs). This study led to the discovery of QTLs *qCT3.12*, *qCT6.7*, and *qCT9.6* that were validated in random BC populations. A QTL for LTG was fine-mapped by Shim et al. (2019) with the help of two introgression lines, TR5 and TR20, which were crossed to common parent Hwaseong to develop $F_{2:3}$ populations. *qLTG1* was located in a 167-kb region between two SSR markers (RM10310 and RM10326) and was found to harbor 18 genes, with nine of them annotated with specific gene functions. The allelic effect at the *qLTG1* locus was contributed by *Oryza rufipogon* that was observed to increase LTG and spikelets per panicle.

5.3   Improving LTS Tolerance by Genetic Transformation

Transgenic approaches can be used to improve cold tolerance in rice. These methods are promising, wherein improvement can be made through the introduction from across trans-species or through the disruption of specific DNA sequences using RNAi technology. Gene expression analysis has made clear the physiological mechanisms responsible for tolerance against various abiotic stresses, including low temperature, in plants (Gao et al. 2008; Pan et al. 2020; Xu et al. 2020). Several chilling stress-responsive genes after their isolation and characterization have been found to encode proteins that act as enzymes for the biosynthesis of osmoprotectants. Transcription factors, especially those from the CBF/DREB1 family, play a more important role (Wang et al. 2008; Zhang et al. 2009; Su et al. 2010). Using the integrated approach of T-DNA-tagged rice plants and inverse PCR, many genes involved with several metabolic pathways were detected and characterized at the molecular level. When plants are exposed to cold stress, they show changes at the gene expression level, and the products of cold-inducible genes may either directly protect against cold stress or further regulate the expression of other genes (Yamaguchi-Shinozaki and Shinozaki 2004; Chen et al. 2008; Su et al. 2010). Several transgenic rice lines or overexpressed lines (OX) have been developed, which are responsible for the overexpression of cold-inducible genes, usually encoding TFs for the transcriptional regulation of these genes. The upregulated or overexpressed important genes/TFs with respect to their function, cold stress condition, the phenotype of OX lines, and side effects in OX lines, if any, in comparison to nontransformed (WT) plants, are described in Table 5. Overexpression of *OsCTZFP8* in transgenic rice exhibited more cold-tolerant phenotypes than nontransgenic control plants by showing higher pollen fertility and seed-setting rate (Jin et al. 2018). The significant overexpression of cold-responsive genes and TFs suggests a practical utility of these genes at the field level requiring a systematic assessment, and this needs further use in crop improvement.

5.4   Improving LTS Tolerance by Genome Editing

Emerging advanced genome editing tools (GETs) such as zinc-finger nucleases (ZFNs), transcription activator-like effector nucleases (TALENS), meganucleases (MNs), and CRISPR/Cas (clustered regularly interspaced short palindromic repeats/CRISPR-associated protein) play a greater role in the understanding of the various biological functions by manipulating the desired target gene of interest and also providing new opportunities to create genetic diversity in various crops (Vats et al. 2019; Wolter et al. 2019). However, these ZFNs, TALENS, and MNs are so expensive in the cloning procedure, have low target specificity, and are time-consuming and laborious, whereas the CRISPR/Cas system is a cheap, easy-to-design, and unique tool for precise and efficient genome editing at the single-base level. CRISPR/Cas9 involves mainly single-guided RNA (sgRNA) in contrast to ZFNs and TALENS and provides mutagenesis at high frequency, thus increasing the number of recombination sites and target specificity in plant genomes (Jaganathan et al. 2018; Molla and Yang 2019). Currently, using this technology, several researchers have targeted genome editing for biotic and abiotic stress tolerance and also grain quality-related genes. Over the past decade, CRISPR/Cas9 has been widely used to produce novel rice varieties with improved target traits such as grain weight, panicle architecture, aroma, high amylose content, and tolerance of bacterial blight and blast, drought, and cold (Mishra et al. 2018; Fiaz et al. 2019; Jun et al. 2019). For instance, Li et al. (2016) edited four grain yield-related genes (*Gn1a*, *DEP1*, *GS3*, and *IPA1*) and Xu et al. (2016) edited three genes related to grain weight (*GW2*, *GW5*, and *TGW6*) using the CRISPR/Cas9 system. They noticed a significant increase in phenotypic traits, such as increasing grain number and size, and dense erect panicles. In response to cold stress in rice, the CRISPR/Cas9 system was used to edit two TFs, TIFY1a (*LOC_Os03g47970*) and TIFY1b (*LOC_Os03g52450*), in rice, revealing single-base-pair insertion and deletion and also long fragment deletion. Thus, employing CRISPR/Cas9 technology to investigate the role of *TIFY1a* and *TIFY1b* might reveal a novel pathway that controls cold adaptation in rice (Huang et al. 2017). CBFs have been studied in response to cold tolerance through genome editing, which has shown that CBF triple mutants were extremely sensitive to cold acclimation-dependent freezing stress (Jia et al. 2016).

Breeding for LTS using conventional methods has met with limited success in improving tolerant rice varieties because of the lack of efficient selection criteria, the time-consuming work, and required expensive facilities for screening, and it also involves a multigenic trait. Therefore, an alternative breeding strategy such as genomic-assisted breeding and gene transformation technologies can provide a viable option to improve LTS tolerance in rice. The ability of the CRISPR/Cas9 system to generate transgene-free genome-modified plants, create site-specific and SNP mutations, and make large-scale changes in chromosome structure at the single and multicellular levels provides a comprehensive target trait mechanism and more helpful breeding strategy to enhance biotic and abiotic stress tolerance in rice. The successful editing of important genes related to improving grain yield and quality traits has proven that this CRISPR/Cas9 technology could be a promising tool for understanding the molecular and physiological functional aspects of genes and TFs that influence target traits.

6 Conclusions and Future Prospects

This chapter has mainly focused on LTS-tolerance traits and associated genetics in rice, which have significant importance in tropical and temperate regions across the globe for increasing yield productivity and sustainability. The LTS tolerance-related QTLs, genes, enzymes, and their interactions are involved in stress-tolerance mechanisms in different growth stages. In order to understand the interaction of genetics, physiological and metabolic responses to plants need to be analyzed using a growth-stage specific and precise phenotypic characterization. The LTS-tolerance trait associated with QTLs and genes will be useful in designing molecular breeding strategies for improving LTS tolerance in rice varieties. The combinational approaches of genomics and metabolomics are more interesting for phenotype prediction and are quite useful in breeding for stress tolerance. Selection based solely on genetic markers is highly biased because the environment profoundly influences most of the economic traits. The development of a metabolomics quantitative trait locus and metabolome-wide association studies could be helpful in crop improvement by overcoming the problems emerging from differing environmental conditions during selection. This field will take advantage of the new plant genomes issued recently and of the modern and more powerful metabolite profiling tools (Dumas 2012; Wei et al. 2018). Some useful results at the metabolome level and the involvement of different metabolites in plant responses, as discussed above, have provided more insight into the complexity of cold-stress response. Even though this has extended our understanding of the molecular mechanisms of plant response to stresses, their reconfiguration under stresses is still quite complex because of the involvement of multiple molecular pathways (Guy et al. 2008).

However, in a breeding program for the development of LTS tolerance, advances in omics-related technologies have led to the development of well-planned phenotypic experiments that offer deeper insight into gene function along with gene effects on the phenotype in a specified biological context. There is a significant positive correlation of LTS-tolerance traits among the different growth stages of rice, and this suggests that rice varieties with high germination rate and early seedling vigor-related traits under LTS conditions are also likely to be more tolerant in the reproductive stage (Ye et al. 2009). It is reasonable to consider that LTS-tolerant genotypes in the GS, SS, and RS may rely on the diverse genetic components of QTLs and genes on different chromosomes that are ensuring LTS tolerance. The rapid development of advanced genome sequencing and larger sets of polymorphic SNPs help in identifying potential candidate genes. Based on the comprehensive literature survey under LTS regarding GS-, SS-, and RS-specific traits and associated with fine-tuned and colocalized stage-specific QTLs (in GS and SS), the highest numbers of QTLs were located on chromosomes 1 and 7 (Pan et al. 2015; Shakiba et al. 2017; Zhang et al. 2018; Sun et al. 2019).

Similarly, in three combinations of stage-specific QTLs (GS, SS, and RS) associated with multiple traits on chromosome 10 (Li et al. 2013; Pan et al. 2015; Schläppi et al. 2017), seven genetic regions are colocalized with candidate genes and QTLs on four chromosomes (1, 3, 5, and 6), and this provides insights into identifying the candidate genes and developing functional allele-specific markers for improving LTS tolerance in breeding programs. The concept of omics studies has amassed a great deal of information at the transcript, protein, and metabolite levels to perceive the tolerance mechanisms of plants under stress. To fully understand the complex regulatory nature of plants against stresses, a highly coordinated approach such as systems biology needs to be identified. Unfortunately, the integration of data outputs from phenomes, transcriptomes, proteomes, and metabolomes has been found to be inefficient and ineffective yet, despite the marvelous progress shown in bioinformatics. However, future approaches need to integrate these through robust computing platforms that should be able to predict the active biochemical and molecular genetic networks to exploit LTS tolerance in rice breeding. This could potentially lead to detecting and identifying master stress regulators to target the right biomarkers for improving rice LTS tolerance effectively.

Acknowledgments The authors would like to thank and acknowledge the Bill & Melinda Gates Foundation for providing a research grant to ZL for the Green Super Rice project under ID OPP1130530. We would also like to thank the Department of Agriculture, Philippines, for providing funds to JA under the Next-Gen project, and also thank and acknowledge IRRI Communications for English language editing and anonymous internal reviewers for their valuable suggestions and constructive comments that helped improve this manuscript.

References

Acharya S (1987) Genetic parameters and their implication in breeding cold tolerant varieties of rice (*Oryza sativa* L.). Crop Improv 14:100–103

Ali AJ, Xu JL, Ismail AM et al (2006) Hidden diversity for abiotic and biotic stress tolerances in the primary gene pool of rice revealed by a large backcross breeding program. Field Crop Res 97:66–76

Ali J, Xu J-L, Gao Y-M et al (2017) Harnessing the hidden genetic diversity for improving multiple abiotic stress tolerance in rice (*Oryza sativa* L.). PLoS One 12:e0172515

Almeida DM, Almadanim MC, Lourenço T et al (2016) Screening for abiotic stress tolerance in rice: salt, cold, and drought. In: Environmental responses in plants. Springer, New York, NY, pp 155–182

Andaya V, Mackill D (2003a) QTLs conferring cold tolerance at the booting stage of rice using recombinant inbred lines from a japonica × indica cross. Theor Appl Genet 106:1084–1090

Andaya VC, Mackill DJ (2003b) Mapping of QTLs associated with cold tolerance during the vegetative stage in rice. J Exp Bot 54:2579–2585. https://doi.org/10.1093/jxb/erg243

Andaya VC, Tai TH (2006) Fine mapping of the *qCTS12* locus, a major QTL for seedling cold tolerance in rice. Theor Appl Genet 113:467–475

Araújo WL, Trofimova L, Mkrtchyan G et al (2013) On the role of the mitochondrial 2-oxoglutarate dehydrogenase complex in amino acid metabolism. Amino Acids 44:683–700

Arbona V, Manzi M, de Ollas C, Gómez-Cadenas A (2013) Metabolomics as a tool to investigate abiotic stress tolerance in plants. Int J Mol Sci 14:4885–4911

Asakura T, Nakaizumi T, Hirose S et al (2004) Proteome analysis of the regulation of rice seed germination. In: Plant and cell physiology. Oxford University Press, Oxford, pp s148–s148

Bai B, Wu J, Sheng W-T et al (2015) Comparative analysis of anther transcriptome profiles of two different rice male sterile lines genotypes under cold stress. Int J Mol Sci 16:11398–11416

Basuchaudhuri P (2014) Cold tolerance in rice cultivation. CRC Press, Boca Raton, FL

Bertin P, Kinet JM, Bouharmont J (1996) Evaluation of chilling sensitivity in different rice varieties: relationship between screening procedures applied during germination and vegetative growth. Euphytica 89:201–210

Bevilacqua CB, Basu S, Pereira A et al (2015) Analysis of stress-responsive gene expression in cultivated and weedy rice differing in cold stress tolerance. PLoS One 10:e0132100

Bierlen R, Wailes EJ, Cramer GL (1997) The mercosur rice economy. Ark Exp Stat Bull 954:1–58

Biswas PS, Khatun H, Das N et al (2017) Mapping and validation of QTLs for cold tolerance at seedling stage in rice from an indica cultivar Habiganj Boro VI (Hbj.BVI). 3 Biotech 7:359. https://doi.org/10.1007/s13205-017-0993-1

Blum A (1988) Plant breeding for stress environments. CRC Press, Boca Raton, FL. 223 p

Board JE, Peterson ML, Ng E (1980) Floret sterility in rice in a cool environment. 1. Agron J 72:483–487

Bonnecarrère V, Borsani O, Díaz P et al (2011) Response to photoxidative stress induced by cold in japonica rice is genotype dependent. Plant Sci 180:726–732

Bonnecarrere V, Quero G, Monteverde E et al (2015) Candidate gene markers associated with cold tolerance in vegetative stage of rice (*Oryza sativa* L.). Euphytica 203:385–398

Bowne JB, Erwin TA, Juttner J et al (2012) Drought responses of leaf tissues from wheat cultivars of differing drought tolerance at the metabolite level. Mol Plant 5:418–429

Chen L, Qiaojun L, Sun ZX et al (2006a) QTL mapping of low temperature germinability in rice. Zhongguo Shuidao Kexue 20:159–164

Chen L, Qiaojun L, Sun ZX et al (2006b) QTL mapping of low temperature on germination rate of rice. Rice Sci 13:93–98

Chen JQ, Meng XP, Zhang Y et al (2008) Over-expression of OsDREB genes lead to enhanced drought tolerance in rice. Biotechnol Lett 30:2191–2198. https://doi.org/10.1007/s10529-008-9811-5

Chen NA, Xu Y, Wang XIN et al (2011) *OsRAN2*, essential for mitosis, enhances cold tolerance in rice by promoting export of intranuclear tubulin and maintaining cell division under cold stress. Plant Cell Environ 34:52–64

Chen J, Lin T, Xu H et al (2012) Cold-induced changes of protein and phosphoprotein expression patterns from rice roots as revealed by multiplex proteomic analysis. Plant Omics 5:194

Cheng L, Wang J, Uzokwe V et al (2012) Genetic analysis of cold tolerance at seedling stage and heat tolerance at anthesis in rice (*Oryza sativa* L.). J Integr Agric 11:359–367

Chinnusamy V, Ohta M, Kanrar S et al (2003) ICE1: a regulator of cold-induced transcriptome and freezing tolerance in *Arabidopsis*. Genes Dev 17:1043–1054

Chinnusamy V, Zhu J, Zhu J-K (2007) Cold stress regulation of gene expression in plants. Trends Plant Sci 12:444–451

Chinnusamy V, Zhu JK, Sunkar R (2010) Gene regulation during cold stress acclimation in plants. In: Plant stress tolerance. Springer, New York, NY, pp 39–55

Chou WL, Huang LF, Fang JC et al (2014) Divergence of the expression and subcellular localization of CCR4-associated factor 1 (CAF1) deadenylase proteins in *Oryza sativa*. Plant Mol Biol 85:443–458. https://doi.org/10.1007/s11103-014-0196-7

Colebrook EH, Thomas SG, Phillips AL, Hedden P (2014) The role of gibberellin signalling in plant responses to abiotic stress. J Exp Biol 217:67–75

da Cruz RP, Milach SCK (2000) Breeding for cold tolerance in irrigated rice. Ciência Rural 30:909–917

da Cruz RP, Milach SCK (2004) Cold tolerance at the germination stage of rice: methods of evaluation and characterization of genotypes. Sci Agric 61:1–8

da Cruz RP, Milach SCK, Federizzi LC (2006a) Rice cold tolerance at the reproductive stage in a controlled environment. Sci Agric 63:255–261

da Cruz RP, Milach SCK, Federizzi LC et al (2006b) Inheritance of rice cold tolerance at the germination stage. Genet Mol Biol 29:314–320

da Cruz RP, Duarte ITL, Cabreira C (2010) Inheritance of rice cold tolerance at the seedling stage. Sci Agric (Piracicaba, Braz) 67:669–674. https://doi.org/10.1590/S0103-90162010000600008

da Cruz RP, Sperotto RA, Cargnelutti D et al (2013) Avoiding damage and achieving cold tolerance in rice plants. Food Energ Secur 2:96–119. https://doi.org/10.1002/fes3.25

Cui S, Huang F, Wang J et al (2005) A proteomic analysis of cold stress responses in rice seedlings. Proteomics 5:3162–3172

Cui Y, Zhang F, Zhou Y (2018) The application of multi-locus GWAS for the detection of salt-tolerance loci in rice. Front Plant Sci 9:1464

Dai L, Lin X, Ye C et al (2004) Identification of quantitative trait loci controlling cold tolerance at the reproductive stage in Yunnan landrace of rice, Kunmingxiaobaigu. Breed Sci 54:253–258

Datta D, Siddiq EA (1983) Genetic analysis of cold tolerance at seedling phase in rice. Indian J Genet Plant Breed 43:345–349

Ding Y, Shi Y, Yang S (2019) Advances and challenges in uncovering cold tolerance regulatory mechanisms in plants. New Phytol 222:1690–1704

Doherty CJ, Van Buskirk HA, Myers SJ, Thomashow MF (2009) Roles for *Arabidopsis* CAMTA transcription factors in cold-regulated gene expression and freezing tolerance. Plant Cell 21:972–984

Donoso G, Cabas P, Paredes M et al (2015) Cold tolerance evaluation of temperate rice (*Oryza sativa* L. ssp. *japonica*) genotypes at seedling stage. Gayana Bot 72:1–13

Dumas M-E (2012) Metabolome 2.0: quantitative genetics and network biology of metabolic phenotypes. Mol BioSyst 8:2494–2502

Farrell T, Fox KM, Williams RL et al (2001) Temperature constraints to rice production in Australia and Laos: a shared problem. In: Increased lowland rice production in the mekong region. Australian Centre for International Agricultural Research, Canberra, ACT, pp 129–137

Feng B, Chen K, Cui Y et al (2018) Genetic dissection and simultaneous improvement of drought and low nitrogen tolerances by designed QTL pyramiding in rice. Front Plant Sci 9:306. https://doi.org/10.3389/fpls.2018.00306

Fiaz S, Ahmad S, Noor MA et al (2019) Applications of the CRISPR/Cas9 system for rice grain quality improvement: perspectives and opportunities. Int J Mol Sci 20:888

Flowers TJ, Yeo AR (1995) Breeding for salinity resistance in crop plants: where next? Funct Plant Biol 22:875–884

Fournier ML, Gilmore JM, Martin-Brown SA, Washburn MP (2007) Multidimensional separations-based shotgun proteomics. Chem Rev 107:3654–3686

Fujino K, Matsuda Y (2010) Genome-wide analysis of genes targeted by qLTG3-1 controlling low-temperature germinability in rice. Plant Mol Biol 72:137

Fujino K, Sekiguchi H (2011) Origins of functional nucleotide polymorphisms in a major quantitative trait locus, qLTG3-1, controlling low-temperature germinability in rice. Plant Mol Biol 75:1–10

Fujino K, Sekiguchi H, Sato T et al (2004) Mapping of quantitative trait loci controlling low-temperature germinability in rice (Oryza sativa L.). Theor Appl Genet 108:794–799

Fujino K, Sekiguchi H, Matsuda Y et al (2008) Molecular identification of a major quantitative trait locus, qLTG3-1, controlling low-temperature germinability in rice. Proc Natl Acad Sci U S A 105:12623–12628. https://doi.org/10.1073/pnas.0805303105

Fursova OV, Pogorelko GV, Tarasov VA (2009) Identification of ICE2, a gene involved in cold acclimation which determines freezing tolerance in Arabidopsis thaliana. Gene 429:98–103

Futuhara Y, Toriyama K (1966) Genetic studies on cool tolerance in rice: III. Linkage relations between genes controlling cool tolerance and marker genes of NAGAO and TAKAHASHI. Jpn J Breed 16:231–242

Gao JP, Chao DY, Lin HX (2008) Toward understanding molecular mechanisms of abiotic stress responses in rice. Rice 1:36–51

Garbarino JE, Hurkman WJ, Tanaka CK, DuPont FM (1991) In vitro and in vivo phosphorylation of polypeptides in plasma membrane and tonoplast-enriched fractions from barley roots. Plant Physiol 95:1219–1228

Garg AK, Kim J-K, Owens TG et al (2002) Trehalose accumulation in rice plants confers high tolerance levels to different abiotic stresses. Proc Natl Acad Sci U S A 99:15898–15903

Gautam A, Suresh J, Gudade BA et al (2018) Rice: grappling with cold under climatic changes, global impact and counter strategies. Adv Res 13:1–9

Ge L-F, Chao D-Y, Shi M et al (2008) Overexpression of the trehalose-6-phosphate phosphatase gene OsTPP1 confers stress tolerance in rice and results in the activation of stress responsive genes. Planta 228:191–201

Ghosh T, Rai M, Tyagi W, Challam C (2016) Seedling stage low temperature response in tolerant and susceptible rice genotypes suggests role of relative water content and members of OsSNAC gene family. Plant Signal Behav 11:1–5. https://doi.org/10.1080/15592324.2016.1138192

Gilmour SJ, Sebolt AM, Salazar MP et al (2000) Overexpression of the Arabidopsis CBF3 transcriptional activator mimics multiple biochemical changes associated with cold acclimation. Plant Physiol 124:1854–1865

Gilmour SJ, Fowler SG, Thomashow MF (2004) Arabidopsis transcriptional activators CBF1, CBF2, and CBF3 have matching functional activities. Plant Mol Biol 54:767–781

Gothandam KM, Nalini E, Karthikeyan S, Shin JS (2010) OsPRP3, a flower specific proline-rich protein of rice, determines extracellular matrix structure of floral organs and its overexpression confers cold-tolerance. Plant Mol Biol 72:125

Greco M, Chiappetta A, Bruno L, Bitonti MB (2012) In Posidonia oceanica cadmium induces changes in DNA methylation and chromatin patterning. J Exp Bot 63(2):695–709

Gupta BB, Salgotra RK, Bali AS (2009) Status paper on rice in Jammu and Kashmir. Rice Knowledge Management Portal (RKMP), Hyderabad. http://www.rkmp.co.in/

Gururani MA, Venkatesh J, Ganesan M et al (2015) In vivo assessment of cold tolerance through chlorophyll-a fluorescence in transgenic zoysiagrass expressing mutant phytochrome A. PLoS One 10:1–17. https://doi.org/10.1371/journal.pone.0127200

Guy C, Kaplan F, Kopka J et al (2008) Metabolomics of temperature stress. Physiol Plant 132:220–235

Han LZ, Qiao YL, Cao GL et al (2004) QTLs analysis of cold tolerance during early growth period for rice. Rice Sci 11:245–250

Han LZ, Zhang YY, Qiao YL et al (2006) Genetic and QTL analysis for low-temperature vigor of germination in rice. Acta Genet Sin 33:998–1006

Hashimoto M, Komatsu S (2007) Proteomic analysis of rice seedlings during cold stress. Proteomics 7:1293–1302

Horisue N, Kunihiro Y, Higashi T et al (1988) Screening for cold tolerance of Chinese and Japanese rice varieties and selection of standard varieties. Trop Agric Res Ser 21:76–87

Hou MY, Jiang L, Wang CM, Wan JM (2003) Detection and analysis of QTLs for low temperature germinability in rice (*Oryza sativa* L.). Rice Genet Newsl 20:52–55

Hu H, You J, Fang Y et al (2008) Characterization of transcription factor gene *SNAC2* conferring cold and salt tolerance in rice. Plant Mol Biol 67:169–181

Huang J, Sun S, Xu D et al (2012) A TFIIIA-type zinc finger protein confers multiple abiotic stress tolerances in transgenic rice (*Oryza sativa* L.). Plant Mol Biol 80:337–350

Huang X, Zeng X, Li J, Zhao D (2017) Construction and analysis of *tify1a* and *tify1b* mutants in rice (*Oryza sativa*) based on CRISPR/Cas9 technology. J Agric Biotechnol 25:1003–1012

Hussain HA, Hussain S, Khaliq A et al (2018) Chilling and drought stresses in crop plants: implications, cross talk, and potential management opportunities. Front Plant Sci 9:393

Ito Y, Katsura K, Maruyama K et al (2006) Functional analysis of rice DREB1/CBF-type transcription factors involved in cold-responsive gene expression in transgenic rice. Plant Cell Physiol 47:141–153. https://doi.org/10.1093/pcp/pci230

Jaganathan D, Ramasamy K, Sellamuthu G et al (2018) CRISPR for crop improvement: an update review. Front Plant Sci 9:985

Jena KK, Kim SM, Suh JP et al (2012) Identification of cold-tolerant breeding lines by quantitative trait loci associated with cold tolerance in rice. Crop Sci 52:517–523

Jeong EG, Baek MK, Yea JD et al (1998) Screening of characters related to cold tolerance of Korean leading rice varieties. RDA J Crop Sci 40:25

Jewel Z, Ali J, Pang Y et al (2018) Developing green super rice varieties with high nutrient use efficiency by phenotypic selection under varied nutrient conditions. Crop J 7:368–377. https://doi.org/10.1016/j.cj.2019.01.002

Ji SL, Jiang L, Wang YH et al (2008) QTL and epistasis for low temperature germinability in rice. Acta Agron Sin 34:551–556

Ji H, Kim SR, Kim YH et al (2010a) Inactivation of the CTD phosphatase-like gene *OsCPL1* enhances the development of the abscission layer and seed shattering in rice. Plant J 61:96–106. https://doi.org/10.1111/j.1365-313X.2009.04039.x

Ji ZJ, Zeng YX, Zeng DI et al (2010b) Identification of QTLs for rice cold tolerance at plumule and 3-leaf-seedling stages by using QTLNetwork software. Rice Sci 17:282–287. https://doi.org/10.1016/S1672-6308(09)60028-7

Ji X, Dong B, Shiran B et al (2011) Control of abscisic acid catabolism and abscisic acid homeostasis is important for reproductive stage stress tolerance in cereals. Plant Physiol 156:647–662

Jia Y, Ding Y, Shi Y et al (2016) The *cbfs* triple mutants reveal the essential functions of *CBFs* in cold acclimation and allow the definition of CBF regulons in *Arabidopsis*. New Phytol 212:345–353

Jiang L, Xun M, Wang J, Wan J (2008) QTL analysis of cold tolerance at seedling stage in rice (*Oryza sativa* L.) using recombination inbred lines. J Cereal Sci 48:173–179

Jiang Y, Cai Z, Xie W et al (2012) Rice functional genomics research: progress and implications for crop genetic improvement. Biotechnol Adv 30:1059–1070. https://doi.org/10.1016/j.biotechadv.2011.08.013

Jiang N, Shi S, Shi H et al (2017) Mapping QTL for seed germinability under low temperature using a new high-density genetic map of rice. Front Plant Sci 8:1–9. https://doi.org/10.3389/fpls.2017.01223

Jin YM, Piao R, Yan YF et al (2018) Overexpression of a new zinc finger protein transcription factor OsCTZFP8 improves cold tolerance in rice. Int J Genom 23:5480617

Joo J, Lee YH, Kim Y et al (2013) Abiotic stress responsive rice *ASR1* and *ASR3* exhibit different tissue-dependent sugar and hormone-sensitivities. Mol Cell 35:421–435. https://doi.org/10.1007/s10059-013-0036-7

Jun REN, Xixun HU, Kejian W, Chun W (2019) Development and application of CRISPR/Cas system in rice. Rice Sci 26:69–76

Kanneganti V, Gupta AK (2008) Overexpression of *OsiSAP8*, a member of stress associated protein (SAP) gene family of rice confers tolerance to salt, drought and cold stress in transgenic tobacco and rice. Plant Mol Biol 66:445–462

Kasuga M, Miura S, Shinozaki K, Yamaguchi-Shinozaki K (2004) A combination of the *Arabidopsis* DREB1A gene and stress-inducible *rd29A* promoter improved drought- and low-temperature stress tolerance in tobacco by gene transfer. Plant Cell Physiol 45:346–350

Kaw RN, Khush GS (1986) Combining ability for low-temperature tolerance in rice. In: Rice genetics I (in two parts). World Scientific, Singapore, pp 593–612

Kawamura Y, Uemura M (2003) Mass spectrometric approach for identifying putative plasma membrane proteins of *Arabidopsis* leaves associated with cold acclimation. Plant J 36:141–154

Khatun H, Biswas PS, Hwang HG, Kim K-M (2016) A quick and simple in-house screening protocol for cold tolerance at seedling stage in rice. Plant Breed Biotechnol 4:373–378

Khush GS (2005) What it will take to feed 5.0 billion rice consumers in 2030. Plant Mol Biol 59:1–6

Kim S-I, Tai TH (2011) Evaluation of seedling cold tolerance in rice cultivars: a comparison of visual ratings and quantitative indicators of physiological changes. Euphytica 178:437–447

Kim S-I, Tai TH (2013) Identification of SNPs in closely related Temperate Japonica rice cultivars using restriction enzyme-phased sequencing. PLoS One 8:e60176

Kim S-H, Kim J-Y, Kim S-J et al (2007) Isolation of cold stress-responsive genes in the reproductive organs, and characterization of the *OsLti6b* gene from rice (*Oryza sativa* L.). Plant Cell Rep 26:1097–1110

Kim S-J, Lee S-C, Hong SK et al (2009) Ectopic expression of a cold-responsive *OsAsr1* cDNA gives enhanced cold tolerance in transgenic rice plants. Mol Cell 27:449–458

Kim S-MM, Suh J-PP, Lee C-KK et al (2014) QTL mapping and development of candidate gene-derived DNA markers associated with seedling cold tolerance in rice (*Oryza sativa* L.). Mol Gen Genomics 289:333–343. https://doi.org/10.1007/s00438-014-0813-9

Koh S, Lee SC, Kim MK et al (2007) T-DNA tagged knockout mutation of rice OsGSK1, an orthologue of *Arabidopsis* BIN2, with enhanced tolerance to various abiotic stresses. Plant Mol Biol 65:453–466. https://doi.org/10.1007/s11103-007-9213-4

Komatsu S, Karibe H, Hamada T, Rakwal R (1999) Phosphorylation upon cold stress in rice (*Oryza sativa* L.) seedlings. Theor Appl Genet 98:1304–1310

Komatsu S, Yang G, Khan M et al (2007) Over-expression of calcium-dependent protein kinase 13 and calreticulin interacting protein 1 confers cold tolerance on rice plants. Mol Gen Genomics 277:713

Koseki M, Kitazawa N, Yonebayashi S et al (2010) Identification and fine mapping of a major quantitative trait locus originating from wild rice, controlling cold tolerance at the seedling stage. Mol Gen Genomics 284:45–54

Kuk YI, Shin JS, Burgos NR et al (2003) Antioxidative enzymes offer protection from chilling damage in rice plants. Crop Sci 43:2109–2117

Kumar K, Rao KP, Sharma P, Sinha AK (2008) Differential regulation of rice mitogen activated protein kinase kinase (MKK) by abiotic stress. Plant Physiol Biochem 46:891–897. https://doi.org/10.1016/j.plaphy.2008.05.014

Kuroki M, Saito K, Matsuba S et al (2007) A quantitative trait locus for cold tolerance at the booting stage on rice chromosome 8. Theor Appl Genet 115:593–600

Kuroki M, Saito K, Matsuba S (2009) Quantitative trait locus analysis for cold tolerance at the booting stage in a rice cultivar, Hatsushizuku. Jpn Agric Res Q 43:115–121

Kwak TS (1984) Inheritance of seedling cold tolerance in rice. SABRAO J 16:83–86

Lee M-H (2001) Low temperature tolerance in rice: the Korean experience. In: ACIAR Proceedings 1998. ACIAR, Canberra, ACT, pp 109–117

Lee T, Lur H, Chu C (1993) Role of abscisic acid in chilling tolerance of rice (*Oryza sativa* L.) seedlings. I. Endogenous abscisic acid levels. Plant Cell Environ 16:481–490

Lee DG, Ahsan N, Lee SH et al (2009a) Chilling stress-induced proteomic changes in rice roots. J Plant Physiol 166:1–11

Lee SJ, Jeon US, Lee SJ et al (2009b) Iron fortification of rice seeds through activation of the nicotianamine synthase gene. Proc Natl Acad Sci U S A 106:22014–22019. https://doi.org/10.4324/9780203605912

Lee J, Lee W, Kwon S-W (2015) A quantitative shotgun proteomics analysis of germinated rice embryos and coleoptiles under low-temperature conditions. Proteome Sci 13:27

Lee J, Heath LS, Grene R, Li S (2019) Comparing time series transcriptome data between plants using a network module finding algorithm. Plant Methods 15:61

Li HB, Wang J, Liu AM et al (1997) Genetic basis of low-temperature-sensitive sterility in indica-japonica hybrids of rice as determined by RFLP analysis. Theor Appl Genet 95:1092–1097

Li G, Zhang M, Cai W-M et al (2008) Characterization of OsPIP2;7, a water channel protein in rice. Plant Cell Physiol 49:1851–1858. https://doi.org/10.1093/pcp/pcn166

Li HW, Zang BS, Deng XW, Wang XP (2011a) Overexpression of the trehalose-6-phosphate synthase gene *OsTPS1* enhances abiotic stress tolerance in rice. Planta 234:1007–1018. https://doi.org/10.1007/s00425-011-1458-0

Li M, Sun P, Zhou H (2011b) Identification of quantitative trait loci associated with germination using chromosome segment substitution lines of rice (*Oryza sativa* L.). Theor Appl Genet 123:411–420. https://doi.org/10.1007/s00122-011-1593-9

Li L, Liu X, Xie K et al (2013) *qLTG-9*, a stable quantitative trait locus for low-temperature germination in rice (*Oryza sativa* L.). Theor Appl Genet 126:2313–2322

Li M, Li X, Zhou Z et al (2016) Reassessment of the four yield-related genes *Gn1a*, *DEP1*, *GS3*, and *IPA1* in rice using a CRISPR/Cas9 system. Front Plant Sci 7:377

Li S, Tian Y, Wu K et al (2018) Modulating plant growth–metabolism coordination for sustainable agriculture. Nature 560:595. https://doi.org/10.1038/s41586-018-0415-5

Liang Y, Meng L, Lin X et al (2018) QTL and QTL networks for cold tolerance at the reproductive stage detected using selective introgression in rice. PLoS One 13:e0200846

Liu K, Wang L, Xu Y et al (2007) Overexpression of OsCOIN, a putative cold inducible zinc finger protein, increased tolerance to chilling, salt and drought, and enhanced proline level in rice. Planta 226:1007–1016. https://doi.org/10.1007/s00425-007-0548-5

Liu C, Wu Y, Wang X (2012) bZIP transcription factor *OsbZIP52/RISBZ5*: a potential negative regulator of cold and drought stress response in rice. Planta 235:1157–1169

Liu F, Xu W, Song Q et al (2013) Microarray-assisted fine-mapping of quantitative trait loci for cold tolerance in rice. Mol Plant 6:757–767

Lone JA, Khan MN, Bhat MA et al (2018) Cold tolerance at germination and seedling stages of rice: methods of evaluation and characterization of thirty rice genotypes under stress conditions. Int J Curr Microbiol App Sci 7:1103–1109

Lou Q, Chen L, Sun Z et al (2007) A major QTL associated with cold tolerance at seedling stage in rice (*Oryza sativa* L.). Euphytica 158:87–94. https://doi.org/10.1007/s10681-007-9431-5

Lourenço T, Sapeta H, Figueiredo DD et al (2013) Isolation and characterization of rice (*Oryza sativa* L.) E3-ubiquitin ligase *OsHOS1* gene in the modulation of cold stress response. Plant Mol Biol 83:351–363

Ma Q, Dai X, Xu Y et al (2009) Enhanced tolerance to chilling stress in *OsMYB3R-2* transgenic rice is mediated by alteration in cell cycle and ectopic expression of stress genes. Plant Physiol 150:244–256

Ma Y, Dai X, Xu Y et al (2015) *COLD1* confers chilling tolerance in rice. Cell 160:1209–1221

Maraña C, García Olmedo F, Carbonero Zalduegui P (1990) Differential expression of two types of sucrose synthase-encoding genes in wheat in response to anaerobiosis, cold shock and light. Gene 88:167–172

Matlon P, Randolph T, Guei R (1998) Impact of rice research in West Africa. In: Impact of rice research. International Rice Research Institute, Los Baños, pp 383–404

Mishra R, Joshi RK, Zhao K (2018) Genome editing in rice: recent advances, challenges, and future implications. Front Plant Sci 9:1361

Miura K, Lin SY, Yano M, Nagamine T (2001) Mapping quantitative trait loci controlling low temperature germinability in rice (*Oryza sativa* L.). Breed Sci 51:293–299

Mohanty S, Wassmann R, Nelson A et al (2012) Rice and climate change: significance for food security and vulnerability. IRRI discussion paper series No. 49. International Rice Research Institute, Los Baños. 14 p

Molla KA, Yang Y (2019) CRISPR/Cas-mediated base editing: technical considerations and practical applications. Trends Biotechnol 37:1121–1142

Moraes De Freitas GP, Basu S, Ramegowda V et al (2016) Comparative analysis of gene expression in response to cold stress in diverse rice genotypes. Biochem Biophys Res Commun 471:253–259. https://doi.org/10.1016/j.bbrc.2016.02.004

Morsy MR, Almutairi AM, Gibbons J et al (2005) The *OsLti6* genes encoding low-molecular-weight membrane proteins are differentially expressed in rice cultivars with contrasting sensitivity to low temperature. Gene 344:171–180

Morsy MR, Jouve L, Hausman J-F et al (2007) Alteration of oxidative and carbohydrate metabolism under abiotic stress in two rice (*Oryza sativa* L.) genotypes contrasting in chilling tolerance. J Plant Physiol 164:157–167

Mukhopadhyay A, Vij S, Tyagi AK (2004) Overexpression of a zinc-finger protein gene from rice confers tolerance to cold, dehydration, and salt stress in transgenic tobacco. Proc Natl Acad Sci U S A 101:6309–6314

Murai-Hatano M, Kuwagata T, Sakurai J et al (2008) Effect of low root temperature on hydraulic conductivity of rice plants and the possible role of aquaporins. Plant Cell Physiol 49:1294–1305

Nagamine T (1991) Genic control of tolerance to chilling injury at seedling stage in rice, *Oryza sativa* L. Jpn J Breed 41:35–40

Nagano K (1998) Development of new breeding techniques for cold tolerance and breeding of new rice cultivars with highly cold tolerance, Hitomebore and Jyoudeki. In: Proceedings of the International Workshop on Breeding and Biotechnology for Environmental Stress in Rice, pp 143–148

Nakashima K, Tran LP, Van Nguyen D et al (2007) Functional analysis of a NAC-type transcription factor OsNAC6 involved in abiotic and biotic stress-responsive gene expression in rice. Plant J 51:617–630

Nalley L, Tack J, Durand A et al (2017) The production, consumption, and environmental impacts of rice hybridization in the United States. Agron J 109:193–203. https://doi.org/10.2134/agronj2016.05.0281

Novillo F, Alonso JM, Ecker JR, Salinas J (2004) CBF2/DREB1C is a negative regulator of CBF1/DREB1B and CBF3/DREB1A expression and plays a central role in stress tolerance in *Arabidopsis*. Proc Natl Acad Sci U S A 101:3985–3990

Oh SJ, Kim YS, Kwon CW et al (2009) Overexpression of the transcription factor *AP37* in rice improves grain yield under drought conditions. Plant Physiol 150:1368–1379

Okamoto E, Matsunaga K, Sasaki T (1986) Cold tolerant rice varieties in temperate district of Japan. Tohoku Agric Res 39:35

Oliver SN, Van Dongen JT, Alfred SC et al (2005) Cold-induced repression of the rice anther-specific cell wall invertase gene *OSINV4* is correlated with sucrose accumulation and pollen sterility. Plant Cell Environ 28:1534–1551

Oliver SN, Zhao X, Dennis ES, Dolferus R (2007) The molecular basis of cold-induced pollen sterility in rice. In: Biotechnology and sustainable agriculture 2006 and beyond. Springer, New York, NY, pp 205–207

Pan Y, Zhang H, Zhang D et al (2015) Genetic analysis of cold tolerance at the germination and booting stages in rice by association mapping. PLoS One 10:e0120590. https://doi.org/10.1371/journal.pone.0120590

Pan Y, Liang H, Gao L et al (2020) Transcriptomic profiling of germinating seeds under cold stress and characterization of the cold-tolerant gene *LTG5* in rice. BMC Plant Biol 20:1–17

Pang Y, Chen K, Wang X et al (2017) Simultaneous improvement and genetic dissection of salt tolerance of rice (*Oryza sativa* L.) by designed QTL pyramiding. Front Plant Sci 8:1275. https://doi.org/10.3389/fpls.2017.01275

Park H, Soon I, Han J et al (2009) *OsDEG10* encoding a small RNA-binding protein is involved in abiotic stress signaling. Biochem Biophys Res Commun 380:597–602. https://doi.org/10.1016/j.bbrc.2009.01.131

Park M, Yun K, Mohanty B et al (2010) Supra-optimal expression of the cold-regulated OsMyb4 transcription factor in transgenic rice changes the complexity of transcriptional network with major effects on stress tolerance and panicle development. Plant Cell Environ 33:2209–2230

Park IK, Oh CS, Kim DM et al (2013) QTL mapping for cold tolerance at the seedling stage using introgression lines derived from an intersubspecific cross in rice. Plant Breed Biotech 1:1–8. https://doi.org/10.9787/PBB.2013.1.1.001

Pouramir Dashtmian F, Khajeh Hosseini M, Esfahani M (2013) Methods for rice genotypes cold tolerance evaluation at germination stage. Int J Agric Crop Sci 5:2111

Ranawake AL, Nakamura C (2011) Cold tolerance of an inbred line population of rice (*Oryza sativa* L) at different growth stages. Trop Agric Res Ext 14:25–33

Ranawake A, Nakamura C (2015) QTL analysis of dehydration tolerance at seedling stage in rice (*Oryza sativa* L.). Trop Agric Res 15:98

Ranawake AL, Manangkil OE, Yoshida S et al (2014) Mapping QTLs for cold tolerance at germination and the early seedling stage in rice (*Oryza sativa* L.). Biotechnol Biotechnol Equip 28:989–998. https://doi.org/10.1080/13102818.2014.978539

Saad RB, Fabre D, Mieulet D et al (2012) Expression of the *Aeluropus littoralis AlSAP* gene in rice confers broad tolerance to abiotic stresses through maintenance of photosynthesis. Plant Cell Environ 35:626–643

Saito K, Miura K, Nagano K et al (1995) Chromosomal location of quantitative trait loci for cool tolerance at the booting stage in rice [*Oryza sativa*] variety 'Norin-PL8'. Breed Sci 45:337–340

Saito K, Miura K, Nagano K et al (2001) Identification of two closely linked quantitative trait loci for cold tolerance on chromosome 4 of rice and their association with anther length. Theor Appl Genet 103:862–868

Saito K, Hayano-Saito Y, Maruyama-Funatsuki W et al (2004) Physical mapping and putative candidate gene identification of a quantitative trait locus *Ctb1* for cold tolerance at the booting stage of rice. Theor Appl Genet 109:515–522. https://doi.org/10.1007/s00122-004-1667-z

Sakamoto A, Murata N (2002) The role of glycine betaine in the protection of plants from stress: clues from transgenic plants. Plant Cell Environ 25:163–171

Sales E, Viruel J, Domingo C, Marqués L (2017) Genome wide association analysis of cold tolerance at germination in temperate japonica rice (*Oryza sativa* L.) varieties. PLoS One 12:e0183416

Sanchez-Ballesta MT, Lafuente MT, Zacarias L, Granell A (2000) Involvement of phenylalanine ammonia-lyase in the response of Fortune mandarin fruits to cold temperature. Physiol Plant 108:382–389

Satake T (1976) Sterile-type cool injury in paddy rice plants. In: Proceedings of the Symposium on Climate and Rice. International Rice Research Institute (IRRI), Los Baños, pp 281–300

Sato Y, Masuta Y, Saito K et al (2011) Enhanced chilling tolerance at the booting stage in rice by transgenic overexpression of the ascorbate peroxidase gene, *OsAPXa*. Plant Cell Rep 30:399–406

Satoh T, Tezuka K, Kawamoto T et al (2016) Identification of QTLs controlling low-temperature germination of the East European rice (*Oryza sativa* L.) variety Maratteli. Euphytica 207:245–254. https://doi.org/10.1007/s10681-015-1531-z

Schiller JM, Linquist B, Douangsila K et al (2001) Constraints to rice production systems in Laos. In: ACIAR Proceedings 1998. ACIAR, Canberra, ACT, pp 3–19

Schläppi MR, Jackson AK, Eizenga GC et al (2017) Assessment of five chilling tolerance traits and GWAS mapping in rice using the USDA Mini-Core collection. Front Plant Sci 8:957. https://doi.org/10.3389/fpls.2017.00957

Seck PA, Diagne A, Mohanty S, Wopereis MCS (2012) Crops that feed the world 7: rice. Food Secur 4:7–24

Serrano-Mislata A, Bencivenga S, Bush M et al (2017) *DELLA* genes restrict inflorescence meristem function independently of plant height. Nat Plants 3:749

Shakiba E, Edwards JD, Jodari F et al (2017) Genetic architecture of cold tolerance in rice (*Oryza sativa*) determined through high resolution genome-wide analysis. PLoS One 12:1–22. https://doi.org/10.1371/journal.pone.0172133

Sharma KD, Nayyar H (2016) Regulatory networks in pollen development under cold stress. Front Plant Sci 7:402

Shi J, Cao Y, Fan X (2012) A rice microsomal delta-12 fatty acid desaturase can enhance resistance to cold stress in yeast and *Oryza sativa*. Mol Breed 29:743–757. https://doi.org/10.1007/s11032-011-9587-5

Shim K-C, Kim S, Le AQ et al (2019) Fine mapping of a low-temperature germinability QTL *qLTG1* using introgression lines derived from *Oryza rufipogon*. Plant Breed Biotechnol 7:141–150

Shinada H, Iwata N, Sato T, Fujino K (2013) Genetical and morphological characterization of cold tolerance at fertilization stage in rice. Breed Sci 63:197–204. https://doi.org/10.1270/jsbbs.63.197

Shinozaki K, Yamaguchi-Shinozaki K (2000) Molecular responses to dehydration and low temperature: differences and cross-talk between two stress signaling pathways. Curr Opin Plant Biol 3:217–223

Shirasawa K, Shiokai S, Yamaguchi M et al (2006) Dot-blot-SNP analysis for practical plant breeding and cultivar identification in rice. Theor Appl Genet 113:147–155

Shirasawa S, Endo T, Nakagomi K et al (2012) Delimitation of a QTL region controlling cold tolerance at booting stage of a cultivar,'Lijiangxintuanheigu', in rice, *Oryza sativa* L. Theor Appl Genet 124:937–946

Singh BK, Sutradhar M, Singh AK, Mandal N (2017) Cold stress in rice at early growth stage: an overview. Int J Pure Appl Biosci 5:407–419

Sinha SK, Amitha Mithra SV, Chaudhary S et al (2018) Transcriptome analysis of two rice varieties contrasting for nitrogen use efficiency under chronic N starvation reveals differences in chloroplast and starch metabolism-related genes. Genes (Basel) 9:1–22. https://doi.org/10.3390/genes9040206

Sivapalan S (2013) Cold tolerance of temperate and tropical rice varieties. Frank Wise Institute of Tropical Agriculture, Kununurra, WA. 4 p

Snell P, Johnston D, Ford R (2008) Cold tolerant rice varieties: a matter of need for Australia. IREC Farm Newsl 177:4–5

Song S-Y, Chen Y, Chen J et al (2011) Physiological mechanisms underlying OsNAC5-dependent tolerance of rice plants to abiotic stress. Planta 234:331–345

Song J, Li J, Sun J et al (2018) Genome-wide association mapping for cold tolerance in a core collection of rice (*Oryza sativa* L.) landraces by using high-density single nucleotide polymorphism markers from specific-locus amplified fragment sequencing. Front Plant Sci 9:875

Sthapit BR, Witcombe JR (1998) Inheritance of tolerance to chilling stress in rice during germination and plumule greening. Crop Sci 38:660–665

Su CF, Wang YC, Hsieh TH et al (2010) A novel MYBS3-dependent pathway confers cold tolerance in rice. Plant Physiol 153:145–158. https://doi.org/10.1104/pp.110.153015

Suh JP, Jeung JU, Lee JI et al (2010) Identification and analysis of QTLs controlling cold tolerance at the reproductive stage and validation of effective QTLs in cold-tolerant genotypes of rice (*Oryza sativa* L.). Theor Appl Genet 120:985–995

Suh JP, Lee CK, Lee JH et al (2012) Identification of quantitative trait loci for seedling cold tolerance using RILs derived from a cross between japonica and tropical japonica rice cultivars. Euphytica 184:101–108

Sun Z, Du J, Pu X et al (2019) Near-isogenic lines of japonica rice revealed new QTLs for cold tolerance at booting stage. Agronomy 9(1):40. https://doi.org/10.3390/agronomy9010040

Sweeney M, McCouch S (2007) The complex history of the domestication of rice. Ann Bot 100:951–957

Takahashi N (1984) Differentiation of ecotypes in *Oryza sativa* L. In: Developments in crop science. Elsevier, Amsterdam, pp 31–67

Takesawa T, Ito M, Kanzaki H et al (2002) Over-expression of ζ glutathione S-transferase in transgenic rice enhances germination and growth at low temperature. Mol Breed 9:93–101

Tao Z, Kou Y, Liu H et al (2011) OsWRKY45 alleles play different roles in abscisic acid signalling and salt stress tolerance but similar roles in drought and cold tolerance in rice. J Exp Bot 62:4863–4874. https://doi.org/10.1093/jxb/err144

Teng S, Zeng D, Qian Q et al (2001) QTL analysis of rice low temperature germinability. Chin Sci Bull 46:1800–1803

Urano K, Maruyama K, Ogata Y et al (2009) Characterization of the ABA-regulated global responses to dehydration in *Arabidopsis* by metabolomics. Plant J 57:1065–1078

Vats S, Kumawat S, Kumar V et al (2019) Genome editing in plants: exploration of technological advancements and challenges. Cell 8:1386

Wainaina CM, Makihara D, Nakamura M et al (2018) Identification and validation of QTLs for cold tolerance at the booting stage and other agronomic traits in a rice cross of a Japanese tolerant variety, Hananomai, and a NERICA parent, WAB56-104. Plant Prod Sci 21:132–143

Wang Q, Guan Y, Wu Y et al (2008) Overexpression of a rice *OsDREB1F* gene increases salt, drought, and low temperature tolerance in both *Arabidopsis* and rice. Plant Mol Biol 67:589–602

Wang ZF, Wang JF, Wang FH et al (2009) Genetic control of germination ability under cold stress in rice. Rice Sci 16:173–180. https://doi.org/10.1016/S1672-6308(08)60076-1

Wang Z, Wang F, Zhou R et al (2011) Identification of quantitative trait loci for cold tolerance during the germination and seedling stages in rice (*Oryza sativa* L.). Euphytica 181:405

Wang C, Wei Q, Zhang K et al (2013a) Down-regulation of *OsSPX1* causes high sensitivity to cold and oxidative stresses in rice seedlings. PLoS One 8(12):e81849. https://doi.org/10.1371/journal.pone.0081849

Wang J, Lin X, Sun Q, Jena KK (2013b) Evaluation of cold tolerance for japonica rice varieties from different countries. Adv J Food Sci Technol 5:54–56. https://doi.org/10.19026/ajfst.5.3311

Wang X, Han F, Yang M et al (2013c) Exploring the response of rice (*Oryza sativa*) leaf to gibberellins: a proteomic strategy. Rice 6:17

Wang D, Liu J, Li C et al (2016) Genome-wide association mapping of cold tolerance genes at the seedling stage in rice. Rice 9:61. https://doi.org/10.1186/s12284-016-0133-2

Wei J, Wang A, Li R et al (2018) Metabolome-wide association studies for agronomic traits of rice. Heredity (Edinb) 120:342

Williams ME, Torabinejad J, Cohick E et al (2005) Mutations in the *Arabidopsis* phosphoinositide phosphatase gene *SAC9* lead to overaccumulation of PtdIns(4,5)P$_2$ and constitutive expression of the stress-response pathway. Plant Physiol 138:686–700

Wittmann-Liebold B, Graack H, Pohl T (2006) Two-dimensional gel electrophoresis as tool for proteomics studies in combination with protein identification by mass spectrometry. Proteomics 6:4688–4703

Wolter F, Schindele P, Puchta H (2019) Plant breeding at the speed of light: the power of CRISPR/Cas to generate directed genetic diversity at multiple sites. BMC Plant Biol 19:176

Xiang Y, Huang Y, Xiong L (2007) Characterization of stress-responsive *CIPK* genes in rice for stress tolerance improvement. Plant Physiol 144:1416–1428. https://doi.org/10.1104/pp.107.101295

Xiao N, Huang W, Li A et al (2015) Fine mapping of the *qLOP2* and *qPSR2-1* loci associated with chilling stress tolerance of wild rice seedlings. Theor Appl Genet 128:173–185

Xiao N, Gao Y, Qian H et al (2018) Identification of genes related to cold tolerance and a functional allele that confers cold tolerance. Plant Physiol 177:1108–1123

Xiong L, Yang Y (2003) Disease resistance and abiotic stress tolerance in rice are inversely modulated by an abscisic acid–inducible mitogen-activated protein kinase. Plant Cell 15:745–759

Xu L-M, Zhou L, Zeng Y-W et al (2008) Identification and mapping of quantitative trait loci for cold tolerance at the booting stage in a japonica rice near-isogenic line. Plant Sci 174:340–347

Xu R, Yang Y, Qin R et al (2016) Rapid improvement of grain weight via highly efficient CRISPR/
    Cas9-mediated multiplex genome editing in rice. J Genet Genom 43:529–532
Xu Y, Hu D, Hou X et al (2020) OsTMF attenuates cold tolerance by affecting cell wall properties
    in rice. New Phytol 227:498. https://doi.org/10.1111/nph.16549
Yamaguchi-Shinozaki K, Shinozaki K (2004) Improving abiotic stress tolerance in crops. In The
    8th International Symposium on the Biosafety of Genetically Modified Organisms, Montpellier,
    France, 26–30 September, 2004, International Society for Biosafety Research, pp 134–136
Yamori W, Noguchi KO, Hikosaka K, Terashima I (2010) Phenotypic plasticity in photosynthetic
    temperature acclimation among crop species with different cold tolerances. Plant Physiol
    152:388–399
Yamori W, Sakata N, Suzuki Y et al (2011) Cyclic electron flow around photosystem I via chlo-
    roplast NAD(P)H dehydrogenase (NDH) complex performs a significant physiological role
    during photosynthesis and plant growth at low temperature in rice. Plant J 68:966–976. https://
    doi.org/10.1111/j.1365-313X.2011.04747.x
Yan SP, Zhang QY, Tang ZC et al (2006) Comparative proteomic analysis provides new insights
    into chilling stress responses in rice. Mol Cell Proteomics 5:484–496
Yang T, Poovaiah BW (2003) Calcium/calmodulin-mediated signal network in plants. Trends
    Plant Sci 8:505–512
Yang G, Rhodes D, Joly RJ (1996) Effects of high temperature on membrane stability and chlo-
    rophyll fluorescence in glycinebetaine-deficient and glycinebetaine-containing maize lines.
    Funct Plant Biol 23:437–443
Yang T, Shad Ali G, Yang L et al (2010) Calcium/calmodulin-regulated receptor-like kinase
    CRLK1 interacts with MEKK1 in plants. Plant Signal Behav 5:991–994
Yang A, Dai X, Zhang W-H (2012) A R2R3-type MYB gene, OsMYB2, is involved in salt, cold,
    and dehydration tolerance in rice. J Exp Bot 63:2541–2556
Yang C, Li D, Mao D et al (2013a) Overexpression of microRNA319 impacts leaf morphogenesis
    and leads to enhanced cold tolerance in rice (*Oryza sativa* L.). Plant Cell Environ 36:2207–2218
Yang Z, Huang D, Tang W et al (2013b) Mapping of quantitative trait loci underlying cold toler-
    ance in rice seedlings via high-throughput sequencing of pooled extremes. PLoS One 8:e68433
Yang W, Guo Z, Huang C et al (2014) Combining high-throughput phenotyping and genome-wide
    association studies to reveal natural genetic variation in rice. Nat Commun 5:5087
Yang LM, Liu HL, Lei L et al (2018) Identification of QTLs controlling low-temperature
    germinability and cold tolerance at the seedling stage in rice (*Oryza sativa* L.). Euphytica
    214:13. https://doi.org/10.1007/s10681-017-2092-0
Ye C, Fukai S, Godwin I et al (2009) Cold tolerance in rice varieties at different growth stages.
    Crop Pasture Sci 60:328–338
Yokotani N, Sato Y, Tanabe S et al (2013) WRKY76 is a rice transcriptional repressor playing
    opposite roles in blast disease resistance and cold stress tolerance. J Exp Bot 64:5085–5097.
    https://doi.org/10.1093/jxb/ert298
Yoshida S (1981) Fundamentals of rice crop science. International Rice Research Institute,
    Los Baños
Yu S, Li M, Xiao Y et al (2018) Mapping QTLs for cold tolerance at seedling stage using an *Oryza
    sativa* × *O. rufipogon* backcross inbred line population. Czech J Genet Plant Breed 54:59–64.
    https://doi.org/10.17221/154/2016-CJGPB
Yuan P, Yang T, Poovaiah BW (2018) Calcium signaling-mediated plant response to cold stress.
    Int J Mol Sci 19:3896
Yun KY, Park MR, Mohanty B et al (2010) Transcriptional regulatory network triggered by oxida-
    tive signals configures the early response mechanisms of japonica rice to chilling stress. BMC
    Plant Biol 10:16
Zhang X, Fowler SG, Cheng H et al (2004) Freezing-sensitive tomato has a functional CBF cold
    response pathway, but a CBF regulon that differs from that of freezing-tolerant *Arabidopsis*.
    Plant J 39:905–919
Zhang Y, Chen C, Jin XF et al (2009) Expression of a rice DREB1 gene, *OsDREB1D*, enhances
    cold and high-salt tolerance in transgenic *Arabidopsis*. BMB Rep 42:486–492

Zhang J, Li J, Wang X, Chen J (2011) OVP1, a vacuolar H+-translocating inorganic pyrophospha-tase (V-PPase), overexpression improved rice cold tolerance. Plant Physiol Biochem 49:33–38

Zhang T, Zhao X, Wang W et al (2012a) Comparative transcriptome profiling of chilling stress responsiveness in two contrasting rice genotypes. PLoS One 7:e43274

Zhang Y, Xu Y, Yi H, Gong J (2012b) Vacuolar membrane transporters OsVIT1 and OsVIT2 modulate iron translocation between flag leaves and seeds in rice. Plant J 72:400–410

Zhang K, Liu H, Tao P, Chen H (2014a) Comparative proteomic analyses provide new insights into low phosphorus stress responses in maize leaves. PLoS One 9:e98215. https://doi.org/10.1371/journal.pone.0098215

Zhang Q, Chen Q, Wang S et al (2014b) Rice and cold stress: methods for its evaluation and summary of cold tolerance-related quantitative trait loci. Rice 7:24. https://doi.org/10.1186/s12284-014-0024-3

Zhang S, Zheng J, Liu B et al (2014c) Identification of QTLs for cold tolerance at seedling stage in rice (*Oryza sativa* L.) using two distinct methods of cold treatment. Euphytica 195:95–104

Zhang J, Luo W, Zhao Y et al (2016) Comparative metabolomic analysis reveals a reactive oxygen species-dominated dynamic model underlying chilling environment adaptation and tolerance in rice. New Phytol 211:1295–1310

Zhang Z, Li JJJ, Pan Y et al (2017) Natural variation in *CTB4a* enhances rice adaptation to cold habitats. Nat Commun 8:1–13. https://doi.org/10.1007/BF01337500

Zhang M, Ye J, Xu Q et al (2018) Genome-wide association study of cold tolerance of Chinese indica rice varieties at the bud burst stage. Plant Cell Rep 37:529–539

Zhao WG, Chung JW, Kwon SW et al (2013) Association analysis of physicochemical traits on eating quality in rice (*Oryza sativa* L.). Euphytica 191:9–21

Zhi-Hong Z, Li S, Wei L et al (2005) A major QTL conferring cold tolerance at the early seedling stage using recombinant inbred lines of rice (*Oryza sativa* L.). Plant Sci 168:527–534. https://doi.org/10.1016/j.plantsci.2004.09.021

Zhou L, Zeng Y, Hu G et al (2012) Characterization and identification of cold tolerant near-isogenic lines in rice. Breed Sci 62:196–201

Zhu JK (2016) Abiotic stress signaling and responses in plants. Cell 167:313–324

Zhu Y, Chen K, Mi X et al (2015) Identification and fine mapping of a stably expressed QTL for cold tolerance at the booting stage using an interconnected breeding population in rice. PLoS One 10:e0145704. https://doi.org/10.1371/journal.pone.0145704

Arsenic Stress Responses and Accumulation in Rice

Varunseelan Murugaiyan, Frederike Zeibig, Mahender Anumalla, Sameer Ali Siddiq, Michael Frei, Jayaseelan Murugaiyan, and Jauhar Ali

Abstract Rice (*Oryza sativa* L.) is one of the world's most vital staple grains, and 90% of it is produced and consumed in Asia alone. It plays a significant role in the entry of mineral nutrients into the food chain. Arsenic (As) is a toxic heavy metal that threatens the major rice-growing regions in the world, particularly in Asia. Arsenic is ubiquitously present in moderate concentrations in the environment because of natural geological processes and anthropogenic impacts. However, rapid industrialization and excessive use of arsenic-rich groundwater are further fueling the increased arsenic concentration in agricultural topsoil. Arsenic accumulation in rice plants has a significant adverse effect on plant, human, and livestock health. Although arsenic contamination in rice is well documented, its interaction and accumulation in rice are poorly understood. So far, no candidate genes or QTLs associated with arsenic interaction are used in breeding programs for the development of low-arsenic-accumulating rice varieties. The development and adaptation of new low-arsenic-accumulating rice cultivars resilient to arsenic toxicity constitute safe

V. Murugaiyan · F. Zeibig
Rice Breeding Platform, International Rice Research Institute (IRRI), Los Baños, Philippines

Institute of Crop Sciences and Resource Conservation (INRES), University of Bonn, Bonn, Germany

M. Anumalla
Rice Breeding Platform, International Rice Research Institute (IRRI), Los Baños, Philippines

S. A. Siddiq
University of Minnesota, Minneapolis, MN, USA

M. Frei
Institute of Crop Sciences and Resource Conservation (INRES), University of Bonn, Bonn, Germany

Department of Agronomy and Plant Breeding, Justus Liebig University, Giessen, Germany

J. Murugaiyan
Department of Biotechnology, SRM University-AP, Amaravati, Andhra Pradesh, India

J. Ali (✉)
Hybrid Rice Breeding Cluster, Hybrid Rice Development Consortium (HRDC), Rice Breeding Platform, International Rice Research Institute (IRRI),
Los Baños, Laguna, Philippines
e-mail: J.Ali@irri.org

J. Ali, S. H. Wani (eds.), *Rice Improvement*,
https://doi.org/10.1007/978-3-030-66530-2_9

281

ways to mitigate arsenic contamination in rice. Recent scientific advances in rice genetics, genomics, and physiology have opened up new opportunities to speed up the process of developing low-arsenic-accumulating rice cultivars for the rapidly growing human population.

Keywords Heavy metal · Arsenic contamination · Arsenic speciation · Phytotoxicity · Quantitative trait loci · Genes

1 Introduction

Rice (*Oryza sativa* L.) belongs to the grass family (*Poaceae*). Its domestication is one of the most significant events in the history of human agricultural advancement (Khush 1997; Molina et al. 2011; Huang et al. 2012a). It is one of the world's vital staple grains and plays a crucial role in the entry of mineral nutrients into the food chain (Ali et al. 2018a). The presence of naturally occurring unwanted arsenic (As) metalloid in flooded paddy soil poses a significant threat to rice production and consumers who depend on rice as their primary staple food (Murugaiyan et al. 2019). A total of 23–25% of the total calories consumed by humans come through the consumption of rice alone (Ashikari and Ma 2015; Yu et al. 2020). As the world population is likely to increase further in the coming decades, keeping rice production sustainable remains a crucial challenge (Ali et al. 2018a). Currently, rice is planted on 166 million hectares worldwide, nurturing some four billion people around the world, and the annual harvest of rice is worth ~USD 200 billion (GRiSP 2013). Rice is the primary cereal crop in Asia, where 90% of the world's rice is produced and consumed (Frei and Becker 2005; Molden 2013). Approximately 480 million metric tons of milled rice are produced annually; China and India alone account for ~50% of the rice produced and 90% of this rice production is consumed domestically (Muthayya et al. 2014). Maintaining a favorable rice supply-demand balance in the future depends mostly on the exploitation of the production capacity of the rainfed ecosystem (Li and Ali 2017). On about 60% of the agricultural land in Asia, rice is grown under rainfed conditions (Rao et al. 2015; Ali et al. 2018a). The rainfed ecosystem is the dominant one in the low-income countries of Asia, where demand for rice is projected to remain very high throughout this century (Li and Ali 2017). Rice production in rainfed conditions is susceptible to a combination of various biotic and abiotic stresses. Abiotic stresses are the primary factor undesirably affecting crop growth and yield worldwide (Gao et al. 2007). Various abiotic stresses limit rice production in rainfed environments, which comprise 35–45% of the global rice area (Wu et al. 2014; Li and Ali 2017). Critical abiotic stresses including extreme temperature, drought, submergence, salinity, iron toxicity, nutritional deficiencies, and heavy metal contamination are known to cause severe losses in rice yield and in the quality of the seed produced (Messerschmidt et al. 2002; Li and Ali 2017; Wu et al. 2017).

Large tracts of paddy soil are directly affected by heavy metal (arsenic, lead, and cadmium) contamination, especially in India, China, and Bangladesh (Das et al. 2008; Chakraborti et al. 2013; Wu et al. 2016). Rice is conventionally produced in flooded paddy fields, which can translocate the unwanted class I carcinogenic-arsenic metalloid into the straw and grain (Sayan et al. 2012). An accumulation of arsenic in the rice plant negatively affects plant performance and also threatens the health of consumers and livestock (Carbonell-Barrachina et al. 2015). Himalayan rivers carry arsenic from rock sediments to the densely populated rice-producing regions of South and Southeast Asia, threatening the primary rice-growing belt of Asia. In these areas, rice production is already threatened by climate change with the frequent intrusion of saltwater in the Ganges-Brahmaputra deltaic regions, and extended periods of drought also force farmers to depend on groundwater for irrigating paddy fields (Murugaiyan et al. 2019). This groundwater is naturally enriched because of arsenic-rich aquifers and it acts as an additional source of arsenic in paddy fields. The long-term use of arsenic-rich groundwater for irrigating rice crops has resulted in an increased concentration of arsenic in paddy topsoil, with up to 83 mg As/kg being reported in some parts of Bangladesh and the Indian subcontinent (Suriyagoda et al. 2018). Without intervention, this concentration will tend to increase in the coming decades because of the significant dependence on groundwater for rice production. However, rice varieties suitable for growing in arsenic-rich fields have not been developed. Despite this arsenic threat, the major rice-growing countries in Asia need more food for their rapidly growing populations, leaving scientists with the challenge of developing rice varieties that do not accumulate arsenic in the grain and straw. These varieties also need to withstand other abiotic stresses such as drought, salinity, and flooding. Because of global food security and the increasing health-related concerns associated with arsenic exposure through rice, it has become vital to understand arsenic toxicity, the interaction with rice plants, and the physiological mechanisms associated with arsenic accumulation in rice. Recent scientific advances, particularly in genetics, genomics, and crop physiology, have opened up new opportunities to speed up the process of developing highly adaptable rice cultivars that are safe and nutritious for meeting the future food demand of the growing population.

2 Heavy Metal Contamination

Heavy metal contamination has become a significant limitation to sustainable crop production. Particularly in rice, it poses a severe threat to human nutrition and food security (Murugaiyan 2019). The presence of toxic heavy metals such as arsenic (As), cadmium (Cd), lead (Pb), and mercury (Hg) in irrigated water systems threatens not only the rice plant but also the populations that depend on rice for their dietary supply. Additionally, feeding rice straw from contaminated paddy fields to cattle leads to an additional entry of these elements into the human food chain. Toxic heavy metals ubiquitously persist at moderate concentrations in the environment because of natural weathering of rocks and minerals, and also through anthropogenic impacts, and they tend to translocate in the food chain (Wu et al. 2016). In recent years, heavy metal contamination has increased significantly in agricultural soil because of the frequent use of polluted irrigation water (Clemens and Ma 2016). Various studies have reported that chronic exposure to heavy metals is often associated with cancer and potentially causes neurotoxicity (Vahidnia et al. 2007). Heavy metals cause developmental neurotoxicity, which is concerning since these developmental abnormalities are often irreversible (Mochizuki 2019).

2.1 Heavy Metal Interaction with the Biological System

In a biological system, the toxic effects of these metals may be similar even though their sources are distinct (Fig. 1). Lead and cadmium share similar ionic size and charge, which make them behave similarly in the biological system, especially regarding their toxicity for the same or related molecular targets (Pohl et al. 2011). In principle, lead and cadmium mimic essential ions such as calcium (Ca^{2+}) and zinc (Zn^{2+}) and use their channels to get into the system (Jomova and Valko 2011). Lead and cadmium exposure generally occurs as a result of industrial exposure. At the same time, arsenic has been reported to occur naturally or as a contaminant, mainly with irrigation water, and is often of geographic origin (Abernathy et al. 2001).

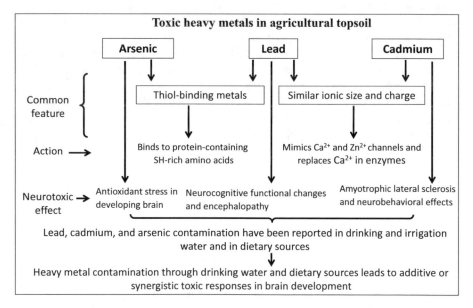

Fig. 1 Similarities between heavy metals and their interaction with the biological system

Arsenic and lead are known as thiol-binding metals. Hence, both are expected to behave similarly and show similar targets and binding sites (Flora et al. 2011). Arsenic and lead could bind to protein-containing sulfhydryl-rich amino acids such as cysteine residue and could have the same toxic effect. When heavy metals are ingested by humans, the nervous system is a primary target for several of them (Sengupta and Bishayi 2002; Nordberg et al. 2005; Thomas 2013).

2.2 Chronic Arsenic Exposure and Its Adverse Effects on Human Health

Arsenic is a metalloid element. It is the 20th most abundant mineral in Earth's crust with an average concentration in sediments ranging from 5 to 10 mg/kg (National Research Council (U.S.) Committee on Medical and Biological Effects of Environmental Pollutants 1977). Chronic exposure to arsenic is linked with myriad possible adverse health effects on humans, including skin lesions, hypertension, cardiovascular disease, pulmonary disease, reproductive and neurological dysfunctions, hematological changes, and malignancies of skin and internal organs (Mandal and Suzuki 2002). Groundwater arsenic contamination and its ill health effects in Southeast Asian countries came into the limelight in 1984 when groundwater used for drinking purposes was directly tainted by arsenic from natural sources (Garai et al. 1984; Duker et al. 2005). A substantial part of the Ganga-Meghna-Brahmaputra plain, with an area of 569,749 km^2 and population surpassing 500 million, was at risk of mass arsenic poisoning (Das et al. 2008; Chakraborti et al. 2013). The World Health Organization (WHO) has recommended <10µg/L concentration of arsenic in drinking water as a safe limit (Muhammad et al. 2010; Kumar and Puri 2012). However, high levels of arsenic contamination through groundwater and their adverse impact on human health have been reported in many countries across the world (Nriagu et al. 2007). The magnitude of this problem is of great concern in Bangladesh, followed by India and China (Nickson et al. 2000; Mohan and Pittman 2007; Guha Mazumder 2015; Tareq et al. 2015). Approximately 85–150 million people in India and Bangladesh are at immediate risk from mass arsenic poisoning through contaminated drinking water (Hossain 2006). Both long- and short-term exposure are hazardous and can lead to skin, bladder, lung, and prostate cancers, cardiovascular diseases, diabetes, and anemia, as well as reproductive, developmental, immunological, and neurological effects (Roy and Saha 2002; Ng 2005; Guha Mazumder 2008). Arsenicosis is a chronic illness linked with drinking water with high concentrations of arsenic over a long period and it commonly occurs in populations exposed to arsenic contamination in their living environment (Mazumder 2003; Sun 2004; Kalia and Flora 2005).

3 Arsenic Contamination in Paddy Soil

Arsenic-containing compounds were used widely in the early twentieth century as pesticides and fungicides, which led to a high-arsenic load in agricultural topsoil and water runoff (Mukherjee et al. 2017). Most arsenic exposures occur through contaminated drinking and irrigation water sources. Arsenic is mainly found in pesticides (lead arsenate, calcium arsenate, and sodium arsenite), herbicides (monosodium arsenate and cacodylic acid and dimethyl arsenic acid), cotton desiccants (arsenic acid), wood preservatives (zinc arsenate and chromium arsenate), and semiconductors (gallium arsenide, indium arsenide, and aluminum gallium arsenide). Arsenic is even used as a desiccant and defoliant in agriculture, and as a by-product in the smelting process, particularly for gold and copper, from coal residues (Järup 2003; Vaughan 2006; Chen et al. 2016). Rivers originating from the higher Himalayas carry arsenic from their rock deposits to the densely inhabited rice-growing regions of South and Southeast Asia, making the primary rice-growing belt of Asia vulnerable to arsenic pollution (Shepherd et al. 2015; Lawson et al. 2016). Climate change is also threatening rice production in these areas. With the frequent occurrence of drought and saltwater intrusion in the Ganges-Brahmaputra deltas of India and Bangladesh, farmers increasingly tap groundwater resources for irrigation (Laha 2017). This groundwater is an additional source of arsenic discharged from the naturally abundant arsenic aquifers (Bondu et al. 2016). Typically, 4–8 mg/kg of arsenic occur in flooded paddy soil, but the concentration increases exponentially in paddy soil throughout the cropping season and can reach 83 mg As/kg in parts of Bangladesh and West Bengal regions of India (Abedin et al. 2002b; Zavala and Duxbury 2008). Evidence has emerged in recent years that arsenic-enriched groundwater occurs commonly in other Asian countries, including Cambodia, Myanmar, Pakistan, Nepal, Vietnam, and Japan (Fig. 2) (Smith et al. 2000; Das et al. 2008; Brammer and Ravenscroft 2009; Jiang et al. 2013). In the early 1970s, the use of surface water was abandoned mainly in the Bengal delta in response to severe health effects caused by pathogens, and this unexpectedly resulted in the extensive use of arsenic-contaminated groundwater (Caldwell et al. 2003). Alluvial and deltaic environments are mainly characterized by reducing conditions that cause a high-arsenic release in groundwater (Abernathy et al. 2001; Lee et al. 2008). Arsenic-abundant groundwater is drawn from shallow (<100 m) depths by domestic and irrigation wells in the Bengal basin aquifer system (Sultana 2013). It has been reported that groundwater from shallow tube wells (12–33 m) contains very high amounts of arsenic. In contrast, the water from deep tube wells (200–300 m) contains lower amounts of arsenic (<50µg/L) (Hossain 2006). Subsurface mobilization of arsenic is mainly caused by a combination of chemical, physical, and microbial factors, and various mechanisms have been proposed to elucidate arsenic mobilization (Anawar et al. 2003; Amini et al. 2008). Among those, the most widely accepted theories are pyrite oxidation and oxy-hydroxide reduction (Hossain 2006). Arsenic dissolution and release in deltaic regions have been modeled considering the contribution of microbes, organic matter, and paleo-sol formation (Gorny et al. 2015).

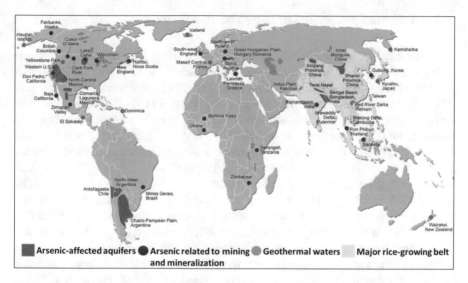

Fig. 2 Worldwide distribution of arsenic-contaminated regions and showing contaminated regions overlapping with major rice-growing belt. (Image modified from the British Geological Survey, Safiuddin et al. 2011)

3.1 Arsenic Contamination in Rice

The chemical characteristics of arsenic in paddy soil are complex, as it exists in both organic and inorganic forms, and it differs distinctly under flooded (anaerobic) and non-flooded (aerobic) conditions (Meharg and Hartley-Whitaker 2002). The toxicity of arsenic is associated with reduced soil conditions (flooded soils), which increase the bioavailability of inorganic arsenic and uptake into rice (Zhao et al. 2010b; Islam et al. 2016). Rice is commonly grown in flooded soil under reduced conditions, in which it assimilates inorganic arsenic into its grain, and this accumulation may adversely affect the nutritional quality of the grain (Islam et al. 2016). Of the total arsenic present in rice grain, inorganic arsenic constitutes approximately 54%. Also, grain arsenic content increases with increased As concentration in paddy soil (Suriyagoda et al. 2018). Total rice grain As concentration can vary from 0.011 to 0.82 mg/kg on average, depending on location, contamination level, and rice cultivars used (Islam et al. 2016). In highly contaminated areas, arsenic concentration can reach 1.7 mg/kg in rice grain, which is ten times more than the limit allowed by the WHO in rice grain (Meharg and Rahman 2003). Arsenic content in the rice plant decreases in the order of roots > leaves > grain, and in the rice grain, husk > bran polish > brown rice > raw rice > polished rice > cooked rice (Suriyagoda et al. 2018). A study with a continuous inorganic arsenic treatment showed an increased enrichment of up to 91.8 mg As/kg in the straw (Abedin et al. 2002c). About 30% of the total arsenic taken up by humans was contributed through rice and rice products (Li et al. 2011). Furthermore, the straw that cattle feed on will also accumulate arsenic in milk and meat, which are in turn consumed by humans, leading to another route of arsenic exposure (Talukder et al. 2011).

3.2 Arsenic Speciation in the Rice Ecosystem

Arsenic metalloid is ubiquitously found in various inorganic and organic forms in paddy soil. Inorganic arsenic compounds are considered to be highly toxic and they enter into paddy soil through both natural and anthropogenic activities (Sturchio et al. 2013). The toxic effects of arsenic in rice plants depend on their species form, with inorganic arsenic species being more extremely toxic than the organic form. The most common inorganic species that occur in the rice ecosystem are arsenate[V] and arsenite[III] (Fig. 3), while the most common organic species are monomethylarsonic acid (MMA) and dimethylarsinic acid (DMA) (Abedin et al. 2002c; Tripathi et al. 2013). Among the inorganic species, trivalent arsenite[III] is considered to be more mobile and toxic than pentavalent arsenate[V]. In both oxidation states, it can combine with methyl groups to form organic arsenic species (Vahter and Concha 2001). However, the existence of organic species in paddy soil is significantly lower than that of inorganic arsenic species. In anaerobic flooded-soil fields (submerged paddy fields), the reduced form arsenite[III] dominates; in aerobic soil conditions, such as upland rice fields, its oxidized form arsenate[V] dominates (Tripathi et al. 2013; Pandey et al. 2015).

3.3 Inorganic Arsenic Interaction with Essential Plant Nutrients

Rice plants do not possess naturally evolved arsenic transporters (Pandey et al. 2015). Instead, arsenic competes with chemically similar essential minerals to enter the plant system (Wenzel and Alloway 2013). Arsenite[III] is physiochemically

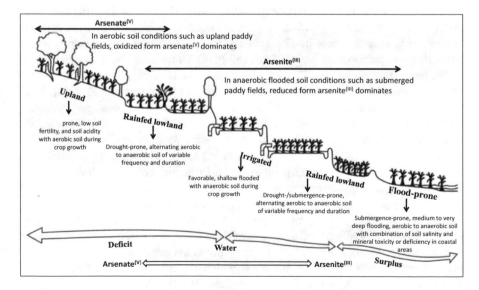

Fig. 3 Inorganic arsenic species found in rice-growing environments. (Adapted and modified from Rice Almanac, 3rd edition, Maclean et al. 2002)

similar to silica (Si), and thus it competes with the Si-uptake pathway. Silica is not considered an essential nutrient in many plant species. However, silica is uptaken actively by the rice plant due to anaerobic respiration, and the silica content in the rice stem and leaf ranges from 10% to 20% (Takahashi 1968; Ma and Takahashi 1990; Dobermann and Fairhurst 2000). Alternatively, arsenate$^{(V)}$ is physiochemically similar to the essential mineral phosphorus (P) and uses P acquisition pathways to enter the root system, and for efflux toward the xylem and various tissues (Clemens 2006; Zhao et al. 2009; Yang et al. 2018). Most rice genotypes possess a mechanism to retain much of the toxic arsenic burden in the roots. However, a genotype-dependent proportion of arsenic is translocated into the shoots and other tissues, including grains of the rice plant (Carey et al. 2010; Pandey et al. 2015). Since 35–55% of rice is produced in irrigated conditions (Ali et al. 2018b), arsenite$^{(III)}$ contributes to the dominant arsenic species loaded into rice plants (Zhao et al. 2010a).

4 Arsenic-Induced Toxicity Symptoms During Different Growth Stages of Rice

Enhanced uptake of arsenic will influence plant growth negatively. Arsenic toxicity in rice plants triggers various symptoms that include lower seed germination rate, poor seed establishment, lower photosynthetic rates, stunted plant growth, low biomass production, sterility-related yield loss, and a physiological disorder referred to as straighthead disease that was associated with arsenic toxicity (Fig. 4) (Rahman et al. 2008; Zhao et al. 2013). Symptoms are often confounded with other soil-related problems associated with rice (Abedin et al. 2002b; Rahman et al. 2008; Zhu et al. 2008).

4.1 Germination Stage

Germination is one of the most delicate stages in a plant's growth cycle. It can be easily affected by abiotic stresses such as drought, salinity, and heavy metals (Rajjou et al. 2012). In general, its starting point is defined with the imbibition of dry seed via water uptake and its termination by radicle or coleoptile appearance (Bewley 1997). A unique feature of rice is the ability to germinate in anaerobic conditions in which the coleoptile occurs first, followed by the radicle. During this process, various complex physiological and biochemical processes take place, for example, the synthesis and degradation of phytohormones, especially abscisic acid (ABA) and gibberellic acid (GA), reactivation of metabolism, and hydrolyzation of starch via enzymes for maintaining a source of energy (Seneviratne et al. 2019). He and Yang (2013) summarized the three phases of germination and their ongoing processes. During phase one, water is taken up rapidly into the seed, where the biosynthesis of mRNA starts, followed by starch degradation. In phase 2, the

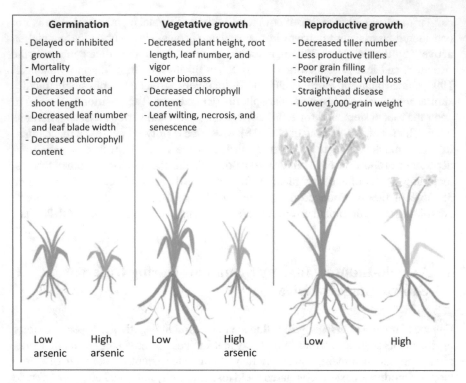

Fig. 4 Arsenic-induced symptoms during the different growth stages in rice

reactivation of metabolism begins, involving starch hydrolysis, glycolysis, and fermentation for mobilizing reserves, accompanied by amino acid biosynthesis, non-functional protein degradation, and mitochondria assembly. In the third phase, the conversion of carbohydrates to sugars begins, and the embryo is activated. The last step is coleoptile emergence, which is accompanied by the onset of aerobic respiration (He and Yang 2013). Especially in phase 1 and 3, water is taken up rapidly (Bewley 1997). With the background of arsenic-contaminated water, this implies that the seed and its emerging seedling confront arsenic stress. This may lead to an interruption or alteration in the normal germination process (Seneviratne et al. 2019). Germination of seeds was delayed or inhibited when the seeds were grown in a nutrient solution with more than 5 ppm arsenic (Begum and Mondal 2019). When seedlings were grown in pots filled with arsenic-contaminated soil, their mortality was observed at 40 mg/kg arsenite[III] in aerobic and anaerobic conditions, but it was higher in the anaerobic treatment (Shah et al. 2014). Those different observations could be explained by different experimental setups, different growth media, or because two different varieties were tested in the studies: IET-4786 and BRRIdhan28. When the seeds were able to germinate, they showed lower dry matter and decreased shoot and root length (Begum and Mondal 2019). A lower number of leaves and a lower leaf blade width were also observed as a consequence of arsenic treatment (Shaibur et al. 2006). Also, the chlorophyll content of seedling leaves decreased, irrespective of whether the plants were treated with arsenate[V] or arsenite[III] (Shaibur et al. 2006; Choudhury et al. 2011; Begum and Mondal 2019). As a general observation, arsenite[III] treatment always caused more toxic symptoms on seedlings than arsenate[V] (Shah et al. 2014; Begum and Mondal 2019).

4.2 Vegetative Growth

The same symptoms as for the seedlings keep occurring during the vegetative growth phase of the rice plant. Decreased plant height was most commonly observed (Abedin et al. 2002b; Shah et al. 2014; Dixit et al. 2016) as well as decreased root length and vigor (Abedin et al. 2002a; Das et al. 2013). Symptoms resulted from a change in IAA biosynthesis and transport, which is responsible for the formation of auxin. Lower auxin content then influenced root development, especially of the lateral roots (Ronzan et al. 2018). The resulting symptom of decreased biomass (Abedin et al. 2002b) was caused by an alteration in photosynthesis (Tuli et al. 2010). Arsenic toxicity degenerated the membrane structure and therefore affected the chloroplast as well as photosynthetic processes (Begum and Mondal 2019). The leaves wilted and turned violet due to increased anthocyanin content (Hossain 2006). As a consequence, leaf tips and margins developed senescence and eventually necrosis (Das et al. 2013). Because of these symptoms, a lower photosynthetically active area occurred, which resulted in lower energy for the plant's metabolism, leading to decreased plant height and stunted growth.

4.3 Reproductive Growth

The changes in the metabolism of the rice plant described in the previous section led to lower yield (Shah et al. 2014). Total yield loss was 80% when the soil contained more than 60 mg As/kg, but a significant loss was observed from 15 mg As/kg (Das et al. 2013). Yield decreased because of lower tiller number, which was observed in a treatment of 40 mg/kg arsenite[III] in a pot study (Das et al. 2013; Shah et al. 2014). This caused a lower number of filled and mature grains per panicle (Das et al. 2013). Another physiological disorder that has been associated with arsenic toxicity is straighthead disease. This physiological disorder produces sterile florets and spikelets and thereby diminishes grain yield (Rahman et al. 2008). A study with arsenate[V]-contaminated irrigation water showed rice plants with decreased plant height and grain yield, explained by a lower number of filled grains and grain weight, and lower root biomass. At the same time, there was a significant increase in arsenic concentration in the root, straw, rice husk, and grain (Abedin et al. 2002b). The leaves can wilt and turn violet because of increased anthocyanin content (Abbas et al. 2018). Also, the chlorophyll content of seedling leaves decreased with both arsenate[(V)] and arsenite[(III)]. This observation explains the lower biomass and growth. Finally, leaf tips and margins develop necrosis (Abedin et al. 2002c; Zhao et al. 2009). Roots show decreased biomass and lower root vigor. Grain yield decreases with decreased tiller number, filled grains, and panicles due to the sterility of florets and spikelets (Rao et al. 2011). Concerning metabolism, arsenic affects carbohydrate, lipid, and protein metabolism (Finnegan and Chen 2012). More importantly, arsenic can cause an increased formation of reactive oxygen species (ROS), exceeding the level that can be scavenged, thus leading to oxidative damage in the plant. Hydrogen peroxide (H_2O_2), and malondialdehyde ($CH_2(CHO)_2$) were the major ROS formed when rice seedlings were exposed to arsenate[(V)] (Rao et al. 2011).

5 Quantitative Trait Loci Associated with Arsenic Stress Tolerance in Rice

Arsenic toxicity, uptake, and accumulation in rice represent a quantitative trait governed by multiple loci (Dasgupta et al. 2004; Zhang et al. 2008; Murugaiyan et al. 2019). Genetic improvement by the selection of rice cultivars with a lower concentration of arsenic in the edible parts is vital in the development of rice varieties accumulating low arsenic (Norton et al. 2012; Duan et al. 2017). Even though rice breeders have been particularly interested in developing cultivars accumulating low arsenic for contaminated ecosystems, no promising loci have been functionally characterized for use in their breeding programs. Identification of appropriate rice cultivars and genetic mapping of chromosomal regions associated with a low-arsenic concentration constitute a practical methodology for diminishing the impact of arsenic in rice.

Arsenic interaction with rice has remained well-documented over the past two decades. Several QTLs have been reported in both vegetative tissues and grains (Table 1). More than a decade ago, Dasgupta et al. (2004) proposed the first QTL related to arsenate[(V)] uptake close to a phosphate (P) uptake QTL on chromosome 6 and named the region AsTol (Dasgupta et al. 2004). This region covers multiple genes with various functions, and it needs to be fine-mapped to identify responsible genes with potential for rice varieties accumulating low arsenic. Nonetheless, chromosome 6 has been identified to harbor several QTLs related to arsenic toxicity, even though these QTLs vary in their physical positions (Dasgupta et al. 2004; Zhang et al. 2008; Norton et al. 2012; Kuramata et al. 2013; Liu et al. 2019; Murugaiyan et al. 2019). QTLs on chromosome 6 were proposed to mediate arsenic concentration in the grain of brown rice (Zhang et al. 2008) and the DNA concentration of the grain (Kuramata et al. 2013). The previously mapped QTLs were closely colocalized to a QTL that mediates leaf arsenic concentration (Norton et al. 2010a). Recently, another QTL responsible for arsenic shoot content was found by Murugaiyan et al. (2019) on chromosome 6. These results indicated that the regions on chromosome 6 were involved in the uptake of arsenate[(V)]. Several genes on chromosome 6 were proposed to be involved in arsenite[(III)] transport: *Lsi6* (OsNIP$_{2;2}$) expressed at the grain-filling stage and plasma membrane intrinsic protein (OsPIPI$_{2;7}$) (Kuramata et al. 2013; Liu et al. 2019). Murugaiyan et al. (2019) also identified a tolerance QTL (*qRChlo1*) on chromosome 1 for relative chlorophyll content in a backcross-selected breeding population of rice, and *qRChlo1* increased the tolerance percentage of rice of arsenite[(III)] stress. However, none of these QTLs and genes has been functionally characterized for use in breeding programs for the development of varieties accumulating low arsenic. Chromosome 5 also contained regions of interest. Norton et al. (2010a) found a QTL associated with arsenic content in leaves on chromosome 5, but the soil at the study site had a low-arsenic concentration (<1 ppm). Also, Murugaiyan et al. (2019) identified the same QTL region on chromosome 5 for shoot arsenic concentration in a breeding population of rice. A heavy metal-associated domain-containing protein (HMA) was suggested as a candidate gene on chromosome 5. Those HMAs are associated with detoxifying heavy metals by containing a metal-binding domain and they have already been shown to help in the metal homeostasis of copper, zinc, lead, and cadmium

Table 1 QTLs reported in rice for arsenic tolerance and accumulation

Reference	Phenotyping under	As concentration	Population used	Putative QTLs on chromosome
Dasgupta et al. (2004)	Hydroponic culture system	1 mg As/kg arsenate[(V)]	Bala (*indica*) × Azucena (*japonica*) recombinant inbred lines	6
Zhang et al. (2008)	Pot culture system	1.27 mg As/kg	CJ06 (*japonica*) × T1 (*indica*) doubled-haploid population	2, 3, 6, 8
Norton et al. (2010a)	Field conditions	0.32 mg As/kg	Bala (*indica*) × Azucena (*japonica*) recombinant inbred lines	1, 3, 5, 6
Norton et al. (2012)	Field conditions	73.8 mg As/kg	Bala (*indica*) × Azucena (*japonica*) recombinant inbred lines	8, 10
Kuramata et al. (2013)	Field conditions	1.4 mg As/kg	69 accessions from World Rice Collection	6, 8
Norton et al. (2014)	Field conditions	14 ± 0.3 mg As/kg	312 accessions from Rice Diversity Panel 1	3, 5
Liu et al. (2019)	Field conditions	3000 mg As/kg	276 accessions from global Rice Diversity Panel	1, 2, 4, 5, 9, 11, 12
Murugaiyan et al. (2019)	Hydroponic culture system	10 mg As/kg arsenite[(III)]	WTR1 (*indica*) × Hao-an-nong (*japonica*) backcross recombinant inbred breeding population	1, 2, 5, 6, 8, 9
Norton et al. (2019)	Field conditions	4.63 mg As/kg	266 Bengal Assam Aus Panel	2, 3, 5, 7, 9

(Murugaiyan et al. 2019). However, their metal-binding ability toward arsenic has not been characterized in rice.

Besides chromosomes 6 and 5, chromosome 8 was found to harbor a few QTLs that influence grain arsenic content, colocalizing with a locus affecting shoot phosphorus concentration at the seedling stage of rice (Zhang et al. 2008; Norton et al. 2012; Kuramata et al. 2013). Also, a QTL related to the trait days to heading was found close to a QTL for grain arsenic content on chromosome 8, leading to the assumption that a prolonged time of the vegetative phase could decrease grain arsenic content (Norton et al. 2012). Furthermore, a QTL that mediated grain DNA content was also localized on this chromosome (Kuramata et al. 2013). Chromosome 3 was also identified several times and was proposed to contain a QTL that encompassed *Lsi2*, a transporter for arsenite[III] (Norton et al. 2012, 2019). Nonetheless, it is also important to identify QTLs associated with arsenic uptake as well as translocation, because, if no arsenic enters the plant or is safely stored, no arsenic will reach the grain. QTLs mediating root arsenic concentration were located on chromosomes 3 and 8 (Zhang et al. 2008; Murugaiyan et al. 2019). In order to estimate arsenic uptake and metabolism, measuring root As content was claimed to be a biased evaluation criterion due to arsenic residuals on the root surface (Zhang et al. 2008). More QTLs contributing to shoot arsenic concentration were found on chromosomes 2, 5, 6, and 9 (Zhang et al. 2008; Murugaiyan et al. 2019), and more precisely for the arsenic concentration in leaves on chromosomes 1, 3, 5, and 6 (Norton et al. 2010a). The QTL on chromosome 2 covered 13 candidate genes, including an H^+ vacuolar pyrophosphatase, which is known to mediate changes in the concentration of essential and toxic ions (Murugaiyan et al. 2019). However, the QTLs mapped for arsenic tolerance in rice are not characterized for use in breeding programs for the development of arsenic-safe varieties.

6   Arsenic Uptake in Rice

As previously stated, arsenite[III] is the predominant form of arsenic in paddy soils. In order to develop rice varieties accumulating low arsenic, it is critical to understand the uptake, translocation, and underlying physiology of arsenic interaction with rice (Fig. 5). The uptake kinetics of both inorganic arsenic species were similar and followed the Michaelis–Menten equation (Lou et al. 2009). At higher substrate concentration, arsenite[III] had a higher uptake and was therefore assumed to have a low-affinity uptake system (Abedin et al. 2002a). Arsenite[III] and arsenate[V] have different ionic size and their uptake pathways differ (Zhao et al. 2013). Arsenite[III] resembles the silica ion in diameter and comparable acid-ionization constant value (pK_a) (Jian et al. 2008). Arsenate[V] has a structure similar to the orthophosphate anion and their second and third pK_a values are similar (O'Day 2006). Dimethylarsinic acid (DMA) and monomethylarsonic acid (MMA) are the most abundant forms of organic arsenic in paddy soil. They were a result of arsenite methylation facilitated by microorganisms (Wu et al. 2012). The uptake of organic arsenic species was the only way for DMA and MMA to enter the rice plant since rice cannot methylate arsenic *in planta*. Instead, methylation occurred in the presence of methylating microorganisms in the rhizosphere (Lomax et al. 2012). Those microorganisms need to contain the *arsM* (*S*-adenosylmethionine methyltransferase) gene in order to methylate arsenic (Jia et al. 2013).

6.1 Arsenite Uptake

Two major transporters were identified for arsenite[III] uptake in rice (Fig. 6). These transporters form part of the aquaporins family, belonging to the group of nodulin 26-like intrinsic membrane proteins (NIPs) in rice (Jian et al. 2008). The aquaporins *Lsi1* (*OsNIP2;1*) and *Lsi2* were permeable only to arsenite[III] and usually mediated silicon influx and efflux. They differed in location and thus in their function. *Lsi1*

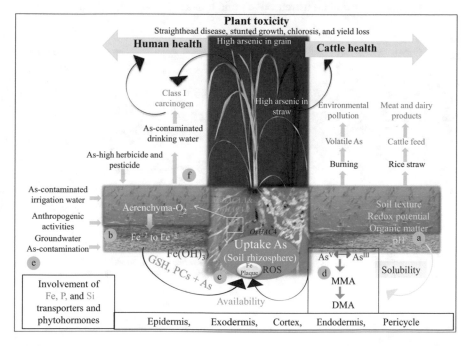

Fig. 5 Factors influencing arsenic availability and transportation from soil to rice plants: (a) the factors involved in solubility and available form of arsenic; (b) anaerobic conditions, rice roots of aerenchyma release a part of the O_2 to the rhizosphere complex; (c) formation of iron (Fe) plaques: Fe^{2+} oxidized to Fe^{3+} on the root surface and subsequent reaction of sulfhydryl-containing short peptides such as glutathione and phytochelatins binding to arsenic; (d) if more oxygen is released, this also may oxidize arsenite[III] to arsenate[V], which is more strongly adsorbed to iron plaque; (e) uptake of arsenate[V] is supported by various micronutrient transporters; (f) As-contaminated paddy soil influenced by different external factors, and effects of consumption of food grain and drinking water on human health

was responsible for the transport of arsenite[(III)] from the external solution to the root cells. It was located and expressed on the outside of the exodermis and endodermis close to the Casparian stripe. The transport was bidirectional, indicating that arsenite[(III)] could also efflux to the external medium via *Lsi1* (Zhao et al. 2010b). However, there were more efflux transporters to be included (Zhao et al. 2010a). *Lsi2* mediated the efflux of arsenite[(III)] toward the xylem and formed the counterpart of *Lsi1* at the expression site. *Lsi2* was expressed on the inner side of the plasma membranes of the exodermis and endodermis. From there, arsenite[(III)] was released into the cortex and stele. *Lsi2* was considered to be more involved in the transport and translocation of arsenite[(III)] from shoot to grain (Jian et al. 2008). In the stele of primary roots and cell layers of lateral roots, more precisely in the plasma membrane, the membrane protein *OsNIP3;2* was expressed, which also belongs to the NIP family. Arsenite[(III)] was also taken up and further transported through this protein (Chen et al. 2017). Besides the NIPs, plasma membrane intrinsic proteins (PIPs) are likely to play a role in arsenic homeostasis in rice. When exposed to elevated arsenite[(III)] concentrations, the transcript concentrations in the roots were downregulated in order to prevent an increased arsenic uptake (Mosa et al. 2012).

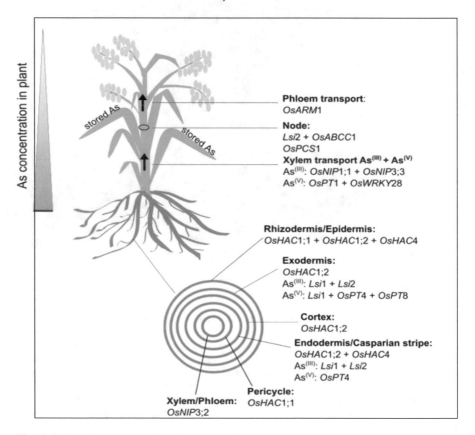

Fig. 6 Schematic representation of arsenic uptake and translocation in rice

6.2 Arsenate Uptake

Arsenate[V] is a chemical analog of phosphate ions and takes advantage of the phosphorus transport pathway. In the presence of phosphate, the uptake of arsenate[V] was actively suppressed, whereas arsenite[III] was not (Abedin et al. 2002a). This finding was a clear indicator that arsenate[V] used phosphorus transporters (*Pht*). On a molecular level, the *Pht* family, comprising 13 genes, was associated with phosphate transport in rice (Wang et al. 2016). Some of the transporter function genes were investigated concerning arsenate[V] uptake. *OsPT8* (*OsPht1;8*) mediated phosphorus homeostasis and was also considered as a critical transporter for arsenate[V] uptake in rice roots (Wang et al. 2016). A role for the uptake and translocation of arsenate[V] in rice was demonstrated by expression analysis of the *OsPT4* transporter gene. The overexpression of this gene resulted in higher arsenic accumulation when grown in arsenate[V] solution (Chao et al. 2014). *Lsi1* also played an essential role in the uptake of arsenate[V] since it was found that arsenite[III] was effluxed through this channel when rice roots were exposed to arsenate[V] (Zhao et al. 2010a).

7 Arsenic Translocation in Rice

The translocation of arsenic depends on its concentration, its capability of complexation, and its sequestration as well as the xylem flow rate (Suriyagoda et al. 2018). In general, the concentration of arsenic in rice decreases in the following order: root > straw > husk > grain (Abedin et al. 2002b). Translocation of arsenic species from root to shoot to grain also depends on the genotype and growth stage. In general, grain arsenic content was found to be affected by genotype, year, and genotype × year interaction effects (Kuramata et al. 2013). A higher-yielding variety was found to be capable of translocating more arsenic from root to shoot and also from shoot to grain than traditionally grown landraces (Bhattacharya et al. 2010). Genotypes that varied in grain arsenic accumulation were found to have different arsenic content in all plant tissues (Duan et al. 2011). Furthermore, there was evidence that translocation mainly took place at the active tillering stage (Das et al. 2013).

7.1 Arsenite Translocation

The phloem plays a significant role in arsenic translocation from the vegetative tissue to the grain. Arsenite[III] was translocated to the vegetative tissue, capable of being remobilized and transported to the grain through the phloem (Fig. 7). An explanation for this phloem loading might be a transfer of arsenic in the xylem vessels into the phloem (Zhao et al. 2012). The phloem transferred 90% of arsenite[(III)] and 55% of DMA to the grain (Carey et al. 2010). Inorganic arsenic was mostly found in the caryopsis and transported during the grain-filling stage. DMA accumulated in the caryopsis before flowering (Zhao et al. 2012). The mechanism by which arsenite[(III)] is remobilized and further transported is not known yet. Arsenic could be stored in the nodes and internodes of rice, where arsenic accumulation was observed in the phloem companion cells. From there, translocation to the grain and flag leaf was restricted. This capability of the nodes was explained by a higher expression of *Lsi2* and the formation of arsenite[(III)]-thiol compounds in the vacuole (Chen et al. 2016). *OsABCC1* was also included in the storage of arsenic in the nodes by sequestering it in the vacuoles of the phloem companion cells (Song et al. 2014). The phytochelatin synthase *OsPCS1* might play a supportive role in this process, as it reduced arsenic grain contents when overexpressed (Hayashi et al. 2017). In the basal and upper nodes, the transcription factor *OsARM1* (Arsenite Responsive MYB 1) was strongly expressed in the phloem region when exposed to high-arsenite[(III)] concentrations. It also showed high expression in the rachis and spikelet and was likely to be involved in mediating uptake and root-to-shoot translocation (Wang et al. 2017). Those findings indicate that translocation through the phloem is more critical than translocation through the xylem.

7.2   Arsenate Translocation

It is widely assumed that arsenate[(V)] is directly reduced to arsenite[(III)] in the root. Seyfferth et al. (2011) showed that oxidized arsenic species were the predominant form in the xylem within a root cross-section. This was supported by the assumption that the uptake of the two inorganic arsenic species occurred at different locations in the root. However, genotype-dependent variation was also observed (Seyfferth et al. 2011). Another fact that disproved the reduction assumption was that straw contained mostly arsenate[(V)] after growing in an arsenate[(V)] solution for 170 days (Abedin et al. 2002a). The high abundance of arsenate[(V)] could be explained by the limited capacity of the reducing enzymes *OsHAC1;1* and *OsHAC1;2*, thus not reducing arsenate[(V)]

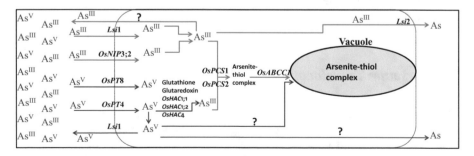

Fig. 7   Schematic representation of arsenic uptake and storage in a cell

anymore (Shi et al. 2016). Nonetheless, as soon as inorganic arsenic was in the xylem, it became transported to the vegetative parts. Nodulin 26-like intrinsic membrane proteins that were included in the translocation of arsenite[(III)] from root to shoot included *OsNIP1;1* and *OsNIP3;3*, which caused a decreased arsenic concentration in shoot and xylem when overexpressed. *OsNIP1;1* was expressed in the basal stem at the seedling and tillering stages. At maturity, it was mostly expressed in the roots, basal stem, nodes, and spikelets. In contrast, *OsNIP3;3* was expressed continuously in different tissues (Sun et al. 2018). For the transport of arsenate[(V)] from root to shoot, the expression level of *OsPT1* was investigated. It was increased after different phosphate (P) and arsenate[(V)] treatments in seedlings, which indicated its facilitating role for the transport (Kamiya et al. 2013). *OsWRKY28*, a transcription factor of the *WRKY* family that is mostly involved in stress tolerance, was found to contribute to the translocation of arsenate[(V)] from root to shoot (Wang et al. 2017).

8 Arsenic Detoxification and Stress Responses in Rice

Arsenic does not have any known biological function within the rice plant, and enhanced uptake triggers various defense mechanisms in the plant to mitigate the negative impact of arsenic stress. Once arsenate[(V)] was taken up into the cell, it was reduced to arsenite[(III)], which was the dominant form inside the plant tissue (Ali et al. 2009). The reduction was a slow process mediated by glutaredoxin (Suriyagoda et al. 2018). Enzymatic reduction of arsenate[(V)] was performed more likely by the arsenate[(V)] reductase gene *High Arsenic Content* (HAC) (Chao et al. 2014). In rice, *OsHAC1;1*, *OsHAC1;2*, and *OsHAC4* played a significant role in this process (Shi et al. 2016). They were most commonly expressed in the roots. *OsHAC1;1* was found in the epidermis, pericycle cells, and root hairs in the mature zone of roots. *OsHAC1;2* was most abundant in the epidermis, exodermis, outer layer of the cortex, and endodermis cells (Shi et al. 2016). *OsHAC4* was expressed in the root epidermis and exodermis (Xu et al. 2017). By overexpression, knocking out, and xylem sap analysis, it was established that they were included in the reduction of arsenate[(V)] to arsenite[(III)] and in the efflux of arsenite[(III)] to the external medium (Shi et al. 2016). Detoxification of the reduced arsenate[(V)] and the readily available arsenite[(III)] was maintained by glutathione-based phytochelatins (PCs). Glutathione acts as a precursor for PCs, and its role in arsenic tolerance was corroborated by a study with a sulfur supplement treatment. It was pointed out that most of the arsenate[(V)] remained in the root, promoting the importance of phytochelatins in binding arsenic (Dixit et al. 2016). The same can be stated for arsenite[(III)], which complexes with PCs, lowering its mobility from shoot to grain (Duan et al. 2011). They are also referred to as arsenite[(III)]-thiol complexes. An essential contribution to this mechanism was suggested for phytochelatin synthase *OsPCS1* and *OsPCS2*, whereby the former was more expressed in the roots under arsenite[(III)] treatment and the latter was identified to be an essential isozyme controlling PC synthesis

(Yamazaki et al. 2018). Arsenite[III]-thiol complexes were transported from the cytoplasm into the vacuole by a member of the C-type ATP-binding cassette (ABC) transporter (*OsABCC*) family, *OsABCC1* (Song et al. 2014). In the vacuole, the complexes could be sequestered efficiently since the acidic pH of 5.5 stabilized them (Ali et al. 2009; Suriyagoda et al. 2018). It was assumed that the same transporters and detoxification mechanisms as in the root cell were used for storing arsenic after relocation, but this is not well investigated yet (Zhao et al. 2010b). Here, the same detoxification mechanism was used again to store arsenic in the vacuole (Suriyagoda et al. 2018). A study with radioactively labeled arsenic during the grain-filling stage showed that arsenite[III] was somewhat immobile in the xylem within the rice plant, probably because of its complexation and sequestration (Zhao et al. 2012). The silica pathway might not be used when translocating and unloading arsenite[III] into the grain since silica mainly accumulated in the husk and not in the grain (Norton et al. 2010b). However, the addition of silica decreased inorganic grain arsenic content but increased DMA, by 59% and 33%, respectively (Li et al. 2009). In contrast, arsenate[V] was proposed to use the phosphorus pathway throughout the whole transport and translocation (Norton et al. 2010b).

8.1   The Oxidative Stress Response in Rice

Concerning metabolism, carbohydrate-, lipid-, protein-, amino acid-, ascorbate-, and aldarate-related pathways were affected by arsenic uptake (Abbas et al. 2018). Membrane transport, cell growth, and death, as well as biodegradation were also altered by showing an up- or down-regulation of gene function (Dubey et al. 2014). When conducting a genome-wide expression study, several gene families were up- and down-regulated when grown in arsenite[III], especially facilitating transporters, stress-related genes, regulatory proteins, growth, and development, as well as secondary metabolism. This finding indicated arsenite[III]-induced oxidative stress (Chakrabarty et al. 2009). More importantly, arsenic triggers an increased formation of reactive oxygen species (ROS) and therefore not all of them could be scavenged, thus leading to plant damage. Hydrogen peroxide (H_2O_2) was an ROS that was formed when rice seedlings were exposed to arsenate[V] (Choudhury et al. 2011). Arsenite[III] was found to promote the development of different ROS, namely, superoxide anion ($O_2^{•-}$) and H_2O_2, thus inducing lipid peroxidation in rice seedlings (Mishra et al. 2011). To prevent an excessive amount of H_2O_2, seedling roots grown in arsenate[V] solution increased ascorbate peroxidase (APX) activity, which reduced H_2O_2 via the ascorbate-glutathione cycle (Dubey et al. 2014). For scavenging the ROS occurring under arsenite[III], the concentrations of enzymatic antioxidants superoxide dismutase (SOD), catalase (CAT), chloroplastic ascorbate peroxidase, guaiacol peroxidase, monodehydroascorbate reductase, and glutathione reductase increased. Higher concentrations of ascorbate and glutathione were found, and the synthesis of phytochelatins and total acid-soluble thiols was enhanced in arsenic-treated seedlings (Mishra et al. 2011). Besides ROS, altered RNase and protease activity were also an indicator of stress. Their activity was inhibited by the presence of arsenite[(III)], thus leading to high abundance of RNA and proteins. However, an increase in proline was also reported (Choudhury et al. 2011). Proline is an osmolyte, and its accumulation was usually found under salinity or drought stress (Amini et al. 2015). Under arsenic stress, its abundance led to the assumption that proline could help to protect enzymes in the presence of arsenite[(III)] (Mishra and Dubey 2006). The same can be reported for heat shock proteins (HSPs), generally formed under a rapid temperature increase, water deficit, or salinity (Ponomarenko et al. 2013). In the presence of arsenic, several HSP genes were upregulated (Chakrabarty et al. 2009).

8.2 Root Plaque Formation as a Scavenger for Arsenic Stress

Despite anaerobic conditions in paddy fields, the rhizosphere can be aerobic due to the semiaquatic characteristic of rice forming an aerenchyma when flooded. From there, oxygen is released by the roots, forming an oxidation barrier. This leads to the formation of iron plaque, visible by the reddish coating around the roots, caused by a reoxidation of $Fe^{(II)}$ to $Fe^{(III)}$ (Becker and Asch 2005). Iron plaque is mostly composed of lepidocrocite, goethite, and ferrihydrite (Seyfferth et al. 2011). It was proposed that the same reoxidation mechanism and an additional formation of iron plaque would decrease the uptake of arsenic (Lee et al. 2013). This hypothesis could not be confirmed since the proportion of the newly formed arsenate[(V)] in a hydroponic solution was very low (Liu et al. 2010). Rice genotypes with a higher radial oxygen loss showed a higher formation of iron plaque. This led to a lower uptake of arsenate[(V)] and hence to a lower concentration of total inorganic arsenic in the plant (Wu et al. 2012). However, iron plaque distribution was not homogeneously covering total roots. Indeed, young roots showed little iron plaque formation. Therefore, iron plaque was assumed to be a bulk scavenger for arsenic since it was absent in the main solute-uptake root area (Seyfferth et al. 2011). In general, arsenate[(V)] was absorbed more than arsenite[(III)], and arsenite[(III)] was desorbed more rapidly (Geng et al. 2017). Surprisingly, the formation of iron plaque enhanced arsenite[(III)] uptake, but, in this experiment, the iron plaque was formed first and then the plants were exposed to arsenite[(III)]/arsenate[(V)]. This practice cannot be transferred to natural field conditions (Liu et al. 2010). The addition of phosphorus increased the uptake of arsenate[(V)] irrespective of iron plaque (Geng et al. 2005; Yang et al. 2020). This could be explained because phosphorus might have occupied the sorption space at the iron plaque, thus releasing more arsenate[(V)].

In conclusion, iron plaque could act as a buffer for arsenate[(V)] uptake (Liu et al. 2010). Concerning a decreased arsenite[(III)] uptake, manganese (Mn) plaque was shown to be more effective in capturing arsenite[(III)] than iron plaque as shown by the lower shoot arsenic concentration. However, the control treatment without any plaque formation also showed lower shoot arsenic concentration. Therefore, the role of manganese plaque needs to be further investigated (Liu et al. 2010). In contrast to the results identified by hydroponic studies, a microcosm study with a contaminated soil from Bangladesh containing 14 mg As/kg dry weight showed that iron plaque indeed prevented uptake, leading to lower shoot arsenic concentration (Huang et al. 2012b).

9 State of Knowledge Gaps for Arsenic Accumulation in Rice

In the past two decades, some progress has been made in understanding the physiological and molecular mechanisms of arsenic metabolism in rice. The regulation and transportation of arsenic from soil to plants are associated with multiple factors: soil texture, water availability, arsenic bioavailability, genotype used, arsenic interactions with other heavy metals and essential minerals, rice ecosystem, microbial interaction, physiological activities, and genotype-by-environment interactions (Norton et al. 2012; Sahoo and Mukherjee 2014) (Fig. 5). Among these factors, the selection of rice cultivars with low-arsenic content in grain and straw is the most efficient and practical approach for decreasing arsenic contamination in rice. The molecular genetic mechanisms of arsenic accumulation in the root, shoot, and grain need to be understood clearly for designing efficient breeding strategies and developing elite breeding materials, which combine arsenic tolerance, low-arsenic uptake, and high-grain yield.

Several experiments were conducted to understand the physiological mechanisms underlying rice varieties accumulating low arsenic (Norton et al. 2012; Murugaiyan et al. 2019). However, some of them show inconsistency and methodological errors, leading to wrong assumptions. Many studies focused on arsenate[V] rather than arsenite[III] although arsenite[III] is the predominant form in paddy soils, where the majority of rice is produced. Abedin et al. (2002a) were the first to elucidate that arsenite[III] and arsenate[V] take advantage of different uptake mechanisms in rice. After this finding, the focus should have been more toward elevated arsenite[III] concentrations in rice experiments, which was not the case. When identifying the *AsTol* QTL on chromosome 6, 1 ppm of arsenate[V] was used (Dasgupta et al. 2004). When investigating the accumulation in widely used cultivars in Bangladesh, the soil was obtained from the Bangladesh Rice Research Institute (BRRI) and enriched with arsenate[V] rather than arsenite[III]. There, they found that a hybrid variety accumulated the most arsenic in the straw at a treatment of 30 mg As/kg, using $Na_2HAsO_4 \cdot 7H_2O$. It was stated that the hybrid variety would show a higher phytotoxicity and accumulation ability (Rahman et al. 2007). This finding needs to be proven when the soil is enriched with arsenite[III] because, otherwise, the hybrid could contain a more efficient phosphorus-uptake pathway, which arsenate[V] took advantage of. When estimating the grain arsenic content determined in different soil arsenic concentrations, arsenate[V] was used to enrich the soil (Rahman et al. 2008). It cannot be clearly stated how important the difference between using arsenate[V] and arsenite[III] really is unless both forms are tested in the same experiment. Another shortcoming, especially when finding QTLs, is when a low-arsenic concentration is used. The concentration in the experiment should simulate field conditions with typical concentrations of 4–8 mg/kg. Therefore, a concentration of less than 1 ppm cannot lead to reliable results (Abedin et al. 2002a; Norton et al. 2010a). Most of the published studies related to arsenic uptake and accumulation were carried out in a hydroponics system or pot culture, and these experiments need to be carried out in a contaminated field for better understanding.

Rice production in the major rice-growing countries is shifting toward more efficient technologies such as direct-seeded rice (DSR) and alternate wetting and drying (AWD) (Pathak et al. 2011; Richards and Sander 2014). It is essential to identify suitable genotypes, genes, and QTLs associated with overcoming arsenic contamination under these technologies. The presence of flooded water influences arsenic speciation. It is crucial to use arsenate$^{(V)}$ for conducting experiments related to dry-DSR conditions and arsenite$^{(III)}$ for wet-DSR conditions. So far, no experiment has been conducted to observe the germination ability of rice under DSR conditions in contaminated soils. AWD technology was proposed to decrease arsenic content in rice. However, no studies have been conducted to observe how arsenic species acted when applying AWD technology. Since arsenic translocation from soil depends on flooding, it is understood that, during the wetting period, arsenite$^{(III)}$ dominates, and, during the drying period, arsenate$^{(V)}$ dominates. However, no data were available to validate this observation.

10 Possible Mitigation Strategies for Arsenic Accumulation in Rice

Several novel methodologies in rice cooking provided promising results for a decrease in arsenic content by using a continual flow of clean water over the raw rice before cooking (Carey et al. 2015). Carey et al. (2015) stated that about 85% of the inorganic arsenic was removed from the raw rice by using a continual flow of clean cooking water. Several researchers also found that using a higher water-to-rice ratio for cooking, followed by removal of the extra water (starch), significantly decreased the arsenic in cooked rice (Sengupta et al. 2006; Stanton et al. 2015). For instance, Sengupta et al. (2006) reported a decrease in arsenic by 57% by washing the rice grain several times until the water looks transparent/clear, and a cooking process with a 6:1 ratio of water to rice grain with draining of the excess water. However, the effect of this procedure on the content of essential water-soluble nutrients such as minerals or B-vitamins remains to be elucidated.

Selection of existing rice cultivars that are biologically restricted in terms of arsenic accumulation in the grain offers great potential for use in breeding programs, as some rice cultivars can accumulate 20–30-fold less arsenic than others (Murugaiyan et al. 2019). The physiological pathways and molecular genetics of arsenic accumulation in rice were linked to many QTLs and genes involved in the uptake and translocation of arsenic in rice (Dasgupta et al. 2004; Tuli et al. 2010).

Still, a gap exists in identifying the candidate genes and allelic variants of the genes governing low-arsenic content in rice grain. Using recent trends of allele mining and haplotype breeding technologies can help to limit arsenic accumulation in the grain, which is a prerequisite for breeding programs to develop low-arsenic-accumulating rice cultivars (Murugaiyan et al. 2019). Hence, several strategies are possible to decrease arsenic in rice grains, including the following:

1. Mitigation strategies related to rice cooking can help to decrease arsenic content in the major staple food crop, rice. The ratio of rice grain for cooking, the amount of water used for washing, cooking vessels, and duration of the cooking process can decrease the arsenic in cooked rice (Carey et al. 2015).
2. Supplementation of mineral nutrient elements such as S, P, Fe, and Si can also significantly diminish the accumulation of arsenic content in rice grain by minimizing its uptake and transport in food crops. For instance, P and Si are complementary in nature to arsenic in competing for uptake. Therefore, the external application of these mineral nutrients decreases the chances of arsenic uptake from the soil (Alkorta et al. 2004; Jian et al. 2008).
3. Water-saving technologies such as direct-seeded rice cultivation are viable, immediate, and sustainable solutions for decreasing arsenic content in rice (Hu et al. 2013; Abedin et al. 2002a). Talukder et al. (2011) reported that arsenic uptake in rice plants (0.23–0.26 mg/kg) is lower with aerobic water management than in anaerobic rice cultivation (0.60–0.67 mg/kg).
4. The identification of donors/selection of rice cultivars with low arsenic can minimize the risk of arsenic associated with human diseases and be eco-friendly to nature (Norton et al. 2012).
5. The development of breeding lines through the introgression of QTLs and genes for low arsenic into elite rice cultivars can ensure high-grain yield and low-arsenic content in cultivars suitable for cultivation in arsenic-contaminated soils (Murugaiyan et al. 2019).
6. CRISPR/Cas9-based genome-editing tools can be useful for modifying crucial-regulated genes (*OsPht1:8*, *Lsi1*, *OsNRAMP1*, and *OsABCC1*) for the interruption or modification of genes that regulate pathways contributing to low-arsenic accumulation in rice grain. Targeting the genes involved in the phloem transport mechanism and nodes could be crucial for transporting to rice grain (Song et al. 2014). Thus, identifying phloem-localized transporters and their manipulation will provide a promising approach toward decreasing arsenic accumulation in grain.

11 Future Directions and Conclusions

The nature of arsenic being redox-active, highly toxic to organisms, and its tendency to be methylated make it a complex and important mineral to study. Arsenic uptake and metabolism in rice need to be placed in the broader context of biogeochemical cycling of arsenic in the environment. Bioavailability and speciation of arsenic in the soil are strongly dependent on environmental conditions, and this knowledge is crucial as it determines the extent of arsenic accumulation by rice and the consequences of arsenic contamination in the food chain. The widespread arsenic contamination in the environment, the mobilization of arsenic into rice grain even in soils with baseline arsenic concentrations, and the realization that excessive arsenic accumulation in rice can present a health risk to humans have provided a recent impetus in research on this subject. Therefore, significant progress has been made in recent years to understand arsenic uptake, speciation, and detoxification in plants. However, substantial knowledge gaps exist on the mechanisms of arsenic sequestration in the vacuoles and for arsenic loading and unloading in xylem and phloem. Still, there is a need to understand the regulation of arsenic accumulation in grain and to unravel the pathways and enzymes responsible for arsenate[V] reduction and methylation. Recent advances in the analytical techniques for arsenic speciation are instrumental in enhancing our understanding of arsenic biogeochemical cycling and plant metabolism. Combining these analytical tools with molecular genetics and functional genomics should provide ample opportunities for unraveling the mechanisms of arsenic transport, metabolism, and regulation.

References

Abbas G, Murtaza B, Bibi I et al (2018) Arsenic uptake, toxicity, detoxification, and speciation in plants: physiological, biochemical, and molecular aspects. Int J Environ Res Public Health 15:59. https://doi.org/10.3390/ijerph15010059

Abedin MJ, Cotter-Howells J, Meharg AA (2002a) Arsenic uptake and accumulation in rice (*Oryza sativa* L.) irrigated with contaminated water. Plant Soil 240:311–319. https://doi.org/10.1023/A:1015792723288

Abedin MJ, Cresser MS, Meharg AA et al (2002b) Arsenic accumulation and metabolism in rice (*Oryza sativa* L.). Environ Sci Technol 36:962–968. https://doi.org/10.1021/es0101678

Abedin MJ, Feldmann J, Meharg AA (2002c) Uptake kinetics of arsenic species in rice plants. Plant Physiol 128:1120–1128. https://doi.org/10.1104/pp.010733

Abernathy C, Chakraborti D, Edmonds JS et al (2001) Environmental health criteria for arsenic and arsenic compounds. Environ health criteria 224. WHO, Geneva

Ali W, Isayenkov SV, Zhao FJ, Maathuis FJM (2009) Arsenite transport in plants. Cell Mol Life Sci 66:2329–2339. https://doi.org/10.1007/s00018-009-0021-7

Ali J, Aslam UM, Tariq R et al (2018a) Exploiting the genomic diversity of rice (Oryza sativa L.): SNP-typing in 11 early-backcross introgression-breeding populations. Front Plant Sci 9:849. https://doi.org/10.3389/fpls.2018.00849

Ali J, Jewel ZA, Mahender A et al (2018b) Molecular genetics and breeding for nutrient use efficiency in rice. Int J Mol Sci 19:1762. https://doi.org/10.3390/ijms19061762

Alkorta I, Hernández-Allica J, Becerril JM et al (2004) Recent findings on the phytoremediation of soils contaminated with environmentally toxic heavy metals and metalloids such as zinc, cadmium, lead, and arsenic. Rev Environ Sci Biotechnol 3:71–90. https://doi.org/10.1023/B:RESB.0000040059.70899.3d

Amini M, Abbaspour KC, Berg M et al (2008) Statistical modeling of global geogenic arsenic contamination in groundwater. Environ Sci Technol 42:3669–3675. https://doi.org/10.1021/es702859e

Amini S, Ghobadi C, Yamchi A (2015) Proline accumulation and osmotic stress: an overview of P5CS gene in plants. Genet Agric Biotechnol Inst Tabarestan 3:44–55

Anawar HM, Akai J, Komaki K et al (2003) Geochemical occurrence of arsenic in groundwater of Bangladesh: sources and mobilization processes. J Geochem Explor 77:109–131. https://doi. org/10.1016/S0375-6742(02)00273-X

Ashikari M, Ma JF (2015) Exploring the power of plants to overcome environmental stresses. Rice 8:10. https://doi.org/10.1186/s12284-014-0037-y

Becker M, Asch F (2005) Iron toxicity in rice:conditions and management concepts. J Plant Nutr Soil Sci 168:558–573

Begum M, Mondal S (2019) Relative toxicity of arsenite and arsenate on early seedling growth and photosynthetic pigments of rice. Curr J Appl Sci Technol 33:1–5. https://doi.org/10.9734/ cjast/2019/v33i430087

Bewley JD (1997) Seed germination and dormancy. Plant Cell 9:1055–1066. https://doi. org/10.1105/tpc.9.7.1055

Bhattacharya P, Samal AC, Majumdar J, Santra SC (2010) Accumulation of arsenic and its distribution in rice plant (*Oryza sativa* L.) in Gangetic West Bengal, India. Paddy Water Environ 8:63–70. https://doi.org/10.1007/s10333-009-0180-z

Bondu R, Cloutier V, Rosa E, Benzaazoua M (2016) A review and evaluation of the impacts of climate change on geogenic arsenic in groundwater from fractured bedrock aquifers. Water Air Soil Pollut 227:296. https://doi.org/10.1007/s11270-016-2936-6

Brammer H, Ravenscroft P (2009) Arsenic in groundwater: a threat to sustainable agriculture in South and South-east Asia. Environ Int 35:647–654. https://doi.org/10.1016/j. envint.2008.10.004

Caldwell BK, Caldwell JC, Mitra SN, Smith W (2003) Searching for an optimum solution to the Bangladesh arsenic crisis. Soc Sci Med 56:2089–2096. https://doi.org/10.1016/ S0277-9536(02)00203-4

Carbonell-Barrachina Á, Munera-Picazo S, Cano-Lamadrid M et al (2015) Arsenic in your food: potential health hazards from arsenic found in rice. Nutr Diet Suppl 7:1. https://doi. org/10.2147/nds.s52027

Carey AM, Scheckel KG, Lombi E et al (2010) Grain unloading of arsenic species in rice. Plant Physiol 152:309–319. https://doi.org/10.1104/pp.109.146126

Carey M, Jiujin X, Farias JG, Meharg AA (2015) Rethinking rice preparation for highly efficient removal of inorganic arsenic using percolating cooking water. PLoS One 10:e0131608. https:// doi.org/10.1371/journal.pone.0131608

Chakrabarty D, Trivedi PK, Misra P et al (2009) Comparative transcriptome analysis of arsenate and arsenite stresses in rice seedlings. Chemosphere 74:688–702. https://doi.org/10.1016/j. chemosphere.2008.09.082

Chakraborti D, Rahman MM, Das B et al (2013) Groundwater arsenic contamination in Ganga-Meghna-Brahmaputra plain, its health effects and an approach for mitigation. Environ Earth Sci 70:1993–2008. https://doi.org/10.1007/s12665-013-2699-y

Chao DY, Chen Y, Chen J et al (2014) Genome-wide association mapping identifies a new arsenate reductase enzyme critical for limiting arsenic accumulation in plants. PLoS Biol 12:e1002009. https://doi.org/10.1371/journal.pbio.1002009

Chen WQ, Shi YL, Wu SL, Zhu YG (2016) Anthropogenic arsenic cycles: a research framework and features. J Clean Prod 139:328–336. https://doi.org/10.1016/j.jclepro.2016.08.050

Chen Y, Sun SK, Tang Z et al (2017) The nodulin 26-like intrinsic membrane protein *OsNIP*3;2 is involved in arsenite uptake by lateral roots in rice. J Exp Bot 68:3007–3016. https://doi. org/10.1093/jxb/erx165

Choudhury B, Chowdhury S, Biswas AK (2011) Regulation of growth and metabolism in rice (*Oryza sativa* L.) by arsenic and its possible reversal by phosphate. J Plant Interact 6:15–24. https://doi.org/10.1080/17429140903487552

Clemens S (2006) Toxic metal accumulation, responses to exposure and mechanisms of tolerance in plants. Biochimie 88:1707–1719

Clemens S, Ma JF (2016) Toxic heavy metal and metalloid accumulation in crop plants and foods. Annu Rev Plant Biol 67:489–512. https://doi.org/10.1146/annurev-arplant-043015-112301

Das B, Nayak B, Pal A et al (2008) Groundwater arsenic contamination and its health effects in the Ganga-Meghna-Brahmaputra plain. In: Groundwater for sustainable development: problems, perspectives and challenges. Routledge, London, pp 257–269

Das I, Ghosh K, Das DK, Sanya SK (2013) Assessment of arsenic toxicity in rice plants in areas of West Bengal. Chem Speciat Bioavailab 25:201–208. https://doi.org/10.318 4/095422913X13785717162124

Dasgupta T, Hossain SA, Meharg AA, Price AH (2004) An arsenate tolerance gene on chromosome 6 of rice. New Phytol 163:45–49. https://doi.org/10.1111/j.1469-8137.2004.01109.x

Dixit G, Singh AP, Kumar A et al (2016) Reduced arsenic accumulation in rice (*Oryza sativa* L.) shoot involves sulfur-mediated improved thiol metabolism, antioxidant system and altered arsenic transporters. Plant Physiol Biochem 99:86–96. https://doi.org/10.1016/j.plaphy.2015.11.005

Dobermann A, Fairhurst T (2000) Rice: nutrient disorders & nutrient management. http://books.irri.org/9810427425_content.pdf

Duan GL, Hu Y, Liu WJ et al (2011) Evidence for a role of phytochelatins in regulating arsenic accumulation in rice grain. Environ Exp Bot 71:416–421. https://doi.org/10.1016/j.envexpbot.2011.02.016

Duan G, Shao G, Tang Z et al (2017) Genotypic and environmental variations in grain cadmium and arsenic concentrations among a panel of high-yielding rice cultivars. Rice 10:9. https://doi.org/10.1186/s12284-017-0149-2

Dubey S, Shri M, Misra P et al (2014) Heavy metals induce oxidative stress and genome-wide modulation in transcriptome of rice root. Funct Integr Genom 14:401–417. https://doi.org/10.1007/s10142-014-0361-8

Duker AA, Carranza EJM, Hale M (2005) Arsenic geochemistry and health. Environ Int 31:631–641

Finnegan PM, Chen W (2012) Arsenic toxicity: the effects on plant metabolism. Front Physiol 3:182. https://doi.org/10.3389/fphys.2012.00182

Flora SJS, Pachauri V, Saxena G (2011) Arsenic, cadmium and lead. In: Reproductive and developmental toxicology. Elsevier, Amsterdam, pp 415–438

Frei M, Becker K (2005) Integrated rice-fish culture: coupled production saves resources. Nat Resour Forum 29:135–143. https://doi.org/10.1111/j.1477-8947.2005.00122.x

Gao JP, Chao DY, Lin HX (2007) Understanding abiotic stress tolerance mechanisms: recent studies on stress response in rice. J Integr Plant Biol 49:742–750

Garai R, Chakraborty AK, Dey SB, Saha KC (1984) Chronic arsenic poisoning from tube-well water. J Indian Med Assoc 82:34–35

Geng CN, Zhu YG, Liu WJ, Smith SE (2005) Arsenate uptake and translocation in seedlings of two genotypes of rice is affected by external phosphate concentrations. Aquat Bot 83:321–331. https://doi.org/10.1016/j.aquabot.2005.07.003

Geng A, Wang X, Wu L et al (2017) Arsenic accumulation and speciation in rice grown in arsanilic acid-elevated paddy soil. Ecotoxicol Environ Saf 137:172–178. https://doi.org/10.1016/j.ecoenv.2016.11.030

Gorny J, Billon G, Lesven L et al (2015) Arsenic behavior in river sediments under redox gradient: a review. Sci Total Environ 505:423–434. https://doi.org/10.1016/j.scitotenv.2014.10.011

GRiSP (Global Rice Science Partnership) (2013) Rice almanac, 4th edn. International Rice Research Institute, Los Baños. 283 p

Guha Mazumder DN (2008) Chronic arsenic toxicity & human health. Indian J Med Res 128:436–447

Guha Mazumder DN (2015) Health effects of chronic arsenic toxicity. In: Handbook of arsenic toxicology. Elsevier, Amsterdam, pp 137–177

Hayashi S, Kuramata M, Abe T et al (2017) Phytochelatin synthase *OsPCS1* plays a crucial role in reducing arsenic levels in rice grains. Plant J 91:840–848. https://doi.org/10.1111/tpj.13612

He D, Yang P (2013) Proteomics of rice seed germination. Front Plant Sci 4:246

Hossain MF (2006) Arsenic contamination in Bangladesh:an overview. Agric Ecosyst Environ 113:1–16. https://doi.org/10.1016/j.agee.2005.08.034

Hu P, Huang J, Ouyang Y et al (2013) Water management affects arsenic and cadmium accumulation in different rice cultivars. Environ Geochem Health 35:767–778. https://doi.org/10.1007/s10653-013-9533-z

Huang X, Kurata N, Wei X et al (2012a) A map of rice genome variation reveals the origin of cultivated rice. Nature 490:497–501. https://doi.org/10.1038/nature11532

Huang Y, Chen Z, Liu W (2012b) Influence of iron plaque and cultivars on antimony uptake by and translocation in rice (*Oryza sativa* L.) seedlings exposed to $Sb^{(III)}$ or $Sb^{(V)}$. Plant Soil 352:41–49. https://doi.org/10.1007/s11104-011-0973-x

Islam S, Rahman MM, Islam MR, Naidu R (2016) Arsenic accumulation in rice: consequences of rice genotypes and management practices to reduce human health risk. Environ Int 96:139–155. https://doi.org/10.1016/j.envint.2016.09.006

Järup L (2003) Hazards of heavy metal contamination. Br Med Bull 68:167–182

Jia Y, Huang H, Zhong M et al (2013) Microbial arsenic methylation in soil and rice rhizosphere. Environ Sci Technol 47:3141–3148. https://doi.org/10.1021/es303649v

Jian FM, Yamaji N, Mitani N et al (2008) Transporters of arsenite in rice and their role in arsenic accumulation in rice grain. Proc Natl Acad Sci U S A 105:9931–9935. https://doi.org/10.1073/pnas.0802361105

Jiang JQ, Ashekuzzaman SM, Jiang A et al (2013) Arsenic-contaminated groundwater and its treatment options in Bangladesh. Int J Environ Res Public Health 10:18–46. https://doi.org/10.3390/ijerph10010018

Jomova K, Valko M (2011) Advances in metal-induced oxidative stress and human disease. Toxicology 283:65–87

Kalia K, Flora SJS (2005) Strategies for safe and effective therapeutic measures for chronic arsenic and lead poisoning. J Occup Health 47:1–21

Kamiya T, Islam MR, Duan G et al (2013) Phosphate deficiency signaling pathway is a target of arsenate and phosphate transporter *OsPT1* is involved in As accumulation in shoots of rice. Soil Sci Plant Nutr 59:580–590. https://doi.org/10.1080/00380768.2013.804390

Khush GS (1997) Origin, dispersal, cultivation and variation of rice. Plant Mol Biol 35:25–34

Kumar M, Puri A (2012) A review of permissible limits of drinking water. Indian J Occup Environ Med 16:40–44. https://doi.org/10.4103/0019-5278.99696

Kuramata M, Abe T, Kawasaki A et al (2013) Genetic diversity of arsenic accumulation in rice and QTL analysis of methylated arsenic in rice grains. Rice 6:1–10. https://doi.org/10.1186/1939-8433-6-3

Laha M (2017) Irrigation and groundwater hazards in India. Trans Inst Indian Geogr 39:237–252

Lawson M, Polya DA, Boyce AJ et al (2016) Tracing organic matter composition and distribution and its role on arsenic release in shallow Cambodian groundwaters. Geochim Cosmochim Acta 178:160–177. https://doi.org/10.1016/j.gca.2016.01.010

Lee JJ, Jang CS, Wang SW et al (2008) Delineation of spatial redox zones using discriminant analysis and geochemical modelling in arsenic-affected alluvial aquifers. Hydrol Process 22:3029–3041. https://doi.org/10.1002/hyp.6884

Lee CH, Hsieh YC, Lin TH, Lee DY (2013) Iron plaque formation and its effect on arsenic uptake by different genotypes of paddy rice. Plant Soil 363:231–241. https://doi.org/10.1007/s11104-012-1308-2

Li Z, Ali J (2017) Breeding green super rice (GSR) varieties for sustainable rice cultivation. In: Sasaki T (ed) Achieving sustainable cultivation of rice, vol 1. Burleigh Dodds Science Publishing, Cambridge, pp 109–130

Li RY, Stroud JL, Ma JF et al (2009) Mitigation of arsenic accumulation in rice with water management and silicon fertilization. Environ Sci Technol 43:3778–3783. https://doi.org/10.1021/es803643v

Li G, Sun GX, Williams PN et al (2011) Inorganic arsenic in Chinese food and its cancer risk. Environ Int 37:1219–1225. https://doi.org/10.1016/j.envint.2011.05.007

Liu J, Cao C, Wong M et al (2010) Variations between rice cultivars in iron and manganese plaque on roots and the relation with plant cadmium uptake. J Environ Sci 22:1067–1072. https://doi.org/10.1016/S1001-0742(09)60218-7

Liu X, Chen S, Chen M et al (2019) Association study reveals genetic loci responsible for arsenic, cadmium and lead accumulation in rice grain in contaminated farmlands. Front Plant Sci 10:61. https://doi.org/10.3389/fpls.2019.00061

Lomax C, Liu WJ, Wu L et al (2012) Methylated arsenic species in plants originate from soil microorganisms. New Phytol 193:665–672. https://doi.org/10.1111/j.1469-8137.2011.03956.x

Lou LQ, Ye ZH, Wong MH (2009) A comparison of arsenic tolerance, uptake and accumulation between arsenic hyperaccumulator, *Pteris vittata* L. and non-accumulator, *P. semipinnata* L.: a hydroponic study. J Hazard Mater 171:436–442. https://doi.org/10.1016/j.jhazmat.2009.06.020

Ma J, Takahashi E (1990) Effect of silicon on the growth and phosphorus uptake of rice. Plant Soil 126:115. https://doi.org/10.1007/BF00041376

Maclean J, Dawe DC, Hardy B, Hettel G (eds) (2002) Rice almanac, 3rd edn. International Rice Research Institute, Los Baños

Mandal BK, Suzuki KT (2002) Arsenic round the world: a review. Talanta 58:201–235. https://doi.org/10.1016/S0039-9140(02)00268-0

Mazumder DNG (2003) Chronic arsenic toxicity: clinical features, epidemiology, and treatment: experience in West Bengal. J Environ Sci Health Part A Toxic Hazardous Subst Environ Eng 38:141–163. https://doi.org/10.1081/ESE-120016886

Meharg AA, Hartley-Whitaker J (2002) Arsenic uptake and metabolism in arsenic-resistant and nonresistant plant species. New Phytol 154:29–43. https://doi.org/10.1046/j.1469-8137.2002.00363.x

Meharg AA, Rahman M (2003) Arsenic contamination of Bangladesh paddy field soils: implications for rice contribution to arsenic consumption. Environ Sci Technol 37:229–234. https://doi.org/10.1021/es0259842

Messerschmidt M, Wagner A, Wong MW, Luger P (2002) Atomic properties of N2O4 based on its experimental charge density. J Am Chem Soc 124:732–733. https://doi.org/10.1021/ja011802c

Mishra S, Dubey RS (2006) Inhibition of ribonuclease and protease activities in arsenic exposed rice seedlings: role of proline as enzyme protectant. J Plant Physiol 163:927–936. https://doi.org/10.1016/j.jplph.2005.08.003

Mishra S, Jha AB, Dubey RS (2011) Arsenite treatment induces oxidative stress, upregulates antioxidant system, and causes phytochelatin synthesis in rice seedlings. Protoplasma 248:565–577. https://doi.org/10.1007/s00709-010-0210-0

Mochizuki H (2019) Arsenic neurotoxicity in humans. Int J Mol Sci 20:3418. https://doi.org/10.3390/ijms20143418

Mohan D, Pittman CU (2007) Arsenic removal from water/wastewater using adsorbents: a critical review. J Hazard Mater 142:1–53. https://doi.org/10.1016/j.jhazmat.2007.01.006

Molden D (2013) Water for food, water for life: a comprehensive assessment of water management in agriculture. Routledge, London

Molina J, Sikora M, Garud N et al (2011) Molecular evidence for a single evolutionary origin of domesticated rice. Proc Natl Acad Sci U S A 108:8351–8356. https://doi.org/10.1073/pnas.1104686108

Mosa KA, Kumar K, Chhikara S et al (2012) Members of rice plasma membrane intrinsic proteins subfamily are involved in arsenite permeability and tolerance in plants. Transgenic Res 21:1265–1277. https://doi.org/10.1007/s11248-012-9600-8

Muhammad S, Tahir Shah M, Khan S (2010) Arsenic health risk assessment in drinking water and source apportionment using multivariate statistical techniques in Kohistan region, northern Pakistan. Food Chem Toxicol 48:2855–2864. https://doi.org/10.1016/j.fct.2010.07.018

Mukherjee A, Kundu M, Basu B et al (2017) Arsenic load in rice ecosystem and its mitigation through deficit irrigation. J Environ Manag 197:89–95. https://doi.org/10.1016/j.jenvman.2017.03.037

Murugaiyan V (2019) Genetic mapping of quantitative trait loci associated with arsenic tolerance and accumulation in rice (*Oryza sativa* L.). Universitäts- und Landesbibliothek, Bonn

Murugaiyan V, Ali J, Mahender A et al (2019) Mapping of genomic regions associated with arsenic toxicity stress in a backcross-breeding populations of rice (*Oryza sativa* L.). Rice 12:61. https://doi.org/10.1186/s12284-019-0321-y

Muthayya S, Sugimoto JD, Montgomery S, Maberly GF (2014) An overview of global rice production, supply, trade, and consumption. Ann N Y Acad Sci 1324:7–14. https://doi.org/10.1111/nyas.12540

National Research Council (US) Committee on Medical and Biological Effects of Environmental Pollutants (1977) Distribution of arsenic in the environment. In: Arsenic: medical and biologic effects of environmental pollutants. National Academies Press, Washington, DC, p 3

Ng JC (2005) Environmental contamination of arsenic and its toxicological impact on humans. Environ Chem 2:146–160

Nickson RT, Mcarthur JM, Ravenscroft P et al (2000) Mechanism of arsenic release to groundwater, Bangladesh and West Bengal. Appl Geochem 15:403–413. https://doi.org/10.1016/S0883-2927(99)00086-4

Nordberg GF, Jin T, Hong F et al (2005) Biomarkers of cadmium and arsenic interactions. Toxicol Appl Pharmacol 206:191–197. https://doi.org/10.1016/j.taap.2004.11.028

Norton GJ, Deacon CM, Xiong L et al (2010a) Genetic mapping of the rice ionome in leaves and grain: identification of QTLs for 17 elements including arsenic, cadmium, iron and selenium. Plant Soil 329:139–153. https://doi.org/10.1007/s11104-009-0141-8

Norton GJ, Islam MR, Duan G et al (2010b) Arsenic shoot-grain relationships in field grown rice cultivars. Environ Sci Technol 44:1471–1477. https://doi.org/10.1021/es902992d

Norton GJ, Pinson SRM, Alexander J et al (2012) Variation in grain arsenic assessed in a diverse panel of rice (*Oryza sativa*) grown in multiple sites. New Phytol 193:650–664. https://doi.org/10.1111/j.1469-8137.2011.03983.x

Norton GJ, Douglas A, Lahner B, Yakubova E, Guerinot ML, et al (2014) Genome wide association mapping of grain arsenic, copper, molybdenum and zinc in rice (*Oryza sativa* L.) grown at four international field sites. PLoS One 9(2): e89685. https://doi.org/10.1371/journal.pone.0089685

Norton GJ, Travis AJ, Talukdar P et al (2019) Genetic loci-regulating arsenic content in rice grains when grown flooded or under alternative wetting and drying irrigation. Rice 12:54. https://doi.org/10.1186/s12284-019-0307-9

Nriagu JO, Bhattacharya P, Mukherjee AB et al (2007) Arsenic in soil and groundwater: an overview. In: Trace metals and other contaminants in the environment. Elsevier, Amsterdam, pp 3–60

O'Day PA (2006) Chemistry and mineralogy of arsenic. Elements 2:77–83. https://doi.org/10.2113/gselements.2.2.77

Pandey S, Rai R, Rai LC (2015) Biochemical and molecular basis of arsenic toxicity and tolerance in microbes and plants. In: Handbook of arsenic toxicology. Elsevier, Amsterdam, pp 627–674

Pathak H, Tewari A, Sankhyan S et al (2011) Direct-seeded rice: potential, performance and problems: a review. Curr Adv Agric Sci Int 3:77

Pohl HR, Roney N, Abadin HG (2011) Metal ions affecting the neurological system. Met Ions Life Sci 8:247–262

Ponomarenko M, Stepanenko I, Kolchanov N (2013) Heat shock proteins. In: Brenner's encyclopedia of genetics, 2nd edn. Elsevier, Amsterdam, pp 402–405

Rahman MA, Hasegawa H, Rahman MM et al (2007) Arsenic accumulation in rice (*Oryza sativa* L.) varieties of Bangladesh: a glass house study. Water Air Soil Pollut 185:53–61. https://doi.org/10.1007/s11270-007-9425-x

Rahman MA, Hasegawa H, Rahman MM et al (2008) Straighthead disease of rice (*Oryza sativa* L.) induced by arsenic toxicity. Environ Exp Bot 62:54–59. https://doi.org/10.1016/j.envexpbot.2007.07.016

Rajjou L, Duval M, Gallardo K et al (2012) Seed germination and vigor. Annu Rev Plant Biol 63:507–533

Rao KP, Vani G, Kumar K et al (2011) Arsenic stress activates MAP kinase in rice roots and leaves. Arch Biochem Biophys 506:73–82. https://doi.org/10.1016/j.abb.2010.11.006

Rao CS, Lal R, Prasad JVNS et al (2015) Potential and challenges of rainfed farming in India. Adv Agron 133:113–181

Richards M, Sander BO (2014) Alternate wetting and drying in irrigated rice. J AHIMA. https://doi.org/10.1016/j.techfore.2006.05.021

Ronzan M, Piacentini D, Fattorini L et al (2018) Cadmium and arsenic affect root development in *Oryza sativa* L. negatively interacting with auxin. Environ Exp Bot 151:64–75. https://doi.org/10.1016/j.envexpbot.2018.04.008

Roy P, Saha A (2002) Metabolism and toxicity of arsenic: a human carcinogen. Curr Sci 82:38–45

Safiuddin M, Shirazi SM, Yussof S (2011) Arsenic contamination of groundwater in Bangladesh: a review. Int J Phys Sci 6:6791–6800. https://doi.org/10.5897/IJPS11.1300

Sahoo PK, Mukherjee A (2014) Arsenic fate and transport in the groundwater-soil-plant system: an understanding of suitable rice paddy cultivation in arsenic-enriched areas. In: Recent trends in modelling of environmental contaminants. Springer, New Delhi, pp 21–44

Sayan B, Kaushik G, Sushanta D et al (2012) Arsenic bioaccumulation in rice and edible plants and subsequent transmission through food chain in Bengal basin: a review of the perspectives for environmental health. Toxicol Environ Chem 94:429

Seneviratne M, Rajakaruna N, Rizwan M et al (2019) Heavy metal-induced oxidative stress on seed germination and seedling development: a critical review. Environ Geochem Health 41:1813–1831. https://doi.org/10.1007/s10653-017-0005-8

Sengupta M, Bishayi B (2002) Effect of lead and arsenic on murine macrophage response. Drug Chem Toxicol 25:459–472

Sengupta MK, Hossain MA, Mukherjee A et al (2006) Arsenic burden of cooked rice: traditional and modern methods. Food Chem Toxicol 44:1823–1829. https://doi.org/10.1016/j.fct.2006.06.003

Seyfferth AL, Webb SM, Andrews JC, Fendorf S (2011) Defining the distribution of arsenic species and plant nutrients in rice (*Oryza sativa* L.) from the root to the grain. Geochim Cosmochim Acta 75:6655–6671. https://doi.org/10.1016/j.gca.2011.06.029

Shah AL, Naher UA, Hasan Z et al (2014) Influence of arsenic on rice growth and its mitigation with different water management techniques. Asian J Crop Sci 6:373–382. https://doi.org/10.3923/ajcs.2014.373.382

Shaibur MR, Kitajima N, Sugawara R et al (2006) Physiological and mineralogical properties of arsenic-induced chlorosis in rice seedlings grown hydroponically. Soil Sci Plant Nutr 52:691–700. https://doi.org/10.1111/j.1747-0765.2006.00085.x

Shepherd K, Hubbard D, Fenton N et al (2015) Policy: development goals should enable decision-making. Nature 523:152–154. https://doi.org/10.1038/523152a

Shi S, Wang T, Chen Z et al (2016) *OsHAC1;1* and *OsHAC1;2* function as arsenate reductases and regulate arsenic accumulation. Plant Physiol 172:1708–1719. https://doi.org/10.1104/pp.16.01332

Smith AH, Lingas EO, Rahman M (2000) Contamination of drinking-water by arsenic in Bangladesh: a public health emergency. Bull World Health Organ 78:1093–1103. https://doi.org/10.1590/S0042-96862000000900005

Song WY, Yamaki T, Yamaji N et al (2014) A rice ABC transporter, *OsABCC1*, reduces arsenic accumulation in the grain. Proc Natl Acad Sci U S A 111:15699–15704. https://doi.org/10.1073/pnas.1414968111

Stanton BA, Caldwell K, Congdon CB et al (2015) MDI Biological Laboratory Arsenic Summit: approaches to limiting human exposure to arsenic. Curr Environ Health Rep 2:329–337. https://doi.org/10.1007/s40572-015-0057-9

Sturchio E, Zanellato M, Minoia C, Bemporad E (2013) Arsenic: environmental contamination and exposure. In: Arsenic: sources, environmental impact, toxicity and human health. A medical geology perspective. Nova Publishers, Hauppauge, NY, pp 3–38

Sultana F (2013) Water, technology, and development: transformations of development technonatures in changing waterscapes. Environ Plan D Soc Sp 31:337–353. https://doi.org/10.1068/d20010

Sun G (2004) Arsenic contamination and arsenicosis in China. Toxicol Appl Pharmacol 198:268–271

Sun SK, Chen Y, Che J et al (2018) Decreasing arsenic accumulation in rice by overexpressing *OsNIP*1;1 and *OsNIP*3;3 through disrupting arsenite radial transport in roots. New Phytol 219:641–653. https://doi.org/10.1111/nph.15190

Suriyagoda LDB, Dittert K, Lambers H (2018) Mechanism of arsenic uptake, translocation and plant resistance to accumulate arsenic in rice grains. Agric Ecosyst Environ 253:23–37

Takahashi E (1968) Silica as a nutrient to the rice plant. Jpn Agric Res Q 3:1

Talukder ASMHM, Meisner CA, Sarkar MAR, Islam MS (2011) Effect of water management, tillage options and phosphorus status on arsenic uptake in rice. Ecotoxicol Environ Saf 74:834–839. https://doi.org/10.1016/j.ecoenv.2010.11.004

Tareq SM, Islam SMN, Rahmam MM, Chowdhury DA (2015) Arsenic pollution in groundwater of Southeast Asia: an overview on mobilization process and health effects. Bangladesh J Environ Res 8:47–67

Thomas DJ (2013) The die is cast: arsenic exposure in early life and disease susceptibility. Chem Res Toxicol 26:1778–1781

Tripathi P, Tripathi RD, Singh RP et al (2013) Arsenite tolerance in rice (*Oryza sativa* L.) involves coordinated role of metabolic pathways of thiols and amino acids. Environ Sci Pollut Res 20:884–896. https://doi.org/10.1007/s11356-012-1205-5

Tuli R, Chakrabarty D, Trivedi PK, Tripathi RD (2010) Recent advances in arsenic accumulation and metabolism in rice. Mol Breed 26:307–323. https://doi.org/10.1007/s11032-010-9412-6

Vahidnia A, Van Der Voet GB, De Wolff FA (2007) Arsenic neurotoxicity: a review. Hum Exp Toxicol 26:823–832

Vahter M, Concha G (2001) Role of metabolism in arsenic toxicity. Pharmacol Toxicol 89:1–5. https://doi.org/10.1111/j.1600-0773.2001.890101.x

Vaughan DJ (2006) Arsenic. Elements 2:71–75. https://doi.org/10.2113/gselements.2.2.71

Wang P, Zhang W, Mao C et al (2016) The role of *OsPT*8 in arsenate uptake and varietal difference in arsenate tolerance in rice. J Exp Bot 67:6051–6059. https://doi.org/10.1093/jxb/erw362

Wang FZ, Chen MX, Yu LJ et al (2017) *OsARM*1, an R2R3 MYB transcription factor, is involved in regulation of the response to arsenic stress in rice. Front Plant Sci 8:1868. https://doi.org/10.3389/fpls.2017.01868

Wenzel WW, Alloway BJ (2013) Chapter 9 Arsenic. Springer, Dordrecht

Wu C, Ye Z, Li H et al (2012) Do radial oxygen loss and external aeration affect iron plaque formation and arsenic accumulation and speciation in rice? J Exp Bot 63:2961–2970. https://doi.org/10.1093/jxb/ers017

Wu LB, Shhadi MY, Gregorio G et al (2014) Genetic and physiological analysis of tolerance to acute iron toxicity in rice. Rice 7:8. https://doi.org/10.1186/s12284-014-0008-3

Wu X, Cobbina SJ, Mao G et al (2016) A review of toxicity and mechanisms of individual and mixtures of heavy metals in the environment. Environ Sci Pollut Res 23:8244–8259. https://doi.org/10.1007/s11356-016-6333-x

Wu LB, Ueda Y, Lai SK, Frei M (2017) Shoot tolerance mechanisms to iron toxicity in rice (*Oryza sativa* L.). Plant Cell Environ 40:570–584. https://doi.org/10.1111/pce.12733

Xu J, Shi S, Wang L et al (2017) *OsHAC*4 is critical for arsenate tolerance and regulates arsenic accumulation in rice. New Phytol 215:1090–1101. https://doi.org/10.1111/nph.14572

Yamazaki S, Ueda Y, Mukai A et al (2018) Rice phytochelatin synthases *OsPCS*1 and *OsPCS*2 make different contributions to cadmium and arsenic tolerance. Plant Direct 2:e00034. https://doi.org/10.1002/pld3.34

Yang Y, Zhang A, Chen Y et al (2018) Impacts of silicon addition on arsenic fractionation in soils and arsenic speciation in *Panax notoginseng* planted in soils contaminated with high levels of arsenic. Ecotoxicol Environ Saf 162:400–407. https://doi.org/10.1016/j.ecoenv.2018.07.015

Yang Y, Hu H, Fu Q et al (2020) Phosphorus regulates As uptake by rice via releasing As into soil porewater and sequestrating it on Fe plaque. Sci Total Environ 738:139869. https://doi.org/10.1016/j.scitotenv.2020.139869

Yu S, Ali J, Zhang C et al (2020) Genomic breeding of green super rice varieties and their deployment in Asia and Africa. Theor Appl Genet 133:1427–1442

Zavala YJ, Duxbury JM (2008) Arsenic in rice: I. Estimating normal levels of total arsenic in rice grain. Environ Sci Technol 42:3856–3860. https://doi.org/10.1021/es702747y

Zhang J, Zhu YG, Zeng DL et al (2008) Mapping quantitative trait loci associated with arsenic accumulation in rice (*Oryza sativa*). New Phytol 177:350–356. https://doi.org/10.1111/j.1469-8137.2007.02267.x

Zhao FJ, Ma JF, Meharg AA, McGrath SP (2009) Arsenic uptake and metabolism in plants. New Phytol 181:777–794. https://doi.org/10.1111/j.1469-8137.2008.02716.x

Zhao FJ, Ago Y, Mitani N et al (2010a) The role of the rice aquaporin *Lsi*1 in arsenite efflux from roots. New Phytol 186:392–399. https://doi.org/10.1111/j.1469-8137.2010.03192.x

Zhao FJ, McGrath SP, Meharg AA (2010b) Arsenic as a food chain contaminant: mechanisms of plant uptake and metabolism and mitigation strategies. Annu Rev Plant Biol 61:535–559. https://doi.org/10.1146/annurev-arplant-042809-112152

Zhao FJ, Stroud JL, Khan AA, McGrath SP (2012) Arsenic translocation in rice investigated using radioactive 73As tracer. Plant Soil 350:413–420. https://doi.org/10.1007/s11104-011-0926-4

Zhao FJ, Zhu YG, Meharg AA (2013) Methylated arsenic species in rice: geographical variation, origin, and uptake mechanisms. Environ Sci Technol 47:3957–3966. https://doi.org/10.1021/es304295n

Zhu YG, Williams PN, Meharg AA (2008) Exposure to inorganic arsenic from rice: a global health issue? Environ Pollut 154:169–171. https://doi.org/10.1016/j.envpol.2008.03.015

Molecular Approaches for Disease Resistance in Rice

Mohammed Jamaloddin, Anumalla Mahender, C. Guru Gokulan,
Chintavaram Balachiranjeevi, A. Maliha, Hitendra Kumar Patel,
and Jauhar Ali

Abstract Rice production needs to be sustained in the coming decades, with
changing climatic conditions becoming more conducive to the prevalence of disease
outbreaks. Major rice diseases collectively cause enormous economic damage and
yield instability. Breeding for disease-resistant rice varieties could be one of the best
options to counter these disease outbreaks. Disease-screening protocols and newer
technologies are essential for effective phenotyping and would aid in gene discovery and function. Understanding the genetics of disease mechanisms and stacking
of broad-spectrum disease-resistance genes could lead to faster development of rice
varieties with multiple disease resistance. New molecular breeding approaches are
discussed for the development of these varieties. The molecular biology of disease
resistance is now better understood and could be well manipulated for improved
resilience. Transgenic approaches for disease resistance are discussed. Genome-
editing tools for the development of disease-resistant rice varieties are thoroughly
discussed. The use of bioinformatics tools to speed up the process and to obtain a
better understanding of molecular genetics mechanisms of disease resistance is
explained.

Keywords Rice · Biotic diseases · Phenotypic screening · QTLs and genes ·
Breeding strategies · Genome editing

M. Jamaloddin · C. G. Gokulan · A. Maliha · H. K. Patel
Centre for Cellular and Molecular Biology (CCMB), Hyderabad, India

A. Mahender · C. Balachiranjeevi
Rice Breeding Platform, International Rice Research Institute (IRRI),
Los Baños, Laguna, Philippines

J. Ali (✉)
Hybrid Rice Breeding Cluster, Hybrid Rice Development Consortium (HRDC), Rice
Breeding Platform, International Rice Research Institute (IRRI),
Los Baños, Laguna, Philippines
e-mail: J.Ali@irri.org

1   Introduction

Rice (*Oryza sativa* L.) is a staple and the most crucial food security crop in the world. It plays a vital role in the human diet and feeds more than 50% of the world's population (Rathna Priya et al. 2019). By 2050, global demand for rice is projected to rise more than 40% to feed the rapidly growing world population (Milovanovic and Smutka 2017). Despite impressive global increases in production from 289 million tons in 1968 to 782 million tons in 2018, this quantum jump still has to keep pace with demand for rice from the rising population (FAOSTAT 2020). At present, rice cultivation throughout South Asia and in ASEAN countries is facing significant threats because of a few major biotic stresses (Yugander et al. 2017). Approximately 52% of the global productivity of rice grain yield is severely damaged by biotic factors, of which nearly 31% is due to various diseases such as bacterial blight (caused by *Xanthomonas oryzae*), blast (caused by *Magnaporthe grisea*), sheath blight (caused by *Rhizoctonia solani*), and tungro disease (tungro bacilliform virus and tungro spherical virus) (Park et al. 2008). Detailed information about the symptoms caused by these major diseases, along with the favorable conditions required by these pathogens and yield losses incurred, is presented in Table 1. The severity of biotic stresses in rice production is increasing at a startling pace of late because of rapid changes in climate (Jamaloddin et al. 2020). Changing climatic conditions are contributing to the emergence of new virulent races and the occurrence of diseases in new localities. Many diseases considered as minor thus far have become economically significant in many rice-cultivating areas and are exacerbating their impact (Anderson et al. 2004). According to Zhang et al. (2009), rice crops are affected by around 70 pathogens, especially viruses, bacteria, fungi, and nematodes. Estimated yield loss because of pathogens globally and as per hotspot range for rice is 30% (24.6–40.9%) (Savary et al. 2019). Over the past 150 years, many rice diseases have caused outbreaks and spread rapidly in different parts of the world. Rice diseases were observed for the first time in different locations, such as bacterial blight and sheath blight in Japan during 1884–1885 and 1910, respectively; false smut in the United States during 1906; blast in Africa during 1922; rice tungro in the Philippines during 1940; rice brown spot in India during 1942; bacterial leaf streak in India during 1963; and rice yellow mottle disease in Kenya during 1966. These diseases, along with a few newly emerging epidemics, are becoming a significant threat to rice production.

Despite so many alternatives for crop protection, plant pathogens still pose a challenge to agriculture. Several management practices have been adopted to decrease their impact, such as chemical control, biological control, optimum fertilizer application, appropriate planting dates, and disease forecasting. However, not all of these methods are environment-friendly and alone are not enough to control the diseases completely. The present situation thus requires environment-friendly and cost-effective modern technologies such as the development and cultivation of disease-resistant cultivars. The development of these varieties using only conventional breeding methods consumes a lot of time, land, and labor. In this context,

Table 1 Key features of major diseases in rice

Disease	Pathogen	Symptoms	Favorable conditions	Yield loss	Reference
Blast	Caused by the fungus *Magnaporthe oryzae* (*Mo*)	Early-stage symptoms appear as white to gray-green lesions with dark green specks. These soon enlarge and spindle-shaped lesions appear with a gray center and dark brown margin.	Prolonged period free from moisture. High humidity conditions. Gentle or no wind at night. Night temperatures from 17 to 22 °C. High rate of fertilizer.	70–80%	Jamaloddin et al. (2020)
Bacterial blight (BB)	Caused by bacterium *Xanthomonas oryzae* pv. *oryzae* (*Xoo*)	Normally, disease appears at heading stage, but can occur early in severe conditions. Infected plants' young leaves change from pale green to gray-green and roll up. As the disease progresses, the entire leaf may eventually be affected, becoming whitish or grayish and then dying.	Suitable temperature is 25–30 °C. High humidity (above 70%), rain, and deep water. Severe winds, which cause wounds. High rate of fertilizer.	Up to 50%	Liu et al. (2014)
Bacterial leaf streak (BLS)	Caused by bacterium *Xanthomonas oryzae* pv. *oryzicola* (*Xoc*)	Plants can be affected from maximum tillering to panicle initiation. Symptoms appear on leaf blades as narrow, dark greenish water-soaked interveinal streaks of various lengths. Later, these streaks become light brown to yellowish gray.	There is a higher probability of developing it in areas having weeds and stubbles harboring infection. Temperatures from 25 to 34 °C with relative humidity >70% are more congenial.	8–32%	Liu et al. (2014)

(continued)

Table 1 (continued)

Disease	Pathogen	Symptoms	Favorable conditions	Yield loss	Reference
Tungro	Caused by Rice tungro bacilliform virus (RTBV) and rice tungro spherical virus (RTSV)	Infection can occur during all growth stages but mostly during the vegetative phase. The tillering stage is the most vulnerable. Leaves of infected plants become yellow or orange-yellow and may also have rust-colored spots. Most of the panicles are entirely or partially sterile with ill-filled grains.	Viruses are transmitted by leafhoppers that feed on tungro-infected plants. Leafhoppers are capable of transmitting viruses to other plants within 5–7 days.	Up to 100%	Bunawan et al. (2014)
False smut	Caused by the fungus *Villosiclava virens* (anamorph: *Ustilaginoidea virens*)	False smut can infect individual rice grains. Only a few panicle grains are usually infected, and the remaining grains are normal. A smut ball appears at first and grows gradually up to 1 cm. As fungi growth intensifies, the smut balls burst and become orange and then later yellowish green/ greenish black in color.	The disease can occur in areas with high relative humidity (>90%) and temperature ranging from 25 to 35 °C. Rain and soils with high nitrogen content also favor false smut. Wind can spread the fungal spores from plant to plant.	In severe cases, tillers will be affected 85–100%.	Huang et al. (2019)

(continued)

Table 1 (continued)

Disease	Pathogen	Symptoms	Favorable conditions	Yield loss	Reference
Sheath blight	Caused by the fungus *Rhizoctonia solani*	The fungus attacks the plants from tillering to heading stage. Initial symptoms appear on leaf sheaths near the water line in the form of oval or irregular greenish gray lesions. Later, lesions extend to the upper parts of the plants and rapidly coalesce, covering entire tillers from the water line to the flag leaf.	Temperature from 28 to 32 °C, high rates of N fertilizer, high seed rate or low spacing, dense canopy, inoculum in soil or floating on the water, and continuous cultivation of high-yielding varieties favor disease development. The crop is more vulnerable during the rainy season.	20–60%	Molla et al. (2020)
Sheath rot	Caused by fungus *Sarocladium oryzae*	The sheath rot lesion starts at the uppermost leaf sheath consisting of young panicles within. Early symptoms are oblong to irregular lesions on the leaves with dark reddish brown margins and brownish gray throughout. The disease can cause partial emergence of panicles present in the infected sheaths. The unmerged panicles rot and turn dark brown with a whitish powdery growth inside the sheaths. Infected panicles and grains look sterile, ill-filled, shriveled, and discolored.	More prevalent during wet seasons than dry seasons. High relative humidity and temperatures from 20 to 28 °C from heading to crop maturity. High rates of N fertilizer application. Plant injuries and wounds caused by insects such as stem borers at the panicle initiation stage.	20–85%	Peeters et al. (2020)

molecular markers come to the rescue of plant breeders by helping them decrease the time between breeding and achieving the desired product. The discovery of DNA markers led to a new tool in plant breeding called marker-assisted selection (MAS), which is one of the widely used components of a discipline called molecular breeding. The application of DNA markers to plant breeding significantly increased its efficiency and precision. MAS is now one of the most advanced methodologies on hand for the transfer of one or more desired genes/genomic regions into elite rice varieties in more durable combinations. Deploying a single *R*-gene often leads to resistance breakdown in a short period as the pathogen evolves and makes itself resistant to the action of the gene. Therefore, pyramiding of multiple *R*-genes imparting resistance against different races of a pathogen through MAS is an efficient way to attain long-term and broad-spectrum resistance. Although this is an advanced method, it has some disadvantages. The main drawback of this approach is that one parent, or even both parents, used in the breeding program may carry quantitative trait loci (QTL) alleles that are either similar or exact to the ones present in the elite germplasm accessions used in other breeding programs. In such a situation, the QTL being introgressed may contribute only partially to the trait improvement. In other cases, the impact of a QTL may differ based on the genetic background as a result of interactions with other loci or epistasis (Holland 2001). Moreover, there are many more important traits for which no genes have been reported so far. In such situations in which a gene is not available in the gene pool, researchers are forced to look outside the gene pool toward other genera or sometimes toward another kingdom to find the desired gene.

Genetic modification (GM) technology has been developed to make changes to an organism's genes to give it new traits that would not occur in nature or to eliminate undesirable characteristics. GM technology using recombinant DNA technology is useful for developing disease-resistant varieties but still has not reached farmers because of a lack of public acceptance and political issues in many countries. Under these situations, researchers are left with an option to create mutations in the gene pool with an expectation to generate variation for a trait not naturally present in the gene pool.

Mutation breeding is helpful in creating novel mutants with genetic variations for plant breeding and functional genomics. It could be used for rice crop improvement programs. Mutation induction can be of advantage to produce cultivars with desired characteristics within defined germplasm pools. Normally, gamma-rays (γ-rays) and ethyl methane sulfonate (EMS) have been used extensively to develop rice mutants. In rice, there have been reports of some important mutant collections developed to carry out functional studies and Hirochika et al. (2004) made available a list of the mutant libraries. Madamba et al. (2009) found a gamma-ray-induced IR64 mutant, G978, that gave enhanced resistance to blast and bacterial blight. The resistance was found to be quantitative and nonrace-specific against bacterial and fungal pathogens. The mutation was shown to be inherited as a single recessive gene, *Bsdr1*, and it caused a shorter stature relative to IR64 and was mapped as a QTL to a 3.8-Mb region on chromosome 12. Comparison of the gene expression profiles of the mutant and wild type showed the candidate gene to encode a U-box domain-containing protein. The disrupted gene exhibited a loss of expression in the mutant and cosegregated with the mutant phenotype (Madamba et al. 2009). These techniques of causing mutations have a problem of creating more undesirable than desirable phenotypes. In other words, these techniques result in random mutations in the genome. The frequency of variations can be controlled but not in the genomic region where they are desired to occur. To achieve a desirable outcome from these experiments, a large population of mutants has to be screened, and this requires a lot of time, space, and resources.

Ultimately, the new era of genome engineering technologies offers vast potential for crop improvement as they allow site-specific modifications of DNA sequences to be executed under laboratory conditions. The accessibility to vast genomic resources and an easy-to-handle genome size make rice more amenable for GM technologies. Advances in genomics and the development of various genome-editing technologies using engineered site-specific nucleases (SSNs) have made the application of genome engineering to crops easy. Among various SSNs, the CRISPR/Cas9 system is commonly applied because of its simplicity, robustness, and high efficiency (Wang et al. 2018). In comparison with other genome-editing tools such as zinc-finger nucleases and transcription activator-like effector nucleases (TALENs), this technique is versatile and simple (Ma et al. 2015b). This technology has been applied to agricultural crop plants with the aim of crop improvement. Oliva et al. (2019) used the CRISPR-Cas9 system to introduce mutations in three *SWEET* gene promoters to make robust and broad-spectrum bacterial blight-resistant lines. There is still much scope for its use and application.

In the future, the challenge for scientists is not only to develop rice varieties for specific diseases but also to select for horizontal resistance without altering other desirable traits of elite rice varieties. A systematically designed experiment involving highly efficient molecular tools would make it possible to achieve this outcome. Hence, the current chapter amalgamates details on the present status of the key diseases that affect rice production, various molecular strategies for attaining disease resistance, and prospects of molecular breeding for disease resistance in rice.

2 Phenotypic Screening Techniques for Major Diseases of Rice: Pathogen Inoculum, Plant Infection Assays, and Disease Scoring

2.1 Bacterial Blight

Bacterial blight (BB), caused by *Xanthomonas oryzae* pv. *oryzae* (*Xoo*), targets the seedling stage of rice, resulting in leaves turning grayish green and rolling up. Usually, BB inoculation can be done in two ways, either in the field or in the greenhouse. Many techniques are available to infect the plant with inoculum such as clipping, needle prick, paint-brush, and spray methods. But the most preferable, efficient, and feasible for inoculation is the clipping method (Jabeen et al. 2011). Individually, collected *Xoo* strains are multiplied and stored on modified Wakimoto'sagar (Sundaram et al. 2009) and the selected rice plants at 45-days-old stage are clip-inoculated with a freshly prepared bacterial suspension (~10^{8-9} cfu/mL) by the following method given by Kauffman et al. (1973). In this method, 1–2-cm tips of five leaves are clipped with scissors, after they were already dipped in bacterial suspension culture, and disease score is recorded 2 weeks post-inoculation both by visual scoring and by measuring the lesion length (LL) as per the Standard Evaluation System (SES) scale of the International Rice Research Institute (IRRI 1996) (0–3 = resistant, 3–5 = moderately resistant (MR), 5–7 = moderately susceptible (MS), and 7–9 = susceptible).

2.2 Blast Disease

The causal organism for blast disease is a fungus, *Magnaporthe oryzae* (*Mo*). Symptoms of blast can appear during any developmental stage and on all parts of the rice plant, including leaves, leaf collars, necks, panicles, pedicels, and seeds. Standard screening protocols of rice varieties for susceptibility to rice blast are usually carried out by spraying the plant with conidial suspensions under greenhouse and field conditions using local isolates of the pathogen (Takahashi et al. 2009). However, for screening against exotic strains, quarantine restrictions are frequently applied to control any escape of the pathogen into the surrounding environment (Jia et al. 2003). In field conditions, artificial leaf blast disease screening usually takes place in a Uniform Blast Nursery (UBN) (Jamaloddin et al. 2020). Applying an excess rate of nitrogen fertilizer (150 kg N/ha) makes rice more vulnerable to spreading blast infection. Artificial inoculation is done with a highly virulent blast race (fungal conidial suspension at a concentration of 1×10^5 spores/mL) by spraying on UBN beds 25–30 days after sowing (DAS). Later, the beds are covered with polythene sheets during the night to create humid conditions for disease development. The disease score is collected 10–15 days after infection, depending on the severity of the infection on the susceptible check using the SES (IRRI 1996). In in vitro conditions, spot inoculation and filter paper inoculation methods are used for inoculation at the vegetative and reproductive stages of rice plants (Jia et al. 2003; Takahashi et al. 2009).

2.3 Sheath Blight

Sheath blight (ShB) disease is caused by a fungus, *Rhizoctonia solani*. The fungus attacks the rice plant from tillering to heading stage. The early symptoms of sheath blight involve oval circles on leaves just above the waterline. Various screening methods have been developed to screen for ShB in greenhouse and field conditions.

Eizenga et al. (2002) delineated a growth-chamber screening technique for sheath blight on *Oryza* spp. Later, Jia et al. (2003) developed the detached-leaf method. For screening ShB under greenhouse conditions, three inoculation methods have been described: liquid-cultured mycelia ball, mycelia suspension, and agar block. Out of these, the liquid culture mycelia ball is a more efficient and better method for successful inoculation (Park et al. 2008). Field screening at the reproductive stage is the most commonly used method. But field trials require a lot of labor and a large amount of seed material, inoculum, and high-humidity conditions for up to 3–5 months to complete the evaluations (Jia et al. 2007). Normally, screening of selected material for ShB tolerance/susceptibility is done using a highly virulent isolate of *Rhizoctonia solani*. Initially, ShB isolate is maintained on a potato dextrose agar (PDA, extract from 200 g/L of potato, 20 g/L of dextrose, and 20 g/L of agar) plate and incubated at 28 °C in darkness. For plant inoculations, *Typha* stem pieces (3–4 cm) are cut and autoclaved in plastic covers. This sterile *Typha* is inoculated with a 5-mm mycelial plug of *R. solani* from a 3-day-old PDA plate and incubated in the dark for 10 days at 28 °C. The colonized *Typha* pieces will be used for inoculating the rice plants at a rate of three to four pieces per hill.

Disease phenotype will be scored 2 weeks after inoculation by measuring the relative lesion height (RLH) as per the following formula:

$$RLH(\%) = (Lesion\ length\ /\ Plant\ height) \times 100$$

The IRRI (1996) phenotype scale is used to classify the plants based on their disease severity index from 0 to 9.

2.4   Sheath Rot of Rice

Sheath rot (ShR) is a symptom that is observed in rice plants when infected by any of the following pathogens: *Sarocladium oryzae*, *Fusarium* sp., *Pseudomonas* sp., and *Cochliobolus lunatus*. Other pathogens have been reported to cause similar symptoms in rice (Bigirimana et al. 2015). Multiple screening techniques that are being used for sheath rot disease resistance in rice include the mealybug inoculation method, rice grain/hull inoculum, leaf piece inoculum, cotton swabbing of conidial spores, spraying or injecting conidial suspension on the sheath, and detached tiller–based assays (Mahadevaiah et al. 2015; Samiyappan et al. 2003). The established screening methods differ depending on the causal agent of ShR as well as the growth stage of the plant. The pathogen is cultured on PDA plates for up to 14 days at 28 °C (Panda and Mishra 2019). A study by Mahadevaiah et al. (2015) compared multiple inoculation methods for *Sarocladium oryzae* during different growth stages and observed that seed inoculation is a suitable screening method for screening for disease resistance in young plants or early infection. In this method, the seeds are soaked overnight in conidial suspension (10^5 spores/mL) and then germinated. The number of germinated plants and lesion lengths 14 days after inoculation are scored.

For screening plants in peak vegetative to booting stage, foliar inoculation methods were able to provide reliable and conclusive results 15 days after infection. For a faster in vitro screening, the detached-tiller assay is recommended, in which the tillers are cut and placed on moist paper and inoculated with mycelial mats. Visible lesions are observed as early as 3 days postinoculation (Samiyappan et al. 2003; Mahadevaiah et al. 2017). Disease severity is estimated by measuring the area of the sheath and/or leaf affected (Mahadevaiah et al. 2017). *Pseudomonas fuscovaginae* also causes ShR in rice (Bigirimana et al. 2015). Adorada et al. (2013) reported and recommended multiple screening techniques for screening ShR caused by *P. fuscovaginae*. The bacteria are cultured using King's medium B initially for about 24 h. For plant inoculations, the following methods were found to be effective: (1) pin-pricking the upper leaf sheath using a needle dipped in bacterial suspension (10^7 cfu/mL) and measuring disease severity 14 days postinoculation in plants at the booting stage; (2) spraying the inoculum was found effective and is recommended for mass screening for ShR resistance in plants at the booting stage; (3) for early-stage resistance, soaking seeds in bacterial inoculum before germination is recommended, followed by measuring the decrease in seedling height 10 days later (Adorada et al. 2013).

2.5 False Smut

The fungus *Ustilaginoidea virens* causes false smut of rice. This fungus attacks the developing panicles and leads to the formation of smutted balls (cottony flakes around the grains). The fungus is generally cultured on potato sucrose agar plates or potato sucrose broth for mass production of conidial suspension (Panguluri and Kumar 2013). Screening for false smut is done during the booting stage of the plants through the following methods. Spraying conidial suspension (5×10^4 spores/mL) at the booting stage is one of the recommended ways for screening for false smut (Kaur and Singh 2017). Another method involves injecting the conidial suspension into the boot (Panguluri and Kumar 2013; Kaur 2014). It has been observed that spraying spores has produced a higher disease incidence and this is suitable for screening for resistant varieties (Kaur 2014). Disease severity is scored by calculating the percentages of infected tillers and infected grains per panicle and a score is assigned as recommended by Rice SES (IRRI 2013; Chaudhari et al. 2019).

2.6 Tungro Disease of Rice

Rice tungro is caused by two viruses, RTBV (rice tungro bacilliform virus) and RTSV (rice tungro spherical virus), and is transmitted by green leafhopper (GLH: *Nephotettix virescens* (Dist.)). The viral infection is manifested by the stunted growth of rice plants and yellowing of leaves (Anjaneyulu et al. 1982; Panguluri and

Kumar 2013). Nursery screening for tungro resistance in rice is carried out by letting three to five viruliferous GLH per plant (20–30 days old) feed in a closed environment and scoring the disease symptoms 14 days later as recommended by Rice SES (Anjaneyulu et al. 1982; Sebastian et al. 1996).

2.7   Bacterial Leaf Streak

Bacterial leaf streak (BLS) of rice is caused by the bacterium *Xanthomonas oryzae* pv. *oryzicola* (*Xoc*). BLS is manifested as water-soaked lesions on the leaf surface, which can result in decreased photosynthesis and hence diminished yield (He et al. 2012). Screening for BLS resistance is mainly performed using either of two methods. For screening seedlings, bacteria are initially grown in peptone sucrose broth, and a bacterial suspension from 10^8 to 10^9 cfu/mL is used for infiltrating the expanded leaves using a needleless syringe. The disease symptoms are scored 14 days postinoculation (Ju et al. 2017). For field screening or screening older plants, matured leaves are pin-pricked with needles that are dipped in bacterial inoculum on either side of the leaves. The lesions caused are measured 20 days postinoculation (Tang et al. 2000; Chen et al. 2006a, b; He et al. 2012). Disease severity is scored as per Rice SES.

3   Genetics of Disease Resistance

Deployment of genes conferring host-plant resistance provides an economical, durable, effective, and environmentally safe approach to combat plant diseases and decrease yield losses (Fig. 1). Major resistance genes from different resistance donors have been reported for various rice diseases. So far, more than 44 resistance genes have been identified against bacterial blight (Kim and Reinke 2019). More than 100 distinctive blast-resistance genes have been reported on different rice chromosomes and, out of these, 21 genes have been cloned (Devi et al. 2020). Two major sheath blight QTLs (*qShB9-2* and *qSBR11-1*) have been reported (Channamallikarjuna et al. 2010). But, thus far, genetic diversity for high resistance to/tolerance of ShB has not been reported in either cultivated rice or its wild relatives; thus, cloning of genes for ShB resistance is straggling (Bonman 1992). For bacterial leaf streak (BLS), no major resistance genes (*R*-genes) have been identified and only a few QTLs have been mapped. Out of these, *qXO-2-1*, *qXO-4-1*, and *qXO-11-2* were showing resistance to more than nine *Xoc* and *Xoo* strains (Bossa-Castro et al. 2018). In the case of tungro disease, a resistance QTL was found in Indian landrace ARC 11554 and was localized on chromosome 4 (Wang et al. 2016). False smut resistance in several rice cultivars has been identified as a quantitative trait controlled by multiple genes (Andargie et al. 2018; Han et al. 2020). But, to date, no rice variety has been identified to show complete resistance to false smut, whereas many cultivars exhibit

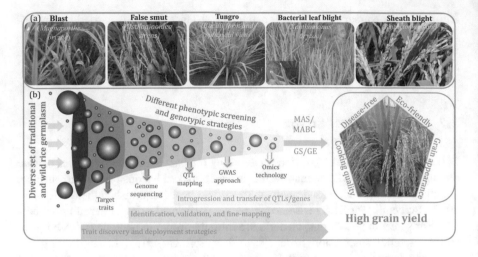

Fig. 1 (**a**) An illustration of the different biotic diseases in rice and (**b**) a funnel diagram representing the sources of valuable traits (tolerance of biotic diseases, yield components, and superior grain quality) that exist in traditional and wild rice germplasm. Using various phenotypic screening techniques and genome sequencing technologies can enable us to understand the molecular genetics and physiological mechanisms of stress tolerance. The identified genomic regions of QTLs and genes associated with the key traits play a vital role in understanding the interactions and further improving disease tolerance and superior grain quality traits with the help of marker-assisted selection and genomic selection approaches for crop improvement

considerable differences in quantitative field resistance to the pathogen (Huang et al. 2019). According to previous studies, the genetics of sheath rot disease resistance was dissected by studying the segregating pattern in an F$_2$ population (Rajashekara et al. 2014; Mvuyekure et al. 2017) and recombinant inbred lines (Graichen et al. 2010; Mahadevaiah et al. 2017). Some of these *R*-genes or loci have been extensively used in MAS breeding programs, and some of them have been fine-mapped and are undergoing cloning efforts. Detailed information on resistance genes/QTLs for economically important rice diseases (i.e., bacterial blight, blast, and sheath blight) appears in Tables 2, 3, and 4, respectively.

4   Breeding for Disease Resistance

Rice breeders have come up with many disease-resistant cultivars adapted to different rice-growing regions worldwide by applying conventional breeding approaches. Because of the dominance and epistasis effects of genes conferring resistance to a few diseases, gene pyramiding through conventional breeding methods becomes a challenge. Also, genes having similar responses to two or more races of a pathogen are difficult to recognize and transfer by conventional approaches (Joseph et al. 2004; Sundaram et al. 2009; Rajpurohit et al. 2011). The exercise of breeding for

Table 2 List of bacterial blight-resistance genes/QTLs

S. No.	Genes/ QTLs	Chr.	Position (bp)	Donor parent	Inheritance	Origin	Resistance to Xoo race	Linked marker	Marker type	References
1	Xa-1	4	31,638,099– 31,644,795	Kogyoku, IRBB1	Dominant, cloned, and characterized	Japan	Japanese race-I	Npb235	RFLP	Yoshimura et al. (1998)
2	Xa-2	4		IRBB2	Dominant	Vietnam	Japanese race-II	HZR950-5	SSR	Kurata and Yamazaki (2006)
3	Xa-3/ Xa-26	11	28,399,360– 28,402,773	WaseAikoku 3, Minghui 63, IRBB3	Dominant, cloned, and characterized	Japan	Chinese, Philippine, and Japanese races	C481S	RFLP	Xiang et al. (2006)
4	Xa-4	11	–	TKM6, IRBB4	Dominant	India	Philippine race-I	Npb181 and RM224	RFLP and SSR	Wang et al. (2001)
5	xa-5	5	437,010– 443,270	IRBB5	Recessive, cloned, and characterized	Bangladesh	Philippine races I, II, and III	RG556 and RM122	CAPS and SSR	Petpisit et al. (1977)
6	Xa-6/ xa-3	11	–	Zenith	Dominant	U.S.	Philippine race-I	Y68SSRA	RFLP	Sidhu et al. (1978)
7	Xa-7	6	–	DZ78	Dominant	Bangladesh	Philippine races	G1091, RM205S2	RFLP, SSR	Chen et al. (2008)
8	xa-8	7	–	PI231128	Recessive	U.S.	Philippine races	RM500, RM533	SSR	Vikal et al. (2014)
9	Xa-9	11	–	Khao Lay Nhay and Sateng	Dominant	Laos	Philippine races	C4S1S	RFLP	Ogawa (1988)
10	Xa-10	11	22,203,734– 22,204,676	Cas 209	Dominant, cloned, and characterized	Senegal	Philippine and Japanese races	M491/M419	RFLP, GAPS	Kurata and Yamazaki (2006)

(continued)

Table 2 (continued)

S. No.	Genes/ QTLs	Chr.	Position (bp)	Donor parent	Inheritance	Origin	Resistance to Xoo race	Linked marker	Marker type	References
11	Xa-11	3	–	IR8	Dominant	Philippines	Japanese races IB, II, IIIA, and V	–	–	Kurata and Yamazaki (2006)
12	Xa-12	4	–	Kogyoku, Java14	Dominant	Japan	Indonesian race-V	–	–	Ogawa et al. (1987)
13	xa-13	8	–	BJ1, IRBB13	Recessive, cloned, and characterized	India	Philippine race-6	RG136, xal3p	STS, SSR	Kurata and Yamazaki (2006)
14	Xa-14	4	–	TN1	Dominant	Taiwan	Philippine race 5	VAZ190B/ RG163	RFLP	Kurata and Yamazaki (2006)
15	xa-15		–	M41 mutant	Recessive	ND	Japanese races	–	–	Ogawa (2008)
16	Xa-16		–	Tetep	Dominant	Vietnam	Japanese races		–	Kurata and Yamazaki (2006)
17	Xa-17		–	Asominori	Dominant	South Korea	Japanese races		–	Kurata and Yamazaki (2006)
18	Xa-18		–	IR24, Miyang23, Toyonishiki	Dominant	Philippines, Japan	Burmese races		–	Kurata and Yamazaki (2006)
19	Xa-19	3	–	XM5 (mutant of IR24)	Recessive	–	Japanese races		–	Kurata and Yamazaki (2006)

S. No.	Genes/QTLs	Chr.	Position (bp)	Donor parent	Inheritance	Origin	Resistance to Xoo race	Linked marker	Marker type	References
20	Xa-20		–	XM6 (mutant of IR24)	Recessive	–	Japanese races	–	–	Kurata and Yamazaki (2006)
21	Xa-21	11	20,802,924–20,806,518	O. longistaminata, IRBB21	Dominant, cloned, and characterized	Africa, Mali	Philippine and Japanese races	pTA248	STS	Song et al. (1995)
22	Xa-22(t)	11		Zhach anglong	Dominant	China	Chinese races	L363B/P143	RFLP	Kurata and Yamazaki (2006)
23	Xa-23	11	22,203,734–22,204,676	O. ruipogon (CBB23)	Dominant, cloned, and characterized	China/Cambodia	Indonesian races		–	Zhang et al. (2001)
24	xa-24	2		DV86	Recessive	Bangladesh	Philippine and Chinese races		–	Khush and Angeles (1999)
25	xa-25/Xa25(t)	12	17,302,073–17,305,326	Minghui 63, HX-3 (somaclonal mutant of Minghui 63)	Recessive, cloned, and characterized	China	Chinese and Philippine races		–	Liu et al. (2011)
26	xa-26(t)	11	–	Nep Bha Bong	Recessive	China	Philippine races	C4S1S/Y6855RA	RFLP	Lee et al. (2003)
27	Xa-27(t)	6	–	O. minuta IRGC 101141, IRBB27	Dominant, cloned, and characterized	Philippines	Chinese strains and Philippine races 2–6	M10S1, M1095	RFLP	Gu et al. (2005)
28	xa-28(t)	11	–	Lota sail	Recessive	Bangladesh	Philippine race 2	–	–	Lee et al. (2003)
29	Xa-29(t)	1	–	O. officinalis (B5)	Dominant	–	Chinese races	–	–	Tan et al. (2004)

(continued)

Table 2 (continued)

S. No.	Genes/QTLs	Chr.	Position (bp)	Donor parent	Inheritance	Origin	Resistance to Xoo race	Linked marker	Marker type	References
30	Xa-30(t)	11	–	O. ruipogon (Y238)	Dominant	India	Indonesian races	–	–	Cheema et al. (2008)
31	xa-31(t)	4	–	Zhach anglong	Recessive	China	Chinese races	–	–	Wang et al. (2009)
32	xa-32(t)	11	–	O. australiensis (introgression line C4064)	Recessive	–	Philippine races	RM27256, 27274	SSRs	Zheng et al. (2009)
33	Xa-33	7	–	Ba7 O. nivara	Dominant	Thailand	Thai races	RMWR7.1 and 7.6	SSRs	Hari et al. (2013)
34	Xa-33(t)	6	–	Ba7O. nivara	Dominant	Thailand	Thai race	RM20590	SSRs	Korinsak et al. (2009)
35	xa-34(t)	1	–	Pin Kaset O. brachyantha	Recessive	Sri Lanka	Thai race	RM493, RM446, RM10927, RM10591	SSRs	Chen et al. (2011b)
36	Xa-35(t)	11	–	Oryza minuta (Acc. No. 101133)	Dominant	Philippines	Philippine races	–	–	Guo et al. (2010)
37	Xa-36(t)	11	–	C4059	Dominant	China	Philippine races	–	–	Miao et al. (2010)
38	Xa-38	4	–	O. nivara IRGC81825	Dominant	–	Indian Punjab races	RM17499, RM459, RM317	STS/SSR	Bhasin et al. (2012)
39	Xa39	11	–	FF329	Dominant	–	Chinese and Philippine races	RM21, RM206	STS/SSR	Zhang et al. (2015)
40	Xa40(t)	11	–	IR65482-7-216-1-2	Dominant	–	Korean BB races	RM27320, ID55, WA18-5	STS/SSR	Kim et al. (2015)

S. No.	Genes/ QTLs	Chr.	Position (bp)	Donor parent	Inheritance	Origin	Resistance to Xoo race	Linked marker	Marker type	References
41	xa41(t)	–	–	Rice germplasm	Dominant	–	Various Xoo strains	–	–	Hutin et al. (2015)
42	xa42	3	–	XM14, a mutant of IR24	Dominant	–	Japanese Xoo races	RM15189	SSR	Busungu et al. (2016)
43	Xa-43 (t)	11	11,92,907– 11,943,779	IR36(P8)	Dominant	–	Korean BB races	–	–	Kim and Reinke (2019)
44	Xa-44 (t)	11	11,964,077– 11,985,463	R73571-3B-11-3- K3(P6)	Dominant	–	Korean BB races	–	–	Kim (2018)
45	Xa-45(t)	8	26,737,175– 26,818,765	O. glaberrima IRGC 102600B	Recessive	–	Indian Punjab races	xa13 prom	STS	(Neelam et al. 2020)
46	Xa46(t)	11	–	Mutant line H 120	Dominant	China	Chinese Xoo races	RM26981– RM26984	SSR	Chen et al. (2020)

Source: Revised and updated from Kou and Wang (2013)

Table 3 List of blast resistance genes/QTLs

S. No.	Gene/QTL	Chr.	Position (bp)	Position (cM)	Donor rice variety	Method of identification	References
1	*Pit*	1	2,270,216–3,043,185	9.08–12.17	Tjahaja	Cloned	Hayashi and Yoshida (2009)
2	*Pi27(t)*	1	6,230,045–6,976,491	24.29–27.90	IR64 (I)	Mapped within 21.6 cM	Sallaud et al. (2003)
3	*Pi24(t)*	1	5,242,654–5,556,378	20.97–22.22	Azuenca (J)	QTL mapping	Zhuang et al. (2002)
4	*Pitp(t)*	1	25,135,400–28,667,306	100.54–117.49	Tetep	Cosegregation marker was identified	Barman et al. (2004)
5	*Pi35(t)*	1	33,000,000–34,150,000	132.0–136.6	Hokkai 188 (J)	Cloned	Xu et al. (2014)
6	*Pi37*	1	33,110,281–33,489,931	132.44–133.95	St. No. 1 (J)	Cloned	Lin et al. (2007)
7	*Pi64*	1	–	–	Yangmaogu (J)	Cloned	Ma et al. (2015a)
8	*Pid1(t)*	2	21,875,000–22,475,000	87.5–89.9	Digu	Mapped within 11.8 cM	Chen et al. (2004)
9	*Pig(t)*	2	34,346,727–35,135,783	137.38–140.54	Guangchang zhan (I)	Mapped within 11.8 cM	Zhou et al. (2004)
10	*Pitq5*	2	37,625,000–39,475,000	150.5–157.9	Teqing	QTL mapping	Tabien et al. (2002)
11	*Piy1(t)*	2	38,300,000–38,525,000	153.2–154.1	Yanxian No. 1	Mapped within 1.6 cM	Lei et al. (2005)
12	*Piy2(t)*	2	38,300,000–38,525,001	153.2–154.1	Yanxian No. 1	Mapped within 3.0 cM	Lei et al. (2005)
13	*Pib*	2	38,300,000–38,525,000	153.2–154.1	Tohoku IL9	Cloned	Wang et al. (1999)
14	*Pi25(t)*	2	34,360,810–37,725,160	137.44–150.90	IR64 (I)	QTL mapping	Wu and Tanksley (1993)
15	*Pi14(t)*	2	1–6,725,831	1.00–26.90	Maowangu	Linkage analysis using isozyme markers	Pan et al. (1996)
16	*Pi16(t)*	2	1–6,725,831	1.00–26.91	Aus373 (I)	Linkage analysis using isozyme markers	Pan and Tanisaka (1997)

(continued)

Table 3 (continued)

S. No.	Gene/ QTL	Chr.	Position (bp)	Position (cM)	Donor rice variety	Method of identification	References
17	Pi68(t)	3	14,738– 14,761	9.30– 9.70	INGR15002	QTL mapping	Devi et al. (2020)
18	Pi63/ Pikahei-1(t)	4	–	–	Kahei	Cloned	Xu et al. (2014)
19	pi21	4	5,242,654– 5,556,378	20.97– 22.22	Owarihatamochi	Cloned	Fukuoka et al. (2009)
20	Pikur1	4	24,611,955– 33,558,479	98.44– 134.23	Kuroka (J)	Linkage analysis using phenotypic marker	Goto (1988)
21	Pi39(t)	4	26,850,000– 27,050,000	107.4– 108.2	Chubu 111 (J)	Mapped within 0.3 cM	Liu et al. (2007)
22	Pi(t)	4	2,270,216– 3,043,185	9.08– 12.17	Tjahaja	Linkage analysis using phenotypic marker	Causse et al. (1994)
23	Pi26(t)	5	8,751,256– 11,676,579	35.00– 46.70	Gumei 2 (I)	QTL mapping	Wu and Tanksley (1993)
24	Pi23(t)	5	10,755,867– 19,175,845	43.02– 76.70	Sweon 365	QTL mapping	Ahn et al. (1997)
25	Pi10	5	14,521,809– 18,854,305	58.08– 75.41	Tongil	Mapped within 6.7 cM	Naqvi et al. (1995)
26	Pi2	6	–	–	C101A51	Cloned	Zhou et al. (2006)
27	Pi22(t)	6	4,897,048– 6,023,472	19.50– 24.09	Suweon365 (J)	QTL mapping	Ahn et al. (1997)
28	Pi26(t)	6	8,751,256– 11,676,579	35.00– 46.70	Azucena (J)	QTL mapping	Wu et al. (2005)
29	Pi27(t)	6	5,556,378– 744,329	22.22– 2.97	IR64 (I)	Mapped within 21.6 cM	Sallaud et al. (2003)
30	Pi40(t)	6	16,274,830– 17,531,111	65.09– 70.12	O. australiensis (W)	Mapped within 1.8 cM	Jeung et al. (2007)
31	Piz	6	10,155,975– 10,517,612	40.60– 42.07	Zenith (J)	Mapped within 0.43 cM	Ahn et al. (1996)
32	Piz-t	6	14,675,000	58.70	Toride 1	Cloned	Hayashi et al. (2006)
33	Pi9	6	10,386,510– 10,389,466	41.50– 41.55	O. minuta (W)	Cloned	Qu et al. (2006)

(continued)

Table 3 (continued)

S. No.	Gene/ QTL	Chr.	Position (bp)	Position (cM)	Donor rice variety	Method of identification	References
34	*Pi25*	6	18,080,056– 19,257,588	72.32– 77.03	Gumei 2	Cloned	Chen et al. (2011a)
35	*Pid2*	6	17,159,337– 17,163,868	68.63– 68.65	Digu	Cloned	Chen et al. (2006b)
36	*Pigm(t)*	6	10,367,751– 10,421,545	41.47– 41.68	Gumei 4	Mapped within 70 kb	Deng et al. (2017)
37	*Pi50*	6	–	–	Er-Ba-zhan (EBZ)	Cloned	Su et al. (2015)
38	*Pid3-I1*	6	–	–	MC276	Cloned	Inukai et al. (2019)
39	*Pi17(t)*	7	22,250,443– 24,995,083	89.00– 99.90	DJ 123	Mapped within 1.8 cM	Pan et al. (1996)
40	*Pi36*	8	2,870,061– 2,884,353	11.48– 11.53	Q61 (I)	Cloned	Liu et al. (2005)
41	*Pi33*	8	5,915,858– 6,152,906	23.66– 24.61	IR64 (I)	Mapped within 1.6 cM	Berruyer et al. (2003)
42	*Pizh*	8	4,372,113– 21,012,219	17.48– 84.04	Zhai-Ya-Quing8 (I)	QTL mapping	Sallaud et al. (2003)
43	*Pi29(t)*	8	9,664,057– 16,241,105	38.65– 64.96	IR64 (I)	Mapped within 0.7 cM	Sallaud et al. (2003)
44	*Pii2(t)*	9	1,022,662– 7,222,779	4.09– 28.89	Azucena	Linkage analysis using phenotypic markers	Kinoshita and Kiyosawa (1997)
45	*Pi5*	9	7,825,000– 8,250,000	31.30– 33.00	RIL125, RIL249, RIL260 (Moroberekan)	Mapped within 170 kb	Lee et al. (2009)
46	*Pi3(t)*	9	7,825,000– 8,250,001	31.3– 33.1	Kan-Tao	Linkage analysis using RFLP markers	Causse et al. (1994)
47	*Pi15*	9	9,641,358– 9,685,993	38.56– 38.74	GA25 (J)	Mapped within 0.7 cM	Pan et al. (1996)
48	*Pii*	9	–	–	Hitomebore	Cloned	Takagi et al. (2013a)
49	*Pi28(t)*	10	19,565,132– 22,667,948	78.26– 90.67	IR64 (I)	QTL mapping	Sallaud et al. (2003)
50	*Pia*	11	–	–	Aichi Asahi (J)	Cloned	Okuyama et al. (2011)

(continued)

Table 3 (continued)

S. No.	Gene/ QTL	Chr.	Position (bp)	Position (cM)	Donor rice variety	Method of identification	References
51	PiCO39(t)	11	6,304,007–6,888,870	25.21–27.55	CO39 (I)	Cloned	Cesari et al. (2013)
52	Pilm2	11	13,635,033–28,377,565	54.54–113.50	Lemont	QTL mapping	Tabien et al. (2002)
53	Pi30(t)	11	441,392–6,578,785	1.76–26.31	IR64 (I)	QTL mapping	Sallaud et al. (2003)
54	Pi7(t)	11	17,850,000–21,075,000	71.40–84.30	RIL29 (Moroberekan)	QTL mapping	Wang et al. (1994)
55	Pi34	11	19,423,000–19,490,000	77.69–77.96	Chubu32 (J)	QTL mapping	Zenbayashi et al. (2002)
56	Pi38	11	19,137,900–21,979,485	76.55–87.91	Tadukan (I)	Mapped within 20 cM	Gowda et al. (2006)
57	PBR	11	20,125,000–30,075,000	80.5–120.3	St. No. 1	Mapped within 22.9 cM	Fujii et al. (1995)
58	Pb1	11	–	–	Modan	Cloned	Hayashi et al. (2010)
59	Pi44(t)	11	22,850,000–29,475,000	91.40–117.90	RIL29 (Moroberekan)	–	Chen et al. (1999)
60	Pik-h/ Pi54	11	24,761,902–24,762,922	99.0–99.05	Tetep	Cloned	Sharma et al. (2005b)
61	Pi1	11	26,498,854–28,374,448	105.99–113.49	LAC23 (J)	Mapped within 11.4 cM	Hua et al. (2012)
62	Pik-m	11	27,314,916–27,532,928	109.25–110.13	Tsuyuake (J)	Cloned	Ashikawa et al. (2008)
63	Pi18(t)	11	26,796,917–28,376,959	107.18–113.50	Suweon365 (J)	Mapped using RFLP markers	Ahn et al. (1996)
64	Pik	11	27,314,916–27,532,928	109.25–110.13	Kusabue (I)	Cloned	Zhai et al. (2011)
65	Pik-p	11			K60	Cloned	Yuan et al. (2011)
66	Pik-s	11	27,314,916–27,532,929	109.25–110.15	Shin 2 (J)	Mapped within 2.7 cM	Fjellstrom et al. (2004)

(continued)

Table 3 (continued)

S. No.	Gene/QTL	Chr.	Position (bp)	Position (cM)	Donor rice variety	Method of identification	References
67	*Pik-g*	11	27,314,916–27,532,930	109.25–110.16	GA20 (J)	Linkage analysis to other resistance genes	Pan et al. (1996)
68	*Pise1*	11	5,740,642–16,730,739	22.96–66.92	Sensho	Linkage analysis using phenotypic markers	Goto (1970)
69	*Pif*	11	24,695,583–28,462,103	98.78–113.84	Chugoku 31-1 (St. No. 1)	QTL mapping	Shinoda et al. (1971)
70	*Mpiz*	11	4,073,024–16,730,739	16.29–66.92	Zenith (J)	Linkage analysis using phenotypic markers	Goto (1970)
71	*Pikur2*	11	2,840,211–18,372,685	11.36–73.49	Kuroka (J)	Linkage analysis using phenotypic markers	Goto (1988)
72	*Piisi*	11	2,840,211–19,029,573	11.36–76.11	Imochi Shirazu (J)	Linkage analysis using phenotypic markers	Goto (1970)
73	*Pike*	11			Xiangzao 143	Cloned	Chen et al. (2015)
74	*Pi24(t)*	12	5,242,654–5,556,378	20.97–22.22	Azuenca (J)	QTL mapping	Zhuang et al. (2002)
75	*Pi62(t)*	12	2,426,648–18,050,026	9.70–77.00	Yashiro-mochi (J), Tsuyuake	Mapped within 1.9 cM	Wu et al. (2008)
76	*Pitq6*	12	5,758,663–7,731,471	23.00–30.92	Tequing (I)	QTL mapping	Tabien et al. (2002)
77	*Pi6(t)*	12	1–6,725,831	1–1.68	Apura (I)	–	McCouch et al. (1994)
78	*Pi12*	12	6,988,220–15,120,464	27.95–60.48	Moroberekan (J)	Linkage analysis using RFLP markers	Inukai et al. (1996)
79	*Pi21(t)*	12	5,242,654–5,556,378	20.94–22.22	Owarihata mochi (J)	–	Ahn et al. (1997)
80	*Pi31(t)*	12	7,731,471–11,915,469	30.92–47.66	IR64 (I)	QTL mapping	Sallaud et al. (2003)

(continued)

Table 3 (continued)

S. No.	Gene/ QTL	Chr.	Position (bp)	Position (cM)	Donor rice variety	Method of identification	References
81	*Pi32(t)*	12	13,103,039– 18,867,450	52.41– 75.46	IR64 (I)	QTL mapping	Sallaud et al. (2003)
82	*Pi157*	12	12,375,000– 15,550,000	49.5– 62.2	Moroberekan	Mapped within 9.5 cM	Causse et al. (1994)
83	*Pita*	12	10,603,772– 10,609,330	42.41– 42.43	Tadukan (I)	Cloned	Hayashi et al. (2006)
84	*Pita-2*	12	10,078,620– 13,211,331	40.31– 52.84	Shimokita (J)	Mapped within 4.0 cM	Nakamura et al. (1997)
85	*Pi19(t)*	12	8,826,555– 13,417,088	35.30– 53.67	Aichi Asahi (J)	Linkage analysis to other resistance genes	Iwata (1996)
86	*Pi39(t)*	12	–	–	Chubu 111 (J),	Mapped within 37 kb	Liu et al. (2007)
87	*Pi20(t)*	12	12,875,000– 12,950,000	51.50– 51.80	IR24 (I)	Mapped within 0.6 cM	Liu et al. (2008)
88	*PiGD-3(t)*	12	13,950,000	55.80	Sanhuangzhan 2	QTL mapping	Liu et al. (2005)
89	*Ptr*	12			Katy	Cloned	Zhao et al. (2018)

Source: Revised and updated from Tanweer et al. (2015)

disease resistance could never obtain a break because of the emergence of new pathotypes, which could overcome the resistance. Advances in rice genomics provided tools such as molecular markers for plant breeders to effectively develop cultivars with resistance against various diseases, which is an environment-friendly alternative vis-à-vis the use of agrochemicals (Miah et al. 2013). Molecular markers can be used to map and introgress one or more desired genes for biotic and abiotic stress resistance from diverse gene pools (Suh et al. 2009). Marker-assisted selection for pyramiding desired genes without altering other quality characteristics of a rice cultivar is crucial in rice improvement (Sundaram et al. 2008; Suh et al. 2009; Shanti et al. 2010). As an added advantage, the availability of gene-linked molecular markers for the resistance genes eases the identification of plants harboring two or more *R*-genes at any growth stage without a bioassay (Sundaram et al. 2008; Shanti et al. 2010; Bainsla and Meena 2016).

Three bacterial blight-resistance genes (*xa5*, *xa13*, and *Xa21*) were pyramided into susceptible cultivar PR106 using MAS. The introgression lines were tested against 17 *Xoo* isolates under both glasshouse and field conditions. The trials suggested that the combination of genes provided broad-spectrum resistance against

Table 4 List of sheath blight-resistance genes/QTLs

S. No.	QTLs	Chr.	Linked markers	Mapping population	Type of marker	LOD	Associated character	Remarks	Reference
1	qshb1.1	1	RM151–RM12253	210 F$_2$ (ARC10531[I] × BPT-5204[I])	SSR (70)	10.7	Percentage relative lesion height	32 candidate genes identified in the region qShB9.2	Yadav et al. (2015)
2	qshb7.1	7	RM81–RM615			8.8			
3	qshb7.2	7	RM10–RM2169			6.7			
4	qshb8.1	8	RM21792–RM310			4.2			
5	qSBL7 (E2)	7	D760–RM248	190 F$_2$ (Yangdao 4[I] × Lemont[I])	SSR (52) and InDel (128)	3.12	DR (disease rating)	Sheath blight resistance is correlated with plant height	Wen et al. (2015)
6	qSBPL-7 (E2)	7	D760–RM248			5.07	LH (lesion height)		
7	qSBD-12-2 (E1)	12	RM1246–D1252			3.74	PL (percentage of lesion height), DR, LH		
8	qHZaLH3	3	RM143–RM514	116 DHs (TN1[I] × CJ06[J])	SSR (214)			No correlation was found between LH and PH	Zeng et al. (2015)
10	qHZaLH6	6	WX–RM587						
11	qHZaDR8	8	RM1376–RM4085						
13	qHZaDR9	9	RM444–AGPSMA						
14	qHZbDR5	5	RM3321–RM3616						

S. No.	QTLs	Chr.	Linked markers	Mapping population	Type of marker	LOD	Associated character	Remarks	Reference
15	qSB-9^{TQ}	9	CY-85 and Y86	235CSSLs (BC$_6$F$_3$) (Teqing^l (TQ) × Lemont^l)	InDel and CAPS (22)	–	ShB resistance	Fine-mapped (146 kb covering region), 12 genes were annotated	Zuo et al. (2014)
16	qDR-4	4	RM1155–RM 5757	155 RILs F$_{8:11}$ (RSB02 × HH1B)	SSR (163)	2.71	DR	Epistasis and QTL × environment (QE) interaction were studied	Liu et al. (2014)
							LL (lesion length)		
17	qRLL-4	4	RM1155–RM5757			5.84	LH		
18	qRLH-4	4	RM1155–RM5757			4.77	Relative LL		
							Relative LH		
19	qSB-11^{LE}	11	Z22-27C and Z23-33C	112CSSLs (Teqing^l × Lemont^l (LE))	STS and CAPS	–	ShB resistance	Fine-mapped (79 kb covering region), 11 genes were annotated	Zuo et al. (2013)
20	qRTL3	3	RM570	BILs (Jarjan^l × Koshihikari^l)	SSR (151)	3.5	RTL (rate of tillers with lesions)		Taguchi-Shiobara et al. (2013)
21	qRTL5	5	RM5784			4.3			
22	qRTL6	6	RM1161			7.7			
23	qRTL9	9	RM6251			3.1			
24	qRTL3	3	RM16200			5.9			
25	qRTL6	6	RM2615			2.9			
26	qRTL12	12	RM7025			3.2			
27	qRTL5	5	RM3286			3.1			
28	qRTL6	6	RM6395			5.8			

(continued)

Table 4 (continued)

S. No.	QTLs	Chr.	Linked markers	Mapping population	Type of marker	LOD	Associated character	Remarks	Reference
29	qRTL9	9	RM3533			3.8			
30	qShB2-1-	2	RM279–RM71	216 RILs (Jasmine 85[1] × Lemont[J])	SSR (199)	3.7	ShB resistance	The major QTL qShB9-2 was reconfirmed based on the field data	Liu et al. (2013)
31	ARqShB7-AR	7	RM5711–RM2			4.0			
32	qShB7 LA	7	RM5711–RM2			6.0			
33	qShB11-1-	11	RM7203–RM536			3.2			
34	TXqShB11-2-TX	11	RM536–RM229			3.3			
35	qShB6 (wild 1-field 2009)	6	RM3431–RM3183	252 Wild-1 and 253 Wild-2 BC$_2$F$_2$	SSR (131)	7.8	ShB resistance	Colocalization of qShB6 with qDH1 and qShB1 with qPH1 revealed the influence of heading date and plant height on resistance	Eizenga et al. (2013)
36	qShB6 (wild 2-field 2009)	6	RM253–RM3431	(Oryzanivara × Bengal[J] (O. sativa))		21.2			
37	qShB6 (wild2-field 2008)	6	RM253–RM3431			11.1			
38	qShB1 (wild 2-field 2008)	1	RM431–RM1361			4.7			
39	qShB6-mc (wild-1 microchamber)	6	RM3183–RM541			3.3			

S. No.	QTLs	Chr.	Linked markers	Mapping population	Type of marker	LOD	Associated character	Remarks	Reference
40	*qsbr_2.1*	2	RM8254–RM8252	197 DHs (MCR10277ⁱ × Cocodrieʲ)	SSR (111)	3.4–29.7	SBF (sheath blight disease		Nelson et al. (2012)
41	*qsbr_2.2*	2	RM3857–RM5404			2.9–37.8	severity in field), SBI (disease severity in microchamber)		
42	*qsbr_12.1*	12	RM3747–RM27608			49,1	SBM (disease severity in mist chamber)		
43	*qSBR1*	1	RM11229	217 core collection of USDA	SSR (154) and InDel (1)	9.5%	Sheath blight resistance		Jia et al. (2012)
44	*qSBR11*	11	RM7203			1.9%			
45	*qSBR1-1, qSBR2-1*	1, 2	RM5389–RM3825, RM5340–RM521	121 RIL (RSB03 × HH1B)	SSR (123)	3.2, 3.1	DR, LL, LH		Fu et al. (2011)
46	*qSBR2-2*	2	RM110–osr14			5.2	Relative LL		
47	*qSBR2-3*	2	RM7245–RM5303			3.3	Relative LH		
48	*qSBR4*	4	RM3288–RM7187			3.8			

(continued)

Table 4 (continued)

S. No.	QTLs	Chr.	Linked markers	Mapping population	Type of marker	LOD	Associated character	Remarks	Reference
49	*qSBR5-2*	5	RM7446–RM3620			4.8			
50	*qSBR7*	7	RM1132–RM473			3.3			
51	*qSBR8*	8	RM8264–RM1109			4.2			
52	*qSBR9*	9	RM23869–RM3769			5.0			
53	*qShB1 (2007/2008)*	1	RM431–RM12017	251 DHs (Baiyeqiu[I] × Maybelle[J])	SSR (227)	5.18–8.03	Sheath blight resistance		Xu et al. (2011)
54	*qShB2 (2008)*	2	RM174–RM145			3.96			
55	*qShB3 (2007)*	3	RM135–RM186			3.42			
56	*qShB5 (2007)*	5	RM18872–RM421			4.35			
57	*qShB1-2 (2020)*	1		184 RILs (SH × DGWG[I]; BHA × DGWG[I])		5.71	Blast resistance		Goad et al. (2020)
58	*qShB4 (2020)*	4				4.50			
59	*qSBR3.2*	3	D311 or RM282	219 RILs (Lemont[J] × Yangdao4[I])		3.3	Sheath blight resistance		Yuan et al. (2019)
60	*qSBR7.1*	7	D709 or D715			3.7			
61	*qSBR8.1*	8	D804		SSR	1.8			

S. No.	QTLs	Chr.	Linked markers	Mapping population	Type of marker	LOD	Associated character	Remarks	Reference
62	*qSBR9.2*	9	D947 or D948			3.0			
63	*qSBR9.3*	9	D949			2.4			
64	*qSBR12.1*	12	D1211			2.3			
65	*qSBR12.2*	12	D1239 or D1246			3.6			

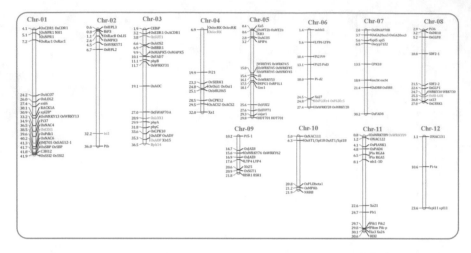

Fig. 2 Physical position of major biotic disease-resistance genes in rice. The chromosome left side indicates the location of genes and the right side shows the names of genes collected from the Q-TARO and Oryza base databases. The color indicates the genes related to various diseases such as bacterial leaf blight (red color) and blast (blue color), and green color indicates insect resistance in rice

the pathogen races predominant in the region (Singh et al. 2001). Recent advances in DNA sequencing have made fine-mapping and characterization of the mapped genes easier, thus contributing significantly to the use of MAS for the development of resistant cultivars. The complete list of cloned genes was collected from Q-TARO (http://qtaro.abr.affrc.go.jp/) and OryGenesDB (https://orygenesdb.cirad.fr/data.html) (Fig. 2). These genes were mainly associated with bacterial blight and blast resistance in rice. Interestingly, the regions on chromosomes 1, 4, and 5 were associated with multiple resistance genes, and these genes were colocalized in the same regions. These genomic regions play a major role in the resistance/tolerance mechanisms for diseases. To date, there are 46 BB R-genes mapped from different sources, out of which 29 are dominant, 12 are recessive, nine cloned, and nine fine-mapped (Chen et al. 2020). More than 100 R-genes (Pi) have been reported, and around 500 QTLs were associated with blast resistance. However, only 25 Pi genes were cloned and characterized (Sharma et al. 2012; Ashkani et al. 2015). Several R QTLs were reported against bacterial leaf streak, but their study was limited to inheritance analysis (He et al. 2012).

4.1 MAS/MABB Foreground/Background Selection

To address the limitations of conventional breeding, molecular breeding through MAS is among the most precise tools used to introgress multiple resistance genes into an elite varietal background at one time. Plant breeders were already successful in using this tool in developing resistant rice cultivars by deploying broad-spectrum multiple-*R*-genes with the help of MAS (Huang et al. 1997; Sanchez et al. 2000; Sundaram et al. 2008; Hari et al. 2013; Hajira et al. 2016; Balachiranjeevi et al. 2018; Swathi et al. 2019; Jamaloddin et al. 2020). In marker-assisted backcross breeding (MABB), a combination of foreground selection and background selection followed by continuous backcrossing can recover up to 99% of the recurrent parent genome (RPG) (Tanksley et al. 1989). In foreground selection, gene-linked markers, or functional markers (SSRs, InDels, and SNPs), are applied to detect the associated *R*-genes in the target population at any stage of plant growth. In contrast, background selection applied by using polymorphic information (SSRs, SNPs) between the donor and recurrent parents can estimate RPG recovery in each backcross generation at any plant growth stage (Singh et al. 2001). Recently, the Green Super Rice (GSR) breeding strategy proved that one backcross followed by selfing could recover more than 90% of the recurrent parent genome (Balachiranjeevi et al. 2019).

4.2   Pyramiding Disease-Resistance Genes

Pyramiding of various biotic disease-resistance genes into a rice cultivar makes it a good candidate for breeders to introgress the resistance into locally adapted varieties that produce higher yield but are susceptible to diseases. The process of gene pyramiding through conventional breeding alone becomes difficult because the linkage between some undesirable traits is difficult to break even after repeated backcrossing (Tanksley et al. 1989). Pyramiding of two or more resistance genes renders the phenotypic assessment of rice genotypes ineffective as distinguishing the effect of each individual gene precisely becomes difficult since each gene imparts resistance to more than one race of the pathogen. Moreover, when a dominant and a recessive allele are present, the effect of the recessive gene is concealed. The availability of tightly linked markers for each of the resistance genes thus eases the recognition of plants with multiple genes. Initially, in rice, Huang et al. (1997) successfully introgressed four major BB resistance genes (*Xa4*, *Xa5*, *Xa13*, and *Xa21*) and developed breeding lines with combinations of two, three, and four genes. In an extension of this work, several research institutions in India and other countries have studied the effectiveness of the pyramided genes against BB disease, to which most of the popular varieties were susceptible. This research has opened the gates in India to address the susceptibility of popular rice varieties such as PR106 (Singh et al. 2001), Pusa Basmati-1 (Joseph et al. 2004), and Samba Mahsuri (Sundaram et al. 2008) by pyramiding the BB *R*-genes (*xa5*, *xa13*, and *Xa21*) in the initial phase of improvement. Later, the improvement of popular rice varieties and parental lines continued mainly against BB (*Xa21*, *Xa23*, *xa5*, *xa13*, *Xa4*, *Xa7*, *Xa33*, and *Xa38*) and blast disease (*Pi* genes *Pi2*, *Pi9*, *Pi40*, *Pi54*, *Piz*, and *Pi1*) separately or by combining *R*-genes for both diseases (Gopalakrishnan et al. 2008; Sundaram et al. 2009; Hari et al. 2013; Balachiranjeevi et al. 2015; Yugander et al.

2018; Rekha et al. 2018; Swathi et al. 2019; Jamaloddin et al. 2020). The possibility of recombination between the gene of interest and the linked marker has led to the selection of false-positive rice genotypes in the marker-assisted selection process, which could be overcome by using gene-specific functional markers (Ingvardsen et al. 2008). Many genetic markers, also called functional markers, have been identified for different disease-resistance genes in rice, such as BB-resistance genes *xa5* (Iyer-Pascuzzi and McCouch 2007), *xa13* (Chu et al. 2006), and *Xa21* (Song et al. 1995). The gene-pyramided lines enable the conducting of quantitative analysis to assess the effect of each gene and interactions between them and, most importantly, enhancing the performance, stability, and longevity of genetic resistance.

4.3 Varieties Improved and Developed

Highly accepted varieties and parental lines were improved against multiple diseases through MAS. For the first time, Huang et al. (1997) developed lines pyramided with two, three, and four genes through MAS and tested their resistance against BB. The resistance levels of introgressed lines showed an elevated resistance compared with lines containing a single gene. Later, Singh et al. (2001) improved Indian rice cultivar PR106 against BB through MAS by pyramiding *xa5*, *xa13*, and *Xa21* genes, followed by Joseph et al. (2004), who improved popular basmati variety Pusa Basmati-1, and Sundaram et al. (2008) improved popular variety Samba Mahsuri for BB and reported more than 95% RPG recovery through MABB. Through MAS, three blast genes (*Pi1*, *Pi2*, and *Pi33*) were introduced in the background of popular Russian rice variety Kuboyar. The improved lines of Kuboyar were used to develop blast-resistant hybrids by using them as hybrid parental lines. Similarly, Hari et al. (2011) improved restorer line KMR3R for resistance against BB by transferring the *Xa21* gene along with *Rf3* and *Rf4* (restorer of fertility) genes through MABB. Balachiranjeevi et al. (2015) imparted resistance to a maintainer line (DRR17B) by introgressing *Xa21* and *Pi54* genes against BB and blast disease, respectively.

4.4 Multiple Disease-Resistance Breeding Strategies

In breeding for disease resistance, multiple methodologies such as pedigree, modified bulk, single seed descent (SSD), doubled-haploid (DH), and MAB have been used to develop resistant rice varieties (Mackill et al. 1996; Khush 2005; Collard et al. 2013). In addition to these strategies, the GSR breeding program was one of the successful strategies that involved vigorous phenotypic screening at early backcross stages (BC_1F_2 to BC_1F_4) combined with three successive rounds of stringent selection for the best plant type to come up with climate-resilient rice varieties. This strategy could develop homozygous inbred cultivars within a short span of 4–5 years vis-à-vis 9–10 years with a conventional breeding program (Yu et al. 2020). The GSR breeding strategy is carried out in three steps. The first is to develop early backcross BC_1F_2 populations by crossing a widely adapted recipient variety with a diverse set of donors. Second is to simultaneously do phenotypic screening of early backcross-derived lines of BC_1F_2, BC_1F_3, and BC_1F_4 generations under the different abiotic and biotic stress conditions in a rigorous manner to identify and select intro-gression lines (ILs) tolerant of different stresses as compared to the tolerant and susceptible checks. The third step is the mapping of genomic regions influenced by particular climate fluctuations and their characterization to decode the molecular and physiological basis of the identified genomic regions (Ali et al. 2017). Three rounds of screening of populations from BC_1F_2 to BC_1F_4 for different diseases simultaneously could help in the development of varieties with tolerance of multiple biotic stresses. The GSR breeding strategy led to successful mapping of the *Xa39* gene and deploying it in the background of Huang-Hua-Zhang (Zhang et al. 2015). Further, through the designed QTL pyramiding approach, one could combine selec-tive ILs carrying different biotic and abiotic stress-tolerance genes/QTLs derived from different donors but having a common recipient parent. Similar to the GSR breeding program, breeders have simultaneously pyramided multiple disease-resistance genes (BB + blast) with different combinations such as *Xa21* + *Pi54*, *Xa21* + *Pi54* + *Pi2*, and *xa5* + *xa13* + *Xa21* + *Pi54* + *Pi2* into the background of an elite cultivar by employing MAS and MABB (Jiang et al. 2015; Jamaloddin et al. 2020). Recently, one of the successful breeding strategies (the GSR breeding pro-gram) revealed lots of hidden genetic diversity for disease resistance through MAS and also proved that RPG recovery could surpass 90% with one backcross followed by selfing (Balachiranjeevi et al. 2019). Furthermore, Feng et al. (2018) reported that pyramiding the detected QTLs effectively broadened the genetic base. Research is being extended to dissect the detected QTLs in order to identify candidate genes through functional validation using a map-based cloning approach.

5 Molecular Mechanisms of Disease Resistance

A wide variety of pathogens, including bacteria, fungi, and viruses, attacks crop plants. Either a pathogen can successfully invade, leading to the development of disease, or the plant can resist the pathogen using an active or passive form of resis-tance. Different strategies have been developed by various pathogens to enter, infect, and reproduce in plants. Pathogens are mainly classified as necrotrophs and bio-trophs based on the method they use to invade, infect, and attack a plant (Oliver and Ipcho 2004). Necrotrophic pathogens kill the host-plant tissue soon after they estab-lish infection and then develop and feed on the dead tissue. Unlike these, biotrophic pathogens require a live-host tissue for their growth and reproduction.

Specific defense mechanisms work effectively against biotrophs through a hypersensitive response developed by rapid local cell death surrounding infection, and this serves to hinder the growth and invasion of pathogens into other plant parts.

This mechanism arises when the first level of the defense mechanism is breached by the pathogen (Zipfel and Felix 2005). Usually, most pathogens that infect plants, such as fungi, harbor secretory proteins, which disrupt these barriers (Serrano et al. 2014). After the entry of the pathogen into the host cell, it is recognized by special molecules called microbe-/pathogen-associated molecular patterns (MAMPs or PAMPs), which include ergosterol, peptidoglycan, lipopolysaccharide, and bacterial flagella in proteins. The innate immune system recognizes these proteins with the help of host plasma membrane-bound receptors called pattern recognition receptors (PRRs) to further obstruct the growth of infection, providing MAMP-triggered immunity (MTI). PRRs also detect molecules that become released in the host when the pathogens cause damage (damage-associated molecular patterns, DAMPs). The binding of these components also triggers pattern-triggered immunity (PTI) and downstream defense responses (Tena et al. 2011). Overall, the recognition of PAMP/MAMP or DAMP results in the activation of PTI, triggering the production of different reactive oxygen species (ROS), initiation of mitogen-activated protein (MAP) kinase activity, and various transcription factor activation, thus limiting the spread of pathogens completely (Nürnberger and Kemmerling 2009).

The widely accepted model of plant disease resistance is explained by a two-level innate immune system. The two levels include PTI, which is usually a weak, basal, and generic immune response, and the other is effector-triggered immunity (ETI), which is a potent response and is specific to the pathogen in question (Jones and Dangl 2006). PTI is mediated by the PRRs that recognize molecular patterns associated with the pathogens or the resulting damage products (PAMPs or DAMPs). On the other hand, ETI includes recognition of a pathogen-specific factor and results in a severe and rapid form of immune response leading to localized cell death (also known as a hypersensitive response or HR) to hinder the pathogen from spreading any further. ETI is achieved by a gene-to-gene interaction and is thus specific to the race of the pathogen. While PRRs mediate PTI, ETI is mediated by specific genes that belong to the nucleotide-binding-leucine-rich repeat (NB-LRR) domain-containing proteins, otherwise called resistance (R) genes. The recognition of cognate ligands results in activation of signaling events that in turn results in the generation of different forms of immune response such as callose and lignin deposition, production of antimicrobial compounds, induction of cell death, changes in primary and secondary metabolic flux, and synthesis of secondary metabolites depending on the type of elicitors. Other classifications of genes involved in disease resistance include major resistance (MR) genes and defense-related genes (DR), whose roles cannot be explained by the definition of PTI- and ETI-associated genes (Ke et al. 2017). PTI is considered to be quantitative in nature, that is, multiple genes function together to achieve immunity, also known as a QTL. ETI against a pathogen strain is controlled by a single gene and is specific only to those strains that contain the cognate avirulence (Avr) protein that the R-gene recognizes, thus leading to a qualitative resistance. Studies in the past few decades established a framework of how the resistance mechanisms act using model pathosystems. Along this line, rice resistance to its major pathogens such as *Xanthomonas oryzae* ssp., *Magnaporthe oryzae*, and *Rhizoctonia solani* has been studied to a reasonable extent. However, rice resistance to other pathogens still needs more investigation to come to a consensus. This section summarizes the established mechanisms of disease resistance in rice.

Plant resistance is dictated by the type of resistance genes and a network of signaling pathways (Chisholm et al. 2006). Broadly, the plant defense system can be categorized into two classes: basal defense and specific defense. The basal defense system is much more effective against necrotrophic pathogens (Singh et al. 2018). Elicitors are molecules that induce a plant defense response at very low concentrations (Thakur and Sohal 2013). The role of the basal defense system is to check the entry of pathogens and provide immunity at the starting stage of infection. This defense response involves membrane permeability, activating ion fluxes (Ca^{2+}, K^+, H^+), generating ROS, producing nitric oxide (NO), and phosphorylation/dephosphorylation of proteins by protein kinases and phosphatases. It also includes the production of signaling molecules such as jasmonic acid (JA), salicylic acid (SA), and ethylene (ET). These proteins are characteristic players in the regulation of defense signal transduction cascades. These steps further trigger an array of signaling that leads to the regulation of the expression of defense-related genes and the stimulation of defense responses. These responses include cell-wall strengthening (callose and lignin deposition), phytoalexin synthesis, and activation of kinase cascades escorted by a hypersensitive response (Jones and Dangl 2006).

5.1 Resistance to Bacterial Blight

To date, 46 resistance genes have been identified to confer resistance to *Xoo* in rice. Among them, 11 genes were cloned and functionally characterized. Some of the resistance genes are quantitative in nature, whereas others confer qualitative resistance (Ke et al. 2017; Jiang et al. 2020; Chen et al. 2020). The 11 cloned genes fall under different classes of resistance genes: LRR-RLKs (leucine-rich repeat receptor-like kinases), NB-LRR, a wall-associated kinase, executor R proteins, *SWEET* (sugars will eventually be exported transporters) genes, and a transcription factor gamma subunit protein. Three of the cloned resistance genes, *Xa3/Xa26*, *Xa4*, and *Xa21*, code for kinases. *Xa4* is a wall-associated kinase (Ke et al. 2017; Jiang et al. 2020) that provides resistance to certain races of *Xoo* through cell-wall reinforcement. *Xa3/Xa26* and *Xa21* are LRR-RLKs that recognize *Xoo*-associated molecules AvrXa3 and sulphated RaxX, respectively. *Xa21*- and *Xa3/Xa26*-mediated resistance has been found to be positively regulated by *OsSERK2* (rice somatic embryogenesis receptor kinase 2). Nine genes were found to be regulating *Xa21*-mediated resistance positively or negatively. *Xa4*-mediated resistance leads to the accumulation of phytoalexins. *Xa1*, an NB-LRR, recognizes intact transcription activation-like effectors (TALEs) from *Xoo* and thus leads to resistance. *SWEET* genes code for sugar transporters and were identified to be targets of different *Xoo* TALEs, thereby acting as susceptibility factors. Natural polymorphisms were identified in the promoters of three SWEET genes, *OsSWEET11/Os8N3/xa13*, *OsSWEET13/*

xa25, and *OsSWEET14/Os11N3/xa41*, which promote their induction by cognate TALEs, thus providing recessive resistance. Genes, including *Xa10*, *Xa23*, and *Xa27*, are classified as executor *R*-genes as the expression of the respective resistance alleles is induced by *Xoo* TALEs. These genes are characterized by the presence of multiple potential transmembrane domains whose expression induction results in HR and thus resistance to *Xoo*. Another recessive resistance gene, *xa5*, codes for transcription factor IIA gamma subunit 5 (TFIIAγ5) with valine to glutamine mutation in the 39th position. The susceptible allele, *Xa5*, is hijacked by the TALEs to induce the expression of other host susceptibility genes. The mutation disrupts the ability of TALEs to bind to TFIIAγ5, thus leading to resistance (Ke et al. 2017; Jiang et al. 2020).

5.2 Resistance to Bacterial Leaf Streak

To date, no major BLS-resistance genes have been identified. However, the *xa5* gene was mapped to be a major resistance QTL for *Xoc* resistance. It was previously observed that TALEs from *Xoc* also hijack TFIIAγ5 for inducing host susceptibility genes. In another study, *Xo1*, a resistance locus in an American rice variety, was identified to be responsible for resistance to African *Xoc* strains but not to Asian strains. *Xa21* was identified to provide weak resistance to *Xoc* through the recognition of Ax21, a quorum-sensing molecule produced by *Xoc* (Jiang et al. 2020). Three major broad-spectrum resistance QTLs, *qXO-2-1*, *qXO-4-1*, and *qXO-11-2*, were identified to confer resistance to *Xoo* and *Xoc* (Bossa-Castro et al. 2018).

5.3 Resistance to Rice Blast

More than 100 resistance genes and 500 QTLs are known to be associated with blast resistance in rice. However, to date, only 25 genes have been cloned (Li et al. 2019b). These 25 cloned *R*-genes are called *Pi* genes. Of the 25 *Pi* genes, 22 encode NB-LRR family proteins. A majority of these *R*-genes trigger ETI, thus leading to qualitative or race-specific resistance. So far, seven *R*-genes have been identified to confer broad-spectrum resistance to blast: *Pi7*, *Pi9*, *Pi21*, *Pi50*, *Pi57*, *Pigm*, and *Ptr*. Apart from canonical *R*-genes, so far, five defense-related genes were also shown to confer resistance to blast: *bsr-d1*, *bsr-k1*, *spl11*, *spl33*, and *OsBBI1*, *Pi9*, *Pi50*, *Pigm*, *Ptr*, and *OsBBI1* are dominant resistance genes or positive regulators of blast resistance, whereas the rest of them are recessive resistance genes, in other words, their wild-type alleles negatively regulate blast resistance (Li et al. 2019b).

5.4   Resistance to Sheath Blight

Information on the mechanisms that govern ShB resistance in rice is just being uncovered. There are no reports on a single resistance gene that confers resistance to ShB. However, many QTLs have been identified to be associated with ShB resistance. Most of the QTLs were reported to provide a minor contribution to the resistance phenotype, whereas two QTLs (*qShB9-2* and *qShB11-1*) were found to contribute more than 10% to ShB resistance. Sequence analyses revealed the presence of various defense-associated genes in these QTLs. *qShB9-2* was identified in many rice varieties that exhibit resistance to ShB. It was observed that *qShB9-2* contains a β-1,3-glucanase, *OsWAK91*, and 12 other possible candidate genes. On the other hand, the *qShB11-1* interval was shown to have receptor-like kinases, a lipase, and a tandem array of 11 chitinase genes. Tens of minor QTLs were found to be associated with ShB resistance. Nevertheless, no information is available on the gene(s) responsible for the resistance. Studies using resistant cultivars shed light on the possible mechanisms by which rice fights *Rhizoctonia solani*. Various studies showed changes in metabolic pathways, including primary and secondary metabolites. Intermediates of glycolysis and tricarboxylic acid cycle were found to be accumulated in rice post *R.solani* infection, indicating the possible involvement of primary metabolism in response to the pathogen. Also, the accumulation of secondary metabolites such as phytoalexins, chlorogenic acid, polyphenols, and flavonoids was reported to be higher in the tolerant varieties than in the susceptible varieties postchitin treatment (Molla et al. 2020). ROS deregulation has been observed to delay pathogen colonization in resistant cultivars (Oreiro et al. 2019).

5.5   Broad-Spectrum Resistance Genes

From a breeder's point of view, a single locus/gene is more preferred as it would permit easier introgression. Many defense-related genes have been identified to provide broad-spectrum resistance to either multiple races of a pathogen (vertical resistance) or multiple pathogens altogether (horizontal resistance). Such responses are quantitative in nature and hence can be highly durable and practical to keep infectious diseases at bay. Several previous studies have been reported that the expression of defense-response genes (DR genes) such as rice germin-like proteins (*OsGLP*) or a class of DR genes present in a QTL along with *R* genes is also most probably associated with rice resistance, as knockdown of these genes escalated the susceptibility against two major rice fungal diseases, blast and sheath blight (Manosalva et al. 2009). *OsPAL4* is reported to impart broad-spectrum resistance to rice (Tonnessen et al. 2015). A LysM receptor-like kinase (RLS), *OsCERK1*, regulates cytoplasmic *OsRLCK176* and *OsRLCK185* recognizes chitin and peptidoglycans activating immune signaling pathways in rice against blast and bacterial blight diseases. *OsSERK1*, *OsWAK25*, *OsWRKY45-1*, *OsWRKY45-2*, *OsWRKY13*, *OsDR8*,

OsMPK6, *OsPAL4*, *OsNH1*, *OsLYP4*, *OsBSR1*, and *OSK35* have all been shown to regulate resistance to bacterial blight and rice blast positively. *OsPAD4* and *OsPAL4* positively regulate resistance to ShB, whereas *OsWAK25* negatively regulates ShB resistance (Ke et al. 2017). These genes, although identified in different studies, play a highly connected role in helping rice fight the invading pathogens. More comprehensive studies are needed to link the dots to construct a complete map of rice resistance to diseases.

6 Impact of Major Nutrient Fertilizers on Biotic Disease Resistance in Rice

The rapidly increasing world population requires a sustainable nutritional global food supply, which is a significant concern for crop production. Changing climatic scenarios and decreasing natural resources suggest that there is a need to intensify agricultural production using an efficient agronomic nutrient management (ANM) system. Following efficient ANM technologies can enable us to understand and mitigate the adverse impacts of stress, inadequate soil fertility status, pathogens, and pests (Dordas 2008).

Several efficient screening technologies exist, such as smart water irrigation systems, integrated fertilizer applications, and disease biocontrol strategies, that have been developed and adopted in different ecosystems to control various diseases in rice (Bargaz et al. 2018). Among these, the rate of fertilizer used, judicious and timely applications of nutrients, and availability of these nutrients play a crucial role in plant growth and also in developing defense mechanisms against various pests and diseases (Fageria et al. 2008; Sun et al. 2020). The management of nutrient statuses in the soil, especially nitrogen (N), phosphorus (P), and potassium (K), is an eco-friendly strategy to control different biotic stresses instead of frequent application of pesticides. Globally, the efficiency of fertilizer use by the crop and the correct rates of fertilizer applications are poorly studied. Earlier studies have indicated that only 30–35% of N, 10–25% of P, and 35–50% of K are taken up by plants. Particularly in China, the amount of fertilizer used has increased drastically from 270 to 350 kg/ha, which is more than 75% of the global average of fertilizer application. This excessive amount of N fertilizer leads to leaching, which is a significant cause of groundwater pollution and degradation of soil quality (Teng et al. 2016).

Developing sustainable agriculture is one of the major strategies to increase global rice production. Application of nutrient fertilizer at the right rate and stage and also microorganisms are the key factors in disease control. The essential nutrient elements can decrease disease severity but also increase the severity of disease incidence (Dordas 2008). Nitrogen is one of the key elements for plant growth and development, which are involved in the major physiological and metabolic pathways related to N assimilation (Bolton and Thomma 2008; Mur et al. 2017). Plenty of research has been conducted on the role of N and its interaction with disease resistance and the results are inconsistent, with a poor understanding of the resistance mechanisms in physiological and metabolic pathways. These differences may be due to various stress signaling mechanisms caused by the different forms of N (NH_4^+ and NO_3^-), the type of pathogen specificity, and the stage of N application (Dordas 2008). However, several researchers have suggested that the correct time application of fertilizers has been significantly increasing disease resistance and decreasing the use of fungicides (Anderson 2002; Hervieux et al. 2002; Bhat et al. 2013). Recently, Sun et al. (2020) reviewed N applications and their critical role in the defense mechanisms in various diseases such as blast, downy mildew, stem rot, powdery mildew, leaf rust, stem rust, and rice blast diseases in plants. Balancing of these nutrients is imperative to understand the cellular structure and composition, which mainly affect plant defense mechanisms. For instance, high rates of N application lead to a significant impact on susceptibility by decreasing the thickness of cell-wall components (cellulose and lignin), whereas decreasing N applications lead to an increase in lodging resistance by changes in stem lignification and secondary cell-wall synthesis (Zhang et al. 2017b; Sun et al. 2018). Also, decreasing N fertilizer significantly increases the incidence of major insect pests, including brown planthopper, leaffolder, and stem borer, the key insect pests in the major rice-growing areas in Asian countries (Lu et al. 2007). Some reports have suggested that N applications significantly influence the size of leaf blast lesions (Matsuyama 1973; Kaur et al. 1979). Sime et al. (2017) studied the different rates of nutrient fertilizer application and their relation to blast disease. The combination of NPK (20-10-10) at a rate of 200 kg/ha has a remarkable impact on decreasing blast disease in all phases of plant growth. Similarly, Reddy et al. (1979) reported an optimal rate of N application (76 kg/ha) to maximize grain yield and also minimize disease. One of the major diseases is bacterial leaf blight of rice, caused by *Xanthomonas oryzae*, which increased significantly when a higher amount of N fertilizer (>100 kg/ha) was applied, and yield decreased. Begum et al. (2011) reported that a balanced application of nutrient fertilizers, including K, significantly decreased the percentage of BLB. The application of K fertilizer has dramatically decreased the intensity of various infectious diseases such as BLB, sheath blight, and stem rot in rice, and also in other cereal crops (Sharma et al. 2005a). Decreasing BLB severity by applying K topdressing is a viable approach just before disease-occurring stages and this makes it possible to maximize grain yield and have lesser disease development.

Using slag-based silicon (Si) fertilizer in rice fields is an alternative approach to control the major disease brown spot, which is caused by the fungus *Bipolaris oryzae*. This disease causes significant yield losses, mainly in tropical and subtropical areas, where the frequent occurrences of heavy rainfall and high temperature are the main factors in decreasing the Si content in highly weathered soils (Raven 2003). The major role of these Si applications is to mediate resistance mechanisms through the physiological and metabolic pathways that can lead to creating more pronounced cell silicification in rice leaves, and the strong leaf epidermal surface might increase the resistance to fungal penetration (Hayasaka et al. 2008; Sun et al. 2010; Ning et al. 2014). These Si fertilizers provided clear evidence showing the importance of increasing the thickness of the silicon layer in the epidermal cell walls that are supposed to be the main site for conferring resistance to brown spot disease in rice. Interestingly, Wu et al. (2017) experimented with the transcriptional responses in two different nutrient fertilizers, Si and N concentrations, and their relation to BPH infestation. These two elements had a trade-off mechanism in terms of resistance. The interaction of these two elements clearly showed decreases in the expression of Si transporters such as OsLsi1 and OsLsi2 under high rates of N application, whereas, in the N transporters *OsNRT1:1*, *OsGS2*, *OsFd-GOGAT*, *OsNADH-GOGAT2*, and *OsGDH2*, expression increased under a high rate of Si fertilizer. This demonstrated that N and Si had antagonistic interactions in rice (Wu et al. 2017). Similarly, Robichaux (2001) identified a significant decrease in the major disease sheath blight, caused by the fungal pathogen *Rhizoctonia solani*, by adding calcium silicate in greenhouse and field conditions. Rice grain yield is increased by almost 13% from the use of a calcium silicate application rate of 3.3 mg/ha and also a significant decrease in ShB in different soil types. These results have proven that Si fertilizer can diminish fungal disease severity by increasing the Si concentration in rice leaves and boosting grain yield.

7 Genome-Editing Tools for Improving Disease Resistance

Diseases cause a considerable yield loss annually (Heinrichs and Muniappan 2017; Mushtaq et al. 2019). Breeding for disease resistance has been pursued for a long time. The traditional practice is to introgress disease resistance into elite cultivars through breeding techniques. Although a successful method, it has its downside (Zafar et al. 2020). The traditional way is time-, labor-, and resource-consuming (Romero and Gatica-Arias 2019). With the arrival of the genomics era, identifying disease-resistance genes has become highly efficient, and resistance alleles can be identified at a single base resolution. With such a massive potential in hand and constant improvement in various genome-editing (GE) tools such as site-specific mutagenesis (SSM), meganucleases (MNs), zinc-finger nucleases (ZFNs), transcription activator-like effector nucleases (TALENS), and clustered regularly interspaced short palindromic repeat (CRISPR)/CRISPR-associated protein (CRISPR/Cas), this has opened a novel path to achieve the improvement of disease resistance in various crops (Zhang et al. 2017a; Mishra et al. 2018; Zafar et al. 2020). Recently, several researchers have reviewed the various GE tool applications and their limitations in target gene specificity and accuracy (Abdallah et al. 2015; Mishra et al. 2018; Zafar et al. 2020). As compared with SSM, MNs, and ZFNs, the most widely used GE tools such as TALENs and the CRISPR/Cas system have a versatile, fast, and relatively efficient GE method. Over the past several years, these two methods have transformed the field of genome engineering, and they can easily edit and also recognize specific genomic regions (Gaj et al. 2016). These methods have a significant impact on the genomic revolution that has accelerated the discovery of novel sequence variations and breakthroughs in the scientific knowledge to demonstrate the power of these GE tools in establishing resistance to pathogens in various diseases. The rapid progress in the CRISPR-Cas9 system makes it a highly accurate and efficient method that can edit in multiple genes at multiple locations using a single molecular construct (Cong et al. 2013).

7.1 Site-Specific Mutagenesis: The Path So Far

Site-specific mutagenesis is achieved by deploying a class of enzymes called "nucleases" fused with DNA-binding motifs to target specific sequences in the genome. The activity of nuclease results in double-strand breaks at the target site, which are then repaired by the host DNA repair mechanisms via nonhomologous end joining (NHEJ) or homologous recombination (Feng et al. 2013). During this process, small insertions or deletions occur in the genome, thus disrupting the gene sequence (Mishra et al. 2018). SSM is an evolving area of research with newer tools often emerging with improved precision and efficiency.

7.1.1 Meganucleases

Meganucleases (MNs) are endonucleases (enzymes that cut within a strand) that occur naturally and possess sequence-specific DNA-binding and nuclease activities. The application of MNs for site-targeted mutagenesis began in the 1980s. Owing to the recognition of long DNA sequences (18–40 bp), MNs were a good choice. On the other hand, the number of naturally occurring MNs was limited, thus diminishing their wider application. Moreover, custom modification of MNs is a viable but expensive option (Abdallah et al. 2015).

7.1.2 Zinc-Finger Nucleases and TALENs

Zinc-finger nucleases (ZFNs) kick-started the wider application of site-specific mutagenesis in 1996 (Kim et al. 1996). The zinc finger motif is one of the most copious DNA-binding motifs present in eukaryotes (Klug and Schwabe 1995). Each ZF motif recognizes a specific 3-bp sequence in the major groove of DNA. Thus, tandemly placing multiple ZF motifs of different base specificity and fusing them to a nuclease can result in the generation of a molecular scissor that can precisely cut the target site. Modular assembly-based methods enabled the construction of ZFNs that can virtually target any DNA sequence (Gaj et al. 2013). However, limitations of using ZFNs exist. The modular assembly is a complex and expensive process that requires many optimizations. Off-target cleavage is another challenge that many SSM techniques face (Ramirez et al. 2008; Gupta and Musunuru 2014).

Transcription activator-like effectors (TALEs) are proteins that naturally occur in the genus *Xanthomonas*, which predominantly consists of phytopathogenic bacteria (Boch et al. 2009; Moscou and Bogdanove 2009). TALEs are employed by

Xanthomonas oryzae pv. *Oryzae* (*Xoo*) to target and activate the expression of specific host genes to increase the susceptibility of the host. The specific binding to DNA is achieved by a 33–35-amino-acid-long tandem repeat domain, each of which targets a specific base. The base specificity is conferred by the amino acids that are located in the 12th and 13th positions of the series. These positions are called repeat variable di-residues (RVDs) (Boch et al. 2009; Moscou and Bogdanove 2009; Gaj et al. 2013; Abdallah et al. 2015). Exploiting this brought in a revolution in the field of genome editing called TALENs. TALENs are TALE nucleases wherein the DNA-binding motif of a TALE is fused with a catalytic domain of a nuclease, thus allowing the domain to target and cleave a specific sequence in the genome. By modifying the RVDs, one could define the target site and thus assemble a custom TALEN to target any region of interest in the genome (Christian et al. 2010; Boch 2011). The design and delivery of TALENs, however, pose a setback for the technique owing to their large size (Abdallah et al. 2015).

7.1.3   CRISPR-Based Genome Editing

Clustered regularly interspaced short palindromic repeats (CRISPR) and CRISPR-associated (Cas), collectively called CRISPR-Cas, is a microbial adaptive immune strategy that works based on an RNA-guided nuclease complex to cleave foreign genetic elements. A CRISPR-Cas locus is a cluster of Cas genes, noncoding RNA, and an array of repetitive elements. The repeated elements are interspaced with protospacers (short variable repeats that are derived from foreign DNA targets). Together, the noncoding RNA and protospacers constitute the CRISPR RNA (crRNA). Each protospacer is associated with a protospacer adjacent motif (PAM) that differs between the types of CRISPR systems. Depending on the organization and composition of the nuclease genes, the CRISPR-Cas system is classified into Class I and Class II. Each class has three types of CRISPR-Cas system each. The Class I CRISPR system is less used owing to its limited knowledge and associated complexities. The Class II system, on the other hand, is a well-characterized and highly used genome-editing system. Class II is further subclassified into three types (types II, V, and VI) based on the specificities for nucleotide substrates, PAM, and the Cas genes that affect the substrate cleavage (Koonin and Makarova 2019; Moon et al. 2019). A brief overview is discussed below.

CRISPR-Cas9 from *Streptomyces pyogenes* is the founding system for CRISPR-based genome editing, which is economical, easier, and more efficient. Cas9 is an RNA-guided nuclease that causes double-strand breaks in the genomic region that is complementary to the crRNA provided that the 3′ of the DNA sequence is 5′-NGG-3′ (G-rich PAM). In addition, the CRISPR-Cas9 system needs a *trans*-acting crRNA (tracrRNA) to be functional (Deltcheva et al. 2011; Jinek et al. 2012). Cas9 proteins of different bacterial origins have different PAM specificities, spacer lengths, and sizes. Improvements in the techniques made it possible to multiplex genome editing with the use of polycistronic tRNA-gRNA (PTG), wherein the tRNA processing system is used to construct a tandem array of tRNA and gRNA (attached to the spacer), which would be transcribed as a full primary transcript that is later processed and cleaved to release individual single-guide RNAs (sgRNAs), each of which targets a unique region in the genome (Zafar et al. 2020).

Cpf1, also known as Cas12a, belongs to the type V CRISPR-Cas system. Unlike Cas9, Cas12a does not need tracrRNA for the complex to be functional and it recognizes a 5′ T-rich PAM sequence (5′-TTTN-3′ or 5′-TTN-3′) (Zetsche et al. 2015). Also, Cpf1 exhibits RNase activity that can cleave pre-crRNAs to mature crRNA, thus enabling the possibility of including multiple crRNAs in a single cassette. Cpf1 allows the use of longer gRNAs of up to 100 nt (Zetsche et al. 2015; Mishra et al. 2018). This system is gaining usage because of its higher specificity and enhanced efficiency. Advances are made in terms of increasing the range of targets by engineering the complex to target other PAMs.

Base editing can be used to modify single bases in the genome, thus opening an avenue to increase the allelic diversity of the genes and also to create specific mutations to alter a gene function. The use of the CRISPR-Cas system achieves this in conjunction with base-modifying enzymes such as cytidine deaminase (to induce C:G to A:T mutations) or adenine deaminase (to induce A:T to G:C mutations). Base-editing techniques eliminate the need for double-strand breaks and thus the activation of DNA repair pathways (Lu and Zhu 2017; Hao et al. 2019).

Gene knock-in thus far required double-strand breaks and activation of homology-directed repair (HDR), which uses a donor template (carrying the gene copy to be knocked-in) to incorporate the new copy in the genome. This technique is extremely limited because of the less frequent and cell cycle stage-dependent nature of HDR. Also, the effective delivery of the donor template has posed a serious challenge. To overcome this, an elegant method was devised, called prime editing. Prime editing depends on a two-component system that includes (1) Cas9 nickase fused with a reverse transcriptase and (2) a prime editing guide RNA (pegRNA) that carries the desired edit(s) to be incorporated into the genome. Once delivered, the complex is guided to the target site by pegRNA and a nick is created in the genome. The nicked DNA serves as a primer that the reverse transcriptase uses to reverse transcribe the pegRNA, thus incorporating the edit into the genome (Anzalone et al. 2019; Lin et al. 2020). All the above-discussed techniques were successfully applied in rice and other plant species to edit various genes. The following few paragraphs will provide a glimpse of the application of genome editing in rice, focusing mainly on disease resistance.

7.2 Application of Genome Editing in Biotic Stress Tolerance in Rice

Conventional breeding has been successfully employed to date to develop disease-resistant rice varieties by introgressing resistance genes from wild rice varieties or landraces into elite cultivars. Although successful, it is a time-consuming

procedure, and also the wild germplasm does not contain genes/loci for all the economically important traits that are of concern for breeders. With the advent of
genome editing, several research groups started testing the possibility of using
genome editing in rice and were by and large successful.

7.2.1 Resistance to Biotic Stress Factors

Genome editing has been successfully employed to generate rice plants resistant to
various biotic stress factors, including bacterial blight, bacterial leaf streak, blast,
and tungro virus. The application of genome-editing techniques for BB resistance
began with modifying the promoter of a BB susceptibility gene, *OsSWEET14*
(*Os11N3*), using TALENs (Li et al. 2012). *SWEET* genes are sugar transporters, and
the expression of certain *SWEET* genes is induced upon *Xoo* infection by the action
of TALEs. So far, the *SWEET* genes, including *OsSWEET11*, *OsSWEET12*,
OsSWEET13, *OsSWEET14*, and *OsSWEET15*, have been shown to be induced by
Xoo and could act as susceptibility factors (Streubel et al. 2013). TAL effectors bind
to the effector-binding elements (EBEs) in the target promoter and activate the
expression of the downstream gene, which tends to be a susceptibility factor in
many cases. Different TAL effectors induce many such susceptibility genes, and
their cognate EBEs were also deciphered. Li et al. (2012) have successfully
employed TALENs to modify the EBEs of *OsSWEET14*. This study established the
possibility of using genome-editing techniques to generate disease-resistant varieties as well as to understand the targets of different TAL effectors. Jiang et al. (2020)
have conducted a proof-of-concept study to confirm the applicability of the CRISPR-
Cas9 system in rice by editing the promoters of *OsSWEET11* and *OsSWEET14*
genes (Jiang et al. 2013). In a study using CRISPR-Cas9, Zhou et al. (2015) created
a null mutant of *OsSWEET13* to show that *PthXo2* (an *Xoo* TAL effector)-depen-
dent disease occurrence needs intact *OsSWEET13*. Xu et al. (2019) used CRISPR-
Cas9 to edit *OsSWEET11* and *OsSWEET14* to engineer broad-spectrum resistance
to BB in rice variety Kitaake. In addition to that, they identified new EBEs in the
promoter of *OsSWEET13* and successfully used CRISPR-Cas9 to disrupt the EBE,
thus generating a rice line that was resistant to all the tested *Xoo* isolates ($n = 133$)
(Xu et al. 2019). Oliva et al. (2019) generated five mutations in the promoters of
OsSWEET11, *OsSWEET13*, and *OsSWEET14* in three rice lines, Kitaake, IR64, and
Ciherang-Sub1. All the lines were reported to show robust and broad-spectrum
resistance in the paddy trials (Oliva et al. 2019). Zhou et al. (2018) used CRISPR-
Cas9 to create a knockout of a susceptibility gene called *BsrK-1* (broad-spectrum
resistance Kitaake-1), which resulted in resistance to BB as well as blast. *BsrK-1* is
a tetratricopeptide domain-containing protein that was shown to bind to the mRNAs
of multiple *OsPAL* (phenylalanine ammonia-lyase) genes and promote their turnover. In *BsrK-1* knockouts, the accumulation of *OsPAL* mRNA was observed along
with increased resistance to diseases (Zhou et al. 2018). The feasibility of the

transgene-free method of genome editing was tested by mutating *Os8N3/ OsSWEET11* (Kim et al. 2019). Cai et al. (2017) demonstrated that a TAL effector (*Tal7*) from *X. oryzae* pv. *oryzicola* RS105 targets the promoter of rice *Cyclin-D4-1* and induces its expression. They have successfully applied TALEN-based genome editing to disrupt the EBE in the promoter of *Cyclin-D4-1*, which leads to resistance to RS105 infection.

The applicability of CRISPR-Cas9 for generating blast-resistant rice lines has been demonstrated by performing both single-site and multisite-targeted mutagenesis of *OsERF922*, a negative regulator of blast resistance, to produce knockouts. All the mutants showed blast resistance while not having any adverse effect on other agronomic traits (Wang et al. 2016). Rice tungro disease (RTD) is a disease caused by rice tungrospherical virus (RTSV) and rice tungrobacilliform virus (RTBV) and is transmitted by green leafhoppers. RTD results in yellowing of leaves, decreased tiller numbers, and stunted growth (Azzam and Chancellor 2002; Lee et al. 2010). Macovei et al. (2018) generated RTSV-resistant lines in the background of IR64 using the CRISPR-Cas9 system. In this study, the *eIF4G* gene was successfully mutated independently, using three different gRNAs, and the mutant plants showed heritable resistance to RTSV (Macovei et al. 2018).

7.2.2 Summary of Nonbiotic Stress-Related Phenotypes

The application of genome editing in rice is rising with time. Other than for biotic stress tolerance, genome editing has been successfully applied to edit several genes having various roles, including nutritional value, yield, and abiotic stress tolerance (Shan et al. 2015; Li et al. 2016; Sun et al. 2016; Shen et al. 2017; Tang et al. 2017; Abe et al. 2018; Endo et al. 2019; Romero and Gatica-Arias 2019).

7.3 Improvements in the Techniques

Currently, transgene-free methods are being tested and employed for genome editing wherein the mutant plants do not contain any of the CRISPR-Cas9 components. This is achieved in several ways, including using Cas9-gRNA ribonucleoproteins (RNPs). This RNP complex is directly delivered into plant cells by transfection or particle bombardment. The RNP complex can perform the editing and will be degraded by the cellular types of machinery. Another approach is to transiently express CRISPR-Cas9 from DNA or RNA in plants from regenerated calli. Both methods suffer from the possibility of component degradation, which might lead to less-efficient editing. To eliminate this disadvantage, He et al. (2018) came up with the suicide gene-based method of a transgene-free CRISPR-Cas9 approach in rice. In this method, a pair of suicidal genes, encoding toxic proteins that kill plant cells, are incorporated into the CRISPR-Cas9 cassette. Therefore, no plant with a CRISPR-Cas9 construct will survive, thus eliminating the plants containing the transgenes. Among other surviving plants, the true mutants can be screened and identified using appropriate techniques. In addition to protein-coding genes, miR-NAs are being targeted for editing owing to their involvement in various growth, development, and stress-response pathways. The use of the CRISPR-Cas system to edit miRNAs has been functionally validated in rice (Zhou et al. 2017; Mangrauthia et al. 2017).

8   Bioinformatics Tools for Disease Resistance and Management

Bioinformatics is an interdisciplinary field that uses computational tools to capture and interpret the function of various genes. The advent of genomics has revolutionized every aspect of life science. The availability of a large amount of data has necessitated better ways to analyze, interpret, and organize the results for the scientific community (Bayat 2002; Vassilev et al. 2005). Thousands of databases and repositories are available for various datasets such as for the genome, gene and protein sequences, expression and coexpression of genes, and genomic variations such as SNPs and InDels, to name a few (Garg and Jaiswal 2016). With time, sequencing platforms have seen an astounding revolution and are becoming more efficient and affordable day by day. Since the first report on the whole-genome sequence of rice in 2005, many varieties were further sequenced as a part of the 3000 Rice Genomes Project (Matsumoto et al. 2005; Li et al. 2014). The data obtained from the project resulted in establishing rice variation databases and these data have provided invaluable insights into rice evolution and domestication (Chen et al. 2019). Moreover, the readily available data can guide breeders to wisely choose varieties and markers for breeding various traits from one cultivar to another. Procedures to score the expression of genes have also undergone an overwhelming transformation from methods such as serial analysis of gene expression (SAGE) to microarrays to RNA-sequencing (Perez-de-Castro et al. 2012). As a result, other than genome databases, gene and protein expression databases play an important role in elucidating the various mechanisms that control a given trait, such as days to flowering, growth and development, abiotic stress tolerance, and disease resistance, among others. Multiple other tools and databases are available to study and acquire information on different aspects, including phylogenomics, protein-protein interaction, promoter analysis, gene and QTL information, marker-trait association, and metabolite profiles (Garg and Jaiswal 2016). This section aims to provide an overview of the application of bioinformatics in breeding for disease resistance in rice.

8.1   The Role of Bioinformatics in Mapping Genomic Loci

8.1.1   Mapping QTLs and Genes Associated with Disease Resistance

Mapping the loci responsible for a desired trait has been successfully carried out for years using a conventional method such as simple sequence repeat (SSR)-based genotyping of a mapping population. With the arrival of affordable sequencing techniques, QTLs and genes can now be mapped at a gene-level resolution. Methods such as MutMap and QTLseq have made it possible not only to identify the genomic locus responsible for a trait but also to pinpoint the causal variation within a gene that led to the phenotype (Abe et al. 2012; Takagi et al. 2013b). Thus, SNP markers that are highly associated with a trait can be identified and employed in breeding programmed for efficient introgression of the trait. In a proof-of-concept study, Takagi et al. (2013a) had identified a QTL conferring partial resistance to blast disease of rice. Following this attempt, multiple studies have successfully used this procedure to map QTLs for various traits in rice and other species. A rice blast resistance gene called *Pii* was mapped by another method called MutMap-Gap (Takagi et al. 2013b).

Sequencing data have been successfully used to compare the genomes of different cultivars and obtain the resistance alleles of cloned rice blast resistance genes (Mahesh et al. 2016). Genes with highly repetitive sequences pose a challenge in accurately characterizing them in context with short-read sequences. A recent study by Read et al. (2020) addressed this challenge by using a long-read sequencing approach called nanopore sequencing in combination with Illumina sequencing to assemble the genome of American rice variety Carolina Gold Select and identify 529 complete or partial NB-LRR domain-containing protein genes that are highly repetitive in nature. The study identified a major disease resistance locus called *Xo1* that confers resistance to *Xanthomonas oryzae* pv. *oryzae* (the causal agent of bacterial blight of rice) and *X. oryzae* pv. *oryzicola* (the causal agent of bacterial leaf streak of rice). Also, a blast resistance gene called *Pi63* at the *Xo1* locus was identified (Read et al. 2020). Another study compared the genomes of 13 domesticated and wild rice relatives and shed light on the complex phylogeny of the *Oryza* genus and identified many haplotypes of disease-resistance genes that can be of potential use for breeding (Stein et al. 2018). Using the genome sequence of rice variety Tetep, an extensive set of molecular markers was designed for breeding novel resistant varieties (Wang et al. 2019).

8.1.2   Genome-Wide Association Studies

Genome-wide association studies (GWAS) exploit the natural variation among different cultivars to identify trait-associated genes (Hu et al. 2018). This is one of the preferred methods for the identification of gene targets for breeding. The availability of genome sequences and phenotype data, along with powerful statistical and bioinformatics tools, have made it possible to analyze hundreds or thousands of genomes in one go and identify genes and haplotypes that are associated with given traits. A GWAS on 373 *indica* rice sequences identified SNPs associated with 14 different agronomic characteristics (Huang et al. 2010). A GWAS with a panel of 584 rice accessions led to the identification of a gene called *PiPR1* that confers partial resistance to blast disease of rice (Liu et al. 2020). Other GWAS have identified tens of QTLs and new alleles of known blast-resistance genes (Li et al. 2019a).

8.1.3 Speeding Up Breeding

The main setback with conventional breeding methods is the time taken for developing a new variety. Also, for durable disease resistance and other complex traits, it is essential to efficiently identify minor-effect QTLs and use the associated markers in breeding programs. The existing methods, including QTL mapping from biparental crossing and GWAS, are not up to the mark to efficiently identify such minor-effect QTLs (Bhat et al. 2016). To address both of these concerns, genomic selection (GS) was proposed (Meuwissen et al. 2001). Unlike MAS, GS does not necessarily need QTL information before selection. GS uses reference population data containing phenotype and high-density marker data to predict breeding values for all the markers. Based on the predicted values, the breeding population data will be analyzed to select the individual that possesses the desirable phenotype (Perez-de-Castro et al. 2012; Hu et al. 2018). In this way, it is possible to introgress even minor-effect QTLs efficiently, as there are no biased marker effects, unlike with MAS. Studies on other plant species have shown the higher prediction accuracy of GS in genetic gains and a significant decrease in the time taken for breeding (Hu et al. 2018). Although proposed two decades ago, the implementation of GS in crop breeding has just begun, mainly because of the advent of high-throughput and affordable genotyping methods that produce dense marker information such as genotyping-by-sequencing and automated phenotyping (Bhat et al. 2016). Given the importance of disease-resistance breeding in rice, the application of GS could be of tremendous benefit.

8.1.4 Using Machine Learning and Artificial Intelligence

The field of computational biology is advancing at an unprecedented rate with the arrival of machine learning (ML) and artificial intelligence (AI). In ML, the machine gains experience by identifying patterns in given datasets and using that experience to interpret the data in question. ML has applications in various aspects of plant sciences, including phenotyping and increasing the accuracy of sequence analysis pipelines, such as differentiating true SNPs from spurious SNPs (Hu et al. 2018). ML was successfully used to phenotype and categorize foliar stresses in soybean with high accuracy (Ghosal et al. 2018). Various parameters such as yield, developmental stage, weed status, crop quality, water and soil management, and disease occurrence were successfully predicted using ML (Liakos et al. 2018). In rice, bakanae disease (caused by *Fusarium fujikuroi*) was predicted with an accuracy of 87.9% using support vector machine classifiers, a popular ML tool that is often used to overcome classification and regression problems (Chung et al. 2016). Although few in number, these studies have put forth the applicability of advanced computational strategies to improve agriculture.

9 Future Prospects and Conclusions

The rice crop plays an essential role in ensuring global food security and providing nutritional security for the rapidly growing world population. Increasing grain yield is a significant target for plant breeders apart from identifying resistance to/tolerance of biotic and abiotic stresses. Enhancing genetic gain is also a primary concern to meet the food demand of the ever-increasing world population, especially with global climate change. In recent years, the innovations in rice breeding programs and advanced genomics technologies such as next-generation sequencing and high-throughput genotyping have been fully exploited to understand trait interactions and select promising rice genotypes for use in breeding programs. The genetic improvements in yield component traits and increasing yield significantly under biotic and abiotic stresses have not been achieved to a great extent due to the complex nature of these stresses. The knowledge of integrated genomics and high-throughput phenomics technologies has laid the foundation to understand these complex traits and also associated molecular genetics and physiological mechanisms that can enable breeders to find better rice genotypes and to move forward as knowledge-based rice breeding is the most acceptable approach in developing climate-smart stress-tolerant and high-yielding rice genotypes. This approach has advanced at a fast pace with low-cost, efficiency, and high resolution of genetic mapping for QTLs and genes and also haplotype blocks to find allelic variations for the target trait of interest. The current advances in CRISPR/Cas9 genome-editing tools have led to significant targeted changes in specific trait-associated genes and changes in single base levels that promise to accelerate crop improvement. These genomic-assisted breeding tools are breeder-friendly, and smart decisions in breeding programs can enhance the efficiency of the selection of rice genotypes in a short period.

Acknowledgments The authors would like to thank and acknowledge the Bill & Melinda Gates Foundation for providing a research grant to ZL for the Green Super Rice project under ID OPP1130530. We would like to thank the Department of Agriculture, Philippines, for providing funds to JA under the Next-Gen project and also thank and acknowledge IRRI Communications for English-language editing and the anonymous internal reviewer's valuable suggestions and constructive comments that helped improve this chapter.

References

Abdallah NA, Prakash CS, McHughen AG (2015) Genome editing for crop improvement: challenges and opportunities. GM Crops Food 6:183–205

Abe A, Kosugi S, Yoshida K et al (2012) Genome sequencing reveals agronomically important loci in rice using MutMap. Nat Biotechnol 30:174–178. https://doi.org/10.1038/nbt.2095

Abe K, Araki E, Suzuki Y et al (2018) Production of high oleic/low linoleic rice by genome editing. Plant Physiol Biochem 131:58–62. https://doi.org/10.1016/j.plaphy.2018.04.033

Adorada DL, Stodart BJ, Cruz CV et al (2013) Standardizing resistance screening to *Pseudomonas fuscovaginae* and evaluation of rice germplasm at seedling and adult plant growth stages. Euphytica 192:1–16. https://doi.org/10.1007/s10681-012-0804-z

Ahn S, Kim Y, Han S et al (1996) Molecular mapping of a gene for resistance to a Korean isolate of rice blast. Rice Genet Newsl 13:74–76

Ahn SN, Kim YK, Hong HC et al (1997) Mapping of genes conferring resistance to Korean isolates of rice blast fungus using DNA markers. Korean J Breed 29:416–423

Ali J, Xu JL, Gao YM et al (2017) Harnessing the hidden genetic diversity for improving multiple abiotic stress tolerance in rice (*Oryza sativa* L.). PLoS One 12:e0172515

Andargie M, Li L, Feng A et al (2018) Mapping of the quantitative trait locus (QTL) conferring resistance to rice false smut disease. Curr Plant Biol 15:38–43. https://doi.org/10.1016/j.cpb.2018.11.003

Anderson S (2002) The relationship between nutrients and other elements to plant diseases. Spectrum Analytic Inc, Washington CH, OH, pp 26–32

Anderson PK, Cunningham AA, Patel NG et al (2004) Emerging infectious diseases of plants: pathogen pollution, climate change and agrotechnology drivers. Trends Ecol Evol 19:535–544

Anjaneyulu A, Singh SK, Shenoi MM (1982) Evaluation of rice varieties for tungro resistance by field screening techniques. Trop Pest Manag 28:147–155. https://doi.org/10.1080/09670878209370692

Anzalone AV, Randolph PB, Davis JR et al (2019) Search-and-replace genome editing without double-strand breaks or donor DNA. Nature 576:149–157. https://doi.org/10.1038/s41586-019-1711-4

Ashikawa I, Hayashi N, Yamane H et al (2008) Two adjacent nucleotide-binding site-leucine-rich repeat class genes are required to confer *Pikm*-specific rice blast resistance. Genetics 180:2267–2276. https://doi.org/10.1534/genetics.108.095034

Ashkani S, Rafii MY, Shabanimofrad M et al (2015) Molecular breeding strategy and challenges towards improvement of blast disease resistance in rice crop. Front Plant Sci 6:886

Azzam O, Chancellor TCB (2002) The biology, epidemiology, and management of rice tungro disease in Asia. Plant Dis 86:88–100

Bainsla NK, Meena HP (2016) Breeding for resistance to biotic stresses in plants. In: Yadav P, Kumar S, Jain V (eds) Recent advances in plant stress physiology. Daya Publishing House, New Delhi

Balachiranjeevi C, Bhaskar NS, Abhilash V et al (2015) Marker-assisted introgression of bacterial blight and blast resistance into DRR17B, an elite, fine-grain type maintainer line of rice. Mol Breed 35:151. https://doi.org/10.1007/s11032-015-0348-8

Balachiranjeevi CH, Bhaskar Naik S, Abhilash Kumar V et al (2018) Marker-assisted pyramiding of two major, broad-spectrum bacterial blight-resistance genes, *Xa21* and *Xa33* into an elite maintainer line of rice, DRR17B. PLoS One 13:e0201271. https://doi.org/10.1371/journal.pone.0201271

Balachiranjeevi CH, Prahalada GD, Mahender A et al (2019) Identification of a novel locus, BPH38 (t), conferring resistance to brown planthopper (*Nilaparvata lugens* Stal.) using early backcross population in rice (*Oryza sativa* L.). Euphytica 215:185

Bargaz A, Lyamlouli K, Chtouki M et al (2018) Soil microbial resources for improving fertilizers efficiency in an integrated plant nutrient management system. Front Microbiol 9:1606

Barman SR, Gowda M, Venu RC, Chattoo BB (2004) Identification of a major blast resistance gene in the rice cultivar "Tetep". Plant Breed 123:300–302. https://doi.org/10.1111/j.1439-0523.2004.00982.x

Bayat A (2002) Bioinformatics. Br Med J 324:1018–1022

Begum N, Rahman MM, Bashar MA et al (2011) Effect of potassium fertilizer on development of bacterial blight of rice. Bangladesh J Sci Ind Res 46:69–76

Berruyer R, Adreit H, Milazzo J et al (2003) Identification and fine-mapping of *Pi33*, the rice resistance gene corresponding to the *Magnaporthe grisea* avirulence gene *ACE1*. Theor Appl Genet 107:1139–1147. https://doi.org/10.1007/s00122-003-1349-2

Bhasin H, Bhatia D, Raghuvanshi S et al (2012) New PCR-based sequence-tagged site marker for bacterial blight-resistance gene *Xa38* of rice. Mol Breed 30:607–611. https://doi.org/10.1007/s11032-011-9646-y

Bhat ZA, Ahangar MA, Sanghera GS, Mubarak T (2013) Effect of cultivar, fungicide spray and nitrogen fertilization on management of rice blast under temperate ecosystem. Int J Sci Environ Technol 2:410–415

Bhat JA, Ali S, Salgotra RK et al (2016) Genomic selection in the era of next-generation sequencing for complex traits in plant breeding. Front Genet 7:221

Bigirimana VP, Hua GKH, Nyamangyoku OI, Höfte M (2015) Rice sheath rot: an emerging ubiquitous destructive disease complex. Front Plant Sci 6:1066

Boch J (2011) TALEs of genome targeting. Nat Biotechnol 29:135–136

Boch J, Scholze H, Schornack S et al (2009) Breaking the code of DNA binding specificity of TAL-type III effectors. Science 326:1509–1512. https://doi.org/10.1126/science.1178811

Bolton MD, Thomma BPHJ (2008) The complexity of nitrogen metabolism and nitrogen-regulated gene expression in plant pathogenic fungi. Physiol Mol Plant Pathol 72:104–110

Bonman JM (1992) Durable resistance to rice blast disease-environmental influences. Euphytica 63:115–123. https://doi.org/10.1007/BF00023917

Bossa-Castro AM, Tekete C, Raghavan C et al (2018) Allelic variation for broad-spectrum resistance and susceptibility to bacterial pathogens identified in a rice MAGIC population. Plant Biotechnol J 16:1559–1568. https://doi.org/10.1111/pbi.12895

Bunawan H, Dusik L, Bunawan SN, Amin NM (2014) Rice tungro disease: from identification to disease control. World Appl Sci J 31:1221–1226. https://doi.org/10.5829/idosi.wasj.2014.31.06.610

Busungu C, Taura S, Sakagami JI, Ichitani K (2016) Identification and linkage analysis of a new rice bacterial blight-resistance gene from XM14, a mutant line from IR24. Breed Sci 66:636–645. https://doi.org/10.1270/jsbbs.16062

Cai L, Cao Y, Xu Z et al (2017) A transcription activator-like effector Tal7 of *Xanthomonas oryzae* pv. *oryzicola* activates rice gene *Os09g29100* to suppress rice immunity. Sci Rep 7:1–13. https://doi.org/10.1038/s41598-017-04800-8

Causse MA, Fulton TM, Cho YG et al (1994) Saturated molecular map of the rice genome based on an interspecific backcross population. Genetics 138:1251–1274

Cesari S, Thilliez G, Ribot C et al (2013) The rice resistance protein pair RGA4/RGA5 recognizes the *Magnaporthe oryzae* effectors AVR-Pia and AVR1-CO39 by direct binding. Plant Cell 25:1463–1481. https://doi.org/10.1105/tpc.112.107201

Channamallikarjuna V, Sonah H, Prasad M et al (2010) Identification of major quantitative trait loci *qSBR11-1* for sheath blight resistance in rice. Mol Breed 25:155–166. https://doi.org/10.1007/s11032-009-9316-5

Chaudhari AK, Rakholiya KB, Baria TT (2019) Screening of rice cultivars against false smut [*Ustilaginoidea virens* (Cooke) Takahashi] of rice. Int J Curr Microbiol App Sci 8:2786–2793. https://doi.org/10.20546/ijcmas.2019.806.336

Cheema KK, Grewal NK, Vikal Y et al (2008) A novel bacterial blight-resistance gene from *Oryza nivara* mapped to 38 kb region on chromosome 4L and transferred to *Oryza sativa* L. Genet Res (Camb) 90:397–407. https://doi.org/10.1017/S0016672308009786

Chen DH, Dela Viña M, Inukai T et al (1999) Molecular mapping of the blast resistance gene, *Pi44(t)*, in a line derived from a durably resistant rice cultivar. Theor Appl Genet 98:1046–1053. https://doi.org/10.1007/s001220051166

Chen XW, Li SG, Xu JC et al (2004) Identification of two blast resistance genes in a rice variety, Digu. J Phytopathol 152:77–85. https://doi.org/10.1046/j.1439-0434.2003.00803.x

Chen C, Zheng W, Huang X et al (2006a) Major QTL conferring resistance to rice bacterial leaf streak. Agric Sci China 5:216–220. https://doi.org/10.1016/S1671-2927(06)60041-2

Chen X, Shang J, Chen D et al (2006b) A B-lectin receptor kinase gene conferring rice blast resistance. Plant J 46:794–804. https://doi.org/10.1111/j.1365-313X.2006.02739.x

Chen S, Huang Z, Zeng L et al (2008) High-resolution mapping and gene prediction of *Xanthomonas oryzae* pv. *oryzae* resistance gene *Xa7*. Mol Breed 22:433–441. https://doi.org/10.1007/s11032-008-9187-1

Chen J, Shi Y, Liu W et al (2011a) A *Pid3* allele from rice cultivar Gumei2 confers resistance to *Magnaporthe oryzae*. J Genet Genom 38:209–216. https://doi.org/10.1016/j.jgg.2011.03.010

Chen S, Liu X, Zeng L et al (2011b) Genetic analysis and molecular mapping of a novel recessive gene *xa34*(t) for resistance against *Xanthomonas oryzae* pv. *oryzae*. Theor Appl Genet 122:1331–1338. https://doi.org/10.1007/s00122-011-1534-7

Chen J, Peng P, Tian J et al (2015) *Pike*, a rice blast resistance allele consisting of two adjacent NBS–LRR genes, was identified as a novel allele at the *Pik* locus. Mol Breed 35:1–15. https://doi.org/10.1007/s11032-015-0305-6

Chen E, Huang X, Tian Z et al (2019) The genomics of *Oryza* species provides insights into rice domestication and heterosis. Annu Rev Plant Biol 70:639–665. https://doi.org/10.1146/annurev-arplant-050718-100320

Chen S, Wang C, Yang J et al (2020) Identification of the novel bacterial blight-resistance gene *Xa46*(t) by mapping and expression analysis of the rice mutant H120. Sci Rep 10:12642. https://doi.org/10.1038/s41598-020-69639-y

Chisholm ST, Coaker G, Day B et al (2006) Host-microbe interactions: shaping the evolution of the plant immune response. Cell 124:803–814

Christian M, Cermak T, Doyle EL et al (2010) Targeting DNA double-strand breaks with TAL effector nucleases. Genetics 186:756–761. https://doi.org/10.1534/genetics.110.120717

Chu Z, Fu B, Yang H et al (2006) Targeting *xa13*, a recessive gene for bacterial blight-resistance in rice. Theor Appl Genet 112:455–461. https://doi.org/10.1007/s00122-005-0145-6

Chung CL, Huang KJ, Chen SY et al (2016) Detecting Bakanae disease in rice seedlings by machine vision. Comput Electron Agric 121:404–411. https://doi.org/10.1016/j.compag.2016.01.008

Collard BC, Ismail AM, Hardy B (eds) (2013) EIRLSBN: twenty years of achievements in rice breeding. International Rice Research Institute, Los Baños. 145 p

Cong L, Ran FA, Cox D et al (2013) Multiplex genome engineering using CRISPR/Cas systems. Science 339:819–823

Deltcheva E, Chylinski K, Sharma CM et al (2011) CRISPR RNA maturation by trans-encoded small RNA and host factor RNase III. Nature 471:602–607. https://doi.org/10.1038/nature09886

Deng Y, Zhai K, Xie Z et al (2017) Epigenetic regulation of antagonistic receptors confers rice blast resistance with yield balance. Science 355:962–965. https://doi.org/10.1126/science.aai8898

Devi SJSR, Singh K, Umakanth B et al (2020) Identification and characterization of a large effect QTL from *Oryza glumaepatula* revealed *Pi68*(t) as putative candidate gene for rice blast resistance. Rice 13:17. https://doi.org/10.1186/s12284-020-00378-4

Dordas C (2008) Role of nutrients in controlling plant diseases in sustainable agriculture. A review. Agron Sustain Dev 28:33–46

Eizenga GC, Lee FN, Rutger JN (2002) Screening *Oryza* species plants for rice sheath blight resistance. Plant Dis 86:808–812

Eizenga GC, Prasad B, Jackson AK, Jia MH (2013) Identification of rice sheath blight and blast quantitative trait loci in two different *O. sativa/O. nivara* advanced backcross populations. Mol Breed 31:889–907. https://doi.org/10.1007/s11032-013-9843-y

Endo M, Mikami M, Endo A et al (2019) Genome editing in plants by engineered CRISPR–Cas9 recognizing NG PAM. Nat Plants 5:14–17. https://doi.org/10.1038/s41477-018-0321-8

Fageria NK, Baligar VC, Li YC (2008) The role of nutrient efficient plants in improving crop yields in the twenty first century. J Plant Nutr 31:1121–1157

FAOSTAT (2020) Statistics data. Food and Agriculture Organization of the United Nations, Rome. http://www.fao.org/faostat/en/#dat

Feng Z, Zhang B, Ding W et al (2013) Efficient genome editing in plants using a CRISPR/Cas system. Cell Res 23:1229–1232

Feng B, Chen K, Cui Y et al (2018) Genetic dissection and simultaneous improvement of drought and low nitrogen tolerances by designed QTL pyramiding in rice. Front Plant Sci 9:306. https://doi.org/10.3389/fpls.2018.00306

Fjellstrom R, Conaway-Bormans CA, McClung AM et al (2004) Development of DNA markers suitable for marker-assisted selection of three *Pi* genes conferring resistance to multiple *Pyricularia grisea* pathotypes. Crop Sci 44:1790–1798. https://doi.org/10.2135/cropsci2004.1790

Fu D, Chen L, Yu G et al (2011) QTL mapping of sheath blight resistance in a deep-water rice cultivar. Euphytica 180:209–218. https://doi.org/10.1007/s10681-011-0366-5

Fujii K, Hayano-Saito Y, Shumiya A et al (1995) Genetical mapping based on the RFLP analysis for the panicle blast resistance derived from a rice parental line St. No. 1. Breed Sci 45:209

Fukuoka S, Saka N, Koga H et al (2009) Loss of function of a proline-containing protein confers durable disease resistance in rice. Science 325:998–1001. https://doi.org/10.1126/science.1175550

Gaj T, Gersbach CA, Barbas CF (2013) ZFN, TALEN, and CRISPR/Cas-based methods for genome engineering. Trends Biotechnol 31:397–405

Gaj T, Sirk SJ, Shui S, Liu J (2016) Genome-editing technologies: principles and applications. Cold Spring Harb Perspect Biol 8:a023754

Garg P, Jaiswal P (2016) Databases and bioinformatics tools for rice research. Curr Plant Biol 7–8:39–52. https://doi.org/10.1016/j.cpb.2016.12.006

Ghosal S, Blystone D, Singh AK et al (2018) An explainable deep machine vision framework for plant stress phenotyping. Proc Natl Acad Sci U S A 115:4613–4618. https://doi.org/10.1073/pnas.1716999115

Goad DM, Jia Y, Gibbons A et al (2020) Identification of novel QTL conferring sheath blight resistance in two weedy rice mapping populations. Rice 13:21. https://doi.org/10.1186/s12284-020-00381-9

Gopalakrishnan S, Sharma RK, Anand Rajkumar K et al (2008) Integrating marker-assisted background analysis with foreground selection for identification of superior bacterial blight-resistant recombinants in Basmati rice. Plant Breed 127:131–139. https://doi.org/10.1111/j.1439-0523.2007.01458.x

Goto I (1970) Genetic studies on the resistance of rice plant to the blast fungus. Jpn J Phytopathol 36:304–312. https://doi.org/10.3186/jjphytopath.36.304

Goto I (1988) Genetic studies on resistance of rice plant to blast fungus (VII): blast resistance genes of Kuroka. Ann Phytopathol Soc Jpn 54:460–465

Gowda M, Roy-Barman S, Chattoo BB (2006) Molecular mapping of a novel blast resistance gene *Pi38* in rice using SSLP and AFLP markers. Plant Breed 125:596–599. https://doi.org/10.1111/j.1439-0523.2006.01248.x

Graichen FAS, Martinelli JA, Federizzi LC et al (2010) Inheritance of resistance to oat crown rust in recombinant inbred lines. Sci Agric 67:435–440. https://doi.org/10.1590/s0103-90162010000400010

Gu K, Yang B, Tian D et al (2005) R-gene expression induced by a type-III effector triggers disease resistance in rice. Nature 435:1122–1125. https://doi.org/10.1038/nature03630

Guo SB, Zhang DP, Lin XH (2010) Identification and mapping of a novel bacterial blight-resistance gene *Xa35*(t) originated from *Oryza minuta*. Scientia Agricultura Sinica 43:2611–2618

Gupta RM, Musunuru K (2014) Expanding the genetic editing tool kit: ZFNs, TALENs, and CRISPR-Cas9. J Clin Invest 124:4154–4161

Hajira SK, Sundaram RM, Laha GS et al (2016) A single-tube, functional marker-based multiplex PCR assay for simultaneous detection of major bacterial blight-resistance genes *Xa21*, *xa13* and *xa5* in rice. Rice Sci 23:144. https://doi.org/10.1016/j.rsci.2015.11.004

Han Y, Li D, Yang J et al (2020) Mapping quantitative trait loci for disease resistance to false smut of rice. Phytopathol Res 2:20. https://doi.org/10.1186/s42483-020-00059-6

Hao L, Ruiying Q, Xiaoshuang L et al (2019) CRISPR/Cas9-mediated adenine base editing in rice genome. Rice Sci 26:125–128

Hari Y, Srinivasarao K, Viraktamath BC et al (2011) Marker-assisted improvement of a stable restorer line, KMR-3R and its derived hybrid KRH2 for bacterial blight resistance and grain quality. Plant Breed 130:608–616. https://doi.org/10.1111/j.1439-0523.2011.01881.x

Hari Y, Srinivasarao K, Viraktamath BC et al (2013) Marker-assisted introgression of bacterial blight and blast resistance into IR 58025B, an elite maintainer line of rice. Plant Breed 132:586–594. https://doi.org/10.1111/pbr.12056

Hayasaka T, Fujii H, Ishiguro K (2008) The role of silicon in preventing appressorial penetration by the rice blast fungus. Phytopathology 98:1038–1044

Hayashi K, Yoshida H (2009) Refunctionalization of the ancient rice blast disease resistance gene *Pit* by the recruitment of a retrotransposon as a promoter. Plant J 57:413–425. https://doi.org/10.1111/j.1365-313X.2008.03694.x

Hayashi K, Yoshida H, Ashikawa I (2006) Development of PCR-based allele-specific and InDel marker sets for nine rice blast-resistance genes. Theor Appl Genet 113:251–260. https://doi.org/10.1007/s00122-006-0290-6

Hayashi N, Inoue H, Kato T et al (2010) Durable panicle blast-resistance gene *Pb1* encodes an atypical CC-NBS-LRR protein and was generated by acquiring a promoter through local genome duplication. Plant J 64:498–510. https://doi.org/10.1111/j.1365-313X.2010.04348.x

He W, Huang D, Li R et al (2012) Identification of a resistance gene *bls1* to bacterial leaf streak in wild rice *Oryza rufipogon* Griff. J Integr Agric 11:962–969. https://doi.org/10.1016/S2095-3119(12)60087-2

He Y, Zhu M, Wang L et al (2018) Programmed self-elimination of the CRISPR/Cas9 construct greatly accelerates the isolation of edited and transgene-free rice plants. Mol Plant 11:1210–1213

Heinrichs EA, Muniappan R (2017) IPM for tropical crops: rice. CAB Rev 12:1

Hervieux V, Yaganza ES, Arul J, Tweddell RJ (2002) Effect of organic and inorganic salts on the development of *Helminthosporium solani*, the causal agent of potato silver scurf. Plant Dis 86:1014–1018

Hirochika H, Guiderdoni E, An G et al (2004) Rice mutant resources for gene discovery. Plant Mol Biol 54:325–334. https://doi.org/10.1023/B:PLAN.0000036368.74758.66

Holland JB (2001) Epistasis and pant breeding. Plant Breed Rev 21:27–92

Hu H, Scheben A, Edwards D (2018) Advances in integrating genomics and bioinformatics in the plant breeding pipeline. Agriculture 8:75

Hua L, Wu J, Chen C et al (2012) The isolation of *Pi1*, an allele at the *Pik* locus which confers broad spectrum resistance to rice blast. Theor Appl Genet 125:1047–1055. https://doi.org/10.1007/s00122-012-1894-7

Huang N, Angeles ER, Domingo J et al (1997) Pyramiding of bacterial blight-resistance genes in rice: marker-assisted selection using RFLP and PCR. Theor Appl Genet 95:313–320. https://doi.org/10.1007/s001220050565

Huang X, Wei X, Sang T et al (2010) Genome-wide asociation studies of 14 agronomic traits in rice landraces. Nat Genet 42:961–967. https://doi.org/10.1038/ng.695

Huang S, Liu L, Wang L, Hou Y (2019) Research on advance of rice false smut *Ustilaginoidea virens* (Cooke) Takah worldwide: Part I. Research status of rice false smut. J Agric Sci 11:240. https://doi.org/10.5539/jas.v11n15p240

Hutin M, Sabot F, Ghesquière A et al (2015) A knowledge-based molecular screen uncovers a broad-spectrum *OsSWEET14* resistance allele to bacterial blight from wild rice. Plant J 84:694–703. https://doi.org/10.1111/tpj.13042

Ingvardsen CR, Schejbel B, Lübberstedt T (2008) Functional markers in resistance breeding. Springer, Berlin, pp 61–87

Inukai T, Nelson RJ, Zeigler RS et al (1996) Genetic analysis of blast resistance in tropical rice cultivars using near-isogenic lines. In: Khush GS (ed) Rice genetics III. International Rice Research Institute, Los Baños, pp 447–455

Inukai T, Nagashima S, Kato M (2019) *Pid3-I1* is a race-specific partial-resistance allele at the *Pid3* blast resistance locus in rice. Theor Appl Genet 132:395–404. https://doi.org/10.1007/s00122-018-3227-y

IRRI (1996) Standard evaluation system for rice, 4th edn. International Rice Research Institute, Los Baños

IRRI (2013) Standard evaluation system for rice, 5th edn. International Rice Research Institute, Los Baños

Iwata N (1996) Registration of new gene symbols. Rice Genet 13:12–18

Iyer-Pascuzzi AS, McCouch SR (2007) Functional markers for *xa5*-mediated resistance in rice (*Oryza sativa*, L.). Mol Breed 19:291–296. https://doi.org/10.1007/s11032-006-9055-9

Jabeen R, Rahman SU, Rais A (2011) Evaluating BLB resistance/aggressiveness in rice through best inoculum concentration, inoculation and application methods. Pak J Bot 43:2635–2638

Jamaloddin M, Durga Rani CV, Swathi G et al (2020) Marker assisted gene pyramiding (MAGP) for bacterial blight and blast resistance into mega rice variety "Tellahamsa". PLoS One 15:e0234088. https://doi.org/10.1371/journal.pone.0234088

Jeung JU, Kim BR, Cho YC et al (2007) A novel gene, *Pi40*(t), linked to the DNA markers derived from NBS-LRR motifs confers broad spectrum of blast resistance in rice. Theor Appl Genet 115:1163–1177. https://doi.org/10.1007/s00122-007-0642-x

Jia Y, Valent B, Lee FN (2003) Determination of host responses to *Magnaporthe grisea* on detached rice leaves using a spot inoculation method. Plant Dis 87:129–133. https://doi.org/10.1094/PDIS.2003.87.2.129

Jia Y, Correa-Victoria F, McClung A et al (2007) Rapid determination of rice cultivar responses to the sheath blight pathogen *Rhizoctonia solani* using a micro-chamber screening method. Plant Dis 91:485–489. https://doi.org/10.1094/PDIS-91-5-0485

Jia L, Yan W, Zhu C et al (2012) Allelic analysis of sheath blight resistance with association mapping in rice. PLoS One 7:e32703. https://doi.org/10.1371/journal.pone.0032703

Jiang W, Zhou H, Bi H et al (2013) Demonstration of CRISPR/Cas9/sgRNA-mediated targeted gene modification in *Arabidopsis*, tobacco, sorghum and rice. Nucleic Acids Res 41:e188–e188

Jiang J, Yang D, Ali J, Mou T (2015) Molecular marker-assisted pyramiding of broad-spectrum disease resistance genes, *Pi2* and *Xa23*, into GZ63-4S, an elite thermo-sensitive genic male-sterile line in rice. Mol Breed 35:1–12. https://doi.org/10.1007/s11032-015-0282-9

Jiang N, Yan J, Liang Y et al (2020) Resistance genes and their interactions with bacterial blight/leaf streak pathogens (*Xanthomonas oryzae*) in rice (*Oryza sativa* L.)—an updated review. Rice 13:3. https://doi.org/10.1186/s12284-019-0358-y

Jinek M, Chylinski K, Fonfara I et al (2012) A programmable dual-RNA-guided DNA endonuclease in adaptive bacterial immunity. Science 337:816–821. https://doi.org/10.1126/science.1225829

Jones JDG, Dangl JL (2006) The plant immune system. Nature 444:323–329

Joseph M, Gopalakrishnan S, Sharma RK et al (2004) Combining bacterial blight resistance and Basmati quality characteristics by phenotypic and molecular marker-assisted selection in rice. Mol Breed 13:377–387. https://doi.org/10.1023/B:MOLB.0000034093.63593.4c

Ju Y, Tian H, Zhang R et al (2017) Overexpression of *OsHSP18.0-CI* enhances resistance to bacterial leaf streak in rice. Rice 10:12. https://doi.org/10.1186/s12284-017-0153-6

Kauffman HE, Reddy APK, Hsieh SPY, Merca SD (1973) An improved technique for evaluating resistance of rice varieties to *Xanthomonas oryzae*. Plant Dis Rep 57:537–541

Kaur Y (2014) Development of screening technique for false smut of rice. Dissertation. Punjab Agricultural University, Ludhiana

Kaur T, Singh I (2017) Standardization of screening technique for false smut of rice. Progress Agric 17:10. https://doi.org/10.5958/0976-4615.2017.00007.2

Kaur S, Padmanabhan SY, Kaur P (1979) Effect of nitrogen on the intracellular spread of *Pyricularia oryzae*. Indian Phytopathol 32:285–286

Ke Y, Deng H, Wang S (2017) Advances in understanding broad-spectrum resistance to pathogens in rice. Plant J 90:738–748. https://doi.org/10.1111/tpj.13438

Khush GS (2005) What it will take to feed 5.0 billion rice consumers in 2030. Plant Mol Biol 59:1–6

Khush GS, Angeles ER (1999) A new gene for resistance to race 6 of bacterial blight in rice, *Oryza sativa*. Rice Genet Newsl 16:92–93

Kim SM (2018) Identification of novel recessive gene *xa44*(t) conferring resistance to bacterial blight races in rice by QTL linkage analysis using an SNP chip. Theor Appl Genet 131:2733–2743. https://doi.org/10.1007/s00122-018-3187-2

Kim SM, Reinke RF (2019) A novel resistance gene for bacterial blight in rice, *Xa43*(t) identified by GWAS, confirmed by QTL mapping using a bi-parental population. PLoS One 14:e0211775. https://doi.org/10.1371/journal.pone.0211775

Kim YG, Cha J, Chandrasegaran S (1996) Hybrid restriction enzymes: zinc finger fusions to Fok I cleavage domain. Proc Natl Acad Sci U S A 93:1156–1160. https://doi.org/10.1073/pnas.93.3.1156

Kim SM, Suh JP, Qin Y et al (2015) Identification and fine-mapping of a new resistance gene, *Xa40*, conferring resistance to bacterial blight races in rice (*Oryza sativa* L.). Theor Appl Genet 128:1933–1943. https://doi.org/10.1007/s00122-015-2557-2

Kim YA, Moon H, Park CJ (2019) CRISPR/Cas9-targeted mutagenesis of *Os8N3* in rice to confer resistance to *Xanthomonas oryzae* pv. *oryzae*. Rice 12:67. https://doi.org/10.1186/s12284-019-0325-7

Kinoshita T, Kiyosawa S (1997) Some considerations on linkage relationships between *Pii* and *Piz* in the blast resistance of rice. Rice Genet Newsl 14:57–59

Klug A, Schwabe JWR (1995) Zinc fingers. FASEB J 9:597–604. https://doi.org/10.1096/fasebj.9.8.7768350

Koonin EV, Makarova KS (2019) Origins and evolution of CRISPR-Cas systems. Philos Trans R Soc B Biol Sci 374:20180087

Korinsak S, Sriprakhon S, Sirithanya P et al (2009) Identification of microsatellite markers (SSR) linked to a new bacterial blight-resistance gene *xa33*(t) in rice cultivar "Ba7". Maejo Int J Sci Technol 3:235–247

Kou Y, Wang S (2013) Bacterial blight resistance in rice. In: Translational genomics for crop breeding. John Wiley & Sons Ltd., Chichester, pp 11–30

Kurata N, Yamazaki Y (2006) Oryzabase:an integrated biological and genome information database for rice. Plant Physiol 140:12–17. https://doi.org/10.1104/pp.105.063008

Lee KS, Rasabandith S, Angeles ER, Khush GS (2003) Inheritance of resistance to bacterial blight in 21 cultivars of rice. Phytopathology 93:147–152. https://doi.org/10.1094/PHYTO.2003.93.2.147

Lee SK, Song MY, Seo YS et al (2009) Rice *Pi5*-mediated resistance to *Magnaporthe oryzae* requires the presence of two coiled-coil-nucleotide-binding-leucine-rich repeat genes. Genetics 181:1627–1638. https://doi.org/10.1534/genetics.108.099226

Lee JH, Muhsin M, Atienza GA et al (2010) Single nucleotide polymorphisms in a gene for translation initiation factor (eIF4G) of rice (*Oryza sativa*) associated with resistance to rice tungro spherical virus. Mol Plant-Microbe Interact 23:29–38. https://doi.org/10.1094/MPMI-23-1-0029

Lei C, Huang D, Li W et al (2005) Molecular mapping of a blast resistance gene in an indica rice cultivar Yanxian No. 1. Rice Genet Newsl 22:76–77

Li T, Liu B, Spalding MH et al (2012) High-efficiency TALEN-based gene editing produces disease-resistant rice. Nat Biotechnol 30:390

Li Z, Fu BY, Gao YM et al (2014) The 3,000 rice genomes project. Giga Sci 3:7. https://doi.org/1 0.1186/2047-217X-3-7

Li M, Li X, Zhou Z et al (2016) Reassessment of the four yield-related genes *Gn1a*, *DEP1*, *GS3*, and *IPA1* in rice using a CRISPR/Cas9 system. Front Plant Sci 7:377

Li C, Wang D, Peng S et al (2019a) Genome-wide association mapping of resistance against rice blast strains in South China and identification of a new *Pik* allele. Rice 12:47. https://doi.org/10.1186/s12284-019-0309-7

Li W, Chern M, Yin J et al (2019b) Recent advances in broad-spectrum resistance to the rice blast disease. Curr Opin Plant Biol 50:114–120

Liakos KG, Busato P, Moshou D et al (2018) Machine learning in agriculture: a review. Sensors 18:2674

Lin F, Chen S, Que Z et al (2007) The blast resistance gene *Pi37* encodes a nucleotide binding site-leucine-rich repeat protein and is a member of a resistance gene cluster on rice chromosome 1. Genetics 177:1871–1880. https://doi.org/10.1534/genetics.107.080648

Lin Q, Zong Y, Xue C et al (2020) Prime genome editing in rice and wheat. Nat Biotechnol 38:582–585. https://doi.org/10.1038/s41587-020-0455-x

Liu XQ, Wang L, Chen S et al (2005) Genetic and physical mapping of *Pi36*(t), a novel rice blast resistance gene located on rice chromosome 8. Mol Gen Genomics 274:394–401. https://doi.org/10.1007/s00438-005-0032-5

Liu X, Yang Q, Lin F et al (2007) Identification and fine-mapping of *Pi39*(t), a major gene conferring the broad-spectrum resistance to *Magnaporthe oryzae*. Mol Gen Genomics 278:403–410. https://doi.org/10.1007/s00438-007-0258-5

Liu W, Jin S, Zhu X et al (2008) Improving blast resistance of a thermo-sensitive genic male sterile rice line GD-8S by molecular marker-assisted selection. Rice Sci 15:179–185. https://doi.org/10.1016/S1672-6308(08)60040-2

Liu Q, Yuan M, Zhou Y et al (2011) A paralog of the MtN3/saliva family recessively confers race-specific resistance to *Xanthomonas oryzae* in rice. Plant Cell Environ 34:1958–1969. https://doi.org/10.1111/j.1365-3040.2011.02391.x

Liu G, Jia Y, McClung A et al (2013) Confirming QTLs and finding additional loci responsible for resistance to rice sheath blight disease. Plant Dis 97:113–117. https://doi.org/10.1094/PDIS-05-12-0466-RE

Liu W, Liu J, Triplett L et al (2014) Novel insights into rice innate immunity against bacterial and fungal pathogens. Annu Rev Phytopathol 52:213–241. https://doi.org/10.1146/annurev-phyto-102313-045926

Liu MH, Kang H, Xu Y et al (2020) Genome-wide association study identifies an NLR gene that confers partial resistance to *Magnaporthe oryzae* in rice. Plant Biotechnol J 18:1376–1383. https://doi.org/10.1111/pbi.13300

Lu Y, Zhu JK (2017) Precise editing of a target base in the rice genome using a modified CRISPR/Cas9 system. Mol Plant 10:523–525

Lu Z, Yu X, Heong K, Cui HU (2007) Effect of nitrogen fertilizer on herbivores and its stimulation to major insect pests in rice. Rice Sci 14:56–66

Ma J, Lei C, Xu X et al (2015a) *Pi64*, encoding a novel CC-NBS-LRR protein, confers resistance to leaf and neck blast in rice. Mol Plant-Microbe Interact 28:558–568. https://doi.org/10.1094/MPMI-11-14-0367-R

Ma X, Zhang Q, Zhu Q et al (2015b) A robust CRISPR/Cas9 system for convenient, high-efficiency multiplex genome editing in monocot and dicot plants. Mol Plant 8:1274–1284. https://doi.org/10.1016/j.molp.2015.04.007

Mackill DJ, Coffman WR, Garrity DP (1996) Rainfed lowland rice improvement. International Rice Research Institute, Los Baños. 242 p

Macovei A, Sevilla NR, Cantos C et al (2018) Novel alleles of rice *eIF4G* generated by CRISPR/Cas9-targeted mutagenesis confer resistance to *Rice tungro spherical virus*. Plant Biotechnol J 16:1918–1927. https://doi.org/10.1111/pbi.12927

Madamba MRS, Sugiyama N, Bordeos A et al (2009) A recessive mutation in rice conferring non-race-specific resistance to bacterial blight and blast. Rice 2:104–114. https://doi.org/10.1007/s12284-009-9027-x

Mahadevaiah C, Hittalmani S, Uday G, Kumar MKP (2015) Standardization of disease screening protocol for sheath rot disease in rice. Int J Agric Sci Res 5:129–137

Mahadevaiah C, Kumar MP, Hittalmani S (2017) Dissecting parameters associated with sheath rot (Sarocladium oryzae [(Sawada) W. Gams &D. Hawksw.]) disease in rice (Oryza sativa L.). Curr Sci 112:151–155. https://doi.org/10.18520/cs/v112/i01/151-155

Mahesh HB, Shirke MD, Singh S et al (2016) Indica rice genome assembly, annotation and mining of blast disease-resistance genes. BMC Genomics 17:242. https://doi.org/10.1186/s12864-016-2523-7

Mangrauthia SK, Maliha A, Prathi NB, Marathi B (2017) MicroRNAs: potential target for genome editing in plants for traits improvement. Indian J Plant Physiol 22:530–548

Manosalva PM, Davidson RM, Liu B et al (2009) A germin-like protein gene family functions as a complex quantitative trait locus conferring broad-spectrum disease resistance in rice. Plant Physiol 149:286–296. https://doi.org/10.1104/pp.108.128348

Matsumoto T, Wu J, Kanamori H et al (2005) The map-based sequence of the rice genome. Nature 436:793–800. https://doi.org/10.1038/nature03895

Matsuyama N (1973) Effect of nitrogenous fertilizer on biochemical processes that could affect lesion size of rice blast. Phytopathology 63:1202–1203

McCouch S, Nelson R, Tohme J, Zeigler R (1994) Mapping of blast resistance genes in rice. In: Zeigler RS, Leong SA, Teng PS (eds) Rice blast disease. CAB International, Wallingford, UK, pp 167–186

Meuwissen THE, Hayes BJ, Goddard ME (2001) Prediction of total genetic value using genome-wide dense marker maps. Genetics 157:1819–1829

Miah G, Rafii M, Ismail M et al (2013) A review of microsatellite markers and their applications in rice breeding programs to improve blast disease resistance. Int J Mol Sci 14:22499–22528

Miao LL, Wang CL, Zheng CK et al (2010) Molecular mapping of a new gene for resistance to rice bacterial blight. Sci Agric Sin 43:2611

Milovanovic V, Smutka L (2017) Asian countries in the global rice market. Acta Univ Agric Silvic Mendel Brun 65:679–688

Mishra R, Joshi RK, Zhao K (2018) Genome editing in rice: recent advances, challenges, and future implications. Front Plant Sci 9:1361

Molla KA, Karmakar S, Molla J et al (2020) Understanding sheath blight resistance in rice: the road behind and the road ahead. Plant Biotechnol J 18:895–915. https://doi.org/10.1111/pbi.13312

Moon SB, Kim DY, Ko JH, Kim YS (2019) Recent advances in the CRISPR genome editing tool set. Exp Mol Med 51:1–11

Moscou MJ, Bogdanove AJ (2009) A simple cipher governs DNA recognition by TAL effectors. Science 326:1501. https://doi.org/10.1126/science.1178817

Mur LAJ, Simpson C, Kumari A et al (2017) Moving nitrogen to the centre of plant defense against pathogens. Ann Bot 119:703–709

Mushtaq M, Sakina A, Wani SH et al (2019) Harnessing genome editing techniques to engineer disease resistance in plants. Front Plant Sci 10:550

Mvuyekure SM, Sibiya J, Derera J et al (2017) Genetic analysis of mechanisms associated with inheritance of resistance to sheath rot of rice. Plant Breed 136:509–515. https://doi.org/10.1111/pbr.12492

Nakamura S, Asakawa S, Ohmido N et al (1997) Construction of an 800-kb contig in the near-centromeric region of the rice blast resistance gene Pi-ta2 using a highly representative rice BAC library. Mol Gen Genet 254:611–620. https://doi.org/10.1007/s004380050459

Naqvi NI, Bonman JM, Mackill DJ et al (1995) Identification of RAPD markers linked to a major blast resistance gene in rice. Mol Breed 1:341–348. https://doi.org/10.1007/BF01248411

Neelam K, Mahajan R, Gupta V et al (2020) High-resolution genetic mapping of a novel bacterial blight- resistance gene xa-45(t) identified from Oryza glaberrima and transferred to Oryza sativa. Theor Appl Genet 133:689–705. https://doi.org/10.1007/s00122-019-03501-2

Nelson JC, Oard JH, Groth D et al (2012) Sheath-blight resistance QTLs in japonica rice germplasm. Euphytica 184:23–34. https://doi.org/10.1007/s10681-011-0475-1

Ning D, Song A, Fan F et al (2014) Effects of slag-based silicon fertilizer on rice growth and brown-spot resistance. PLoS One 9:e102681

Nürnberger T, Kemmerling B (2009) PAMP-triggered basal immunity in plants. Adv Bot Res 51:1–38

Ogawa T (1988) Near-isogenic lines as international differentials for resistance to bacterial blight of rice. Rice Genet Newsl 5:106

Ogawa T (2008) Monitoring race distribution and identification of genes for resistance to bacterial leaf blight. Rice Genet III 3:456–459

Ogawa T, Lin L, Tabien RE et al (1987) A new recessive gene for resistance to bacterial blight of rice. Rice Genet Newsl 4:98

Okuyama Y, Kanzaki H, Abe A et al (2011) A multifaceted genomics approach allows the isolation of the rice *Pia*-blast resistance gene consisting of two adjacent NBS-LRR protein genes. Plant J 66:467–479. https://doi.org/10.1111/j.1365-313X.2011.04502.x

Oliva R, Ji C, Atienza-Grande G et al (2019) Broad-spectrum resistance to bacterial blight in rice using genome editing. Nat Biotechnol 37:1344–1350. https://doi.org/10.1038/s41587-019-0267-z

Oliver RP, Ipcho SVS (2004) *Arabidopsis* pathology breathes new life into the necrotrophs-vs.-biotrophs classification of fungal pathogens. Mol Plant Pathol 5:347–352

Oreiro EG, Grimares EK, Atienza-Grande G et al (2019) Genome-wide associations and transcriptional profiling reveal ROS regulation as one underlying mechanism of sheath blight resistance in rice. Mol Plant-Microbe Interact 33:212–222. https://doi.org/10.1094/MPMI-05-19-0141-R

Pan Q, Tanisaka T (1997) Studies on the genetics and breeding of blast resistance in rice. VII. Gene analysis for the blast resistance of Indian native cultivar, Aus 373. Breed Sci 47:35

Pan Q, Wang L, Ikehashi H, Tanisaka T (1996) Identification of a new blast resistance gene in the indica rice cultivar Kasalath using Japanese differential cultivars and isozyme markers. Phytopathology 86:1071–1075

Panda KK, Mishra MK (2019) Studies on physiological characteristics of *Sarocladium oryzae* causing sheath rot of rice. Int J Curr Microbiol App Sci 8:1767–1774. https://doi.org/10.20546/ijcmas.2019.808.209

Panguluri SK, Kumar AA (2013) Phenotyping for plant breeding: applications of phenotyping methods for crop improvement. Springer, New York

Park DS, Sayler RJ, Hong YG et al (2008) A method for inoculation and evaluation of rice sheath blight disease. Plant Dis 92:25–29. https://doi.org/10.1094/PDIS-92-1-0025

Peeters KJ, Haeck A, Harinck L et al (2020) Morphological, pathogenic and toxigenic variability in the rice sheath rot pathogen *Sarocladium oryzae*. Toxins (Basel) 12:109. https://doi.org/10.3390/toxins12020109

Perez-de Castro A, Vilanova S, Canizares J et al (2012) Application of genomic tools in plant breeding. Curr Genom 13:179–195. https://doi.org/10.2174/138920212800543084

Petpisit V, Khush GS, Kauffman HE (1977) Inheritance of resistance to bacterial blight in rice. 1. Crop Sci 17:551–554. https://doi.org/10.2135/cropsci1977.0011183x001700040018x

Qu S, Liu G, Zhou B et al (2006) The broad-spectrum blast resistance gene *Pi9* encodes a nucleotide-binding site-leucine-rich repeat protein and is a member of a multigene family in rice. Genetics 172:1901–1914. https://doi.org/10.1534/genetics.105.044891

Rajashekara H, Ellur RK, Khanna A et al (2014) Inheritance of blast resistance and its allelic relationship with five major *R*-genes in a rice landrace "Vanasurya". Indian Phytopathol 67:365–369

Rajpurohit D, Kumar R, Kumar M et al (2011) Pyramiding of two bacterial blight resistance and a semidwarfing gene in Type 3 Basmati using marker-assisted selection. Euphytica 178:111–126. https://doi.org/10.1007/s10681-010-0279-8

Ramirez CL, Foley JE, Wright DA et al (2008) Unexpected failure rates for modular assembly of engineered zinc fingers. Nat Methods 5:374–375

Rathna Priya TS, Eliazer Nelson ARL, Ravichandran K, Antony U (2019) Nutritional and functional properties of coloured rice varieties of South India: a review. J Ethn Foods 6:11

Raven JA (2003) Cycling silicon: the role of accumulation in plants. New Phytol 158:419–421

Read AC, Moscou MJ, Zimin AV et al (2020) Genome assembly and characterization of a complex zfBED-NLR gene-containing disease resistance locus in Carolina Gold Select rice with Nanopore sequencing. PLoS Genet 16:e1008571. https://doi.org/10.1371/journal.pgen.1008571

Reddy APK, Katyal JC, Rouse DI, MacKenzie DR (1979) Relationship between nitrogen fertilization, bacterial leaf blight severity and yield of rice. Phytopathology 69:970–973

Rekha G, Abhilash Kumar V, Viraktamath BC et al (2018) Improvement of blast resistance of the popular high-yielding, medium slender-grain type, bacterial blight-resistant rice variety, Improved Samba Mahsuri by marker-assisted breeding. J Plant Biochem Biotechnol 27:463–472. https://doi.org/10.1007/s13562-018-0455-9

Robichaux CR (2001) The effect of calcium silicate on rice yield and sheath blight disease. Fitopatol Bras 30:457–469

Romero FM, Gatica-Arias A (2019) CRISPR/Cas9: development and application in rice breeding. Rice Sci 26:265–281

Sallaud C, Lorieux M, Roumen E et al (2003) Identification of five new blast resistance genes in the highly blast-resistant rice variety IR64 using a QTL mapping strategy. Theor Appl Genet 106:794–803. https://doi.org/10.1007/s00122-002-1088-9

Samiyappan R, Amutha G, Kandan A et al (2003) Purification and partial characterization of a phytotoxin produced by *Sarocladium oryzae*, the rice sheath rot pathogen (Reinigung und partielle charakterisierung eines vom reisblattscheidenfäule-pathogen *sarocladium oryzae* erzeugten phytoto). Arch Phytopathol Plant Protect 36:247–256. https://doi.org/10.1080/03235400310001617879

Sanchez AC, Brar DS, Huang N et al (2000) Sequence tagged site marker-assisted selection for three bacterial blight-resistance genes in rice. Crop Sci 40:792–797. https://doi.org/10.2135/cropsci2000.403792x

Savary S, Willocquet L, Pethybridge SJ et al (2019) The global burden of pathogens and pests on major food crops. Nat Ecol Evol 3:430–439. https://doi.org/10.1038/s41559-018-0793-y

Sebastian LS, Ikeda R, Huang N et al (1996) Molecular mapping of resistance to rice tungro spherical virus and green leafhopper. Phytopathology 86:25–30. https://doi.org/10.1094/Phyto-86-25

Serrano M, Coluccia F, Torres M et al (2014) The cuticle and plant defense to pathogens. Front Plant Sci 5:274. https://doi.org/10.3389/fpls.2014.00274

Shan Q, Zhang Y, Chen K et al (2015) Creation of fragrant rice by targeted knockout of the *OsBADH2* gene using TALEN technology. Plant Biotechnol J 13:791–800. https://doi.org/10.1111/pbi.12312

Shanti ML, Kumar Varm CM, Premalatha P et al (2010) Understanding the bacterial blight pathogen-combining pathotyping and molecular marker studies. Int J Plant Pathol 1:58–68. https://doi.org/10.3923/ijpp.2010.58.68

Sharma S, Duveiller E, Basnet R et al (2005a) Effect of potash fertilization on *Helminthosporium* leaf blight severity in wheat, and associated increases in grain yield and kernel weight. Field Crop Res 93:142–150

Sharma TR, Madhav MS, Singh BK et al (2005b) High-resolution mapping, cloning and molecular characterization of the *Pi-kh* gene of rice, which confers resistance to *Magnaporthe grisea*. Mol Gen Genomics 274:569–578. https://doi.org/10.1007/s00438-005-0035-2

Sharma TR, Rai AK, Gupta SK et al (2012) Rice blast management through host-plant resistance: retrospect and prospects. Agric Res 1:37–52

Shen C, Que Z, Xia Y et al (2017) Knock out of the annexin gene *OsAnn3* via CRISPR/Cas9-mediated genome editing decreased cold tolerance in rice. J Plant Biol 60:539–547

Shinoda H, Toriyama K, Yunoki T et al (1971) Studies on the varietal resistance of rice to blast. 6. Linkage relationship of blast resistance genes. Jpn Chugoku Nogyo Shikengo Fukuyama Bull Ser A 20:1

Sidhu GS, Khush GS, Mew TW (1978) Genetic analysis of bacterial blight-resistance in seventy-four cultivars of rice, *Oryza sativa* L. Theor Appl Genet 53:105–111. https://doi.org/10.1007/BF00272687

Sime HD, Mbong GA, Malla DK, Suh C (2017) Effect of different doses of NPK fertilizer on the infection coefficient of rice (*Orysa sativa* L.) blast in Ndop, North West of Cameroon. Agron Afr 29:245–255

Singh S, Sidhu JS, Huang N et al (2001) Pyramiding three bacterial blight-resistance genes (*xa5*, *xa13* and *Xa21*) using marker-assisted selection into indica rice cultivar PR106. Theor Appl Genet 102:1011–1015. https://doi.org/10.1007/s001220000495

Singh PK, Nag A, Arya P et al (2018) Prospects of understanding the molecular biology of disease resistance in rice. Int J Mol Sci 19:1141

Song WY, Wang GL, Chen LL et al (1995) A receptor kinase-like protein encoded by the rice disease resistance gene, *Xa21*. Science 270:1804. https://doi.org/10.1126/science.270.5243.1804

Stein JC, Yu Y, Copetti D et al (2018) Genomes of 13 domesticated and wild rice relatives highlight genetic conservation, turnover and innovation across the genus *Oryza*. Nat Genet 50:285–296. https://doi.org/10.1038/s41588-018-0040-0

Streubel J, Pesce C, Hutin M et al (2013) Five phylogenetically close rice *SWEET* genes confer TAL effector-mediated susceptibility to *Xanthomonas oryzae* pv. *oryzae*. New Phytol 200:808–819. https://doi.org/10.1111/nph.12411

Su J, Wang W, Han J et al (2015) Functional divergence of duplicated genes results in a novel blast resistance gene *Pi50* at the *Pi2/9* locus. Theor Appl Genet 128:2213–2225

Suh J-P, Noh T-H, Kim K-Y et al (2009) Expression levels of three bacterial blight-resistance genes against K3a race of Korea by molecular and phenotype analysis in japonica rice (*O. sativa* L.). J Crop Sci Biotechnol 12:103–108. https://doi.org/10.1007/s12892-009-0103-y

Sun W, Zhang J, Fan Q et al (2010) Silicon-enhanced resistance to rice blast is attributed to silicon-mediated defense resistance and its role as physical barrier. Eur J Plant Pathol 128:39–49

Sun Y, Zhang X, Wu C et al (2016) Engineering herbicide-resistant rice plants through CRISPR/Cas9-mediated homologous recombination of acetolactate synthase. Mol Plant 9:628–631

Sun Q, Liu X, Yang J et al (2018) MicroRNA528 affects lodging resistance of maize by regulating lignin biosynthesis under nitrogen-luxury conditions. Mol Plant 11:806–814

Sun Y, Wang M, Mur LAJ et al (2020) Unravelling the roles of nitrogen nutrition in plant disease defenses. Int J Mol Sci 21:572

Sundaram RM, Vishnupriya MR, Biradar SK et al (2008) Marker-assisted introgression of bacterial blight resistance in Samba Mahsuri, an elite indica rice variety. Euphytica 160:411–422. https://doi.org/10.1007/s10681-007-9564-6

Sundaram RM, Vishnupriya MR, Laha GS et al (2009) Introduction of bacterial blight resistance into Triguna, a high-yielding, mid-early duration rice variety. Biotechnol J 4:400–407. https://doi.org/10.1002/biot.200800310

Swathi G, Durga Rani CV, Md J et al (2019) Marker-assisted introgression of the major bacterial blight- resistance genes, *Xa21* and *xa13*, and blast resistance gene, *Pi54*, into the popular rice variety, JGL1798. Mol Breed 39:1–12. https://doi.org/10.1007/s11032-019-0950-2

Tabien RE, Li Z, Paterson AH et al (2002) Mapping QTLs for field resistance to the rice blast pathogen and evaluating their individual and combined utility in improved varieties. Theor Appl Genet 105:313–324. https://doi.org/10.1007/s00122-002-0940-2

Taguchi-Shiobara F, Ozaki H, Sato H et al (2013) Mapping and validation of QTLs for rice sheath blight resistance. Breed Sci 63:301–308. https://doi.org/10.1270/jsbbs.63.301

Takagi H, Abe A, Yoshida K et al (2013a) QTL-seq: rapid mapping of quantitative trait loci in rice by whole- genome resequencing of DNA from two bulked populations. Plant J 74:174–183

Takagi H, Uemura A, Yaegashi H et al (2013b) MutMap-Gap: whole-genome resequencing of mutant F2 progeny bulk combined with *de novo* assembly of gap regions identifies the rice blast-resistance gene *Pii*. New Phytol 200:276–283. https://doi.org/10.1111/nph.12369

Takahashi W, Miura Y, Sasaki T (2009) A novel inoculation method for evaluation of grey leaf spot resistance in Italian ryegrass. J Plant Pathol 91:171–176. https://doi.org/10.4454/jpp.v91i1.638

Tan GX, Ren X, Weng QM et al (2004) Mapping of a new resistance gene to bacterial blight in rice line introgressed from *Oryza officinalis*. Acta Genet Sin 31:724–729

Tang D, Wu W, Li W et al (2000) Mapping of QTLs conferring resistrance to bacterial leaf streak in rice. Theor Appl Genet 101:286–291. https://doi.org/10.1007/s001220051481

Tang L, Mao B, Li Y et al (2017) Knockout of *OsNramp5* using the CRISPR/Cas9 system produces low Cd-accumulating indica rice without compromising yield. Sci Rep 7:14438. https://doi.org/10.1038/s41598-017-14832-9

Tanksley SD, Young ND, Paterson AH, Bonierbale MW (1989) RFLP mapping in plant breeding: new tools for an old science. Bio/Technology 7:257–264. https://doi.org/10.1038/nbt0389-257

Tanweer FA, Rafii MY, Sijam K et al (2015) Current advance methods for the identification of blast resistance genes in rice. C R Biol 338:321–334. https://doi.org/10.1016/j.crvi.2015.03.001

Tena G, Boudsocq M, Sheen J (2011) Protein kinase signaling networks in plant innate immunity. Curr Opin Plant Biol 14:519–529

Teng Q, Hu X-F, Chang Y-Y et al (2016) Effects of different fertilisers on rice resistance to pests and diseases. Soil Res 54:242–253

Thakur M, Sohal BS (2013) Role of elicitors in inducing resistance in plants against pathogen infection: a review. Int Sch Res Not 2013:762412

Tonnessen BW, Manosalva P, Lang JM et al (2015) Rice phenylalanine ammonia-lyase gene *OsPAL4* is associated with broad spectrum disease resistance. Plant Mol Biol 87:273–286. https://doi.org/10.1007/s11103-014-0275-9

Vassilev D, Leunissen J, Atanassov A et al (2005) Application of bioinformatics in plant breeding. Biotechnol Biotechnol Equip 19:139. https://doi.org/10.1080/13102818.2005.10817293

Vikal Y, Chawla H, Sharma R et al (2014) Mapping of bacterial blight-resistance gene *xa8* in rice (*Oryza sativa* L.). Indian J Genet Plant Breed 74:589. https://doi.org/10.5958/0975-6906.2014.00895.5

Wang GL, Mackill DJ, Bonman JM et al (1994) RFLP mapping of genes conferring complete and partial resistance to blast in a durably resistant rice cultivar. Genetics 136:1421–1434

Wang ZX, Yano M, Yamanouchi U et al (1999) The *Pib* gene for rice blast resistance belongs to the nucleotide binding and leucine-rich repeat class of plant disease resistance genes. Plant J 19:55–64. https://doi.org/10.1046/j.1365-313X.1999.00498.x

Wang W, Zhai W, Luo M et al (2001) Chromosome landing at the bacterial blight-resistance gene *Xa4* locus using a deep coverage rice BAC library. Mol Gen Genet 265:118–125. https://doi.org/10.1007/s004380000382

Wang C, Wen G, Lin X et al (2009) Identification and fine-mapping of the new bacterial blight-resistance gene, *Xa31*(t), in rice. Eur J Plant Pathol 123:235–240. https://doi.org/10.1007/s10658-008-9356-4

Wang F, Wang C, Liu P et al (2016) Enhanced rice blast resistance by CRISPR/Cas9-targeted mutagenesis of the ERF transcription factor gene *OsERF922*. PLoS One 11:e0154027

Wang M, Wang S, Liang Z et al (2018) From genetic stock to genome editing: gene exploitation in wheat. Trends Biotechnol 36:160–172

Wang L, Zhao L, Zhang X et al (2019) Large-scale identification and functional analysis of NLR genes in blast resistance in the Tetep rice genome sequence. Proc Natl Acad Sci U S A 116:8479–18487. https://doi.org/10.1073/pnas.1910229116

Wen ZH, Zeng YX, Ji ZJ, Yang CD (2015) Mapping quantitative trait loci for sheath blight disease resistance in Yangdao 4 rice. Genet Mol Res 14:1636–1649. https://doi.org/10.4238/2015.March.6.10

Wu KS, Tanksley SD (1993) Abundance, polymorphism and genetic mapping of microsatellites in rice. Mol Gen Genet 241:225–235. https://doi.org/10.1007/BF00280220

Wu JL, Fan YY, Li DB et al (2005) Genetic control of rice blast resistance in the durably resistant cultivar Gumei 2 against multiple isolates. Theor Appl Genet 111:50–56. https://doi.org/10.1007/s00122-005-1971-2

Wu KS, Martinez C, Lentini Z et al (2008) Cloning a blast resistance gene by chromosome walking. Rice Genet III:669. https://doi.org/10.1142/9789812814 2789_0082

Wu X, Yu Y, Baerson SR et al (2017) Interactions between nitrogen and silicon in rice and their effects on resistance toward the brown planthopper *Nilaparvata lugens*. Front Plant Sci 8:28

Xiang Y, Cao Y, Xu C et al (2006) *Xa3*, conferring resistance for rice bacterial blight and encoding a receptor kinase-like protein, is the same as *Xa26*. Theor Appl Genet 113:1347–1355. https://doi.org/10.1007/s00122-006-0388-x

Xu Q, Yuan X, Yu H et al (2011) Mapping quantitative trait loci for sheath blight resistance in rice using double haploid population. Plant Breed 130:404–406. https://doi.org/10.1111/j.1439-0523.2010.01806.x

Xu X, Hayashi N, Wang CT et al (2014) Rice blast resistance gene *Pikahei-1*(t), a member of a resistance gene cluster on chromosome 4, encodes a nucleotide-binding site and leucine-rich repeat protein. Mol Breed 34:691–700. https://doi.org/10.1007/s11032-014-0067-6

Xu Z, Xu X, Gong Q et al (2019) Engineering broad-spectrum bacterial blight resistance by simultaneously disrupting variable TALE-binding elements of multiple susceptibility genes in rice. Mol Plant 12:1434–1446. https://doi.org/10.1016/j.molp.2019.08.006

Yadav S, Anuradha G, Kumar RR et al (2015) Identification of QTLs and possible candidate genes conferring sheath blight resistance in rice (*Oryza sativa* L.). Springerplus 4:175. https://doi.org/10.1186/s40064-015-0954-2

Yoshimura S, Yamanouchi U, Katayose Y et al (1998) Expression of *Xa1*, a bacterial blight-resistance gene in rice, is induced by bacterial inoculation. Proc Natl Acad Sci U S A 95:1663–1668. https://doi.org/10.1073/pnas.95.4.1663

Yu S, Ali J, Zhang C et al (2020) Correction to: Genomic breeding of green super rice varieties and their deployment in Asia and Africa. Theor Appl Genet 133:1337

Yuan B, Zhai C, Wang W et al (2011) The *Pik-p* resistance to *Magnaporthe oryzae* in rice is mediated by a pair of closely linked CC-NBS-LRR genes. Theor Appl Genet 122:1017–1028. https://doi.org/10.1007/s00122-010-1506-3

Yuan C, Yuxiang Z, Zhijuan J et al (2019) Identification of stable quantitative trait loci for sheath blight resistance using recombinant inbred line. Rice Sci 26:331–338. https://doi.org/10.1016/j.rsci.2019.08.007

Yugander A, Sundaram RM, Ladhalakshmi D et al (2017) Virulence profiling of *Xanthomonas oryzae* pv. *oryzae* isolates, causing bacterial blight of rice in India. Eur J Plant Pathol 149:171–191. https://doi.org/10.1007/s10658-017-1176-y

Yugander A, Sundaram RM, Singh K et al (2018) Incorporation of the novel bacterial blight-resistance gene *Xa38* into the genetic background of elite rice variety Improved Samba Mahsuri. PLoS One 13:e0198260. https://doi.org/10.1371/journal.pone.0198260

Zafar K, Sedeek KEM, Rao GS et al (2020) Genome editing technologies for rice improvement: progress, prospects, and safety concerns. Front Genome Ed 2:5. https://doi.org/10.3389/fgeed.2020.00005

Zenbayashi K, Ashizawa T, Tani T, Koizumi S (2002) Mapping of the QTL (quantitative trait locus) conferring partial resistance to leaf blast in rice cultivar Chubu 32. Theor Appl Genet 104:547–552. https://doi.org/10.1007/s00122-001-0779-y

Zeng YX, Xia LZ, Wen ZH et al (2015) Mapping resistant QTLs for rice sheath blight disease with a doubled haploid population. J Integr Agric 14:801–810. https://doi.org/10.1016/S2095-3119(14)60909-6

Zetsche B, Gootenberg JS, Abudayyeh OO et al (2015) Cpf1 is a single RNA-guided endonuclease of a Class 2 CRISPR-Cas system. Cell 163:759–771. https://doi.org/10.1016/j.cell.2015.09.038

Zhai C, Lin F, Dong Z et al (2011) The isolation and characterization of *Pik*, a rice blast-resistance gene which emerged after rice domestication. New Phytol 189:321–334. https://doi.org/10.1111/j.1469-8137.2010.03462.x

Zhang Q, Wanf CL, Zhao KJ et al (2001) The effectiveness of advanced rice lines with new resistance gene *Xa23* to rice bacterial blight. Rice Genet Newsl 18:71–72

Zhang H, Li G, Li W, Song F (2009) Transgenic strategies for improving rice disease resistance. Afr J Biotechnol 8:1750–1757

Zhang F, Zhuo DL, Zhang F et al (2015) *Xa39*, a novel-dominant gene conferring broad-spectrum resistance to *Xanthomonas oryzae* pv. *oryzae* in rice. Plant Pathol 64:568–575. https://doi.org/10.1111/ppa.12283

Zhang H, Zhang J, Lang Z et al (2017a) Genome editing: principles and applications for functional genomics research and crop improvement. CRC Crit Rev Plant Sci 36:291–309

Zhang W, Wu L, Ding Y et al (2017b) Nitrogen fertilizer application affects lodging resistance by altering secondary cell wall synthesis in japonica rice (*Oryza sativa*). J Plant Res 130:859–871

Zhao H, Wang X, Jia Y et al (2018) The rice blast-resistance gene *Ptr* encodes an atypical protein required for broad-spectrum disease resistance. Nat Commun 9:1–12. https://doi.org/10.1038/s41467-018-04369-4

Zheng CK, Wang CL, Yu YJ et al (2009) Identification and molecular mapping of *Xa32*(t), a novel resistance gene for bacterial blight (*Xanthomonas oryzae* pv. *oryzae*) in rice. Acta Agron Sin 35:1173–1180. https://doi.org/10.1016/S1875-2780(08)60089-9

Zhou JH, Wang JL, Xu JC et al (2004) Identification and mapping of a rice blast-resistance gene *Pi-g*(t) in the cultivar Guangchangzhan. Plant Pathol 53:191–196. https://doi.org/10.1111/j.0032-0862.2004.00986.x

Zhou B, Qu S, Liu G et al (2006) The eight amino-acid differences within three leucine-rich repeats between Pi2 and Piz-t resistance proteins determine the resistance specificity to *Magnaporthe grisea*. Mol Plant-Microbe Interact 19:1216–1228. https://doi.org/10.1094/MPMI-19-1216

Zhou J, Peng Z, Long J et al (2015) Gene targeting by the TAL effector PthXo2 reveals cryptic resistance gene for bacterial blight of rice. Plant J 82:632–643

Zhou J, Deng K, Cheng Y et al (2017) CRISPR-Cas9 based genome editing reveals new insights into microRNA function and regulation in rice. Front Plant Sci 8:1598. https://doi.org/10.3389/fpls.2017.01598

Zhou X, Liao H, Chern M et al (2018) Loss of function of a rice TPR-domain RNA-binding protein confers broad-spectrum disease resistance. Proc Natl Acad Sci U S A 115:3174–3179. https://doi.org/10.1073/pnas.1705927115

Zhuang JY, Ma WB, Wu JL et al (2002) Mapping of leaf and neck blast-resistance genes with resistance gene analog, RAPD and RFLP in rice. Euphytica 128:363–370. https://doi.org/10.1023/A:1021272710294

Zipfel C, Felix G (2005) Plants and animals: a different taste for microbes? Curr Opin Plant Biol 8:353–360

Zuo S, Yin Y, Pan C et al (2013) Fine-mapping of *qSB-11LE*, the QTL that confers partial resistance to rice sheath blight. Theor Appl Genet 126:1257–1272. https://doi.org/10.1007/s00122-013-2051-7

Zuo S, Zhang Y, Yin Y et al (2014) Fine-mapping of *qSB-9* ^{TQ}, a gene conferring major quantitative resistance to rice sheath blight. Mol Breed 34:2191–2203. https://doi.org/10.1007/s11032-014-0173-5

Molecular Approaches for Insect Pest Management in Rice

Jagadish S. Bentur, R. M. Sundaram, Satendra Kumar Mangrauthia, and Suresh Nair

Abstract This chapter focuses on the progress made in using molecular tools in understanding resistance in rice to insect pests and breeding rice for multiple and durable insect resistance. Currently, molecular markers are being extensively used to tag, map, introgress, and clone plant resistance genes against gall midge, planthoppers, and leafhoppers. Studies on cloned insect resistance genes are leading to a better understanding of plant defense against insect pests under different feeding guilds. While marker-assisted breeding is successfully tackling problems in durable and multiple pest resistance in rice, genomics of plants and insects has identified RNAi-based gene silencing as an alternative approach for conferring insect resistance. The use of these techniques in rice is in the developmental stage, with the main focus on brown planthopper and yellow stem borer. CRISPR-based genome editing techniques for pest control in plants has just begun. Insect susceptibility genes (negative regulators of resistance genes) in plants are apt targets for this approach while gene drive in insect populations, as a tool to study rice-pest interactions, is another concept being tested. Transformation of crop plants with diverse insecticidal genes is a proven technology with potential for commercial success. Despite advances in the development and testing of transgenic rice for insect resistance, no insect-resistant rice cultivar is now being commercially cultivated. An array of molecular tools is being used to study insect-rice interactions at transcriptome, proteome, metabolome, mitogenome, and metagenome levels, especially with reference to BPH and gall midge, and such studies are uncovering new approaches for insect pest management and for understanding population genetics and phylogeography of rice pests. Thus, it is evident that the new knowledge being gained through these studies has provided us with new tools and information for

J. S. Bentur (✉)
Agri Biotech Foundation, Hyderabad, Telangana, India

R. M. Sundaram
ICAR-Indian Institute of Rice Research, Hyderabad, Telangana, India

S. K. Mangrauthia
ICAR-Indian Institute of Rice Research, Hyderabad, Telangana, India

S. Nair
Plant-Insect Interaction Group, International Centre for Genetic Engineering and Biotechnology, New Delhi, India

facing future challenges. However, what is also evident is that our attempts to manage rice pests cannot be a one-time effort but must be a continuing one.

Keywords Insect resistance · Molecular markers · Marker-assisted breeding · RNAi · Genome editing · Transgenic rice · Insect-plant interaction

1 Introduction

Insect pests of rice form a formidable biotic stress component and a significant production constraint across the globe. Although more than 200 insect species are reported to feed on rice plants, about a dozen of them are economically important in a specific rice ecosystem at a given time. Several of these have coevolved over thousands of years along with their host and many have no alternate host. The pest complex of rice is represented by insects from all the feeding guilds, from defoliators, tissue borers, and sap-suckers to gall formers, and several of these are occupied by a complex of species (Heinrichs 1994). Most important among these are stem borers: yellow stem borer (YSB), *Scirpophaga incertulas*; striped stem borer (SSB), *Chilo suppressalis*; and pink stem borer (PSB), *Sesamia inferens*; planthoppers: brown planthopper (BPH), *Nilaparvata lugens*; white-backed planthopper (WBPH), *Sogatella furcifera*; and small brown planthopper (SBPH), *Laodelphax striatellus*; leafhoppers: green leafhopper (GLH), *Nephottetix virescens*; green rice leafhopper (GRL/GRH), *Nephotettix cincticeps*; and zigzag leafhopper (ZLH), *Recilia dorsalis*; gall midges: Asian rice gall midge (ARGM), *Orseolia oryzae*; and African rice gall midge (AfRGM), *Orseolia oryzivora*; and leaffolders: *Cnaphalocrocis medinalis* and *Marasmia* spp. Several other insects such as rice hispa, grain bugs, aphids, mealy bug, and stem fly are of minor or regional importance (Bentur 2010).

Several studies reported yield losses due to either a single pest or a complex of pests but most of them end up with either overestimating or underestimating the damage caused by these pests. Savary et al. (2000) critically studied yield losses caused by different pests under varying production environments and suggested that stem borer damage at heading stage accounted for 2.3% loss. They also noted that yield attrition from chronic injuries by stem borer deadheart damage and defoliation is underestimated. Although this study represented the macro-level scenario, micro-level yield losses due to any single or combination of insect pests can be high and deserves to be mitigated. Deutsch et al. (2018) predict future increases in yield loss in rice because of insect pests under the scenario of global warming.

Past experience has clearly shown that any unilateral approach based on chemical control, plant resistance, biocontrol, or behavioral means with pheromones has not provided desirable and sustainable solutions to pest problems. However, an early lead in exploring host-plant resistance taken by the International Rice Research Institute and emulated by various national programs paved the way for breeding for insect resistance with exemplary success against pests such as striped stem borer, gall midge, and planthoppers. But the first wave of success was countered by the rapid evolution of virulent biotypes. With the recent advances in molecular biology and biotechnology, researchers now have a new set of tools with which they can address several problems at the molecular level and identify new strategies to overcome old problems. In this chapter, we examine progress made through the classical approach and how the limitations of classical approach-based insect resistance and breeding for multiple and durable insect resistance in rice are being overcome with molecular marker-based approaches. In addition, we attempt to present the future scenario of genomics-based tools that may provide novel strategies of pest management.

2 Classical Approach Through Host-Plant Resistance Gene Deployment

Following the seminal publication of R.H. Painter (1951), genetic resistance in the host plant has been extensively explored and plant resistance (R) genes have been transferred to elite cultivars of field crops and other economically important plant species. Classical breeding methods and phenotypic selections were followed to achieve this until molecular markers were discovered. Currently, desired R-genes can be transferred and pyramided through marker-assisted selection and breeding. The status of donor sources, genetics of resistance, tagging and mapping of R-genes, and reported gene-linked markers are provided in what follows for the major insect pests of rice.

2.1 Gall Midge

Asian rice gall midge (ARGM) is a serious pest of rice in South and Southeast Asia. In India, gall midge damage is estimated to cause an annual yield loss of about USD 80 million (Bentur et al. 2003). The insect displays a unique life cycle, which is completed in 3–4 weeks. Maggots hatched from eggs laid on the plant surface crawl down between leaf sheaths to reach the apical meristem to feed. The insect feeds by laceration of the apical meristem and secretion of saliva, resulting in hypertrophy and hyperplasia of cells, ultimately leading to the development of a nutritive tissue and a gall chamber in the tiller. The insect also renders the tiller sterile and arrests further differentiation. Maggots cease feeding in the third instar and pupate in the gall. The adult fly emerges from this modified sheath gall called "silver shoot," which is a typical symptom of gall midge damage. ARGM is predominantly a vegetative-stage pest and, in the event of a high percentage of tillers being converted into galls, there will be a proportionate decrease in the number of productive tillers, panicles, and therefore grain yield.

Rice varieties differ in their response to gall midge infestation. A small proportion of varieties is immune to pest attack by effectively killing the maggot within hours of feeding. The resistance mechanism displayed by the varieties is categorized into two distinct types. A majority of the resistant rice genotypes express a hypersensitive reaction (HR), leading to tissue necrosis at the site of maggot feeding, and are referred to as HR +ve (HR+) types, whereas a few of the resistant genotypes do not display HR but still maggot mortality is noticed, and these are termed HR −ve (HR−) types. The role of phenols in HR+ resistance has been reported (Amudhan et al. 1999). Because the nature of resistance is antibiosis in both HR+ and HR− types, host-plant resistance is the most effective way of managing the pest (Bentur et al. 2003).

Field and greenhouse evaluations of more than 50,000 germplasm accessions resulted in the identification of more than 300 primary sources of resistance (Bentur et al. 2011). Studies on the genetics of gall midge resistance in rice have often shown the involvement of a single dominant or recessive gene. To date, 12 genes conferring resistance against the pest have been reported (Himabindu et al. 2010; Leelagud et al. 2020), 10 of which are dominant (*Gm1* through *Gm11*, except *gm3 and gm12*). The presence of gall midge biotypes within India was suspected during the early phase of breeding for resistance. So far, seven distinct biotypes have been characterized based on their reaction pattern against five groups of differential rice varieties (Vijayalakshmi et al. 2006). Similar to the interaction between pathogens and their plant hosts, a gene-for-gene interaction has been reported between rice resistance genes (i.e., R genes) and gall midge biotypes (Nair et al. 2011). Each of the biotypes displays a specific range of virulence against R-genes, and likewise each R-gene confers resistance to specific biotypes, which also implies that none of the R-genes conferred resistance to all biotypes and none of the biotypes showed virulence against all of the R-genes. Hence, it is possible to extend the range of resistance against biotypes by combining diverse resistance genes through gene pyramiding.

Of the 12 gall midge resistance genes identified thus far, 10 (*Gm1*, *Gm2*, *gm3*, *Gm4*, *Gm5*, *Gm6*, *Gm7*, *Gm8*, *Gm11*, and *gm12*) have been tagged and mapped with reported linked markers (Table 1). As it is well known that single gene-conferred resistance against gall midge can break down within a short time, the strategy of pyramiding two or more genes with divergent mechanisms of resistance (i.e., HR+ and HR−) has been advocated for durable resistance against the insect pest (Sundaram et al. 2013). So far, three gall midge resistance genes, *gm3* (Sama et al. 2014), *Gm4* (Divya et al. 2015), and *Gm8* (Divya et al. 2018b), have been cloned and characterized, and another gene, *Gm2*, has been reported to be allelic to *gm3* (Sama et al. 2014; Sundaram 2007). The recessive resistance gene, *gm3*, which displays HR+, encodes an NB-ARC (NBS-LRR) domain-containing protein, while the dominant gene, *Gm4*, which also displays HR+, encodes a leucine-rich repeat (LRR) protein, and *Gm8*, displaying HR−, encodes a proline-rich protein (PRP). It is desirable to deploy two or more previously undeployed genes that differ in their mechanism of resistance, for example, *Gm4* (HR+) and *Gm8* (HR−) genes would meet the above-specified requirements. Another gene, *gm3*, may also be considered

Table 1 List of rice resistance genes against major insects tagged and mapped along with information on linked markers

Gene	Chromosome no.	Primer name	Primer physical position (bp)	F primer	R primer	Reference
Brown planthopper						
Bph 1	12L	BpE18-3 STS	?	CGCTGCGAGAGTGTGACACT	TTGGGTTACACGGGTTTGAC	Kim and Sohn (2005)
Bph 1	12L	pBPH9F	22,858,861	AGCGCTGGTCGTTGGGGTTGTAGT	ATTAAAAGTGATCGCAGCCGTTCG	Cha et al. (2008)
bph2	12L	KAM4STS	?	TAACTGGTGTTAGTGCGAATGC	AATTCACGGCATGTGAAGCCCTAG	Murai et al. (2001)
bph2	12L	RM7102	13,213,987	GGGCGGTTCGGTTTACTTGGTTACTCG	GGCGGCATAGGAGTGTTTAGAGTGC	Sun et al. (2006)
bph2	12L	RM463	22,125,420	GAGGATTAATTAGCGTGTGACC	GTCGTGACATCTACTCAAATGG	Sun et al. (2006)
BPh3	6S	RM190	1,765,637	CTTTGTCTATCTCAAGACAC	TTGCAGATGTTCTTCCTGATG	Jairin et al. (2007a)
BPh3	6S	RM589	1,381,875	GTGGCTTAACCACCACATGAGAAACTACC	TCACATCATTAGGTGGCAATCG	Jairin et al. (2007a)
BPh3	6S	RM588	1,612,412	TCTTGCTGTGTGCTGTTAGTGTACG	GCAGGACATAAATACTAGGCATGG	Jairin et al. (2007a)
BPh3	6S	RM19291	1,216,883	CACTTGCACGTGTCCTCTGTACG	GTGTTTCAGTTCACCTTGCATCG	Jairin et al. (2007b)
BPh3	6S	RM8072	1,409,335	GATCACTCAGGTCATCCATTC	AATCAGAGAGGCTAAAGACAATAAT	Jairin et al. (2007b)
bph4	6S	RM586	1,477,792	TGCCATCTCATAAACCCACTAACC	CTGAGATACGCCAACGAGATACC	Jairin et al. (2010)
bph4	6S	RM589	1,381,875	GTGGCTTAACCACCACATGAGAAACTACC	TCACATCATTAGGTGGCAATCG	Jairin et al. (2010)

(continued)

Table 1 (continued)

Gene	Chromosome no.	Primer name	Primer physical position (bp)	F primer	R primer	Reference
bph4	6S	RM217	?	ATCGCAGCAATGCCTCGT	GGGTGTGAACAAAGACAC	Kawaguchi et al. (2001)
bph4	6S	RM225	3,417,595	TGCCCATATGGTCTGGATG	GAAAGTGGATCAGGAAGGC	Kawaguchi et al. (2001)
Bph6	4L	RM6997	21,272,870	CGGCAGTAAATTTGCATTGACC	AGTGGCCTTGTCAGTCTACATGC	Qiu et al. (2010)
Bph6	4L	RM5742	21,559,553	GATCCTCAAACGGCCTCTGC	CCTTCAAAGTTTACTCACGCTCTGC	Qiu et al. (2010)
Bph9	12L	RM463	22,125,420	GAGGATTAATTAGCGTGTGACC	GTCGTGACATCTACTCAAATGG	Su et al. (2006)
Bph9	12L	RM5341	19,114,746	CATCCGGAGGAAGTTTGAAAGAAGG	CAAGGGCAACCTCTTCCACTACGC	Su et al. (2006)
Bph10[a]	12L	RG457FL/RB STS marker		?	?	Lang and Buu (2003)
Bph10[a]	12L	RG457FL/RL STS marker		?	?	Lang and Buu (2003)
Bph12[a]	4L	RM16459	5,213,984	TCCAGGAGTTTGCCTTGTAGTGC	TAGCGAAGTCAGGATGGCATAGG	Qiu et al. (2012)
Bph12[a]	4L	RM1305	5,659,601	GGTACTACAAAGAAACCTGCATCG	TCCTAGCTCAAATGTGCTATCTGG	Qiu et al. (2012)
Bph13[a]	2L	RM240	31,497,147	CCTTAATGGGTAGTGTGCAC	TGTAACCATTCCTTCCATCC	Liu et al. (2001)
Bph13[a]	2L	RM250	32,774,365	GGTTCAAACCAAGCTGATCA	GATGAAGGCCTTCCACGCAG	Liu et al. (2001)
BPh13[a]	2L	AJ09b StS from RAPD	?	?	?	Renganayaki et al. (2002)
Bph14[a]	3L	SM1	38,751,114	AGCGTTAAGCGCCATTATCA	CGCCGAGGCATTAGAGTAGA	Du et al. (2009)

Table 1 (continued)

Gene	Chromosome no.	Primer name	Primer physical position (bp)	F primer	R primer	Reference
Bph14[b]	3L	NBS-LRR		AAATTCGTGGTTGTTCTGGT	TCGGTAACGATCCATGATGA	Du et al. (2009)
Bph15[a]	4S	MS1 reported earlier by C820 and S11182 RFLP marker	9,278,808	CATGGACCCACTTGTCATCC	AGCATGAGAGACTGCCAAGG	Yang et al. (2004)
Bph15[a]	4S	RM261 reported earlier by C820 and S11182 RFLP marker	6,578,953	CTACTTCTCCCCTTGTGTCG	TGTACCATCGCCAAATCTCC	Yang et al. (2004)
Bph16[a]	12L	RM6732	21,983,109	AATTTTGAAACACCTCAAAGG	TTTTCAGTGCATGTCTTCG	Hirabayashi et al. (2004)
Bph17	4S	RM8213	4,446,064	TGTTGGGTGGGTAAAGTAGATGC	CCCAGTGATACAAAGATGAGTTGG	Sun et al. (2005)
Bph17	4S	RM5953	9,379,510	AAACTTTCTGTGATGGTATC	ATCCTTGTCTAGAATTGACA	Sun et al. (2005)
Bph17[b]	4S	LRK2	6,953,940	CTTTCGCAGGGTGGCAAATAGGGT	CCTTCGCTGCTCACTAGGACCGTGTA	Liu et al. (2015)
Bph18[a]	12L	RM463	22,125,420	GAGGATTAATTAGCGTGTGACC	GTCGTGACATCTACTCAAATGG	Jena et al. (2006)
Bph18[a]	12L	S15552 STS marker cleaved by alu1 enzyme		?	?	Jena et al. (2006)
Bph18[a]	12L	RM511	17,401,530	AACGAAAGCGAAGCTGTCTCC	ATTTGTTCCCTTCCTTCGATCC	Suh et al. (2011)

(continued)

Table 1 (continued)

Gene	Chromosome no.	Primer name	Primer physical position (bp)	F primer	R primer	Reference
Bph18[a]	12L	RM1584	27,104,354	TAGCCTGCAGCCACCCTGATCC	CAATGTGACTTCCGTGCGTAGTGG	Suh et al. (2011)
bph19	3S	RM6308	7,181,156	TCTCGACCTGGCTCTCCTCTAGC	AGTGCACGGACATGTCACTCTCG	Chen et al. (2006)
bph19	3S	RM3134	7,240,409	GCAGGCACAAAAGCAAAGAG	AGGTGAAGGTGCATTGTGTG	Chen et al. (2006)
Bph20[a]	4S	MS10 STS marker	8,071,921	CAATACGAGAAGCCCCTCAC	CTGAAGGAACACGCGGTAGT	Rahman et al. (2009)
Bph20[a]	4S	RM5953	9,379,510	AAACTTTCTGTGATGGTATC	ATCCTTGTCTAGAATTGACA	Rahman et al. (2009)
bph20[a]	4S	RM435 BYL7 closest marker	535,684	AAGCTAGGGAATCAGCGGTTA BYL	TGTGGCATGTCACTCACTCAC	Yang et al. (2012)
bph20[a]	4S	RM540 BYL8 closest marker	468,236	CCCACTTCCACAACCACA BYL8	ATGCTCCTAGCTTCCTATTCC	Yang et al. (2012)
Bph21[a]	12L	RM3726	23,275,187	TACACCCACCCACATACGTCAGC	GTCGTACTCCCGGATCTTCTTCC	Rahman et al. (2009)
Bph21[a]	12L	RM5479	24,412,536	CTCACCATAGCAATCTCCTGTGC	ACTTCGTTCACTTGCATCATGG	Rahman et al. (2009)
bph21	10S	RM222	2,620,380	CTTAAATGGGCCACATGCG	CAAAGCTTCCGGCCAAAAG	Yang et al. (2012)
bph21	10S	RM244		CCGACTGTTCGTCCTTATCA	CTGCTCTCGGGTGAACGT	Yang et al. (2012)
bph22	4	RM8212	99,416	CCACCGCACTTGTCTATG	TCCAATCTCACTCTCGACTC	Hou et al. (2011)

Table 1 (continued)

Gene	Chromosome no.	Primer name	Primer physical position (bp)	F primer	R primer	Reference
bph22	4	RM261	6,574,396	CTACTTCTCCCCTTGTGTCG	TGTACCATCGCCAAATCTCC	Hou et al. (2011)
Bph22[a]	6	RM19429Rm584	3,417,532	TATGTGGTTGGCTTGCCTAGTGG	TGCCCATATGGTCTGGATGTGC	Harini et al. (2010)
Bph22[a]	6	RM584	3,417,532	TATGTGGTTGGCTTGCCTAGTGG	TGCCCATATGGTCTGGATGTGC	Harini et al. (2010)
Bph22[a]	6	RM585	3,169,373	CTAGCTAGCCATGCTCTCGTACC	CTGTGACTGACTTGGTCATAGGG	Harini et al. (2010)
bph23[a]	8	RM2655	2,017,692	TGTCTGTGTTGTCACTGCCTTATCG	TCCGCTCTGTGTGTATCTGATCTGG	Hou et al. (2011)
bph23[a]	8	RM3572	3,928,306	CCATTTGGTAGGTCCATCTTACCC	CTCCCAAGTGAAGTGCTGTCTGG	Hou et al. (2011)
Bph25	6S	S00310	214,278	CAACAAGATGGACGGCAAGG	TTGGAAGAAAGGCAGGCAC	Myint et al. (2012)
Bph25	6S	RM8101	1,705,742	CACTGACATAGCTAAGGTCTCATGTCTTAT	TGGTTAACTCGCTATTATAATGAGTTCG	Myint et al. (2012)
Bph26	12L	RM5479	24,409,241	CTAAGCTCACCATAGCAATC	ATACACTTCTCCCCTCTCTG	Myint et al. (2012)
Bph26	12L	MSSR2	25,033,993	CATGTCGAAGAGGTTGCAGA	GGTTTCATCCAAGTCCACGA	Myint et al. (2012)
Bph26[b]	12L	LRR	22,884,874	TAGCATCAGTCCCTTGCTTGTTTGC	ATTGATTTAATTAGCAGACAAGTTG	Tamura et al. (2014)
Bph27[a]	4	RM16853	19,201,791	CTCCCCATCCTTCATTTCATCTCG	CTTTCTGCAAGACACTGCAAACG	Huang et al. (2013)
Bph27[a]	4	RM16846	19,115,583	CTACAAGCAACACAGTATCACAGC	GGTAACTGGTGCTTATTTAGCC	Huang et al. (2013)

(continued)

Table 1 (continued)

Gene	Chromosome no.	Primer name	Primer physical position (bp)	F primer	R primer	Reference
Bph28	11	RM26656	15,738,361	GCAAAGAACATTGTGGCAAACACC	TGGACACATTGTATCGTGCTTCG	Wu et al. (2014)
BPh28	11	RM26725	17,114,046	ATGTAGGCCCAAACGAGCTCTGACC	ATGCCATAGTAGCGCTTGCGTATCC	Wu et al. (2014)
bph29[a]	6	RM435 BYL7 closest marker	535,684	AAGCTAGGGAATCAGCGGTTA BYL	TGTGGCATGTCACTCACTCAC	Wang et al. (2008)
bph29[a]	6	RM540 BYL8 closest marker	468,236	CCCACTTCCACAACCACA BYL8	ATGCTCCTAGCTTCCTATTCC	Wang et al. (2008)
bph29[b]	6	G5 B3 domain	484,709	ATGGCCACCATTGTTGCATG	TCAAAGCTGCAAATCCAGCG	Wang et al. (2015)
bph30[a]	10	RM222	2,620,380	CTTAAATGGGCCACATGCG	CAAAGCTTCCGGCCAAAAG	Wang et al. (2008)
bph30[a]	10	RM244		CCGACTGTTCGTCCTTATCA	CTGCTCTCGGGTGAACGT	Wang et al. (2008)
BPH31	3	PA26		?	?	Prahalada et al. (2017)
BPH31	3	RM2334	26,740,452	CATGCATCTGATCTGATTAT	TGTGAAGAGTACAAGTAGGG	Prahalada et al. (2017)
Bph32	6	RM19291	1,215,884	CACTTGCACGTGTCCTCTGTACG	GTGTTTCAGTTCACCTTGCATCG	Ren et al. (2016)
Bph32	6	RM8072	1,408,336	GATCACTCAGGTCATCCATTC	AATCAGAGAGGCTAAAGACAATAAT	Ren et al. (2016)
Bph32[b]	6	SCP	1,099,689	TGAGGGAGTTGTAGTAGGAGTA	CGTCGTTGATGAAGTAAAGGT	Ren et al. (2016)
Bph33	1	RM11522	28,071,767	TAACTGCAGTGCTCAACAAAGG	CTAGGTACCGGATTAAGATTCACC	Naik et al. (2018)

Gene	Chromosome no.	Primer name	Primer physical position (bp)	F primer	R primer	Reference
Bph33	1	RM488	24,808,556	AACAAACCAGCGTATGCGTTCTCG	CCCACGGCTTTGTAGGAAGAAGC	Naik et al. (2018)
Bph34[a]	4	RM16994	21,331,444	TGGCAGTACACACTACAGTACATGC	AGAGGGAGGAGAGAAAAGGAAGG	Kumar et al. (2018)
Bph34[a]	4	RM17007	21,484,908	CTTCCCACGCGAAACTTCATGG	TCCGGCCAAGACAATATCAACG	Kumar et al. (2018)
White-backed planthopper						
Wbph7	3L	R1925		?	?	Tan et al. (2004)
Wbph7	3L	G1318		?	?	Tan et al. (2004)
Wbph8	4S	R288		?	?	Tan et al. (2004)
Wbph8	4S	S11182		?	?	Tan et al. (2004)
Wbph9t	6	RM589	1,380,866	ATCATGGTCGGTGGCTTAAC	CAGGTTCCAACCAGACACTG	Ramesh et al. (2014)
Wbph9t	6	RM539		GAGCGTCCTTGTTAAAACCG	AGTAGGGTATCACGCATCCG	Ramesh et al. (2014)
Wbph10t	12	SSR12-17.2				Ramesh et al. (2014)
Wbph10t	12	RM28487	23,142,514	GAGGTGATCTTAATGCCATCTTGACG	TACATGCAACCTGGGTATGAGAGTGC	Ramesh et al. (2014)
Wbph11t	4	RM3643	19,948,112	AGCATGAGCAGGTGCTAGTG	CGTTGCATGTGTGATGGC	Ramesh et al. (2014)
Wbph11t	4	RM1223	25,292,767	CAGCGTCTCCAAGAAACTCC	GCTACCAGGTCAGAGTTGCC	Ramesh et al. (2014)

(continued)

Gene	Chromosome no.	Primer name	Primer physical position (bp)	F primer	R primer	Reference
Wbph12t	4	RM16913	19,646,468	GTGTACGTGTTGGCTCTCTGTACG	GATGTTGCTTGTGTGCTGCAACC	Ramesh et al. (2014)
Wbph12t	4	RM471	19,426,246	CGGATCCAAGAAAACAGCAG	TTCGGTATCCTCCACACCTC	Ramesh et al. (2014)
Ovc	6S	S1520		?	?	Yamasaki et al. (2003)
Ovc	6S	L688		?	?	Yamasaki et al. (2003)
Green rice leafhopper						
Grh1	5	C309		?	?	Tamura et al. (1999)
Grh1	5	R569		?	?	Kadowaki et al. (2003)
Grh2	11L	G144		?	?	Fukuta et al. (1998)
Grh2	11L	G4001		?	?	Kadowaki et al. (2003)
Grh2	11L	G1465		?	?	Kadowaki et al. (2003)
Grh3	6	C288B		?	?	Saka et al. (2006)
Grh3	6	C133A		?	?	Saka et al. (2006)
Grh4	3L	G1465		?	?	Fukuta et al. (1998)
Grh4	3L	C1186		?	?	Kadowaki et al. (2003)

Gene	Chromosome no.	Primer name	Primer physical position (bp)	F primer	R primer	Reference
Grh4	3L	R2982		?	?	Kadowaki et al. (2003)
Grh5	8L	RM502	26,492,117	GCGATCGATGGCTACGAC	ACAACCCAACAAGAAGGACG	Fujita et al. (2006)
Grh5	8L	RM6845	27,560,145	GTGACGGCAAGAGGAAGAAG	GTTCGACAGGAACGCCAC	Fujita et al. (2006)
Grh6	4S	Y3635R		?	??	Tamura et al. (2004)
Grh6	4S	C708		?		Tamura et al. (2004)
Grh6	4S	RM8213	4,441,638	AGCCCAGTGATACAAAGATG	GCGAGGAGATACCAAGAAAG	Fujita et al. (2004)
Grh6	4S	G6-9				Fujita et al. (2004)
QGRH9	9L	RM201	20,174,289	CTCGTTTATTACCTACAGTACC	CTACCTCCTTTCTAGACCGATA	Fujita et al. (2010)
QGRH9	9L	RM205	22,720,624	CTGGTTCTGTATGGGAGCAG	CTGGCCCTTCACGTTTCAGTG	Fujita et al. (2010)
Gall midge						
Gm1	9	RM23941	7,818,607	AGAATCGAACCCTAACACATGC	TATCGCTTGATTCTTGGACAGC	Sundaram (2007)
Gm1	9	RM23956	7,992,661	GTCTCTCCCTCTCTCATCTTGTCG	CCCTATTCATGTGCAATGGAACC	Sundaram (2007)
Gm2	4	RM17473	30,838,175	TCTCTCCAGCTCCCTAAACATTCC	AGCGACACACTGTTCACCTTGC	Sundaram (2007)
Gm2	4	RM17503	31,717,625	CCAGATCATCCAGGCATAACATCACC	CGGCGCTGGTAAACTCCATTCC	Sundaram (2007)

(continued)

Gene	Chromosome no.	Primer name	Primer physical position (bp)	F primer	R primer	Reference
gm3	4	RM17480	31,146,877	GAGTTCGTCCCTGACAAACAGAAACG	GTGAGCGAGCGAGTGAGTGAGC	Sama et al. (2014)
gm3	4	gm3SSR4	32,075,106	AGACACGAGGGAATTATGC	CTCTATATTTGCCGCATCC	Sama et al. (2014)
gm3[b]	4	NB-ARC	31,950,000	CTGCCAGAGATGGGCCTTCCA	CGTACAAATTCCTGTACCACTC	Sama et al. (2014)
Gm4	8	RM22551	5,451,661	CTTCGATCTCCTCGTCCTCTTCC	GAGCATGAGATGATGCATGACG	Divya et al. (2015)
Gm4	8	RM22562	5,782,732	GATCGGAGGAGGGAGGAAGACG	GTCGCATCCACTCATATTCCAAGC	Divya et al. (2015)
Gm4[b]	8	LRR-del	5,583,333	GTGGATCGAGAGAAGACAAG	CTTGAGGACGATATTCAAGC	Divya et al. (2015)
Gm5	12	OPB14		?	?	see Bentur et al. (2003)
Gm6	4	OPM061400		?	?	Katiyar et al. (2001)
Gm6	4	RG214		?	?	Katiyar et al. (2001)
Gm6	4	RG476		?	?	Katiyar et al. (2001)
Gm7	4	SA598		?	?	Sardesai et al. (2002)
Gm7	4	F8		?	?	Sardesai et al. (2002)
Gm8	8	RM22685	8,800,203	ATGGGCTTCCAGGCTCAATCTCG	CCCACTCTCACGTCTCCTCTCTTCC	Sama et al. (2012)
Gm8	8	RM22709	9,226,565	CGCGTGGGCGAGACTAATCG	CCTTGACTCCGAGGATTCATTGTCC	Sama et al. (2012)

Gene	Chromosome no.	Primer name	Primer physical position (bp)	F primer	R primer	Reference
Gm8[b]	8	PRP-Del	9,127,586	TATAAAGAGGACGGTCTAACCTTTA	GCACAGGGAAGTTGTCAGTTCAAGTA	Divya et al. (2018a)
Gm11	12	RM28574	24,292,314	TAGTTTGGTGAAGTGGCATTGG	ATAGTAGGGCAAGGATTCAGAAGAGG	Himabindu et al. (2010)
Gm11	12	RM28706	25,992,979	GGTTCCCGGTCATCATATTTCC	ACTTTACCCACGCGCTTTGC	Himabindu et al. (2010)
gm12	2	RM3340	386,212	GAGAGAGACACCAAATGATCCATCC	ACTGATTTGGCCCTTGTTCTTGG	Leelagud et al. 2020
gm12	2	S2-76222	76,222	CACACATACCCACCCACTAGTGAAGATGAA[C/T]TGAATGCCAATGGGAGAAGAGGAGGAGCAGA		Leelagud et al. 2020

? = sequence information not available in the publication cited

F primer and R primer = forward and reverse primers, respectively

[a]Gene has been introgressed from wild species

[b]Cloned gene

for pyramiding since it is a HR+ type and recessively inherited and has not been deployed so far in any variety.

Using *Gm4* and *Gm8*, the research group at ICAR-IIRR (Abhilash Kumar et al. 2017) has developed gene-pyramided lines in the genetic background of the elite restorer line RPHR1005R (restorer line for the popular rice hybrid DRRH3) through marker-assisted breeding. In another such effort, the high-yielding rice variety Akshayadhan has been improved for its resistance against gall midge by targeted transfer of *Gm4* and *Gm8* genes. Sama et al. (2014) introduced the recessive gene *gm3* into the genetic background of elite rice variety Improved Samba Mahsuri with the help of markers. In a recent report (Venkanna et al. 2018), two major resistance genes, *gm3* and *Gm8*, have been pyramided in the genetic background of the fine-grain-type rice variety Kavya, which already possesses *Gm1*. Now that closely linked markers/functional markers are available for all the major gall midge resistance genes, selected gene combinations can be pyramided into elite genetic backgrounds (Divya et al. 2018c) easily through marker-assisted breeding for developing durable multiARGM biotype-resistant rice cultivars/hybrids.

2.2   Planthoppers and Leafhoppers

Although more than 100 species of planthoppers (Delphacidae) and leafhoppers (Cicadellidae) are reported to feed on rice, three species of planthoppers (BPH, WBPH, and SBPH) have gained high economic importance since 2000 (Bentur and Viratkamath 2008). Likewise, of the leafhoppers, GLH, GRL, and ZLH are important. Although both groups consist of phloem sap feeders, planthoppers cause severe damage by feeding alone, leading to total death of plants, termed hopper burn. Both leafhoppers and planthoppers vector disease-causing viruses and cause indirect damage to the crop. The main virus diseases thus transmitted are rice ragged stunt, rice grassy stunt, and wilted stunt by BPH; southern rice black-streaked dwarf by WBPH; stripe and black-streaked dwarf by SBPH; and rice tungro by several species of leafhoppers. In response to planthoppers and leafhoppers gaining importance, screening of rice germplasm for resistance began at the International Rice Research Institute (IRRI), Philippines, during the 1970s. Such initiatives were also taken up in other Asian rice-growing countries, leading to reports of a large number of resistance sources (IRRI 1979; Heinrichs et al. 1985; Heong and Hardy 2010). Several of these sources were selected for systematic studies on genetics of resistance, resulting in the identification of more than 38 major resistance genes against BPH, 14 against WBPH, 14 against GLH, six against GRL, and three against ZLH (Fujita et al. 2013; Ling and Weillin 2016; Du et al. 2020). Most of these genes are now tagged and mapped on different rice chromosomes and reliable molecular markers linked to these traits are available (Table 1). Marker-assisted selection as a tool for breeding for BPH resistance using a single gene or multiple genes is being reported (Liu et al. 2016; H Wang et al. 2016a; Y Wang et al. 2017b; Jiang et al. 2018). However, a few issues remain to be resolved. Several of the reported BPH Rgenes are not effective on the Indian subcontinent (Horgan et al. 2016); hence, more effective genes for the region need to be characterized from the reported sources. All the BPH Rgenes, except probably *Bph3* (Liu et al. 2015), are not effective against WBPH. Since these two species are sympatric and are often under severe interspecific competition (Srinivasan et al. 2016), selecting for resistance against BPH alone may not be the right approach. But, efforts to tag, map, and clone WBPH resistance genes are few. Another limitation is the ability of BPH to quickly evolve virulent biotypes, especially if a single Rgene is deployed. Hence, gene pyramiding is suggested for durability. Ideally, undeployed genes with different mechanisms of resistance are the choice for pyramiding. Of the 13 BPH Rgenes cloned, eight (*Bph14*, *Bph26*, *Bph18*, *Bph9/1/7/10/21*) represent the NBS-LRR family; proteins coded by these are located in the cytoplasm, while others are reported as lectin receptor kinases (*Bph3*, *Bph15*), B3 DNA-binding domain (*bph29*), or SCR domain protein (*Bph32*) coding (Ren et al. 2016; Y Zhao et al. 2016b; Du et al. 2020), which are membrane bound. It is suggested to combine two genes from these two classes (such as *Bph14* + *Bph15*) to achieve durability (Jing et al. 2017).

2.3 Other Pests

Rice stem borers are ubiquitous insects representing Diopsidae (Diptera), Noctuidae, and Crambidae (Lepidoptera) families. Among several species of rice stem borers, YSB is considered the most economically important insect pest in almost all rice-growing countries of Asia (Makkar and Bentur 2017). Larvae of YSB feed only on rice and can cause damage at both the vegetative (deadheart) and reproductive (whitehead) stages of the rice crop, with the latter being the main cause of yield loss (Savary et al. 2000). Because of the lack of highly resistant sources in rice germplasm explored so for, breeding for resistance or molecular mapping of resistance genes against YSB has not been encouraging (Bentur 2007). Nonetheless, rice varieties such as Vikas, Ratna, and Sasyasree have been developed and relcascd in India with moderate YSB resistance. Most of these varieties have either TKM6 or W1263 as the source of resistance. Because of the lack of the desired level of resistance against YSB in the primary gene pool of rice, the secondary gene pool consisting of wild species of *Oryza* is being explored at IRRI, ICAR-IIRR, and other institutes. Chromosome segment substitution lines (CSSLs) need to be developed for different accessions of wild rice that can be evaluated for YSB resistance. Also, ethyl methanesulfonate (EMS) mutants of rice have been generated and evaluated at ICAR-IIRR and have shown encouraging results in preliminary evaluations. More extensive and concerted efforts in this direction have potential to identify novel sources of resistance that can be used by breeders and entomologists for understanding the resistance mechanisms and developing YSB-resistant rice cultivars.

3   Novel Approaches Through Genomics

3.1   RNAi Approach for Insect Resistance

Transgenic crops harboring *Bt* endotoxin genes or other insecticidal protein-coding genes have shown tremendous potential for managing insect pests. Several of these genes have been used in transforming rice as described in the next section, although none of these have been commercially cultivated. As an alternative to this approach, RNA interference (RNAi) can be exploited, which has been well demonstrated for resistance induction in plants against viruses, bacteria, and nematodes. RNAi is an RNA-driven post-transcriptional homology-based gene-silencing mechanism through the mRNA degradation pathway present in all eukaryotic organisms. The RNAi is triggered by double-stranded RNAs (dsRNA), which are processed by the RNase-III-like Dicer protein to produce small interfering RNAs (siRNAs). The guide strand of siRNA directs an RNA-induced silencing complex (RISC) to the target mRNA (Fig. 1). The most important constituent of RISC is RNase protein Argonaute, which helps in the degradation of target mRNAs sharing homology with the guide strand of siRNA (Zamore et al. 2000). The double-stranded RNAs specific to key insect genes can be stably expressed in plant tissues fed on by the insect and that in turn can trigger the RNAi pathway to degrade the mRNAs transcribed by the key insect genes (Price and Gatehouse 2008; Agarwal et al. 2012).

Key genes in insects are identified as targets of RNAi, that is, genes coding developmental proteins, digestive enzymes, salivary gland proteins, nervous system regulatory proteins, proteins involved in host-insect interaction, hormone receptors, gut enzymes, and proteins involved in metabolism (Gatehouse 2008; Huvenne and Smagghe 2010; Agarwal et al. 2012; Kola et al. 2015).

Initial successes in experiments (Tomoyasu and Denell 2004; Turner et al. 2006) raised hope among researchers that RNAi could be another alternative and effective tool to develop insect resistance in crop plants. Initially, dsRNAs were delivered to the target insects either by injection or through artificial diet. Baum et al. (2007) demonstrated the effectiveness of host-plant-mediated production of dsRNA in crop protection. Transgenic maize plants producing insect-specific vacuolar H⁺ ATPase dsRNAs had decreased root damage by western corn rootworm. In another similar report, Mao et al. (2007) generated transgenic *Nicotiana tabacum* and *Arabidopsis thaliana* targeting RNAi against the cytochrome P450 gene of *Helicoverpa armigera*, resulting in retarded larval growth of insects feeding on these modified hosts. The versatility of the application of RNAi against different insect orders and target genes shows the potential of RNAi for managing diverse crop pests (Terenius et al. 2011; Khajuria et al. 2015; Zhang et al. 2017). Recent reports suggest that the production of dsRNAs in chloroplasts, rather than in cytoplasm, can improve insect resistance significantly as long as dsRNAs can be stably produced in chloroplasts, which are devoid of RNAi machinery (Zhang et al. 2015; Jin et al. 2015; Bally et al. 2018). The first such RNAi-based DvSnf7 dsRNA-expressing maize crop targeting western corn rootworm is scheduled to be commercialized (Khajuria et al. 2018). Several research groups have been working on modifying the technology for its more efficient application.

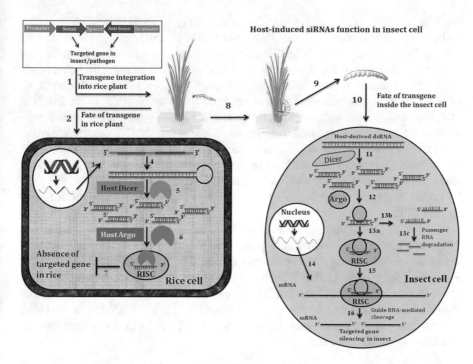

Fig. 1 Schematic representation of host-induced gene silencing in insects through siRNA approach. (**1**) Integration of insect gene-targeted siRNA cassette (transgene) into rice genome; (**2**) fate of transgene in rice cells; (**3**) expression of transgene in rice cell generates the mRNA; (**4**) formations of dsRNA through self-complementation of transgene's sense and antisense strands in rice cell; (**5**) host Dicer-mediated specific cleavage of dsRNA leads to production of siRNAs in rice cell; (**6**) host-generated siRNAs processed by host Argonaute protein (the main component of RNA-induced silencing complex or RISC); (**7**) host-generated siRNAs are nonfunctional in rice cells due to absence of targeted gene; (**8 and 9**) host-synthesized dsRNAs/siRNAs transfer from rice plant to insect through feeding on rice tissues; (**10**) fate of transferred dsRNA/siRNAs in insect cells; (**11**) generation of siRNAs from dsRNA through insect Dicer-mediated cleavage; (**12**) the siRNAs are processed by insect Argonaute proteins/RISC complex; (**13a**) formation of activated RISC along with target-specific guide RNA; (**13b and 13c**) the passenger RNA is separated from guide RNA and degraded; (**14**) transcription of insect DNA resulted in the expression of targeted functional mRNA (transcript); (**15**) guide strand of siRNA helps in identification and binding of activated RISC to the targeted mRNA; (**16**) silencing of targeted gene expression by RISC-mediated cleavage of corresponding mRNA

Application of the RNAi tool for insect resistance in rice is in the developmental phase. Most of the reports on RNAi in rice are centered on BPH (Zha et al. 2011; Zhou et al. 2013; Yu et al. 2014; Wang et al. 2018) and YSB (Renuka et al. 2017). RNAi has been used for functional genomics of glutathione *S*-transferase (GST) genes, which are involved in the degradation of toxins produced by host plants and insecticides. Injecting dsRNAs targeting the NlGSTe1 and NlGSTm2 genes into nymphs of BPH enhanced their sensitivity to chlorpyrifos but not to beta-cypermethrin. Through feeding assays and stable expression of NlEcR (ecdysone receptor gene) targeting dsRNAs in rice, Yu et al. (2014) showed significant down-regulation of target gene expression and a decrease in the number of offspring produced by BPH adults. Likewise, Zha et al. (2011) targeted three midgut genes, carboxypeptidase (Nlcar), hexose transporter (NlHT1), and trypsin-like serine protease (Nltry). L Zhao et al. (2016a) aimed at trehalase genes involved in chitin biosynthesis and degradation. Wang et al. (2018) aimed at the calmodulin gene, Waris et al. (2018) aimed at the chemosensory protein 8 (*CSP8*) gene, and Zhu et al. (2017b) studied the function of the ribosomal protein gene (*NlRPL5*) using an RNAi tool. Pan et al. (2018) used RNAi to knock down 135 CP (chitin and cuticular protein) genes by injecting specific dsRNAs and showed that 32 CPs are necessary for normal egg production and development. Li et al. (2015) suggested that dsRNAs are stable under diverse environments and can be absorbed by roots of crop plants. This study provides scope to use dsRNAs as biopesticides. The above-cited studies are laying the foundation for the development of RNAi as a tool for managing rice pests such as BPH.

Kola et al. (2016) showed by feeding YSB larvae with dsRNA of cytochrome P450 derivative (*CYP6*) and amino peptidase N (*APN*) that expression of target genes decreased and resulted in increased mortality of larvae after 12–15 days. Similarly, Zeng et al. (2018) knocked down three chemosensory protein (CSP) genes in rice leaffolder (*C. medinalis*) through injection of dsRNAs, which downregulated insect response to the specific chemicals. He et al. (2018), in contrast, overexpressed striped stem borer-derived miR-14 microRNA in rice, which resulted in a high resistance against the pest.

3.2 Genome Editing Approach for Insect Resistance

Genome editing tools enable us to edit the genome or specific genes of an organism by addition/deletion or replacement of nucleotides with high precision and with few off-target effects. Because of its simplicity and wider applicability, genome editing is being practiced in many laboratories for functional genomics and trait improvement. In agriculture, the technology has immense potential to improve yield and abiotic and biotic stress tolerance of crops. Also, the technology does not attract many concerns regarding biosafety regulators, specifically in the case of deletion of nucleotides. Most of the research on genome editing has been focused on functional genomics, trait discovery, and improvement in plants (Arora and Narula 2017; Yin et al. 2017; Aglawe et al. 2018).

The use of genome editing techniques for pest control in plants has just been hypothesized. Zuo et al. (2017) created a mutation in *Spodoptera exigua* with CRISPR/Cas9 technology, which resulted in a mutant insect with high resistance against chlorantraniliprole, cyantraniliprole, and flubendiamide insecticides. To demonstrate the role of the cadherin gene in developing resistance against *Bt* toxin, the gene was edited by CRISPR/Cas9 in *Helicoverpa armigera*. The mutant strain of the insect showed high resistance to Cry1Ac (J Wang et al. 2016b). The pheromone-binding protein 1 (PBP1) gene of *H. armigera* was mutated and the mutant male adults showed impaired responses to sex pheromone (ZF Ye et al.

2017b). Dong et al. (2017) mutated PBP1 and PBP3 genes in striped stem borer to demonstrate their function. Biogenesis of the lysosome-related organelles complex 1 subunit 2 (BLOS2) gene of *Spodoptera litura* was edited, which resulted in the disappearance of the yellow strips and white spots on the larval integument (Zhu et al. 2017a). Similarly, when the abdominal-A (Slabd-A) gene of *S. litura* was mutated, it resulted in ectopic pigmentation and anomalous segmentation during the larval stage (Bi et al. 2016). Recently, Xue et al. (2018) edited two eye pigmentation genes in BPH, resulting in bright red compound eyes.

Although most of these reports showed successful application of CRISPR-based genome editing technology for functional genomics of insect genes, its use for incorporating and enhancing pest resistance in crops is yet to be realized. It is possible to derive genome editing-mediated resistance against insects by targeting either the host genes or the gene drive in insect populations. There is a dearth of information with regard to insect-susceptibility genes of host plants, specifically in rice. The recessive resistance genes identified so far are likely to represent nonfunctional susceptibility genes and hence the need for more studies to characterize such candidate genes, which represent ideal targets for genome editing to develop new sources of resistance. Alternatively, novel resistance alleles can be created in susceptible rice cultivars either by replacement of a few nucleotides/motifs/domains or by editing of specific bases or transfer of a complete gene. However, precise replacement of nucleotides, base editing, and insertion of a gene through CRISPR technology are relatively complex at this stage and it may require a few more years for researchers in not-so-sophisticated laboratories to be able to use this technology.

Gene drive is another highly potential technology that can be exploited to promote the inheritance of CRISPR-generated mutated alleles or any other DNA sequence by sexual reproduction, which allows a rapid spread of genes among the insect population. Even the whole CRISPR machinery, that is, Cas9 mRNA and specific sgRNAs, can be spread into insect populations via a gene drive. Besides controlling the insect population, it can decrease vector-borne virus diseases such as rice tungro disease. However, its application requires a thorough public debate among scientists, policymakers, and regulators and other stakeholders (Courtier-Orgogozo et al. 2017). In addition, the rapid advancement in genome editing technologies will facilitate the functional analysis of insect genes, which would indirectly help in developing more effective strategies for achieving effective and durable biotic stress resistance in crops.

4 Transgenic Approach Through Gene Transfer

As naturally occurring resistance to lepidopteran pests of rice is yet to be identified/discovered (e.g., stem borers and leaffolders; Schuler et al. 1998), transgenesis offers a potent, immediate, cost-effective, and environment-friendly option for control of these pests through access and use of resistance from unrelated sources (i.e., nonrice sources). Fortunately, tissue culture and genetic transformation protocols are well established in rice and many bacterial-derived insecticidal proteins have been deployed in rice through transgenic breeding (Sundaram et al. 2013). *Bt* genes derived from the soil bacterium *Bacillus thuringiensis* have been the most successful group of related genes used commercially for genetic transformation of many crop plants, including rice. *Bt* genes encode for insecticidal proteins that are filled in crystalline inclusion bodies produced by the bacterium upon sporulation (e.g., Cry protein) or expressed during bacterial growth (e.g., Vip proteins). In addition, several research groups are assessing the potential of using non-*Bt* insecticidal proteins such as lectins (carbohydrate-binding proteins), proteinase inhibitors, ribosome-inactivating proteins, secondary plant metabolites, small RNA viruses, etc. (Makkar and Bentur 2017).

4.1   Development of Transgenic Rice for Insect Resistance

The crystal insecticidal proteins (Cry toxins or delta-endotoxins) encoded by *Bt* genes are known to possess high toxicity to lepidopteran pests (Cohen et al. 2000), Dipterans (Andrews et al. 1987), and Coleopterans (Krieg et al. 1983; Herrnstadt et al. 1986) but are nontoxic to other groups of insects, other animals, and humans. Fujimoto et al. (1993) reported the first transformation of rice with a *Bt* gene. Many reports on the development and evaluation of *Bt* rice lines have since appeared (see review by High et al. 2004; Chen et al. 2006). Rice lines expressing *Cry1Aa*, *Cry1Ab*, *Cry1Ac*, *Cry1Ab/Cry1Ac* fusion gene, *Cry1B*, *Cry1C*, *Cry2A*, and a pyramid of *Cry1Ac* with *Cry2A*, under the control of various constitutive and conditional promoters, have been shown to confer resistance to stem borers, leaffolders, and other foliage-feeding lepidopteran insects (Table 2). Several rice lines expressing insecticidal genes with anti-lepidopteran activity using *Cry* genes (*Cry1Aa*, *Cry1Ab*, *Cry1Ac*, *Cry1Ab/Ac*, *Cry1C*, *Cry2A*), CpTI (cowpea trypsin inhibitor), Vip (vegetative insecticidal protein), etc., have been reported. Various transgenic *Bt* (*Cry1Ab*, *Cry1Ac*) rice varieties (IR64, Karnal Local, etc.) resistant to YSB have been produced in India (Khanna and Raina 2002; Ramesh et al. 2004a). Pradhan et al. (2016) deployed a vegetative insecticidal protein (*vip*) in the genetic background of Swarna and demonstrated that the transgenic rice is resistant to multiple lepidopteran pests such as yellow stem borer, leaffolder, and rice horny caterpillar. Field evaluation and validation of transgenic rice possessing *Cry1A* (Shu et al. 2000; Tu et al. 2000) and synthetic *Cry1Ab* (Shu et al. 2002) have been reported from China. Field trials of *Bt* rice have also been conducted in Pakistan (Bashir et al. 2005; Mahmood-ur-Rahman et al. 2007), Spain (Breitler et al. 2004), Iran (James 2005), and India (Bunsha 2006). Iran was the first country to release *Bt* rice for commercial cultivation in 2004 (Makkar and Bentur 2017). China permitted the commercial production of *Bt* rice lines Huahui No. 1 (CMS restorer line) and *Bt* Shanyou 63 (a hybrid of Huahui No. 1 and Zhenshan 97A, a CMS line), both lines expressing *Cry1Ab/Ac* fusion gene (Chen et al. 2011), but cultivation was discontinued afterward. Currently, no *Bt* rice is grown in any country across the world, including China, although

Table 2 List of transgenes used in rice transformation to provide insect resistance

S. No.	Transgene(s) deployed	Recipient rice variety/ genotype	Promoter deployed	Reported resistant against	Reference(s)
Lepidopteran pests					
1	*cry1Ab*	IR58 (indica rice)	CaMV35S	Yellow stem borer, striped stem borer, leaffolder	Wunn et al. (1996)
2	*PINII* (potato proteinase inhibitor)	Japonica rice	–	Pink stem borer	Duan et al. (1996)
3	*cry1Ab*	Japonica, Taipei 309	Rice actin-1 promoter	Yellow stem borer	Wu et al. (1997a)
4	*cry1A*, cowpea proteinase inhibitor gene	Japonica, Taipei 309, and Taipei 85-93. Indica, Minghui 63, and Qingliu Rai	–	Yellow stem borer	Wu et al. (1997b)
5	*cry1AC*	IR64 (indica rice)	Maize ubiquitin 1 promoter	Yellow stem borer	Nayak et al. (1997)
6	*cry1Aa, cry 1Ac, cry2A, cry1C*	Indica, japonica	–	Yellow stem borer	Lee et al. (1997)
7	*cry1Ab*	Aromatic rice, Tarom molaii	–	Yellow stem borer	Ghareyazie et al. (1997)
8	*cry2A*	Basmati 370 and M7 (indica rice)	CaMV35S promoter	Yellow stem borer, leaffolder	Maqbool et al. (1998)
9	*cry1Ab*	Indica and japonica rice	–	Yellow stem borer	Datta et al. (1998)
10	*cry1Ab, cry1Ac, hph, gus* genes	Japonica rice	Maize ubiquitin promoter, the CaMV 35S promoter, and the Brassica Bp10 gene promoter	Yellow stem borer, striped stem borer	Cheng et al. (1998)
11	*cry1Ab*	Maintainer line IR68899B	35S constitutive promoter	Yellow stem borer	Alam et al. (1999)
12	*cry1Ab*	Vaidehi (indica rice)	–	Yellow stem borer	Alam et al. (1998)
13	*cry1Ab*	PR16 and PR18	Maize ubiquitin promoter	Yellow stem borer	Ye et al. (2000)

(continued)

Table 2 (continued)

S. No.	Transgene(s) deployed	Recipient rice variety/ genotype	Promoter deployed	Reported resistant against	Reference(s)
14	*cry1Ab, cry1Ac*	Minghui 63 (indica CMS restorer line) and its derived hybrid rice Shanyou 63	Rice actin-1 promoter	Yellow stem borer and leaffolder	Tu et al. (2000)
15	*cry1Ab*	KMD1 (japonica elite line)	–	Yellow stem borer	Shu et al. (2000)
16	*cry1A, cry1Ab, cry1Ac, cry1c* and *cry2A*	Indica rice	–	Yellow stem borer	Intikhab et al. (2000)
17	*cry1Ab, Xa21*	Pusa Basmati 1 (indica rice)	–	Yellow stem borer, Bacterial blight disease	Gosal et al. (2000)
18	CRY1AB	KMD1 and KMD2	–	Yellow stem borer, striped stem borer	Ye et al. (2001)
19	*cry1Ac, cry2A,* snowdrop lectin *gna*	M7 and Basmati 370 (indica rice varieties)	Maize ubiquitin-1 promoter, CaMV 35S promoter	Yellow stem borer, leaffolder, and BPH	Maqbool et al. (2001)
20	*cry1Ab*	IR64 (indica rice)	–	Yellow stem borer	Maiti et al. (2001)
21	Spider insecticidal gene	Xiushuill and Chunjiang 11	–	Leaffolder and striped stem borer	Huang et al. (2001)
22	*cry1Ac* gene	Minghui 81	Maize ubiquitin-1 promoter	Striped stem borer	Zeng et al. (2002)
23	*cry1Ac* gene	Pusa Basmati-1, IR64, and Karnal Local (indica rice)	Maize ubiquitin-1 promoter	Yellow stem borer	Khanna and Raina (2002)
24	*Bt* fusion gene (for insect resistance), *Xa21* gene (for BLB), chitinase gene (sheath blight)	IR72 (indica rice)	–	Yellow stem borer, bacterial blight disease, sheath blight disease	Datta et al. (2002)
25	Chimeric *Bt* gene, *cry1Ab*; *cry1Ab/cry1Ac* fusion gene	IR68899B and IR68897B (maintainer lines), MH63 and BR827-35R (restorer lines)	35S and PEPC promoters; actin 1 promoter	Yellow stem borer, leaffolder	Balachandran et al. (2002)

(continued)

Table 2 (continued)

S. No.	Transgene(s) deployed	Recipient rice variety/ genotype	Promoter deployed	Reported resistant against	Reference(s)
26	*cry1Ab*, snowdrop lectin *gna*	Rajalele (javanica progenies)	–	Yellow stem borer, planthoppers	Slamet et al. (2003)
27	*cry1Ac*	IR64, Pusa Basmati-1, and Karnal Local (indica rice)	Maize ubiquitin promoter	Yellow stem borer	Raina et al. (2003)
28	*cry1Ac, cry2A*	Basmati (indica rice)	PEPC promoter and PB 10 (pollen-specific) promoter	Yellow stem borer	Husnain et al. (2003)
29	*cry1Ac, Xa21*	Pusa Basmati-1 (indica rice)	–	Yellow stem borer, bacterial blight	Gosal et al. (2003)
30	*CRY1AB, CRY1AC* genes; *bar* gene for herbicide resistance	IR58025A, IR58025B, and Vajram (indica rice)	Maize ubiquitin promoter, CaMV 35S promoter (for BAR gene)	Yellow stem borer	Ramesh et al. (2004b)
31	*cry1B, cry1Aa*	Ariete and Senia	ubi 1 promoter or mpi promoter	Striped stem borer	Breitler et al. (2004)
32	*cry1Ab, cry1Ac, cry1C, cry2A, cry9C*	Indica rice	–	Yellow stem borer, Striped stem borer	Alcantara et al. (2004)
33	*mpi* gene (maize proteinase inhibitor)	Senia and Ariete	Maize ubiquitin 1 promoter	Striped stem borer	Vila et al. (2005)
34	*cry2A*	Minghui 63 (indica restorer line)	Maize ubiquitin promoter	Yellow stem borer	Chen et al. (2005)
35	*cry1Ac, cry2A*	Basmati line B-370 (indica rice)	–	Yellow stem borer, leaffolder	Bashir et al. (2005)
36	*cry1Ac, cry2A*	Basmati 370 (indica rice)	Ubiquitin promoter and CaMV 35S promoter	Yellow stem borer	Riaz et al. (2006)

(continued)

Table 2 (continued)

S. No.	Transgene(s) deployed	Recipient rice variety/ genotype	Promoter deployed	Reported resistant against	Reference(s)
37	*cry1Ab-1B* (translationally fused gene) and *cry1A/cry1Ac* (hybrid *Bt* gene)	Elite Vietnamese cultivars	Maize ubiquitin promoter and rice actin-1 promoter	Yellow stem borer	Ho et al. (2006)
38	*PINII* (potato proteinase inhibitor)	Pusa basmati-1 and Tarori Basmati (indica rice) and TNG67 (japonica rice)	Pin2 wound-inducible promoter	Yellow stem borer	Bhutani et al. (2006)
39	*cry2Ab* gene	Minghui 63 (indica restorer line)/T(1Ab)-10	–	Yellow stem borer, leaffolder	Tang and Lin (2007)
40	*cry1Ab*	Korean varieties, P-I, P-II, P-III	Maize ubiquitin promoter	Yellow stem borer	Kim et al. (2008)
41	*cry1Ab* gene	Khazar, Neda, and Nemat	–	Striped stem borer	Kiani et al. (2008)
42	Ten transgenic lines (two *cry1Ac* lines, three *cry2A* lines, five *cry9C* lines)	Minghui 63 (elite indica restorer line)	–	Yellow stem borer, striped stem borer	Chen et al. (2008)
43	*cry1C*	Zhonghua 11 (*Oryza sativa* L. subsp. japonica)/RJ5 line.	rbcS promoter	Yellow stem borer, striped stem borer, leaffolder	Ye et al. (2009)
44	*cry1Ia5*	*Oryza sativa*	–	Yellow stem borer	Moghaieb (2010)
45	*cry1b* and *cry1Aa* fusion gene	*Oryza sativa*	PEPC promoter	Yellow stem borer	Kumar et al. (2010)
46	*cry1Ab* and *Vip3H* fusion gene	G6H1, G6H2, G6H3, G6H4, G6H5, G6H6	–	Striped stem borer, pink stem borer	Chen et al. (2010)
47	*cry1Ab, cry1Ac* fusion gene	Bt-SY63	–	Striped stem borer	Zhang et al. (2011)
48	*cry1Ac, CpTI* genes	Bt-KF6	–	Striped stem borer	Zhang et al. (2011)
49	*cry1Ab*	Bt-DL	–	Striped stem borer	Zhang et al. (2011)

(continued)

Table 2 (continued)

S. No.	Transgene(s) deployed	Recipient rice variety/ genotype	Promoter deployed	Reported resistant against	Reference(s)
50	cry1Ab, cry1Ac, cry1C, cry2A	Minghui 63 (elite indica restorer line)	Maize ubiquitin promoter	Yellow stem borer, striped stem borer, leaffolder	Yang et al. (2011)
51	Cry1Ac, cry1I-like gene	Rice	pGreen	Striped stem borer, leaffolder	Yang et al. (2014)
52	cry1Ab gene	Mfb-MH86	Ubiquitin promoter	Striped stem borer and other lepidopteran pests	Wang et al. (2014)
53	mpi-pci fusion gene	Ariete	mpi promoter	Striped stem borer	Quilis et al. (2014)
54	Ds-Bt	Zhejing-22, Kongyu-131	–	Striped stem borer	Gao et al. (2014)
55	cry1Ac, cry1Ig, G10 (EPSPS gene)	Xiushui 134	Maize ubiquitin promoter (pUBi)/ modified cauliflower 35S promoter	Striped stem borer, leaffolder, and glyphosate	Zhao (2015)
Sucking pests					
56	GNA (Galanthus nivalis agglutinin)	?	Phloem-specific rice-sucrose-synthase	BPH	Rao et al. (1998)
57	GNA	ASD16/M12	Rice sucrose synthase/ maize ubiquitin	BPH and GLH	Foissac et al. (2000)
58	GNA	?	?	SBPH	Wu et al. (2002)
59	GNA	Chaitanya and Phalguna, indica cultivars	Phloem-specific rice-sucrose-synthase	BPH, GLH, and WBPH	Nagadhara et al. (2003, 2004)
60	GNA			BPH, GLH, and WBPH	Ramesh et al. (2004a, b)
61	GNA	Zhuxian B, an indica rice		BPH	Li et al. (2005)

(continued)

Table 2 (continued)

S. No.	Transgene(s) deployed	Recipient rice variety/ genotype	Promoter deployed	Reported resistant against	Reference(s)
62	*ASAL* (*Allium sativum* agglutinin)	IR64	CaMV35S	BPH and GLH	Saha et al. (2006)
63	*ASAL*	Chaitanya and BPT5204, indica cultivars	CaMV35S	BPH, GLH, and WBPH	Yarasi et al. (2008)
64	*ASAL*	IR64	CaMV35S	BPH and GLH	Sengupta et al. (2010)
65	*DB1/G95A-mALS* (*Dioscoria batata* tuber lectin)	Tachisugata	Phloem-specific rice-sucrose-synthase	BPH	Yoshimura et al. (2012)
66	*ASAL*	?	Phloem-specific rice-sucrose-synthase	BPH	Chandrasekhar et al. (2014)
67	Loop replacements with gut-binding peptides in *Cry1Ab* domain II	In vitro assay	–	BPH	Shao et al. (2016)
68	*Cry64Ba* and *Cry64Ca*	report		Effective against sap-sucking insects	Liu et al. (2018)

several such transgenic rice lines have been deregulated by the respective regulatory authorities of these countries due to various policy-related issues.

Genetic engineering for the control of planthopper and leafhopper pests of rice has begun with the use of plant-derived lectin genes. The snowdrop lectin gene, *Galanthus nivalis* agglutinin (*GNA*), has been transferred to several rice varieties and has been shown to provide partial to complete resistance to planthoppers and leafhoppers. Partial resistance to leafhoppers and planthoppers was demonstrated by rice transformation with a lectin gene from garlic (*Allium sativum* leaf agglutinin gene, *ASAL*; Saha et al. 2006). Bala et al. (2013) reported that *ASAL* interacts with NADH quinone oxidoreductase (NQO), a key player in the electron transport chain, and results in toxicity and increased mortality of BPH in transgenic rice lines. This study also demonstrated that, among all the transgenes available for control of sucking pests, *ASAL* holds significant promise, particularly against BPH. Yoshimura et al. (2012) developed transgenic rice possessing lectin1 gene from *Dioscorea batatas* under the control of a phloem-specific promoter (i.e., promoter of sucrose synthase-1 gene) that showed a 30% decrease in the survival rate of BPH. Even though, in general, it is known that Cry proteins are ineffective against sucking pests, through loop replacements with gut-binding peptides in *Cry1AB* domain II, enhanced toxicity against BPH has been demonstrated (Shao et al. 2016). Liu et al. (2018) have shown the effectivity of Cry64Ba and Cry64Ca, two ETX/MTX2-type *Bt* proteins, against hemipteran pests. Boddupally et al. (2018) recently demonstrated that the expression of hybrid fusion protein (Cry1Ac::ASAL) in transgenic rice plants imparted resistance against multiple insect pests: BPH, stem borer, and leaffolder. The list of transgenes deployed for the control of sucking pests such as BPH is summarized in Table 2.

5 Insect-Plant Interactions at the Genomic Level

5.1 Planthopper Genomes

The genome of BPH and its endosymbionts have been sequenced (Xue et al. 2014). It is a large genome (1141 Mb) with 27,571 protein-coding genes, of which 16,330 are specific to this species. In comparison, the WBPH genome is relatively smaller (720 Mb) with 21,254 protein-coding genes (L Wang et al. 2017a), while the SBPH genome size is 541 Mb with 17,736 protein-coding genes (Zhu et al. 2017c). Mitochondrial (mt) genomes of these three planthopper species have also been sequenced (Zhang et al. 2013, 2014). These studies are now providing insights into the genetic plasticity of this group, possible causes of rapid evolution of virulent biotypes, and resistance against a wide range of synthetic insecticides. In addition, the role of endosymbionts such as yeast-like symbiont (YLS) and *Wolbachia* spp. in enhancing insect fitness is being studied. Additional genetic markers are being developed for studying population genetics, individual differences, and the phylogeography of planthoppers. Several key genes of the insects have been identified, which can be targeted for RNAi-based genetic tools of pest management. Transcriptomics of the salivary gland has revealed more than 350 secretory proteins, of which several, such as NlSEF1 (W Ye et al. 2017a), act as effectors modulating plant defense response. Likewise, muscin-like protein of the salivary gland secretion of BPH (Huang et al. 2017; Shangguan et al. 2018) and WBPH (Miao et al. 2018) is likely to be an effector. Such genes can be suitable targets for their control using an RNAi-based approach described above. A high number of cytochrome P450 genes and their functional diversification are attributed to drive the evolution of insecticide resistance and virulence against host-plant resistance (Peng et al. 2017; Zimmer et al. 2018). In spite of efforts to map virulence loci onto the BPH genome (Jing et al. 2014; Kobayashi et al. 2014), no aviR gene has yet been cloned and characterized. Although mitochondrial markers based on mt genes COI and ND4 have been screened for population differentiation, the results have not been encouraging over large populations across countries. Further, Zhang et al. (2013) suggest that markers based on the control region of the mt genome might provide more reliable markers for studying population genetics and the phylogeography of planthoppers.

5.2   Rice-Planthopper Interactions

Using both candidate gene cloning and a characterization-based approach and functional genomics-based omics approaches, attempts are being made to understand
planthopper and rice interactions. Based on initial information on the nature of R
genes as being members of the NBS-LRR class or receptor kinase class, the rice
resistance mechanism against BPH is, rather hurriedly, aligned to rice resistance
against pathogens under two-tier immunity involving pattern-triggered immunity
(PTI) and effector-triggered immunity (ETI) and the involvement of both JA- and
SA-mediated pathways (Jing et al. 2017). Even the cloned genes are assigned to PTI
(*Bph3*, *Bph15*) or ETI (*Bph14*, *Bph1/10/18*) tiers. However, what is not accounted
for is the lack of documented evidence of hypersensitive reaction (HR) and systemic
acquired resistance (SAR), which are hallmarks of plant response to biotrophic
pathogens. Further, resistance in rice against planthoppers is not at the immune level
but with moderate antibiosis coupled with antifeeding and antixenosis components.
It is generally understood that SA- and JA-mediated plant defenses act reciprocally
antagonistic to each other with adaptive significance (Thaler et al. 2012). Such
antagonism has not been convincingly illustrated in the case of planthopper resistance in rice. Thus, greater understanding of planthopper-rice interactions is needed.

5.3   Rice-Gall Midge Interactions

Although genome sequence data for ARGM are yet to be published (Nair et al.
unpublished), the mitogenome has been sequenced (Atray et al. 2015) and the
microbiome analyzed in different stages of the life cycle of the insect (Ojha et al.
2017). Based on identification and functional analysis of candidate genes, global
gene expression profiles and differential gene expressions detected through SSH
cDNA libraries, microarray studies, and the pyrosequencing approach in both the
plants and the insect rice-gall midge interactions have been fairly well studied. In
essence, with results from these studies indicating strategies used by both the pest
and the host to defeat each other, defense ploys can be termed as a battle for survival
(Bentur et al. 2016; Sinha et al. 2017).

During the infestation process and subsequent feeding on the host, the larvae
inject substance(s) into the host. As in the case of pathogenic bacteria and fungi,
these products could be determinants (effectors) of the avirulence/virulence phenomenon. Extending this idea further, the genes that encode these molecules could
be determinants of gall midge biotypes. Further, the genes that encode such molecules could be those that encode secreted salivary gland proteins (SSGPs). Therefore,
characterizing genes that encode SSGPs could provide a handle to study this interaction and also gain valuable insight into the process of infestation of rice by this
pest. The expression patterns of some of these SSGPs in larvae interacting with a
susceptible host (SH; compatible interaction) or resistant host (RH; incompatible
interaction) indicated that some of the SSGPs such as gamma subunit of

oligosaccharyl transferase (*OoOST*) and nucleoside diphosphate kinase (*OoNDPK*) overexpress when interacting with SH compared with those in maggots when feeding on RH (Sinha et al. 2011a, 2012a). Furthermore, NDPK protein has been demonstrated to influence the host physiology. In contrast, two genes, *OoprotI* and *OoprotII*, homologous to serine proteases, and *OoDAD1* (defender against death) overexpress in midgut of the maggots feeding on RH when compared with those feeding on SH (Sinha et al. 2011b, 2015). Although the former interactions represent effector-induced susceptibility, the latter set forms neutralizers attempting to overcome plant-secreted defensive toxins. Earlier studies also brought out similarities in rice defense expression against gall midge with those seen against plant pathogens (Rawat et al. 2010, 2013), complete with HR and SAR. Subsequent analysis of SSH-generated cDNA libraries and microarray data brought out differences in the defense pathways underlying HR+-type and HR−-type resistance (Rawat et al. 2012b), among the two HR-type resistances conferred by *Gm1* and *Gm8* genes (Divya et al. 2016, 2018b), and also the diversity in susceptibility pathways in rice genotypes with ineffective R-genes against virulent biotypes (Rawat et al. 2012a). Generally, in the three gall midge-susceptible rice varieties studied, the insect-challenged plants tend to step up metabolism and transport of nutrients to their feeding site and have suppressed defense responses. However, one of the rice varieties mounted an elevated defense response during early hours of infestation, only to be overpowered later, eventually resulting in host-plant susceptibility.

Pyrosequencing-based transcriptome analysis of ARGM revealed a differential response of the midge depending on whether it is in a compatible or incompatible interaction with its host (Sinha et al. 2012b). A recent study with sequencing of 16S rRNA bacterial gene (V3-V4 region) revealed differences in the microflora of the gall midge-rice maggots feeding on susceptible or resistant rice hosts. Results revealed that *Wolbachia* was the predominant bacterium in pupae and adults while *Pseudomonas* was predominant in maggots. Further, it was observed that members of proteobacteria were predominant across all the samples. There was high species diversity in maggots isolated from susceptible rice and a high representation of unclassified bacteria in maggots isolated from resistant rice. A first step in this direction is a report that highlights variation in the microbiome of the rice gall midge, based on the host phenotype from which it was isolated, and results suggest that these variations could have an important role in the host's susceptibility/resistance (Ojha et al. 2017).

The availability of the complete sequence of the gall midge mt genome and subsequent sequence analysis revealed the presence of two tandem repeat elements in the noncoding regions of the mt genome. Further, sequencing of the iterated regions demonstrated that the iterations of the repeat elements could not only differentiate different gall midge biotypes present in India but also were able to genetically separate the ARGM from its counterpart, the African rice gall midge (Atray et al. 2015). Thus, this study identified a reliable tool to monitor changes in the insect populations so as to have an "early warning system" in place. Janique et al. (2017) reported that two noncoding repeat motifs observed in the mitogenome of ARGM in India were absent in Thai populations and these were replaced by an 89-bp noncoding sequence.

6   Conclusions and Perspectives

In terms of an evolutionary perspective, survival of neither the host nor the herbivore has ever been under threat. Understandably, however, over the past couple of millennia when half of the human population started depending on this one cereal as its staple food, conflict of interest erupted between these insects and humans. All feasible efforts were made to protect the crop from possible damage by insects during the early phase of domestication and cultivation of the rice crop. With the advent of modern scientific methods of crop husbandry, crop improvement, and synthetic chemicals, insect pests became targets of a frontal attack by humans. With quick development of resistance against a range of synthetic insecticides, insect pests proved their evolutionary superiority, compelling humans to concede defeat and conclude that pest management was the best solution for sustainable productivity rather than pest eradication or control. Rice insect pest management has traversed the same course as that in other crops such as cotton.

Insect pest management is complex and fraught with many variables. From the foregoing account, it is quite clear that we are just beginning to understand and make inroads into the complex interactions between the pest and its host, rice. It is also evident that, although productivity loss due to these insects alone runs into several million US dollars, insects are rapidly overcoming any management strategy that we are able to deploy, whether it is resistance genes or the development of new pesticides. What this review hopes to highlight primarily is that as a central concept we need to know how the rice plant interacts with its several insect enemies from an evolutionary point of view. Against YSB, no high resistance is expressed, probably because it is a "k" strategist and monophagous insect that does not kill the host. Against gall midge with an intermediate population strategy displaying buck and boost cycles, the host plant has a diverse array of immune-level R-genes that are constantly evolving along with the virulence in the pest populations. In stark contrast, the rice plant has stockpiled multiple major and minor R-genes against planthoppers, which are typical "r" strategists. Second, the molecular tools now available have provided novel products for deployment to alleviate pest-induced yield losses. Notable among these are gene-pyramided elite cultivars, derived from marker-assisted selection, to manage multiple pests and their strains/biotypes (Divya et al. 2018c). Also present is the array of transgenic rice lines with potent genes against all the guilds of insect pests. It is unfortunate that these products are not yet available for commercial use. Molecular approaches have also broadened our knowledge and identified unexplored facets for possible use in pest management. Finally, this flush in information has reiterated the evolutionary advantage of insect genome, mitogenome, and metagenome in facing any future challenges. A recent report on quick field selection for dsRNA resistance in western corn rootworm (Khajuria et al. 2018) exemplifies this. As summarized, representatives of each insect guild have evolved their own strategy to overcome plant resistance. Considering that, in the coming years, we are likely to be under pressure to grow more in less area, it is therefore imperative that we cut the losses in productivity due to insect pests.

We have made rapid strides in the past couple of decades toward this goal with emerging new tools and strategies. What is also clear is that the solution to the insect problem is unlikely to come from one area of study but from an amalgamation of information obtained from several different studies that can provide durable, effective, and targeted resistance to insect pests of rice. The caveat is that this is unlikely to be a one-time effort but must be a continuing one.

References

Abhilash Kumar V, Balachiranjeevi C, Naik BS et al (2017) Marker-assisted pyramiding of bacterial blight and gall midge resistance genes into RPHR-1005, the restorer line of the popular rice hybrid DRRH-3. Mol Breed 37:86. https://doi.org/10.1007/s11032-017-0687-8

Agarwal S, Mohan M, Mangrauthia SK (2012) RNAi: machinery and role in pest and disease management. In: Venkateshwarlu B, Shankar AK, Shankar C, Maheshwari M (eds) Crop stress and its management: perspectives and strategies. Springer, New York, pp 447–469

Aglawe SB, Barbadikar KM, Mangrauthia SK, Madhav MS (2018) New breeding technique "genome editing" for crop improvement: applications, potentials and challenges. 3 Biotech 8(8):336

Alam MF, Datta K, Abrigo E et al (1998) Production of transgenic deep water indica rice plants expressing a synthetic *Bacillus thuringiensis cry1Ab* gene with enhanced resistance to yellow stem borer. Plant Sci 135(1):25–30

Alam MF, Abrigo E, Datta K et al (1999) Transgenic insect resistance maintainer line (IR68899B) for improvement of hybrid rice. Plant Cell Rep 18:572–575

Alcantara EP, Aguda RM, Curtiss A et al (2004) *Bacillus thuringiensis* delta endotoxin binding to brush border membrane vesicles of rice stem borers. Arch Insect Biochem Physiol 55(4):169–177. https://doi.org/10.1002/arch.10128

Amudhan S, Prasad Rao U, Bentur JS (1999) Total phenol profile in some rice varieties in relation to infestation by Asian rice gall midge *Orseolia oryzae* (Wood-Mason). Curr Sci 76:1577–1580

Andrews RW, Faust R, Wabiko MH et al (1987) The biotechnology of *Bacillus thuringiensis*. Crit Rev Biotechnol 6:163–232. https://doi.org/10.3109/07388558709113596

Arora L, Narula A (2017) Gene editing and crop improvement using CRISPR-Cas9 system. Front Plant Sci 8:1932. https://doi.org/10.3389/fpls.2017.01932

Atray I, Bentur JS, Nair S (2015) The Asian rice gall midge (*Orseolia oryzae*) mitogenome has evolved novel gene boundaries and tandem repeats that distinguish its biotypes. PLoS One 10(7):e0134625. https://doi.org/10.1371/journal.pone.0134625

Bala A, Roy A, Behura N, Hess D, Das S (2013) Insight to the mode of action of *Allium sativum* leaf agglutinin (ASAL) expressing in T3 rice lines on brown planthopper. Am J Plant Sci 4:400–407

Balachandran S, Chandel G, Alam M et al (2002) Improving hybrid rice through anther culture and transgenic approaches. In: 4th International Symposium on Hybrid Rice. Hanoi, Vietnam. IRRI, Los Baños, pp 105–118

Bally J, Fishilevich E, Bowling AJ, Pence HE, Narva KE, Waterhouse PM (2018) Improved insect-proofing: expressing double-stranded RNA in chloroplasts. Pest Manag Sci 74:1751–1758

Bashir K, Husnain T, Fatima T, Riaz N, Makhdoom R, Riazuddin S (2005) Novel indica basmati line (B-370) expressing two unrelated genes of *Bacillus thuringiensis* is highly resistant to two lepidopteran insects in the field. Crop Prot 24:870–879

Baum JA, Bogaert T, Clinton W et al (2007) Control of coleopteran insect pests through RNA interference. Nat Biotechnol 25(11):1322–1326

Bentur JS (2007) Host-plant resistance to insects as a core of rice IPM. In: Aggarwal PK, Ladha JK, Singh RK, Devkumar C, Hardy B (eds) Science, technology, and trade for peace and prosperity. Proceedings of the 26th International Rice Research Conference, New Delhi, India, 9–12 October 2006. International Rice Research Institute, Indian Council of Agricultural Research, and National Academy of Agricultural Sciences, Los Baños, New Delhi, pp 419–435

Bentur JS (2010) Insect pests of rice in India and their management. In: Reddy DV, Rao NP, Rao KV (eds) Pests and pathogens: management strategies. BS Publications, Hyderabad, pp 1–42

Bentur JS, Viratkamath BC (2008) Rice planthoppers strike back. Curr Sci 95:441–442

Bentur JS, Pasalu IC, Sarma NP, Prasad Rao U, Mishra B (2003) Gall midge resistance in rice. DRR technical bulletin #1. Directorate of Rice Research, Hyderabad. 20 p

Bentur JS, Padmakumari AP, Jhansilakshmi V et al (2011) Insect resistance in rice. DRR technical bulletin #51. Directorate of Rice Research, Hyderabad. 86 p

Bentur JS, Rawat N, Divya D, Sinha DK, Agarrwal R, Atray I, Nair S (2016) Rice-gall midge interactions: battle for survival. J Insect Physiol 84:40–49. dx.doi.org. https://doi.org/10.1016/j.jinsphys.2015.09.008

Bhutani S, Kumar R, Chauhan R et al (2006) Development of transgenic indica rice plants containing potato proteinase inhibitor 2 gene with improved defense against yellow stem borer. Physiol Mol Biol Plants 12(1):43–52

Bi HL, Xu J, Tan AJ, Huang YP (2016) CRISPR/Cas9-mediated targeted gene mutagenesis in *Spodoptera litura*. Insect Sci 23:469–477

Boddupally D, Tamirisa S, Gundra SR, Vudem DR, Rao KV (2018) Expression of hybrid fusion protein (Cry1Ac::ASAL) in transgenic rice plants imparts resistance against multiple insect pests. Sci Rep 8:8458. https://doi.org/10.1038/s41598-018-26881-9

Breitler JC, Vassal JM, Catala MDM et al (2004) *Bt* rice harbouring *cry* genes controlled by a constitutive or wound-inducible promoter: protection and transgene expression under Mediterranean field conditions. Plant Biotechnol J 2:417–430

Bunsha D (2006) Crops on trial. Frontline 23(23) http://www.hinduonnet.com/fline/fl2323/stories/20061201003603000.htm. Accessed 15 Jan 2008

Cha YS, Ji H, Yun DW et al (2008) Fine mapping of the rice *Bph1* gene, which confers resistance to the brown planthopper (*Nilaparvata lugens* Stal), and development of STS markers for marker-assisted selection. Mol Cell 26:146–151

Chandrasekhar K, Reddy MG, Singh J, Vani K, Vijayalakshmi M, Kaul T, Reddy MK (2014) Development of transgenic rice harbouring mutated rice 5-*Enolpyruvylshikimate 3-Phosphate Synthase* (Os-*mEPSPS*) and *Allium sativum* leaf agglutinin (ASAL) genes conferring tolerance to herbicides and sap-sucking insects. Plant Mol Biol Report 32(6):1146. https://doi.org/10.1007/s11105-014-0715-3

Chen H, Tang W, Xu C et al (2005) Transgenic indica rice plants harboring a synthetic *cry2A* gene of *Bacillus thuringiensis* exhibit enhanced resistance against lepidopteran rice pests. Theor Appl Genet 111:1330–1337

Chen M, Zhao JZ, Ye GY, Fu Q, Shelton AM (2006) Impact of insect-resistant transgenic rice on target insect pests and non-target arthropods in China. Insect Sci 13:409–420

Chen H, Zhang G, Zhang Q, Lin Y (2008) Effect of transgenic *Bacillus thuringiensis* rice lines on mortality and feeding behavior of rice stem borers (Lepidoptera: Crambidae). J Econ Entomol 101:182–189. https://doi.org/10.1093/jee/101.1.182

Chen Y, Tian J, Shen Z et al (2010) Transgenic rice plants expressing a fused protein of *Cry1Ab/Vip3H* have resistance to rice stem borers under laboratory and field conditions. J Econ Entomol 103(4):1444–1453. https://doi.org/10.1603/EC10014

Chen M, Shelton A, Ye G (2011) Insect-resistant genetically modified rice in China: from research to commercialization. Annu Rev Entomol 56:81–101

Cheng X, Sardana R, Kaplan H et al (1998) *Agrobacterium*-transformed rice plants expressing synthetic *cry1A*(b) and *cry1A*(c) genes are highly toxic to striped stem borer and yellow stem borer. Proc Natl Acad Sci U S A 95(6):2767–2772. https://doi.org/10.1073/pnas.95.6.2767

Cohen MB, Romena AM, Gould F (2000) Dispersal by larvae of the stem borers *Scirpophaga incertulas* (Lepidoptera: Pyralidae) and *Chilo suppressalis* (Lepidoptera: Crambidae) in plots of transplanted rice. Environ Entomol 29:958–971. https://doi.org/10.1603/0046-225x-29.5.958

Courtier-Orgogozo V, Morizot B, Boëte C (2017) Agricultural pest control with CRISPR-based gene drive: time for public debate. Should we use gene drive for pest control? EMBO Rep 18:878–880

Datta K, Vasquez A, Tu J et al (1998) Constitutive and tissue specific differential expression of *cry1Ab* gene in transgenic rice plants conferring resistance of rice insect pest. Theor Appl Genet 97:20–30

Datta K, Baisakh N, Thet K et al (2002) Pyramiding transgenes for multiple resistance in rice against bacterial blight, yellow stem borer and sheath blight. Theor Appl Genet 106:1–8

Deutsch CA, Tewksbury JJ, Tigchelaar M et al (2018) Increase in crop losses to insect pests in a warming climate. Science 361:916–919

Divya D, Himabindu K, Nair S, Bentur JS (2015) Cloning of a gene encoding LRR protein and its validation as candidate gall midge resistance gene, *Gm4*, in rice. Euphytica 203:185–195

Divya D, Tunginba Singh Y, Nair S, Bentur JS (2016) Analysis of SSH library of rice variety Aganni reveals candidate gall midge resistance genes. Funct Integr Genom 16:153–169

Divya D, Nair S, Bentur JS (2018a) Expression profiles of key genes involved in rice gall midge interactions reveal diversity in resistance pathways. Curr Sci 115:74–82

Divya D, Sahu N, Nair S, Bentur JS (2018b) Map-based cloning and validation of a gall midge resistance gene, *Gm8*, encoding a proline-rich protein in the rice variety Aganni. Mol Biol Rep 45:2075. https://doi.org/10.1007/s11033-018-4364-8

Divya D, Ratna Madhavi K, Dass MA et al (2018c) Expression profile of defense genes in rice lines pyramided with resistance genes against bacterial blight, fungal blast and insect gall midge. Rice 11:40

Dong XT, Liao H, Zhu GH et al (2017) CRISPR/Cas9 mediated PBP1 and PBP3 mutagenesis induced significant reduction in electrophysiological response to sex pheromones in male *Chilo suppressalis*. Insect Sci 26:388. https://doi.org/10.1111/1744-7917.12544

Du B, Zhang W, Liu B et al (2009) Identification and characterization of *Bph14*, a gene conferring resistance to brown planthopper in rice. Proc Natl Acad Sci U S A 106:22163–22168

Du B, Chen R, He J (2020) Current understanding of the genomic, genetic, and molecular control of insect resistance in rice. Mol Breed 40:24. https://doi.org/10.1007/s11032-020-1103-3

Duan X, Li X, Xue Q et al (1996) Transgenic rice plants harboring an introduced potato protein-ase inhibitor II gene are insect resistant. Nat Biotechnol 14:494–498. https://doi.org/10.1038/nbt0496-494

Foissac X, Nguyen TL, Christou P, Gatehouse AMR, Gatehouse JA (2000) Resistance to green leafhopper (*Nephotettix virescens*) and brown planthopper (*Nilaparvata lugens*) in trans-genic rice expressing snowdrop lectin (*Galanthus nivalis* agglutinin; GNA). J Insect Physiol 46:573–583

Fujimoto H, Itoh K, Yamamoto M, Kyozuka J, Shimamoto K (1993) Insect-resistant rice gener-ated by introduction of a modified endotoxin gene of *Bacillus thuringiensis*. Biotechnology 11:1151–1155

Fujita D, Doi K, Yoshimura A, Yasui H (2004) Introgression of a resistance gene for green rice leafhopper from *Oryza nivara* into cultivated rice, *Oryza sativa* L. Rice Genet Newsl 21:64–66

Fujita D, Doi K, Yoshimura A, Yasui H (2006) Molecular mapping of a novel gene, *Grh5*, confer-ring resistance to green rice leafhopper (*Nephotettix cincticeps* Uhler) in rice, *Oryza sativa* L. Theor Appl Genet 113:567–573

Fujita D, Doi K, Yoshimura A, Yasui H (2010) A major QTL for resistance to green rice leafhopper (*Nephotettix cincticeps* Uhler) derived from African rice (*Oryza glaberrima* Steud.). Breed Sci 60:336–341

Fujita D, Kohli A, Horgan FG (2013) Rice resistance to planthoppers and leafhoppers. Crit Rev Plant Sci 32:162–191

Fukuta T, Tamura K, Hirae M, Oya S (1998) Genetic analysis of resistance to green rice leafhopper (*Nephotettix cincticeps* (Uhler)) in rice parental line, Norin-PL6, using RFLP markers. Breed Sci 48:243–249

Gao X, Zhou J, Li J et al (2014) Efficient generation of marker-free transgenic rice plants using an improved transposon-mediated transgene reintegration strategy. Plant Physiol 167:11–24. https://doi.org/10.1104/pp.114.246173

Gatehouse JA (2008) Biotechnological prospects for engineering insect-resistant plants. Plant Physiol 146:881–887

Ghareyazie B, Alinia F, Menguito CA et al (1997) Enhanced resistance to two stem borers in an aromatic rice containing a synthetic *CRY1A(B)* gene. Mol Breed 3:401–404

Gosal SS, Gill R, Sindhu AS et al (2000) Transgenic basmati rice carrying genes for stem borer and bacterial leaf blight resistance. In: Proceedings International Rice Research Conference. IRRI, Los Baños, pp 353–360

Gosal SS, Gill R, Sindhu AS et al (2003) Introducing the *cry1Ac* gene into basmati rice and transmitting transgenes to R3 progeny. In: Proceedings International Rice Research Conference. IRRI, Los Baños, pp 558–560

Harini AS, Lakshmi SS, Kumar SS, Sivaramakrishnan S, Kadirvel P (2010) Validation and fine-mapping of genetic locus associated with resistance to brown planthopper [*Nilaparvata lugens* (Stål.)] in rice (*Oryza sativa* L.). Asian J Biol Sci 5:32–37

He K, Xiao H, Sun Y, Ding S, Situ G, Li F (2018) Transgenic microRNA-14 rice shows high resistance to rice stem borer. Plant Biotechnol J 17:461. https://doi.org/10.1111/pbi.12990

Heinrichs EA (1994) Biology and management of rice insects. Wiley Eastern Limited and New Age International Limited, New Delhi. 779 p

Heinrichs EA, Medrano FG, Rapusas HR (1985) Genetic evaluation for insect resistance in rice. International Rice Research Institute, Los Baños. 352 p

Heong KL, Hardy B (eds) (2010) Planthoppers: new threats to the sustainability of intensive rice production systems in Asia. International Rice Research Institute, Los Baños. 460 p

Herrnstadt C, Soares GG, Wilcox ER et al (1986) A new strain of *Bacillus thuringiensis* with activity against coleopteran insects. Nat Biotechnol 4:305–308

High SM, Cohen MB, Shu QY, Altosaar I (2004) Achieving successful deployment of *Bt* rice. Trends Plant Sci 9:286–292

Himabindu K, Suneetha K, Sama VSAK, Bentur JS (2010) A new rice gall midge resistance gene in the breeding line CR57-MR1523, mapping with flanking markers and development of NILs. Euphytica 174:179–187

Hirabayashi H, Ideta O, Sato H et al (2004) Identification of a resistance gene to brown planthopper derived from *Oryza minuta* in rice. Breed Res 6(Suppl 1):285

Ho NH, Baisakh N, Oliva N et al (2006) Translational fusion hybrid *Bt* genes confer resistance against yellow stem borer in transgenic elite Vietnamese rice cultivars. Crop Sci 46:781–789

Horgan FG, Ramal AF, Bentur JS et al (2016) Virulence of brown planthopper (*Nilaparvata lugens*) populations from South and South East Asia against resistant rice varieties. Crop Prot 78:222–231

Hou LY, Ping YU, Qun XU et al (2011) Genetic analysis and preliminary mapping of two recessive resistance genes to brown planthopper, *Nilaparvata lugens* Stål in rice. Rice Sci 18:238–242

Huang J, Wei Z, An H et al (2001) *Agrobacterium tumefaciens*-mediated transformation of rice with the spider insecticidal gene conferring resistance to leaffolder and striped stem borer. Cell Res 11:149–155. https://doi.org/10.1038/sj.cr.7290080

Huang D, Qiu Y, Zhang Y et al (2013) Fine-mapping and characterization of *BPH27*, a brown planthopper resistance gene from wild rice (*Oryza rufipogon* Griff.). Theor Appl Genet 126:219–229

Huang HJ, Liu CW, Xu HJ, Bao YY, Zhang CX (2017) Mucin-like protein, a saliva component involved in brown planthopper virulence and host adaptation. J Insect Physiol 98:223–230

Husnain T, Bokhari SM, Riaz N et al (2003) Pesticidal genes of *Bacillus thuringiensis* in transgenic rice technology to breed insect resistance. Pak J Biochem Mol Biol 36(3):133–142

Huvenne H, Smagghe G (2010) Mechanisms of dsRNA uptake in insects and potential of RNAi for pest control: a review. J Insect Physiol 56:227–235

Intikhab S, Karim S, Riazuddin S (2000) Natural variation among rice yellow stem borer and rice leaffolder populations to C delta endotoxins. Pak J Biol Sci 3(8):1285–1289

IRRI (International Rice Research Institute) (1979) Brown planthopper: threat to rice production in Asia. IRRI, Los Baños. 369 p

Jairin J, Phengrat K, Teangdeerith S, Vanavichit A, Toojinda T (2007a) Mapping of a broad-spectrum brown planthopper resistance gene, *Bph3*, on rice chromosome 6. Mol Breed 19:35–44

Jairin J, Teangdeerith SN, Leelagud P, Phengrat K, Vanavichit A, Toojinda T (2007b) Physical mapping of *Bph3*, a brown planthopper resistance locus in rice. Maejo Int J Sci Technol 1:166–177

Jairin J, Teangdeerith S, Leelagud P et al (2010) Detection of a brown planthopper resistance gene *bph4* at the same chromosomal position of *Bph3* using two different genetic backgrounds of rice. Breed Sci 60:71–75

James C (2005) Global status of commercialized biotech/GM crops: 2005. ISAAA brief No. 34. International Service for the Acquisition of Agri-Biotech Applications, Ithaca, NY

Janique S, Sriratanasak W, Ketsuwan K, Jairin J, Jeratthitikul E (2017) Phylogeography of the Asian rice gall midge *Orseolia oryzae* (Wood Mason) (Diptera: Cecidomyiidae) in Thailand. Genetica 145:37–49

Jena KK, Jeung JU, Lee JH, Choi HC, Brar DS (2006) High-resolution mapping of a new brown planthopper (BPH) resistance gene, *Bph18*(t), and marker-assisted selection for BPH resistance in rice (*Oryza sativa* L.). Theor Appl Genet 112:288–297

Jiang H, Hu J, Li Z et al (2018) Evaluation and breeding application of six brown planthopper resistance genes in rice maintainer line Jin 23B. Rice 11:22

Jin S, Singh ND, Li L, Zhang X, Daniell H (2015) Engineered chloroplast dsRNA silences cytochrome p450 monooxygenase, V-ATPase and chitin synthase genes in the insect gut and disrupts *Helicoverpa armigera* larval development and pupation. Plant Biotechnol J 13:435–446

Jing S, Zhang L, Ma Y et al (2014) Genome-wide mapping of virulence in brown planthopper identifies loci that break down host-plant resistance. PLoS One 9:e98911

Jing S, Zhao Y, Du B, Chen R, Zhu L, He G (2017) Genomics of interaction between the brown planthopper and rice. Curr Opin Insect Sci 19:82–87

Kadowaki M, Yoshimura A, Yasui H (2003) RFLP mapping of antibiosis to rice green leafhopper. In: Khush GS, Brar DS, Hardy B (eds) Advances in rice genetics. International Rice Research Institute, Los Baños, pp 270–272

Katiyar SK, Tan Y, Huang B et al (2001) Molecular mapping of gene *Gm-6*(t) which confers resistance against four biotypes of Asian rice gall midge in China. Theor Appl Genet 103:953–961

Kawaguchi M, Murata K, Ishii T, Takumi S, Mori N, Nakamura C (2001) Assignment of a brown planthopper (*Nilaparvata lugens* Stål) resistance gene *bph4* to the rice chromosome 6. Breed Sci 51(1):13–18

Khajuria C, Vélez AM, Rangasamy M et al (2015) Parental RNA interference of genes involved in embryonic development of the western corn rootworm, *Diabrotica virgifera virgifera* LeConte. Insect Biochem Mol Biol 63:54–62

Khajuria C, Ivashuta S, Wiggins E et al (2018) Development and characterization of the first dsRNA-resistant insect population from western corn rootworm, *Diabrotica virgifera virgifera* LeConte. PLoS One 13(5):e0197059. https://doi.org/10.1371/journal.pone.0197059. http://www.epa.gov/pesticide-registration/registration-rnai-control-corn-rootworm

Khanna HK, Raina SK (2002) Elite indica transgenic rice plants expressing modified *CryIAc* endotoxin of *Bacillus thuringiensis* show enhanced resistance to yellow stem borer (*Scirpophaga incertulas*). Transgenic Res 11:411–423

Kiani G, Nematzadeh G, Ghareyazie B et al (2008) Evaluation of *Bt* (*Bacillus thuringiensis*) rice varieties against stem borer (*Chilo suppressalis*). Pak J Biol Sci 11:648–651. https://doi. org/10.3923/pjbs.2008.648.651

Kim SM, Sohn JK (2005) Identification of a rice gene (*Bph1*) conferring resistance to brown planthopper (*Nilaparvata lugens* Stal) using STS markers. Mol Cell 20:30–34

Kim S, Kim C, Kim WLT et al (2008) Inheritance and field performance of transgenic Korean *Bt* rice lines resistant to rice yellow stem borer. Euphytica 164:829–839

Kobayashi T, Yamamoto K, Suetsugu Y et al (2014) Genetic mapping of the rice resistance breaking gene of the brown planthopper *Nilaparvata lugens*. Proc R Soc B 281:20140726

Kola VSR, Renuka P, Madhav MS, Mangrauthia SK (2015) Key enzymes and proteins of crop insects as candidate for RNAi-based gene silencing. Front Physiol 6:119. https://doi. org/10.3389/fphys.2015.00119

Kola VSR, Renuka P, Padmakumari AP et al (2016) Silencing of CYP6 and APN genes affects the growth and development of rice Yellow Stem Borer, *Scirpophaga incertulas*. Front Physiol 7:20. https://doi.org/10.3389/fphys.2016.00020

Krieg A, Huger AM, Langenbruch GA et al (1983) *Bacillus thuringiensis* var. *tenebrionis*: a new pathotype effective against larvae of coleopteran. J Appl Entomol 96:500–508

Kumar S, Arul L, Talwar D (2010) Generation of marker-free *Bt* transgenic indica rice and evaluation of its yellow stem borer resistance. J Appl Genet 51:243–257. https://doi.org/10.1007/ bf03208854

Kumar K, Sarao PS, Bhatia D et al (2018) High-resolution genetic mapping of a novel brown planthopper resistance locus, *Bph34* in *Oryza sativa* L. X *Oryza nivara* (Sharma & Shastry) derived interspecific F$_2$ population. Theor Apppl Genet 131:1163. https://doi.org/10.1007/ s00122-018-3069-7

Lang TN, Buu CB (2003) Genetic and physical maps of gene *Bph10* controlling brown planthopper resistance in rice (*Oryza sativa* L.). Omonrice 11:35–41

Lee M, Aguda R, Cohen M et al (1997) Determination of binding of *Bacillus thuringiensis* (delta)-endotoxin receptors to rice stem borer midguts. Appl Environ Microbiol 63:1453–1459

Leelagud P, Kongsila S, Vejchasarn P et al (2020) Genetic diversity of Asian rice gall midge based on mtCOI gene sequences and identification of a novel resistance locus gm12 in rice cultivar MN62M. Mol Biol Rep 47:4273. https://doi.org/10.1007/s11033-020-05546-9

Li G, Xu X, Xing H, Zhu H, Fan Q (2005) Insects resistance to *Nilaparvata lugens* and *Cnephalocrosis medinalis* in transgenic indica rice and the inheritance of gna+ sbti transgenes. Pest Manag Sci 61:390–396

Li H, Guan R, Guo H, Miao X (2015) New insights into an RNAi approach for plant defence against piercing-sucking and stem-borer insect pests. Plant Cell Environ 38:2277–2285. https://doi.org/10.1111/pce.12546

Ling Y, Weillin Z (2016) Genetic and biochemical mechanisms of rice resistance to planthopper. Plant Cell Rep 35:1559–1572

Liu G, Yan H, Fu Q et al (2001) Mapping of a new gene for brown planthopper resistance in cultivated rice introgressed from *Oryza eichingeri*. Chin Sci Bull 46:1459–1462

Liu Y, Wu H, Chen H et al (2015) A gene cluster encoding lectin receptor kinases confers broad-spectrum and durable insect resistance in rice. Nat Biotechnol 33:301–305

Liu Y, Chen L, Liu Y et al (2016) Marker-assisted pyramiding of two brown planthopper resistance genes, *Bph3* and *Bph27* (t), into elite rice cultivars. Rice 9:27

Liu Y, Wang Y, Shu C et al (2018) *Cry64Ba* and *Cry64Ca*, two ETX/MTX2 *Bacillus thuringiensis* insecticidal proteins against hemipteran pests. Appl Environ Microbiol 84:e01996–e01917. https://doi.org/10.1128/AEM.01996-17

Mahmood-ur-Rahman RH, Shahid AA, Bashir K, Husnain T, Riazuddin S (2007) Insect resistance and risk assessment studies of advanced generations of basmati rice expressing two genes of *Bacillus thuringiensis*. Electron J Biotechnol 10(2):15. http://www.ejbiotechnology.info/ content/vol10/issue2/full/3/

Maiti MK, Nayak P, Basu A et al (2001) Performance of *Bt* IR64 rice plants resistant against yellow stem borer in their advanced generations. Food security and environment protection in the new millennium. In: Proceedings Asian Agriculture Congress. IRRI, Los Baños, p 314

Makkar GS, Bentur JS (2017) Breeding for stem borer and gall midge resistance in rice. In: Arora R, Sandhu S (eds) Breeding insect resistant crops for sustainable agriculture. Springer, Singapore, pp 323–352

Mao YB, Cai WJ, Wang JW et al (2007) Silencing a cotton bollworm P450 monooxygenase gene by plant-mediated RNAi impairs larval tolerance of gossypol. Nat Biotechnol 25:1307–1313

Maqbool SB, Husnain T, Riazuddin S et al (1998) Effective control of yellow stem borer and rice leaf folder in transgenic rice indica varieties Basmati 370 and M7 using the novel endotoxin cryIIA *Bacillus thuringiensis* gene. Mol Breed 6:1–7

Maqbool SB, Riazuddin S, Loc NT et al (2001) Expression of multiple insecticidal genes confers broad resistance against a range of different rice pests. Mol Breed 7:85–93

Miao YT, Deng Y, Jia HK, Liu YD, Hou ML (2018) Proteomic analysis of watery saliva secreted by white-backed planthopper, *Sogatella furcifera*. PLoS One 13:e0193831. https://doi.org/10.1371/journal.pone.0193831

Moghaieb R (2010) Transgenic rice plants expressing *cryIIa5* gene are resistant to stem borer (*Chilo agamemnon*). GM Crops 1:288–293. https://doi.org/10.4161/gmcr.1.5.14276

Murai H, Hashimoto Z, Sharma PN, Shimizu T, Murata K, Takumi S, Mori N, Kawasaki S, Nakamura C (2001) Construction of a high-resolution linkage map of a rice brown planthopper (*Nilaparvata lugens* Stål) resistance gene *bph2*. Theor Appl Genet 103(4):526–532

Myint KKM, Fujita D, Matsumura M, Sonoda T, Yoshimura A, Yasui H (2012) Mapping and pyramiding of two major genes for resistance to the brown planthopper (*Nilaparvata lugens* [Stål]) in the rice cultivar ADR52. Theor Appl Genet 124:495–504

Nagadhara D, Ramesh S, Pasalu IC, Rao YK, Krishnaiah NV, Sarma NP et al (2003) Transgenic *indica* rice-resistant to sap sucking insects. Plant Biotechnol J 1:231–240

Nagadhara D, Ramesh S, Pasalu IC, Rao YK, Sarma NP, Reddy VD, Rao KV (2004) Transgenic rice plants expressing the snowdrop lectin gene (*gna*) exhibit high-level resistance to the white-backed planthopper (*Sogatella furcifera*). Theor Appl Genet 109:1399–1405

Naik BS, Divya D, Sahu N et al (2018) A new gene *Bph33(t)* conferring resistance to brown planthopper (BPH), *Nilaparvata lugens* (Stal) in rice line RP2068-18-3-5. Euphytica 214:53

Nair S, Bentur JS, Sama VSAK (2011) Mapping gall midge resistance genes: towards durable resistance through gene pyramiding. In: Muralidharan K, Siddiq EA (eds) Genomics and crop improvement: relevance and reservations. Institute of Biotechnology, ANGR Agricultural University, Hyderabad, pp 256–264

Nayak P, Basu D, Das S et al (1997) Transgenic elite *indica* rice plants expressing CryIAc-endotoxin of *Bacillus thuringiensis* are resistant against yellow stem borer (*Scirpophaga incertulas*). Proc Natl Acad Sci U S A 94:2111–2116

Ojha A, Sinha DK, Padmakumari AP, Bentur JS, Nair S (2017) Bacterial community structure in the Asian rice gall midge reveals a varied microbiome rich in Proteobacteria. Sci Rep 7:9424. https://doi.org/10.1038/s41598-017-09791-0

Painter RH (1951) Insect resistance in crop plants. University of Kansas Press, Lawrence, KS

Pan PL, Ye YX, Lou YH et al (2018) A comprehensive omics analysis and functional survey of cuticular proteins in the brown planthopper. Proc Natl Acad Sci U S A 115:5175. https://doi.org/10.1073/pnas.1716951115

Peng L, Zhao Y, Wang H et al (2017) Functional study of cytochrome P450 enzymes from the brown planthopper (*Nilaparvata lugens* Stål) to analyze its adaptation to BPH-resistant rice. Front Physiol 8:972

Pradhan S, Chakraborty A, Sikdar N et al (2016) Marker-free transgenic rice expressing the vegetative insecticidal protein (Vip) of *Bacillus thuringiensis* shows broad insecticidal properties. Planta 244:789–804

Prahalada GD, Shivakumar N, Lohithaswa HC et al (2017) Identification and fine mapping of a new gene, *BPH31* conferring resistance to brown planthopper biotype 4 of India to improve rice, *Oryza sativa* L. Rice 10:41

Price DR, Gatehouse JA (2008) RNAi-mediated crop protection against insects. Trends Biotechnol 26:393–400

Qiu Y, Guo J, Jing S, Zhu L, He G (2010) High-resolution mapping of the brown planthopper resistance gene *Bph6* in rice and characterizing its resistance in the 9311 and Nipponbare near isogenic backgrounds. Theor Appl Genet 121:1601–1611

Qiu Y, Guo J, Jing S, Zhu L, He G (2012) Development and characterization of *japonica* rice lines carrying the brown planthopper-resistance genes *BPH12* and *BPH6*. Theor Appl Genet 124:485–494

Quilis J, López-García B, Meynard D et al (2014) Inducible expression of a fusion gene encoding two proteinase inhibitors leads to insect and pathogen resistance in transgenic rice. Plant Biotechnol J 12:367–377

Rahman ML, Jiang W, Chu SH, Qiao Y, Ham TH, Woo MO, Lee J, Khanam MS, Chin JH, Jeung JU, Brar DS (2009) High-resolution mapping of two rice brown planthopper resistance genes, *Bph20* (t) and *Bph21* (t), originating from *Oryza minuta*. Theor Appl Genet 119(7):1237–1246

Raina SK, Khanna HK, Talwar D et al (2003) Insect bioassays of transgenic indica rice carrying a synthetic *Bt* toxin gene, *cry1Ac*. In: Advances in rice genetics, Proceedings of the 4th International Rice Research Conference, pp 567–569

Ramesh S, Nagadhara D, Pasalu IC et al (2004a) Development of stem borer- resistant transgenic parental lines involved in the production of hybrid rice. J Biotechnol 111:131–141

Ramesh S, Nagadhara D, Reddy VD, Rao KV (2004b) Production of transgenic *indica* rice-resistant to yellow stem borer and sap-sucking insects, using super-binary vectors of *Agrobacterium tumefaciens*. Plant Sci 166:1077–1085

Ramesh K, Padmavathi G, Dean R, Pandey MK, Lakshmi VJ, Bentur JS (2014) White-backed planthopper *Sogatella furcifera* (Horvath) (Homoptera: Delphacidae) resistance in rice variety Sinna Sivappu. Euphytica 200:139–148

Rao KV, Rathore KS, Hodges TK et al (1998) Expression of snowdrop lectin (GNA) in transgenic rice plants confers resistance to rice brown planthopper. Plant J 15:469–477

Rawat N, Sinha DK, Rajendrakumar P et al (2010) Role of pathogenesis-related genes in rice-gall midge interactions. Curr Sci 99(10):1361–1368

Rawat N, Neeraja CN, Nair S, Bentur JS (2012a) Differential gene expression in gall midge susceptible rice genotypes revealed by suppressive subtraction hybridization (SSH) cDNA libraries and microarray analysis. Rice 5:8

Rawat N, Neeraja CN, Sundaram RM, Nair S, Bentur JS (2012b) A novel mechanism of gall midge resistance in the rice variety Kavya revealed by microarray analysis. Funct Integr Genom 12:249–264

Rawat N, Himabindu K, Neeraja CN, Nair S, Bentur JS (2013) Suppressive subtraction hybridization reveals that rice gall midge attack elicits plant-pathogen like responses in rice. Plant Physiol Biochem 63:122–130

Ren J, Gao F, Wu X et al (2016) *Bph32*, a novel gene encoding an unknown SCR domain-containing protein, confers resistance against the brown planthopper in rice. Sci Rep 6:37645

Renganayaki K, Fritz AK, Sadasivam S et al (2002) Mapping and progress toward map-based cloning of brown planthopper biotype-4 resistance gene introgressed from *Oryza officinalis* into cultivated rice, *O. sativa*. Crop Sci 42:2112–2117

Renuka P, Maganti SM, Padmakumari AP et al (2017) RNA-Seq of rice yellow stem borer, *Scirpophaga incertulas* reveals molecular insights during four larval developmental stages. G3 (Bethesda) 7:3031. https://doi.org/10.1534/g3.117.043737

Riaz N, Husnain T, Fatima T et al (2006) Development of *indica* Basmati rice harboring two insecticidal genes for sustainable resistance against lepidopteran insects. South Afr J Bot 72:217–223

Saha P, Majumder P, Dutta I, Ray T, Roy SC, Das S (2006) Transgenic rice expressing *Allium sativum* leaf lectin with enhanced resistance against sap-sucking insect pests. Planta 223:1329–1343. https://doi.org/10.1007/s00425-005-0182-z

Saka N, Tsuji T, Toyama T, Yano M, Izawa T, Sasaki T (2006) Development of cleaved amplified polymorphic sequence (CAPS) markers linked to a green rice leafhopper resistance gene, *Grh3*(t). Plant Breed 125:140–143

Sama VSAK, Himabindu K, Naik BS, Sundaram RM, Viraktamath BC, Bentur JS (2012) Mapping and MAS breeding of an allelic gene to the *Gm8* for resistance to Asian rice gall midge. Euphytica 187:393–400

Sama VSAK, Rawat N, Sundaram RM et al (2014) A putative candidate for the recessive gall midge resistance gene *gm3* in rice identified and validated. Theor Appl Genet 127:113–124

Sardesai N, Kumar A, Rajyashri KR, Nair S, Mohan M (2002) Identification and mapping of an AFLP marker linked to *Gm7*, a gall midge resistance gene and its conversion to a SCAR marker for its utility in marker aided selection in rice. Theor Appl Genet 105:691–698

Savary S, Willocquet L, Elazegui FA, Castilla NP, Teng PS (2000) Rice pest constraints in tropical Asia: quantification of yield losses due to rice pests in a range of production situations. Plant Dis 84:357–369

Schuler TH, Poppy GM, Kerry BR et al (1998) Insect-resistant transgenic plants. Trends Biotechnol 16:168–175

Sengupta S, Chakraborti D, Mondal HA, Das S (2010) Selectable antibiotic resistance marker gene-free transgenic rice harbouring the garlic leaf lectin gene exhibits resistance to sap-sucking planthoppers. Plant Cell Rep 29:261–271

Shangguan X, Zhang J, Liu B et al (2018) A mucin-like protein of planthopper is required for feeding and induces immunity response in plants. Plant Physiol 176:552–565

Shao E, Lin L, Chen C et al (2016) Loop replacements with gut-binding peptides in Cry1Ab domain II enhanced toxicity against the brown planthopper, *Nilaparvata lugens* (Stål). Sci Rep 6:20106

Shu Q, Ye G, Cui H et al (2000) Transgenic rice plants with a synthetic *Cry1Ab* gene from *Bacillus thuringiensis* were highly resistant to eight lepidopteran rice pest species. Mol Breed 6:433–439

Shu QY, Cui HR, Ye GY et al (2002) Agronomic and morphological characterization of *Agrobacterium*-transformed *Bt* rice plants. Euphytica 127:345–352. https://doi.org/10.1023/a:1020358617257

Sinha DK, Bentur JS, Nair S (2011a) Compatible interaction with its rice host leads to enhanced expression of gamma subunit of oligosaccharyl transferase (OoOST) in the Asian rice gall midge (*Orseolia oryzae*). Insect Mol Biol 20(5):567–575

Sinha DK, Lakshmi M, Anuradha G et al (2011b) Serine proteases-like genes in the rice gall midge show differential expression in compatible and incompatible interactions with rice. Int J Mol Sci 12:2842–2852 .

Sinha DK, Atray I, Bentur JS, Nair S (2012a) Expression of nucleoside diphosphate kinase (OoNDPK) is enhanced in rice gall midge feeding on susceptible rice hosts and its overexpression leads to salt tolerance in *E. coli*. Insect Mol Biol 21(6):593–603

Sinha DK, Nagaraju J, Tomar A, Bentur JS, Nair S (2012b) Pyrosequencing-based transcriptome analysis of the Asian rice gall midge reveals differential response during compatible and incompatible interaction. Int J Mol Sci 13:13079–13103

Sinha D, Atray I, Bentur JS, Nair S (2015) Feeding on resistant rice leads to enhanced expression of defender against apoptotic cell death (*OoDAD1*) in the Asian rice gall midge. BMC Plant Biol 15:235

Sinha DK, Atray I, Agarrwal R, Bentur JS, Nair S (2017) Genomics of the Asian rice gall midge and its interactions with rice. Curr Opin Insect Sci 19:76–81

Slamet LIH, Novalina S, Damayanti D et al (2003) Inheritance of *cry1Ab* and snowdrop lectin *gna* genes in transgenic javanica rice progenies and bioassay for resistance to brown planthopper and yellow stem borer. International Rice Research Institute (IRRI), Los Baños, pp 565–566

Srinivasan TS, Almazan MLP, Bernal CC, Ramal AF, Subbarayalu MK, Horgan FG (2016) Interactions between nymphs of *Nilaparvata lugens* and *Sogatella furcifera* (Hemiptera: Delphacidae) on resistant and susceptible rice varieties. Appl Entomol Zool 51:81–90

Su C-C, Zhai H-Q, Wang C-MG, Sun LH, Wan J-M (2006) SSR mapping of brown planthopper resistance gene *Bph9* in Kaharamana, an indica rice (*Oryza sativa* L.). Acta Genet Sin 33:262–268

Suh JP, Yang SJ, Jeung JU et al (2011) Development of elite breeding lines conferring *Bph18* gene-derived resistance to brown planthopper (BPH) by marker-assisted selection and genome-wide background analysis in japonica rice (*Oryza sativa* L.). Field Crop Res 120:215–222

Sun L, Su C, Wang C, Zhai H, Wan J (2005) Mapping of a major resistance gene to the brown planthopper in the rice cultivar Rathu Heenati. Breed Sci 55:391–396

Sun L-H, Wang C-M, Su C-C, Liu Y-Q, Zhai H-Q, Wan J-M (2006) Mapping and marker-assisted selection of a brown planthopper resistance gene *bph2* in rice (*Oryza sativa* L.). Acta Genet Sin 33:717–723

Sundaram RM (2007) Fine-mapping of rice-gall midge resistance genes *Gm1* and *Gm2* and validation of the linked markers. Dissertation. University of Hyderabad, Hyderabad

Sundaram RM, Balachandran SM, Madhav MS, Viraktamath BC (2013) Biotechnological options for rice improvement. In: Shetty PK, Hegde MR, Mahadevappa M (eds) Innovation in rice production. National Institute of Advanced Studies, Bangalore, pp 167–202

Tamura K, Fukuta Y, Hirae M, Oya S, Ashikawa I, Yagi T (1999) Mapping of the *Grh1* locus for green rice leafhopper resistance in rice using RFLP markers. Breed Sci 49:11–14

Tamura K, Fukuta Y, Hirae M, Oya S, Ashikawa I, Yagi T (2004) RFLP mapping of a new resistance gene for green rice leafhopper in Kanto PL10. Rice Genet Newsl 21:62–64

Tamura Y, Hattori M, Yoshioka H et al (2014) Map-based cloning and characterization of a brown planthopper resistance gene *BPH26* from *Oryza sativa* L. ssp. *indica* cultivar ADR52. Sci Rep 4:5872

Tan GX, Wang QM, Ren X, Huang Z, Zhu LL, He GC (2004) Two white-backed planthopper resistance genes in rice share the same loci with those for brown planthopper resistance. Heredity 92:212–217

Tang W, Lin YJ (2007) Field experiment of transgenic *cry1Ab* insect resistant rice. Hereditas 29:1008–1012

Terenius O, Papanicolaou A, Garbutt JS et al (2011) RNA interference in Lepidoptera: an overview of successful and unsuccessful studies and implications for experimental design. J Insect Physiol 57:231–245

Thaler JS, Humphrey PT, Whiteman NK (2012) Evolution of jasmonate and salicylate signal crosstalk. Trends Plant Sci 17:260–270

Tomoyasu Y, Denell RE (2004) Larval RNAi in *Tribolium* (Coleoptera) for analyzing adult development. Dev Genes Evol 214:575–578. https://doi.org/10.1007/s00427-004-0434-0

Tu J, Zhang G, Datta K et al (2000) Field performance of transgenic elite commercial hybrid rice expressing *Bacillus thuringiensis* delta-endotoxin. Nat Biotechnol 18:1101–1104

Turner CT, Davy MW, MacDiarmid RM, Plummer KM, Birch NP, Newcomb RD (2006) RNA interference in the light brown apple moth, *Epiphyas postvittana* (Walker) induced by double-stranded RNA feeding. Insect Mol Biol 15:383–391

Venkanna V, Hari Y, Rukminidevi K et al (2018) Marker-assisted selection for pyramiding of gall midge resistance genes in Kavya, a popular rice variety. Int J Curr Microbiol App Sci 7:745–753

Vijayalakshmi P, Amudhan S, Himabindu K, Cheralu C, Bentur JS (2006) A new biotype of the Asian rice gall midge *Orseolia oryzae* (Diptera: Cecidomyiidae) characterized from the Warangal population in Andhra Pradesh. India Int J Trop Insect Sci 26:207–211

Vila L, Quilis J, Meynard D et al (2005) Expression of the maize proteinase inhibitor (*mpi*) gene in rice plants enhances resistance against the striped stem borer (*Chilo suppressalis*): effects on larval growth and insect gut proteinases. Plant Biotechnol J 3:187–202

Wang Y, Wang X, Yuan H et al (2008) Responses of two contrasting genotypes of rice to brown planthopper. Mol Plant-Microbe Interact 21:122–132

Wang Y, Zhang L, Li Y et al (2014) Expression of Cry1Ab protein in a marker-free transgenic *Bt* rice line and its efficacy in controlling a target pest, *Chilo suppressalis* (Lepidoptera: Crambidae). Environ Entomol 43:528–536

Wang Y, Cao L, Zhang Y et al (2015) Map-based cloning and characterization of *BPH29*, a B3 domain-containing recessive gene conferring brown planthopper resistance in rice. J Exp Bot 66:6035–6045

Wang H, Ye S, Mou T (2016a) Molecular breeding of rice restorer lines and hybrids for brown planthopper (BPH) resistance using the *Bph14* and *Bph15* genes. Rice 9:53

Wang J, Zhang H, Wang H et al (2016b) Functional validation of cadherin as a receptor of Bt toxin Cry1Ac in *Helicoverpa armigera* utilizing the CRISPR/Cas9 system. Insect Biochem Mol Biol 76:11–17. https://doi.org/10.1016/j.ibmb.2016.06.008

Wang L, Tang N, Gao X et al (2017a) Genome sequence of a rice pest, the white-backed planthopper (*Sogatella furcifera*). Giga Sci 6:1–9

Wang Y, Jiang W, Liu H et al (2017b) Marker-assisted pyramiding of *Bph6* and *Bph9* into elite restorer line 93–11 and development of functional marker for *Bph9*. Rice 10:51

Wang W, Wan P, Lai F, Zhu T, Fu Q (2018) Double-stranded RNA targeting calmodulin reveals a potential target for pest management of *Nilaparvata lugens*. Pest Manag Sci 74:1711–1719. https://doi.org/10.1002/ps.4865

Waris MI, Younas A, ul Qamar MT et al (2018) Silencing of chemosensory protein gene NlugCSP8 by RNAi induces declining behavioral responses of *Nilaparvata lugens*. Front Physiol 9:379. https://doi.org/10.3389/fphys.2018.00379

Wu C, Fan Y, Zhang C et al (1997a) Transgenic fertile japonica rice plants expressing a modified *cry1Ab* gene resistant to yellow stem borer. Plant Cell Rep 17:129–132

Wu C, Zhao R, Fan Y et al (1997b) Transgenic rice plants resistant to yellow stem borer. Rice Biotechnol 9:7

Wu J, Maehara T, Shimokawa T et al (2002) A comprehensive rice-transcript map containing 6591 expressed sequence tag sites. Plant Cell Online 14:525–535

Wu H, Liu Y, He J et al (2014) Fine-mapping of brown planthopper (*Nilaparvata lugens* Stål) resistance gene *Bph28* (t) in rice (*Oryza sativa* L.). Mol Breed 33:909–918

Wunn J, Kloti A, Burkhardt P et al (1996) Transgenic indica rice breeding line IR58 expressing a synthetic *cryIA(b)* gene from *Bacillus thuringiensis* provides effective insect pest control. Nat Biotechnol 14:171–176

Xue J, Zhou X, Zhang CX et al (2014) Genomes of the rice pest brown planthopper and its endosymbionts reveal complex complementary contributions for host adaptation. Genome Biol 15:521

Xue WH, Xu N, Yuan XB et al (2018) CRISPR/Cas9-mediated knockout of two eye pigmentation genes in the brown planthopper, *Nilaparvata lugens* (Hemiptera: Delphacidae). Insect Biochem Mol Biol 93:19–26

Yamasaki M, Yoshimura A, Yasui H (2003) Genetic basis of ovicidal response to white-backed planthopper (*Sogatella furcifera* Horvath) in rice (*Oryza sativa* L.). Mol Breed 12:133–143

Yang H, You A, Yang Z et al (2004) High-resolution genetic mapping at the *Bph15* locus for brown planthopper resistance in rice (*Oryza sativa* L.). Theor Appl Genet 110:182–191

Yang Z, Chen H, Tang W et al (2011) Development and characterization of transgenic rice expressing two *Bacillus thuringiensis* genes. Pest Manag Sci 67:414–422

Yang L, Li RB, Li YR et al (2012) Genetic mapping of *bph20* (t) and *bph21*(t) loci conferring brown planthopper resistance to *Nilaparvata lugens* Stål in rice (*Oryza sativa* L.). Euphytica 183:161–171

Yang Y, Mei F, Zhang W et al (2014) Creation of Bt rice expressing a fusion protein of *cry1Ac* and *Cry1I*-like using a green tissue-specific promoter. J Econ Entomol 107:1674–1679

Yarasi B, Sadumpati V, Pasalu IC, Reddy DV, Rao KV (2008) Transgenic rice expressing *Allium sativum* leaf agglutinin (ASAL) exhibits high-level resistance against major sap-sucking pests. BMC Plant Biol 8:102. https://doi.org/10.1186/1471-2229-8-102

Ye G, Shu Q, Cui H et al (2000) A leaf-section bioassay for evaluating rice stem borer resistance in transgenic rice containing a synthetic *cry1Ab* gene from *Bacillus thuringiensis* Berliner. Bull Entomol Res 90:179–182. https://doi.org/10.1017/s0007485300000298

Ye G, Shu Q, Yao H et al (2001) Field evaluation of resistance of transgenic rice containing a synthetic *cry1Ab* gene from *Bacillus thuringiensis* Berliner to two stem borers. J Econ Entomol 94:271–276

Ye R, Huang H, Yang Z et al (2009) Development of insect-resistant transgenic rice with *Cry1C*-free endosperm. Pest Manag Sci 65:1015–1020

Ye W, Yu H, Jian Y et al (2017a) A salivary EF-hand calcium-binding protein of the brown planthopper *Nilaparvata lugens* functions as an effector for defense responses in rice. Sci Rep 7:40498

Ye ZF, Liu XL, Han Q et al (2017b) Functional characterization of PBP1 gene in *Helicoverpa armigera* (Lepidoptera: Noctuidae) by using the CRISPR/Cas9 system. Sci Rep 7:8470. https://doi.org/10.1038/s41598-017-08769-2

Yin K, Gao C, Qiu JL (2017) Progress and prospects in plant genome editing. Nat Plants 3:17107

Yoshimura S, Komatsu M, Kaku K et al (2012) Production of transgenic rice plants expressing *Dioscorea batatas* tuber *lectin1* to confer resistance against brown planthopper. Plant Biotechnol 29:501–504

Yu R, Xu X, Liang Y et al (2014) The insect ecdysone receptor is a good potential target for RNAi-based pest control. Int J Biol Sci 10:1171–1180. https://doi.org/10.7150/ijbs.9598

Zamore PD, Tuschl T, Shorp PA, Bartel DP (2000) RNAi: double-stranded RNA directs the ATP-dependent cleavage of mRNA at 21 to 23 nucleotide intervals. Cell 101:25–33

Zeng Q, Wu Q, Zhou K et al (2002) Obtaining stem borer-resistant homozygous transgenic lines of Minghui 81 harboring novel *cry1Ac* gene via particle bombardment. Yi Chuan Xue Bao 29:519–524

Zeng F, Liu H, Zhang A et al (2018) Three chemosensory proteins from the rice leaffolder *Cnaphalocrocis medinalis* involved in host volatile and sex pheromone reception. Insect Mol Biol 27:710. https://doi.org/10.1111/imb.12503

Zha W, Peng X, Chen R, Du B, Zhu L, He G (2011) Knockdown of midgut genes by dsRNA-transgenic plant-mediated RNA interference in the Hemipteran insect *Nilaparvata lugens*. PLoS One 6:e20504

Zhang Y, Li Y, Zhang Y et al (2011) Seasonal expression of Bt proteins in transgenic rice lines and the resistance against Asiatic rice borer *Chilo suppressalis* (Walker). Environ Entomol 40:1323–1330

Zhang KJ, Zhu WC, Rong X et al (2013) The complete mitochondrial genomes of two rice planthoppers, *Nilaparvata lugens* and *Laodelphax striatellus*: conserved genome rearrangement in Delphacidae and discovery of new characteristics of atp8 and tRNA genes. BMC Genomics 14:417

Zhang KJ, Zhu WC, Rong X et al (2014) The complete mitochondrial genome sequence of *Sogatella furcifera* (Horváth) and a comparative mitogenomic analysis of three predominant rice planthoppers. Gene 533:100–109

Zhang J, Khan SA, Hasse C, Ruf S, Heckel DG, Bock R (2015) Full crop protection from an insect pest by expression of long double-stranded RNAs in plastids. Science 347:991–994

Zhang L, Qiu LY, Yang HL et al (2017) Study on the effect of wing bud chitin metabolism and its developmental network genes in the brown planthopper, *Nilaparvata lugens*, by knockdown of TRE gene. Front Physiol 8:750

Zhao Q (2015) Generation of insect-resistant and glyphosate-tolerant rice by introduction of a T-DNA containing two *Bt* insecticidal genes and an *EPSPS* gene. J Zhejiang Univ Sci B 16(10):824–831

Zhao L, Yang M, Shen Q et al (2016a) Functional characterization of three trehalase genes regulating the chitin metabolism pathway in rice brown planthopper using RNA interference. Sci Rep 6:27841. https://doi.org/10.1038/srep27841

Zhao Y, Huanga J, Wanga Z et al (2016b) Allelic diversity in an NLR gene *BPH9* enables rice to combat planthopper variation. Proc Natl Acad Sci U S A 113:45

Zhou WW, Liang QM, Xu Y et al (2013) Genomic insights into the Glutathione S-transferase gene family of two rice planthoppers, *Nilaparvata lugens* (Stål) and *Sogatella furcifera* (Horváth) (Hemiptera: Delphacidae). PLoS One 8(2):e56604. https://doi.org/10.1371/journal.pone.0056604

Zhu GH, Peng YC, Zheng MY et al (2017a) CRISPR/Cas9-mediated BLOS2 knockout resulting in disappearance of yellow strips and white spots on the larval integument in *Spodoptera litura*. J Insect Physiol 103:29–35

Zhu J, Hao P, Lu C, Ma Y, Feng Y, Yu X (2017b) Expression and RNA interference of ribosomal protein L5 gene in *Nilaparvata lugens* (Hemiptera: Delphacidae). J Insect Sci 17:73. https://doi.org/10.1093/jisesa/iex047

Zhu J, Jiang F, Wang X et al (2017c) Genome sequence of the small brown planthopper, *Laodelphax striatellus*. Giga Sci 6:1–12

Zimmer CT, Garrood WT, Singh SK et al (2018) Neofunctionalization of duplicated P450 genes drives the evolution of insecticide resistance in the brown planthopper. Curr Biol 28:268–274

Zuo Y, Wang H, Xu Y et al (2017) CRISPR/Cas9-mediated G4946E substitution in the ryanodine receptor of *Spodoptera exigua* confers high levels of resistance to diamide insecticides. Insect Biochem Mol Biol 89:79–85

Doubled Haploids in Rice Improvement: Approaches, Applications, and Future Prospects

Sanghamitra Samantaray, Jauhar Ali, Katrina L. C. Nicolas,
Jawahar Lal Katara, Ram Lakhan Verma, C. Parameswaran, B. N. Devanna,
Awadhesh Kumar, Byomkesh Dash, and Sudhansu Sekhar Bhuyan

Abstract Exploitation of biotechnological tools in conventional breeding strategies is the need of the hour for overcoming limitations in rice production and productivity. In addition, improvement in quantity and quality along with resistance to climatic and disease stress in rice require immediate attention. Anther culture has proven its efficiency by instantaneously fixing homozygosity through diploidization of regenerated haploid plants. Therefore, androgenesis provides an efficient platform for developing inbred lines in a short period of time. Although anther culture shows its efficiency in speeding up breeding in several crop species, including rice, associated limitations still prevent the exploitation of its optimum potential. Although anther culture is well exploited in *japonica* rice breeding, its application in *indica* rice is limited because of inherent recalcitrant genetic backgrounds. The success of anther culture is determined by several factors that limit the efficiency of androgenesis. Identified constraints are early anther necrosis, poor-callus response, and proliferation, and low green-plant regeneration, along with the most frustrating albinism associated with *indica* rice, which has been considerably clarified. This chapter details the method of androgenesis and scope for improving the applicability of anther culture producing doubled haploids of rice in order to use it as a complementary tool for precision breeding.

Keywords Androgenesis · Doubled haploid · Callus · Regeneration · Anther culture

S. Samantaray (✉) · J. L. Katara · R. L. Verma · C. Parameswaran · B. N. Devanna
A. Kumar · B. Dash · S. S. Bhuyan
ICAR-National Rice Research Institute, Cuttack, Odisha, India

J. Ali · K. L. C. Nicolas
IRRI, Metro Manila, Philippines

© The Author(s) 2021
J. Ali, S. H. Wani (eds.), *Rice Improvement*,
https://doi.org/10.1007/978-3-030-66530-2_12

1   Introduction

Doubled-haploid breeding through anther culture has emerged as an exciting and powerful tool, and a convenient alternative to conventional techniques for crop improvement (Purwoko et al. 2010). Doubled haploids have several advantages, such as shortening the breeding cycle by immediate fixation of homozygosity, offering high-selection efficiency, widening genetic variability through the production of gametoclonal variants, and expressing desirable recessive genes suitable for breeding (Devaux and Pickering 2005). Despite all the advantages DH technology offers in several crops, it has not been put to use in rice to sufficient extent in order to take maximum advantage, even though more than half of the world's population depends on rice for consumption as staple food. Further, the development of high-yielding rice cultivars is urgently needed to meet the demand of the increasing population and the challenges of a changing climate since cultivar development is a lengthy and time-consuming process. However, anther culture has been exploited to develop several varieties and improved breeding lines, mostly in *japonica* cultivars (Grewal et al. 2011). Contrastingly, this technique has poor implications in *indica* rice cultivars owing to poor-androgenic response. In addition, early anther necrosis, poor-callus proliferation, and albino-plant regeneration are some of the major problems encountered in the case of *indica* rice at the time of androgenesis, which needs vast improvement. Hence, attempts are being made to overcome low-anther culturability by evaluating several key factors involved in affecting the success of anther culture (Trejo-Tapia et al. 2002; Jacquard et al. 2006). DH technology integrated with phenomics and genomics could accelerate cultivar development and economize plant-breeding operations.

Several approaches in the production of DHs were developed, for which in vitro culture was found to be the most efficient and simplest technique. It is being used to produce haploids/doubled haploids in several crops. The two commonly used in vitro methods for DH production are gynogenesis and androgenesis. However, androgenesis shows its effectiveness and applicability in producing haploids and DHs in numerous cereals (Forster et al. 2007). To alleviate the problems associated with *indica* rice DH breeding, manipulation of the limiting factors related to androgenesis is required to achieve a successful protocol for effective exploitation.

This chapter deals with insights into the practical aspects of anther culture technique for it to be fully exploited for improving rice breeding.

2   Status of Doubled-Haploid Research in the Success of *Indica* Rice Anther Culture

The past and current status of research highlights the usefulness of DH technology in rice improvement. The discovery of haploids in plants led to the use of DH technology in plant breeding. Although there are different methods to generate haploids, they are usually accompanied by chromosome doubling. In general, in vitro methods were found to be the most pertinent in developing haploids and DHs, for which androgenesis shows its effectiveness and applicability in numerous cereals, including rice. These systems allow completely homozygous lines to be developed from heterozygous parents in a single generation.

The first naturally occurring haploids were reported by Blakeslee et al. (1922) in jimson weed (*Datura stramonium* L.) and thereafter natural haploids were documented in several other species. However, the relevance of DHs came into prominence only when Guha and Maheshwari (1964, 1966) reported a breakthrough in the production of haploids from anther culture of *Datura innoxia* (Mill.). Further, their research revolutionized the use of DH technology in plant breeding worldwide. Subsequently, this haploid discovery by anther culture provided several opportunities for applying this technique in crop improvement programs. In rice, the first report on producing haploids through anther culture came from Niizeki and Oono (1968). Subsequently, the doubled-haploidy approach coupled with conventional breeding led to the development of several rice varieties for pest and disease resistance, high yield, and good-quality grains in many countries. Furthermore, anther culture could facilitate other biotechnological approaches such as gene transformation technology and the identification of QTLs.

Despite all the advantages DH technology offers, it has not been put to use in *indica* rice to a great extent for maximizing its advantage. This is mostly because of the lack in expertise and the variable response of different genotypes under in vitro culture. Although the androgenic response to *japonica*-subspecies has led to the release of many varieties, the potential of anther culture for *indica* rice breeding is not fully exploited, in spite of the release of a salt-tolerant *indica* variety through anther culture (Senadhira et al. 2002). Early anther necrosis, poor-callus proliferation, and albino-plant regeneration are some of the problems encountered in *indica* rice at the time of androgenesis, which are researchable concerns. Several determining factors associated with poor-androgenic response were also addressed to obtain success in rice anther culture. Although one-step androgenesis (somatic embryogenesis) is cost-effective vis-à-vis two-step organogenesis, the latter was widely adopted because of its responsiveness in rice.

Physiology of the donor plant is an important contributory factor for the success of rice anther culture. Anthers of panicles collected from field-grown plants have been decided better in their anther culture response than anthers collected from pot plants placed in the greenhouse or near the field (Veeraraghavan 2007). Second, the microspore stage is considered as an important factor for androgenic response. An easily observable morphological trait of the plant that shows good correlation with the pollen development stage is used as a guide to identify the required stage of the microspore (Nurhasanah et al. 2016). Usually, the distance between the collar of the flag leaf and the ligule of the penultimate leaf of the tiller serves as a reliable guide to anther maturity. The most suitable stage of microspore development has been described as the late uninucleate to early binucleate stage. However, the appropriate microspore stage for effective androgenic response was reported as early as mid-uninucleate stage (Fig. 1a).

Fig. 1 Androgenesis in rice hybrids (CRHR32 and BS6444G). (**a**) Early to mid-uninucleate stages of microspore; (**b**) callus induction; (**c**) green spot indicating shoot bud emergence; (**d**) shoot elongation; (**e**) albino shoot regeneration; (**f**) root to microshoots

A wide range of chemical and physical factors influences androgenesis in vitro. The most widely used pretreatment for androgenesis is low-temperature shock with appropriate duration. Mishra et al. (2013) assessed the influence of cold pretreatment at 10 °C for 7–9 days on the anther culture response of Rajalaxmi (CRHR 5) and Ajay (CRHR 7), which showed a positive influence on callus induction frequency irrespective of the media and plant growth regulators (PGRs) employed; prolonged treatment over the optimum duration proved to be inhibitory for androgenesis. However, cold treatment (10 °C) for 8 days was found to be effective for callus induction and green-plant regeneration in a popular *indica* rice hybrid, BS6444G

(Naik et al. 2017). In addition, 2 days of preincubation period at 10 °C was quite interesting for the success of androgenesis in a long-duration *indica* rice hybrid (Rout et al. 2016).

The most commonly used basal media for anther culture are N6, MS, B5, and Potato-2 medium. Subsequently, several media (MSN, SK1, He2, and RZ) were developed from N6 media by modifying the nitrogen rates and sources, carbon content and sources, and changes in vitamins and their concentrations, which were found to be encouraging for anther response in rice. N6 media were found to induce maximum callusing in Taraori Basmati (Grewal et al. 2006). Min et al. (2016) found out the best media for androgenesis in the generation of DHs from F_1s of two intervarietal crosses in terms of callusing (N6) and shoot regeneration (½MS) after a trial with 16 different media. However, two basal media, such as N6 and MS, were found to be effective for callusing and green-shoot regeneration, respectively, in *indica* rice hybrids (Rout et al. 2016; Naik et al. 2017).

Considering the importance of plant growth regulators in tissue culture, the effects of different PGRs were investigated for androgenesis. Even though 2,4-D has proven to be a potent auxin for callus induction from cultured anthers, medium with lower 2,4-D rates was found to be more effective for the regeneration ability of calli induced in *indica* rice than higher 2,4-D rates (Fig. 1b). The calli developed green spots in 7–10 days after transfer to media supplemented with NAA (0.5 mg/L), BAP (1.0 mg/L), and Kn (1.0 mg/L), which adequately supported green-plant regeneration from subcultured calli (Fig. 1c, d). The type and concentration of auxins seem to determine the pathway of microspore development with 2,4-D inducing callus formation and IAA and NAA promoting direct embryogenesis (Ball et al. 1993). The 2,4-D was found effective for callus response while the combinations of NAA, Kn, and BAP showed shoot regeneration in the generation of DHs from F_1s of two intervarietal crosses and hybrid rice (Min et al. 2016; Rout et al. 2016; Naik et al. 2017).

The nitrogen composition supplied in the form of nitrate and/or ammonium ions in culture media plays a significant role in androgenesis. The ratio of nitrate (NO_3^-): ammonium (NH_4^+) has been observed to be an important determinant for the success of anther culture as well as for the in vitro induction of embryogenic calli in *indica* rice (Grimes and Hodges 1990). Ivanova and Van Staden (2009) investigated the elimination of total nitrogen in media, resulting in limited ability of proliferation and shoot growth, but higher ability was observed in media containing NO_3^- as the only sole nitrogen source, whereas replacing the NH_4^+ with NO_3^- decreased the rate of hyperhydricity. Herath et al. (2007) proved that the frequency of callus induction was improved by modifying three different media (N6, B5, and Miller) with one-half the rate of NH_4^+ and double the rate of KNO_3 nitrogen.

A carbohydrate source is essential for androgenesis because of the osmotic and nutritional effects. Naik et al. (2017) aptly demonstrated the superiority of maltose over sucrose as the carbon source in rice anther culture for callusing. Replacing sucrose (146 mM) with maltose (146 mM) in the callus induction medium had a significant positive effect on anther response in both *indica* and *japonica* types, with a greater effect on *indica* rice. With respect to the effect of light quality on anther culture, the embryogenic induction of microspores is inhibited by high-intensity white light, whereas darkness or low-intensity white light is found encouraging (Bjørnstad et al. 1989). The incubation of anthers continuously in the dark has, on occasion, been found to be essential.

Most in vitro morphogenic responses are genotype-dependent. In general, *indica* rice cultivars exhibit poor-androgenic response vis-à-vis *japonica* ones. Even among the *indica* cultivars, a considerable variation in microspore callusing and green-shoot regeneration has been observed (Rout et al. 2016; Naik et al. 2017). The highest callus-responsive cultivars often show the best regeneration frequency and the best responsive genotypes to callusing exhibit low-regeneration ability. Therefore, selection of a single step, either callus improvement or shoot regeneration alone, may not help in establishing an effective androgenic method. It is rather important to identify genotypes carrying the two traits for overall improvement in anther culture efficiency.

The occurrence of a large proportion of albinos among the regenerated plants following anther culture is the most frustrating feature of androgenesis and this remains a formidable obstacle for application to rice breeding. The frequency of albinos can vary from 5% to 100% (Talebi et al. 2007). *Indica* rice cultivars are more prone to this problem than *japonica* cultivars. Several factors, including pretreatment conditions and culture medium, affect the frequency of albinos, which could be considerably decreased by shortening the culture period (Asaduzzaman et al. 2003). The presence of large-scale deletions in some plastid genomes of the albino haploid plants derived from anther culture of *japonica* × *indica* hybrids and absence of such deletions in green regenerants (Yamagishi 2002) appeared to suggest a role for the plastid genome in determining the albino phenotype. The albinos (60–100%) in the regeneration culture proved to be detrimental for optimizing an androgenic response (Fig. 1e). Therefore, ICAR-NRRI, India, attempted to develop a protocol for the suppression of albinism, which led to standardization of a 100% albino-free shoot regeneration method in *indica* rice (patent filed 1355/KOL/2015 titled "Method for albino-free shoot regeneration in rice through anther culture").

The well-developed shoots were transferred to MS media supplemented with 2.0 mg/L NAA, 0.5 mg/L Kn, and 5% sucrose for rooting. Root induction starts 7 days after transfer, followed by profuse rooting after 4 weeks of culture (Fig. 1f).

3   Diploidization of Haploid Genomes

Haploids can be diploidized (duplication of chromosomes) to produce homozygous plants. There are two approaches for haploidy diploidization.

3.1   Artificial Genome Doubling

Artificial genome doubling is the most popular method applied for doubling genomes in large-scale DH production. Colchicine, an antimicrotubule drug, has been widely used and is the most effective genome-doubling agent. Colchicine duplicates the genome by binding to tubulins and inhibits microtubule polymerization (Kleiber et al. 2012; Prasanna et al. 2012; Weber 2014). However, colchicine is highly toxic, which is not only potentially carcinogenic but also hazardous to the environment (Melchinger et al. 2016). The effects of other agents with lower toxicity on chromosome doubling, such as amiprophos-methyl (APM), oryzalin, pronamide, and trifluralin (all of which are herbicides), have been reported (Wan et al. 1991; Häntzschel and Weber 2010; Murovec and Bohanec 2012).

3.2   Spontaneous Genome Doubling

Spontaneous genome doubling is cost-effective and has been reported in several cereals, including rice. The frequency of spontaneous genome doubling is reported to be 50–60% in rice (Seguí-Simarro and Nuez 2008). However, interestingly, a recent report on rice hybrid BS6444G showed 90–99% spontaneous doubling (Naik et al. 2017). The doubling rate fluctuates extremely among genotypes (Chalyk 1994; Kleiber et al. 2012). Endomitosis is one of the phenomena of doubling the number of chromosomes without division of the nucleus. The haploid cells in general are unstable in culture, with a tendency to undergo endomitosis. This property of haploid cells is exploited for diploidization to produce homozygous plants. The procedure involves growing a small segment of haploid plant stem in a suitable medium supplemented with growth regulators (auxin and cytokinin). This induces callus formation followed by differentiation. During callus growth, chromosomal doubling occurs by endomitosis. This results in the production of diploid homozygous cells and ultimately true homozygous plants.

4   Characterization of Regenerants: Ploidy Analysis

After a successful haploid induction and the regeneration procedure, evaluation of regenerants is needed to distinguish doubled haploids from redundant heterozygous diploids grown in the net house (Fig. 2a). During the production of homozygous lines, various undesired heterozygous plantlets can be obtained. In anther culture, these plants can be regenerated from the somatic tissue of inoculated plant organs such as anther wall cells, tapetum, and filaments. Reliable and fast selection of regenerants is therefore necessary before further employment of putative haploids and doubled haploids.

Fig. 2 Ploidy status in green regenerants. (**a**) A_0 plants grown in net house; (**b**) morphological discrimination of ploidy status; (**c**) SSR marker (RM480) differentiating somatic tissue-derived heterozygotes from DHs. M = 100-bp DNA ladder; C = CRMS31A; R = CRL22R; H = CRHR32; 1–3 to 5–20 = DHs; 4 = heterozygote (*arrow*)

Variations in nuclear DNA content of cultured plant tissues resulted in changes in ploidy level, which in turn changes the phenotypic characters of the regenerants, termed as triploid, tetraploid, polyploid, and aneuploid. This kind of variability occurred especially in chromosome number variations of regenerated plants, that is, ploidy changes: deletion, duplication, and rearrangements (d'Amato 1989). The chromosome aberrations influenced agronomic traits and the ploidy level of rice plants (Zhang and Chu 1984). Since the isolated anthers sometimes become contaminated from the anther wall tissues, the developed calli may be of mixed type such as haploid, diploid, triploid, and tetraploid (Dunwell 2010).

Different ploidy status can be evaluated through several approaches.

4.1 Morphological

Fertile diploids can be well-distinguished from other ploids such as haploids, triploids, and polyploids through morphological evaluation (Fig. 2b). The diploids or putative DHs show normal morphological appearance with 70–80% grain fertility, whereas the polyploids are tall, large, and possess broad thick leaves with less than 1% spikelet fertility. Short stature with no spikelet fertility confirms haploids. Sometimes, mixploids are observed in the regenerants, which can be discriminated by observing grain type after grain maturity (Fig. 2b).

4.2   Cytological

Of the several methods tested earlier, the classical cytological approach was thought to be ideal to determine ploidy status. Ploidy status in plants can be determined by counting the chromosomes of root tip cells arrested in metaphase stage (Mishra et al. 2015). However, this approach is tedious and time-consuming and also requires expertise. This is also considered an unambiguous method. Most importantly, tissues containing dividing cells may not always be readily available for analysis.

4.3   Pollen Fertility Analysis

Stomatal density and size, the size of pollen grains, cell size, and plant size are mainly considered to analyze the ploidy level in DHs, out of which a cytology-based approach through pollen size determination was used as an alternative, convenient, rapid, and reliable method to assess the ploidy level in several plant species (Zonneveld and Van Iren 2001). This method also has its own demerits like cytology.

4.4   Flow Cytometry

Since the nuclear DNA content and the ploidy level have close association, flow cytometry is the most reliable method and it has wide application in plant-ploidy analysis (Ochatt 2008; Cousin et al. 2009; X Wang et al. 2016). It is frequently used to analyze the ploidy level of individuals obtained from experiments of haplo-diploidization or chromosome doubling (Grewal et al. 2009; Ochatt et al. 2009). This method was used to discriminate the ploidy status of the regenerants developed from rice hybrids (Mishra et al. 2015).

4.5   High-Throughput Cell Analysis

Most recently, a quick and precise ploidy determination method was developed using high-throughput cell analysis (Sahoo et al. 2019). This is useful to analyze nuclear DNA content-based ploidy validation of a large number of samples.

   None of these approaches could differentiate DHs from heterozygotes (somatic tissue-derived diploids), for which molecular markers might be the best option for the identification of true DHs (Rout et al. 2016; Naik et al. 2017).

4.6   Molecular Markers

Genomic molecular markers are highly heritable, available in high numbers, and often exhibit enough polymorphism to discriminate closely related genotypes. Simple sequence repeats (SSRs) are abundant and well-distributed throughout the rice genome. They are the most commonly used markers for understanding the source of origin (originated from the microspore mother cells or embryogenesis from diploid somatic tissue), parental contribution (allelic frequencies), and homozygosity in plant derivatives. SSRs have been successfully used to identify homozygous spontaneous-doubled haploids in rice hybrids (Fig. 2c).

5   Application of DHs in Rice Improvement

To meet the challenge of food security for the increasing population amid diminishing resources such as cultivable land and irrigation water along with climate change associated with unpredictable and unseasonal weather patterns, the development of high-yielding rice varieties is required to feed the ever-growing population. Hybrid rice is considered as a best option to break through the yield barrier, showing significant yield advantages over conventional cultivars. Although hybrid rice can outyield conventional cultivars by 30–40% in production fields, it does not gain in popularity among Indian farmers because of its complicated seed production system, purchase of nonreplaceable seed every season (as it segregates in consequent seasons), higher seed cost, less-preferred quality, and vulnerability to abiotic and biotic stresses. Therefore, we need to find an alternative way to exploit hybrid potential in fixing heterosis along with the associated problems for which DH technology was found to be efficient in the rapid fixation of favorable alleles for yield and related traits. The DH breeding technique shortens the time required for breeding a new variety from the usual ~8 to ~5 years, thus saving time, labor, and financial resources. Conversely, the DH technique could be more appropriate for developing new varieties from photosensitive rice genotypes. Most of these advances were achieved in *japonica* cultivars, which were amenable to anther culture, rather than in *indica* rice. Significantly, several varieties and improved parental lines have been developed through androgenic approaches, but restricted to only *japonica* rice cultivars. However, the use of anther culture as a routine technique for breeding is extremely limited in *indica* rice due to the poor induction of androgenic calli and subsequent plant regeneration. An efficient androgenic protocol could generate a considerable number of DHs for various applications.

Fig. 3 Promising doubled haploids derived from rice hybrids (CRHR32 and BS6444G). (**a**) DHs of BS6444G showing yield equal to that of the parent; (**b**) grain quality of DH lines of hybrid BS6444G

5.1 Performance of DHs Derived from Hybrid Rice

The DH technique was used to overcome the constraints associated with *indica* rice hybrids: (1) expensive seed, depriving Indian marginal farmers from using the seed year after year; and (2) unpredictable environmental conditions and synchronized flowering. Standardization of DH technology in rice hybrids (PHB71 and KRH2) generated DHs from which two were released as varieties, Satyakrishna in 2008 and Phalguni in 2010. Subsequently, androgenic protocols in two more *indica* rice hybrids, CRHR32 (an elite long-duration *indica* rice hybrid developed at NRRI, Cuttack) and BS6444G (a popular rice hybrid, Bayer Seed Pvt. Ltd., Hyderabad), generated a considerable number of DHs (Rout et al. 2016; Naik et al. 2017). Some of the DHs showed grain yield on par with that of the parent rice hybrids (Fig. 3a); the grain quality of the DHs was also higher than that of the rice hybrids (Fig. 3b).

5.2 Mapping of Genes and QTLs for Biotic and Abiotic Stress

A quantitative trait locus (QTL) is the genomic region that contributes toward discrete traits, governed by multiple genes. In plant species, QTLs are the major determinants of important agronomic traits and mapping these genomic regions onto the specific chromosomal locations is vital for their application for crop improvement. Among the different mapping populations used for mapping QTLs in plants, doubled-haploid populations have their own significance. Since the effects of QTLs are small and are greatly influenced by environmental factors, this necessitates accurate phenotyping with replicated trials under different environmental conditions. Therefore, DH populations are among the few mapping populations with true breeding nature that can facilitate effective mapping of QTLs in plants. In addition, DH-mapping populations are developed in a short time, in one generation, as compared with other nonsegregating, true-breeding populations such as near-isogenic lines (NILs) and recombinant inbred lines (RILs). Therefore, DH populations are an effective and efficient source for mapping of genes and QTLs

(Forster and Thomas 2003; Forster et al. 2007). Subsequently, DH breeding through androgenesis has emerged as a potent tool and a suitable alternative to other techniques for crop improvement (Purwoko et al. 2010; Germana 2011).

A doubled-haploid population was used to map rice blast resistance genes and QTLs in a more efficient way (Z Wang et al. 2001). A DH population from a cross between blast-resistant variety Zhai Ye Qing 8 and susceptible variety Jin Xi17 was used in combination with cDNA-amplified fragment length polymorphism and bulked segregant analysis to identify rice blast resistance genes (Zheng et al. 2004). Subsequently, Fatah et al. (2014) used DH lines developed from IR64 and Azucena to map QTL regions having major blast resistance genes. A QTL responsible for rice blast disease resistance was also mapped using a combination of DH mapping population consisting of 88 lines (DH lines), derived by crossing Joiku No. 462 (a blast-resistant line) and Jokei06214 (a blast-susceptible line). Besides leaf blast, three main QTLs for neck blast resistance were identified and mapped onto rice chromosomes 10 (*qNBL-10*), 9 (*qNBL-9*), and 5 (*qNBL-5*) using a DH population from a cross between IR64 and Azucena, and the DH lines displayed complete neck blast resistance under field conditions (Hittalmani et al. 2000).

Similarly, QTLs linked to resistance to another major rice disease, sheath blight, were identified in a DH population. Kunihiro et al. (2002) employed *indica* rice line Zhai Ye Qing 8 (ZYQ8) and *japonica* rice line Jing Xi 17 (JX17), along with their DH population, to identify four sheath blight resistance (ShBR) QTLs, *qSBR-2*, *qSBR-3*, *qSBR-7*, and *qSBR-11*, and mapped them on chromosomes 2, 3, 7, and 11, respectively. A DH population derived from *japonica* rice line Maybelle and *indica* rice line Baiyeqiu was also used to map a ShBR QTL using marker-assisted selection (MAS) (Xu et al. 2011). Zeng et al. (2015) developed a DH population by crossing *japonica* rice line CJ06 and *indica* line TN1, which underwent field evaluation under three different environmental conditions. They identified a total of eight QTLs each for lesion height (LH) and disease rating (DR) under three environments.

Besides for rice diseases, DH lines are used for mapping QTLs for insect resistance. Using DH lines developed by crossing IR64 and Azucena, six QTLs linked with brown planthopper (BPH) resistance in rice were mapped (Soundararajan et al. 2004). Using the same DH population, Geethanjali et al. (2009) identified QTLs responsible for white-backed planthopper (WBPH) resistance. Later, a DH population obtained from a cross between Cheonhcheong and Nagdong was used to identify and map QTLs for WBPH resistance. Further, the F_1 rice line derived from a cross between JSNDH13 (BPH-resistant) and CNDH32 (WBPH-resistant) was used to develop a DH line having resistance to both insect pests (Yi et al. 2015).

Efforts are also made to map QTLs contributing to viral disease resistance in rice. Among two DH populations derived from crosses between IR64 and Azucena, and IRAT177 and Apura, the DH population derived from the former cross could map a QTL responsible for rice yellow mottle virus resistance to chromosome 12 of the rice genome (Ghesquière et al. 1997).

Further, anther culture was used to develop a mapping population from Savitri (a high-yielding *indica* rice variety) × Pokkali (a salt-tolerant *indica* rice genotype) for the identification of salt tolerance QTLs/genes. A systematic study with 117 DHs

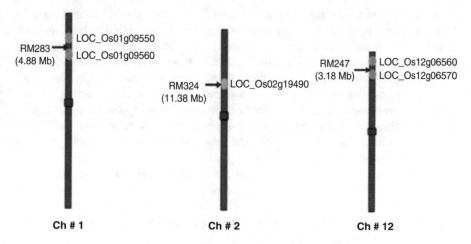

Fig. 4 Identified four candidate genes for salinity tolerance at germination stage using a doubled-haploid mapping population (Savitri × Pokkali)

derived from F_1s of Savitri and Pokkali (developed at ICAR-NRRI, Cuttack) was able to identify four candidate genes, LOC_Os01g09550 (no apical meristem protein), LOC_Os01g09560 (mitochondrial processing peptidase subunit alpha), LOC_Os12g06560 (putative protein), and LOC_Os12g06570 (cyclic nucleotide-gated ion channel), for salinity tolerance at germination stage (Fig. 4).

5.3 Development of Improved Lines and Rice Cultivars

Anther culture technology is able to fix all segregating loci (100%) in one generation (Purwoko et al. 2010) rather than eight to ten inbreeding generations with the conventional approach. Furthermore, it is able to maintain maximum genetic variance and heritability (Prigge et al. 2012), thus providing an opportunity to select transgressive segregants/superior genotypes at a very early stage. Anther culture provides high reproducibility for early-selection results, which is efficient in trait improvement strategies with simplified steps (Dicu and Cristea 2016). Hence, it is found to be an economically viable option for accelerating breeding with more selection accuracy and breeding value (Rober et al. 2005).

Given the scenario of dynamic demographics and diversified quality preferences, breeding varieties combined for higher yield, more enhanced quality traits, and wider acceptability are major concerns. Grain quality in rice hybrids is found to be a major limitation in hybridization of rice area in India (Babu et al. 2013). Using genetically diverse parental lines with similar quality parameters are found to be a workable strategy to address quality concerns in hybrids. The DH approach in turn is a probable means of helping to overcome quality problems in hybrids as it is quite efficient in developing transgressive segregants with superior quality that might be useful as parents for hybrid rice research and for cultivar release. Several varieties and improved parental lines have been developed and exploited for cultivar development, mostly with a *japonica* genetic background (Grewal et al. 2011). This technology is observed to be useful in creating immortalized genetic units from genetic resources to make them amenable to crop improvement. The libraries of DH lines help in making genetic resources accessible to crop improvement by linking molecular inventories of gene banks with meaningful phenotypes. Thus, a decrease in the breeding cycle per se in varietal development has substantial economic impact on rice breeding by saving developmental costs and maximizing genetic gain in a short span of time in a more precise way.

Doubled haploids have been released for commercialization in several crops, including rice. In addition, DHs are used as parents in developing rice hybrids (Mishra and Rao 2016). In rice, more than 100 rice breeding lines or varieties have been developed and released in China, India, Japan, South Korea, Hungary, and the United States (Siddique 2015).

Rice varieties such as Huayu I, Huayu II, Xin Xiu, Late Keng 959, Tunghua 1, Tunghua 2, Tunghua 3, Zhonghua 8, Zhonghua 9, Huahanzao, Huajian 7902, Tanghuo 2, Shanhua 7706, and Huahanzao 77001 are high-yielding varieties with superior grain quality and resistance to blast and bacterial blight diseases (Zhang 1989; Hu and Zeng 1984; Chen et al. 1986). Nanhua 5, Noll, Hua 03, and Guan 18 have early maturity, good quality, and disease resistance (Zhu and Pan 1990). Huayu 15 is resistant to lodging and diseases and has good quality (Shouyi and Shouyin 1991). Milyang 90 has good-grain quality and is resistant to brown planthopper and stripe virus disease (Chung 1987). Hwacheongbyeo, Joryeongbyeo, and Hwajinbyeo are resistant to BPH, rice stripe tenui-virus, blast, and bacterial blight (Lee et al. 1988). Bicoll (IR51500AC11-1) has salt tolerance (Senadhira et al. 2002) in China, South Korea, and the Philippines. Salinity-tolerant DH line AC-1 was developed at IRRI, Philippines, and is commercialized for salinity-prone areas of Bangladesh (Thomson et al. 2010). Table 1 lists the commercialized DH rice varieties.

In India, the National Rice Research Institute (NRRI), Cuttack, began DH work during 1997 to overcome the constraints associated with *indica* rice hybrids, that is, expensive seed, which led to preventing marginal farmers from using the seed over the years, and unpredictable environmental conditions. NRRI released two DH varieties, Satyakrishna (CR Dhan 10) in 2008 and Phalguni (CR Dhan 801) in 2010. In addition, rice variety Parag 401 (Patil et al. 1997) was also bred through DH breeding.

5.4   Biofortification of Rice for Essential Traits

Being a premier food crop, rice serves as the major source of energy, protein, thiamine, riboflavin, niacin, and micronutrients (Fe, Zn, Ca) in the diet (Juliano 1997). However, owing to deficiency for essential micronutrients ("hidden hunger"), it is unable to address the nutritional food security of the country. Among

Table 1   Rice doubled haploids commercialized in different countries

Country	Name of DH	Important features
India	Satyakrishna (CR Dhan 10)	High yield, higher
	Phalguni (CR Dhan 801)	quality
China	Huayu I, Huayu II, Xin Xiu, Late Keng 959, Tunghua 1, Tunghua 2, Tunghua 3, Zhonghua 8, Zhonghua 9, Huahanzao, Huajian 7902, Tanghuo 2, Shanhua 7706, Huahanzao 77001, Tanfeng 1, Huayu III, Ta Be 78, Guan 18	High yield, superior grain quality, resistance to blast and bacterial blight diseases
South Korea	Nanhua 5, Noll, Hua 03, Guan 18	Early maturity, good quality, and disease resistance
Philippines	AC-1	Salinity tolerance
Japan	Joiku N. 394, Hirohikari, Hirohonami AC No. 1, Kibinohana	High yield, quality type, cold resistance
Argentina	Patei and Moccoi	High yield
Hungary	Dama	High yield

the micronutrient malnutrition conditions, Fe and Zn deficiencies are of major concern not only because of their serious health consequences but also of the number of people affected worldwide. Hence, enriching rice grain with iron and zinc is likely to have tremendous health benefits for the rice-eating population. In addition, the presence of a substantial amount of phytate (0.06–2.22%) in rice, which inhibits the absorption of Fe and Zn, needs to be lowered (Liang et al. 2007). Hence, biofortification of rice varieties with enhanced content of protein, iron, and zinc and lower phytate would be an ideal goal. Biofortification is a promising food-based approach for helping to overcome micronutrient malnutrition. Significant efforts have been made over the past decade to biofortify the major cereals targeted to different parts of the world and a lot of progress has already been achieved in this endeavor. There are two distinct ways in which the nutritional value of cereals can be enhanced. The first is by using the genetic variation available through breeding or genetic engineering. Another promising approach is the doubled-haploid method, in which fixed genetic materials can be developed faster relative to other mapping populations and can be evaluated across years and locations readily. This method has less-genetic background noise than in traditional breeding methods, which makes it an important genetic resource for mapping QTLs/genes for various traits related to biofortification. Several reports have shown the utility of DH populations in identifying QTLs for the concentration of micronutrients in rice grain. Recently, two DH populations developed by IRRI were evaluated to map the QTLs related to agronomic traits and grain micronutrients (Swamy et al. 2018). Considerable genetic variation has been observed for all traits in these DH populations. These kinds of DH lines can be used as donors in breeding programs or can be directly tested in multilocation trials to further evaluate their performance.

5.5   Exchange of Cytoplasmic and Nuclear Genomes

In *Arabidopsis*, both maternal and paternal haploids containing wild-type
chromosomes and maternal cytoplasm can be generated using CENH3-mediated
haploid inducers as the male or female parent. Ravi et al. (2010) developed a
CENH3-1 GFP-tailswap haploid inducer with Ler cytoplasm: Ler-cytoplasmic
haploid inducer (HI). When pollinating Ler-cytoplasmic HI with pollen from a wild
type with Col-0 cytoplasm or Col-0 WT, haploids with Col-0 WT chromosomes and
Ler cytoplasm are generated. This method can be used to develop any combination
of cytoplasmic and nuclear genomes by transferring the male nuclear genome into
a heterologous cytoplasm rapidly and conveniently. This facilitates the production
of new cytoplasmic male sterile (CMS) lines for F_1 hybrid seed production. If the
haploid inducer line has a CMS background, pollinating this HI line with different
inbred lines generates paternal haploids, which carry CMS. One or a few paternal
haploids need to be pollinated with pollen from the maternal inbred to produce a
new diploid CMS line. Using paternal haploids for cytoplasmic conversions have
three distinct advantages: (1) only two generations are needed, (2) the new CMS
line has 100% of the genomes of either of the inbred lines, and (3) chromosome
doubling is not required (Weber 2014). This method has been employed in maize
using the *ig1* system for quite a while (Evans 2007). Most recently, knocking out the
MATL ortholog in rice resulted in haploid induction at a rate of 2–6%, suggesting
the functional conservation of *MATL*, and this represents an advance for rice
breeding (Yao et al. 2018).

5.6   Reverse Breeding

Hybrid seed is traditionally produced from a cross between two inbred lines. Dirks
et al. (2009) proposed a novel plant breeding technology, reverse breeding, which
can directly generate parental inbred lines from any hybrid. Three steps are required
for reverse breeding: (1) inhibition of meiotic crossover in F_1 plants to produce
gametes containing combinations of nonrecombinant parental chromosomes, (2)
generation of DH lines via in vitro unfertilized ovule or anther culture, and (3)
regeneration of the original hybrid through crossing DH lines with complementary
sets of parental chromosomes.

5.7   Gene Stacking from Biparental Crosses

Introgression of one or a limited number of genes into elite inbreds by marker-
assisted backcrossing is routine in plant breeding (Lübberstedt and Frei 2012). At
the end of backcross programs, a heterozygous plant is selfed to produce a fixed

line. For single-gene introgression, the expected probability of individuals with the desired homozygous genotype is 1/4. The frequency of expected genotypes decreases exponentially following the formula $\frac{1}{4}^{(n)}$, where n is the number of independently segregating genes (Lübberstedt and Frei 2012; Ravi et al. 2010; Shen et al. 2015). In contrast, haploid target genotypes are generated with a frequency of $\frac{1}{2}^{(n)}$. For example, for five loci, the frequency of the desired homozygous genotype is 1/1024 in selfed diploid progenies and 1/32 in haploid progenies. The application of doubled haploids thus significantly decreases the population size required to find desirable genotypes.

5.8   Accelerating Plant Breeding by MAS and GS

The availability of cheap and abundant molecular markers allows breeders to apply marker-assisted selection (MAS) and genomic selection (GS) in crop improvement. MAS depends on the identification of markers significantly associated with a trait. MAS allows breeders to discard a large number of plants with undesired gene combinations, pyramid beneficial genes in subsequent generations, minimize field testing, and decrease the number of generations (Collard and Mackill 2008; Dwivedi et al. 2015). The combination of MAS and DHs offers new opportunities for increasing genetic gain and shortens the time required for cultivar breeding. MAS and DHs have been successfully used to accelerate resistance breeding in cereal crops (Wessels and Botes 2014), which demonstrates the integration of MAS and DH technology to increase the speed of cultivar development vis-à-vis conventional breeding processes.

5.9   Transgenic Research

In general, the DH technique offers a rapid homozygous state attainment of heterologous loci. This principle can be used in transgenic research for developing homozygous plants within one generation. The selfing of transgenic plants after regeneration is usually the methodology followed to obtain homozygous plants for the transgenic loci. However, this requires two generations of selfing from positive-transformed plants for phenotypic validation. In contrast, anther culture of positive-transformed plants in T1 will result in homozygous conditions with one generation of selfing. Figure 5 gives a schematic representation of the segregation of the transgenic loci. Anther culture was successfully used for developing homozygous transgenic rice with the chitinase gene for enhanced sheath blight resistance (Baisakh et al. 2001). Further employment of anther culture could generate DHs from transgenics-containing genes involved in β-carotene metabolism (Datta et al. 2014).

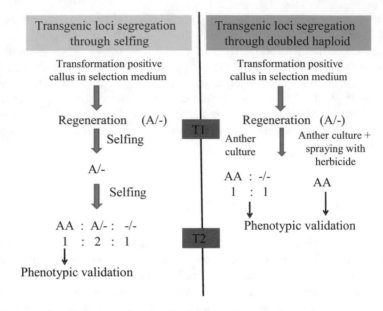

Fig. 5 Segregation of transgenic loci in selfed plants and anther culture plants

6   Knowledge Gaps

The recalcitrant nature of *indica* rice requires optimization of the anther culture method, which is essential to achieve the potential yield of androgenesis technology in *indica* rice lines. In spite of attempts made by Rout et al. (2016) and Naik et al. (2017) with these concerns, additional novel attempts for media manipulation have to be made to increase callusing potential or somatic embryogenesis. In addition, the identification of genomic regions that contribute to promising yield in rice from heterotic F_1 hybrids has to be addressed by analyzing the genetic structure of DHs. Recent studies with *Arabidopsis* and maize (D Wang et al. 2015) suggest that heterosis could be fixed by carefully combining genomic regions from the parents that contribute to higher performance.

7   The Way Forward

Given the importance of DHs in the quick fixation of segregating loci and varietal development, we need to develop novel media composition to increase the callusing potential of anthers from *indica* rice. Simultaneously, we need to focus on direct somatic embryogenesis from microspores as this method is considered cost-effective among all the pathways in tissue culture. Moreover, the mechanism of spontaneous chromosome doubling in androgenesis requires immediate attention. Conceptual understanding of the superior yield of DHs is still not clear. Thus, the scientific basis of superior yield of DHs needs to be comprehensively studied through the generation of a large number of DHs (~200 DHs from each hybrid).

Acknowledgments The authors acknowledge the director general of ICAR for continuous guidance and support for the research on doubled haploids at NRRI, Cuttack. We also greatly acknowledge the director of NRRI for his kind support and approval for bringing out a chapter on doubled haploids.

References

Asaduzzaman M, Bari MA, Rahman MH, Khatun N, Islam MA, Rahman M (2003) In vitro plant regeneration through anther culture of five rice varieties. J Biol Sci 3(2):167–171

Babu VR, Shreya K, Dangi KS, Usharani G, Nagesh P (2013) Evaluation of popular rice (*Oryza sativa* L.) hybrids for quantitative, qualitative and nutritional aspects. Int J Sci Res Publ 3(1):1

Baisakh N, Datta K, Oliva N, Ona I, Rao GJN, Mew TW, Datta SK (2001) Rapid development of homozygous transgenic rice using anther culture-harboring rice chitinase gene for enhanced sheath blight resistance. Plant Biotechnol 18(2):101–108

Ball ST, Zhou H, Konzak CF (1993) Influence of 2,4-D, IAA, and duration of callus induction in anther cultures of spring wheat. Plant Sci 90(2):195–200

Bjørnstad A, Opsahl-Ferstad HG, Aasmo M (1989) Effects of donor plant environment and light during incubation of anther cultures of some spring wheat (*Triticum aestivum* L.) cultivars. Plant Cell Tissue Organ Cult 17:27–37

Blakeslee AF, Belling J, Farnham ME, Bergner AD (1922) A haploid mutant in the jimson weed, "*Daturastramonium*". Science 55(1433):646–647

Chalyk ST (1994) Properties of maternal haploid maize plants and potential application to maize breeding. Euphytica 79:13–18

Chen JJ, Hus YJ, Tsay HS (1986) Effect of iron on rice anther culture. J Agric Res China 35(3):244–252

Chung GS (1987) Application of anther culture technique for rice (*Oryza sativa* L.) improvement. In: Korea-China Plant Tissue Culture Symposium. Academia Sinica Publishers, Beijing, pp 36–56

Collard BC, Mackill DJ (2008) Marker-assisted selection: an approach for precision plant breeding in the twenty-first century. Philos Trans R Soc Lond Biol 363:557–572

Cousin A, Heel K, Cowling WA, Nelson MN (2009) An efficient high-throughput flow cytometric method for estimating DNA ploidy level in plants. Cytom Part A 75(12):1015–1019

d'Amato F (1989) Polyploidy in cell differentiation. Caryologia 42:183–211

Datta K, Sahoo G, Krishnan S, Ganguly M, Datta SK (2014) Genetic stability developed for β-carotene synthesis in BR29 rice line using dihaploid homozygosity. PLoS One 9(6):e100212

Devaux P, Pickering R (2005) Haploids in the improvement of Poaceae. In: Palmer CE, Keller WA, Kasha KJ (eds) Haploids in crop improvement II. Biotechnology in agriculture and forestry, vol 56. Springer, Berlin; New York, NY, pp 215–242

Dicu G, Cristea S (2016) The efficiency use of doubled-haploid technology in maize breeding: obtaining DH parent lines and hybrids. Sci Pap Ser A Agron 59:273–278

Dirks R, Van Dun K, De Snoo CB, Van Den Berg M, Lelivelt CL, Voermans W, Woudenberg L, De Wit JP, Reinink K, Schut JW, Van Der Zeeuw E (2009) Reverse breeding: a novel breeding approach based on engineered meiosis. Plant Biotechnol J 7(9):837–845

Dunwell JM (2010) Haploids in flowering plants: origins and exploitation. Plant Biotechnol J 8:377–424

Dwivedi SL, Britt AB, Tripathi L, Sharma S, Upadhyaya HD, Ortiz R (2015) Haploids: constraints and opportunities in plant breeding. Biotechnol Adv 33:812–829

Evans MM (2007) The indeterminate gametophyte 1 gene of maize encodes a LOB domain protein required for embryo sac and leaf development. Plant Cell 19(1):46–62

Fatah T, Rafii MY, Rahim HA, Meon S, Azhar M, Latif MA (2014) Cloning and analysis of QTL linked to blast disease resistance in Malaysian rice variety PongsuSeribu 2. Int J Agric Biol 16:395–400

Forster BP, Thomas WTB (2003) Doubled haploids in genetic mapping and genomics. In: Doubled haploid production in crop plants. Springer, Dordrecht, pp 367–390

Forster BP, Heberle-Bors E, Kasha KJ, Touraev A (2007) The resurgence of haploids in higher plants. Trends Plant Sci 12(8):368–375

Geethanjali S, Kadirvel P, Gunathilagaraj K, Maheswaran M (2009) Detection of quantitative trait loci (QTL) associated with resistance to white-backed planthopper (*Sogatella furcifera*) in rice (*Oryza sativa*). Plant Breed 128:130–136

Germana MA (2011) Anther culture for haploid and doubled haploid production. Plant Cell Tissue Organ Cult 104(3):283–300

Ghesquière A, Albar L, Lorieux M, Ahmadi N, Fargette D, Huang N, McCouch SR, Nottéghem JL (1997) A major quantitative trait locus for rice yellow mottle virus resistance maps to a cluster of blast resistance genes on chromosome 12. Phytopathology 87(12):1243–1249

Grewal D, Gill R, Gosal SS (2006) Role of cysteine in enhancing androgenesis and regeneration of indica rice (*Oryza sativa* L.). Plant Growth Regul 49(1):43–47

Grewal RK, Lulsdorf M, Croser J, Ochatt S, Vandenberg A, Warkenti TD (2009) Doubled haploid production in chickpea (*Cicer arietinum* L.): role of stress treatments. Plant Cell Rep 28:1289–1299

Grewal D, Manito C, Bartolome V (2011) Doubled haploids generated through anther culture from crosses of elite indica and japonica cultivars and/or lines of rice: large-scale production, agronomic performance, and molecular characterization. Crop Sci 51(6):2544–2553. https://doi.org/10.2135/cropsci2011.04.0236

Grimes HD, Hodges TK (1990) The inorganic NO_3^-: NH_4^+ ratio influences plant regeneration and auxin sensitivity in primary callus derived from immature embryos of indica rice (*Oryza sativa* L.). J Plant Physiol 136:362–367

Guha S, Maheshwari SC (1964) *In vitro* production of embryos from anthers of *Datura*. Nature 204(4957):497

Guha S, Maheshwari SC (1966) Cell division and differentiation of embryos in the pollen grains of *Datura in vitro*. Nature 212:97–98

Häntzschel KR, Weber G (2010) Blockage of mitosis in maize root tips using colchicine-alternatives. Protoplasma 241(1–4):99–104

Herath HMI, Bandara DC, Samarajeewa PK (2007) Effect of culture media for anther culture of indica rice varieties and hybrids of Indica and Japonica. Trop Agric Res Ext 10:17–22. https://doi.org/10.4038/tare.v10i0.1866

Hittalmani S, Parco A, Mew TV, Zeigler RS, Huang N (2000) Fine-mapping and DNA marker-assisted pyramiding of the three major genes for blast resistance in rice. Theor Appl Genet 100(7):1121–1128

Hu H, Zeng JZ (1984) Development of new varieties via anther culture. In: Ammirato PV, Evans DA, Sharp WR et al (eds) Handbook of plant cell culture, vol 3. Macmillan, New York, NY, pp 65–90

Ivanova M, Van Staden J (2009) Nitrogen source, concentration, and NH_4^+:NO_3^- ratio influence shoot regeneration and hyperhydricity in tissue-cultured *Aloe polyphylla*. Plant Cell Tissue Organ Cult 99(2):167–174

Jacquard C, Asakaviciute R, Hamalian AM, Sangwan RS, Devaux P, Clement C (2006) Barley anther culture: effects of annual cycle and spike position on microspore embryogenesis and albinism. Plant Cell Rep 25:375–381

Juliano BO (1997) Rice products in Asia. FAO-RAP publication No. 38. FAO, Rome

Kleiber D, Prigge V, Melchinger AE, Burkard F, San Vicente F, Palomino G, Gordillo GA (2012) Haploid fertility in temperate and tropical maize germplasm. Crop Sci 52:623–630

Kunihiro Y, Qian Q, Sato H, Teng S, Zeng DL, Fujimoto K, Zhu LH (2002) QTL analysis of sheath blight resistance in rice (*Oryza sativa* L.). Chin J Genet 29:50–55

Lee SY, Lee YT, Lee MS (1988) Studies on the anther culture of *Oryza sativa* L. 3. Growing environment of donor plant in anther culture, effects of photoperiod and light intensity. Res Rep Rural Dev Adm Biotechnol 30:7–12

Liang J, Han B-Z, Han L, Robert Nout MJ, Hamer RJ (2007) Iron, zinc and phytic acid content of selected rice varieties from China. J Sci Food Agric 87:504–510

Lübberstedt T, Frei UK (2012) Application of doubled haploids for target gene fixation in backcross programmes of maize. Plant Breed 131(3):449–452

Melchinger AE, Molenaar WS, Mirdita V, Schipprack W (2016) Colchicine alternatives for chromosome doubling in maize haploids for doubled-haploid production. Crop Sci 56:559–569

Min ZY, Li H, Zou T, Tong L, Cheng J, Sun XW (2016) Studies of in vitro culture and plant regeneration of unfertilized ovary of pumpkin. Chin Bull Bot 51(1):74–80

Mishra R, Rao GJN (2016) In-vitro androgenesis in rice: advantages, constraints and future prospects. Rice Sci 23:57–68

Mishra R, Rao GJN, Rao RN (2013) Effect of cold pretreatment and phytohormones on anther culture efficiency of two *indica* rice (*Oryza sativa* L.) hybrids, Ajay and Rajalaxmi. J Exp Biol Agric Sci 1(2):69–76

Mishra R, Rao GJN, Rao RN, Kaushal P (2015) Development and characterization of elite-doubled haploid lines from two *indica* rice hybrids. Rice Sci 22(6):290–299

Murovec J, Bohanec B (2012) Haploids and doubled haploids in plant breeding. In: Abdurakhmonov I (ed) Plant breeding. InTech, Rijeka, pp 87–106

Naik N, Rout P, Umakanta N, Verma RL, Katara JL, Sahoo KK et al (2017) Development of doubled haploids from an elite indica rice hybrid (BS6444G) using anther culture. Plant Cell Tissue Organ Cult 128(3):679–689

Niizeki H, Oono K (1968) Induction of haploid rice plant from anther culture. Proc Jpn Acad 44(6):554–557

Nurhasanah, Pratama AN, Sunaryo W (2016) Anther culture of local upland rice varieties from East Kalimantan: effect of panicle cold pre-treatment and putrescine-enriched medium. Biodiversitas 17(1):148–153

Ochatt SJ (2008) Flow cytometry in plant breeding. Cytom Part A 73(7):581–598

Ochatt S, Pech C, Grewal R, Conreux C, Lulsdorf M, Jacas L (2009) Abiotic stress enhances androgenesis from isolated microspores of some legume species (Fabaceae). J Plant Physiol 166:1314–1328

Patil VD, Nerkar YS, Misal MB, Harkal SR (1997) Parag 401, a semidwarf rice variety developed through anther culture. Int Rice Res Notes 22(2):19

Prasanna BM, Chaikam V, Mahuku G (2012) Doubled haploid technology in maize breeding: theory and practice. International Maize and Wheat Improvement Center (CIMMYT), Mexico

Prigge V, Xu XW, Li L, Babu R, Chen SJ, Atlin GN, Melchinger AE (2012) New insights into the genetics of *in vivo* induction of maternal haploids, the backbone of doubled haploid technology in maize. Genetics 111:781–793

Purwoko BS, Dewi IS, Khumaida K (2010) Rice anther culture to obtain doubled-haploids with multiple tolerances. Asia Pac J Mol Biol Biotechnol 18(1):55–57

Ravi M, Kwong PN, Menorca RM, Valencia JT, Ramahi JS, Stewart JL, Tran RK et al (2010) The rapidly evolving centromere-specific histone has stringent functional requirements in *Arabidopsis thaliana*. Genetics 186:461–471

Rober FK, Gordillo GA, Geiger HH (2005) In vivo haploid induction in maize: performance of new inducers and significance of doubled haploid lines in hybrid breeding. Maydica 50:275–283

Rout P, Naik N, Ngangkham U, Verma RL, Katara JL, Singh ON, Samantaray S (2016) Doubled haploids generated through anther culture from an elite long-duration rice hybrid, CRHR32: method optimization and molecular characterization. Plant Biotechnol 33(3):177–186

Sahoo SA, Jha Z, Verulkar SB, Srivastava AK, Suprasanna P (2019) High-throughput cell analysis based protocol for ploidy determination in anther-derived rice callus. Plant Cell Tissue Organ Cult 137(1):187–192

Seguí-Simarro JM, Nuez F (2008) How microspores transform into haploid embryos: changes associated with embryogenesis induction and microspore-derived embryogenesis. Physiol Plant 134:1–12

Senadhira D, Zapata-Arias FJ, Gregorio GB, Alejar MS, De La Cruz HC, Padolina TF, Galvez AM (2002) Development of the first salt-tolerant rice cultivar through *indica/indica* anther culture. Field Crop Res 76:103–110

Shen Y, Pan G, Lubberstedt T (2015) Haploid strategies for functional validation of plant genes. Trends Biotechnol 33:611–620

Shouyi L, Shouyin H (1991) Huayu 15, a high-yielding rice variety bred by anther culture. In: Bajaj YPS (ed) Biotechnology in agriculture and forestry, Volume 14: rice. Springer, Berlin; New York, NY, pp 230–247

Siddique R (2015) Impact of different media and genotypes in improving anther culture response in rice (*Oryza sativa*) in Bangladesh. Eur Sci J 11(6)

Soundararajan RP, Kadirvel P, Gunathilagaraj K, Maheswaran M (2004) Mapping of quantitative trait loci associated with resistance to brown plant hopper in rice by means of a doubled haploid population. Crop Sci 44:2214–2220

Swamy BM, Descalsota GIL, Nha CT, Amparado A, Inabangan-Asilo MA, Manito C et al (2018) Identification of genomic regions associated with agronomic and biofortification traits in DH populations of rice. PLoS One 13(8):e0201756

Talebi R, Rahemi MR, Arefi H, Nourozi M, Bagheri N (2007) In vitro plant regeneration through anther culture of some Iranian local rice (*Oryza sativa* L.) cultivars. Pak J Biol Sci 10(12):2056–2060

Thomson MJ, de Ocampo M, Egdane J, Rahman MA, Sajise AG, Adorada DL, Tumimbang-Raiz E, Blumwald E, Seraj ZI, Singh K, Gregorio GB, Ismail AM (2010) Characterizing the *Saltol* quantitative trait locus for salinity tolerance in rice. Rice 3(2):148–160

Trejo-Tapia G, Amaya UM, Morales GS, Sanchez ADJ, Bonfil BM, Rodriguez-Monroy M, Jimenez-Aparicio A (2002) The effects of cold-pretreatment, auxins and carbon source on anther culture of rice. Plant Cell Tissue Organ Cult 71:41–46. https://doi.org/10.1023/A:1016558025840

Veeraraghavan R (2007) A study on the comparison of anther culture response in different varieties of rice (*Oryza sativa* L.) subspecies *indica*. University of Colombo, Colombo. PMCid: PMC1891355

Wan Y, Duncan DR, Rayburn AL, Petolino JF, Widholm JM (1991) The use of antimicrotubule herbicides for the production of doubled haploid plants from anther-derived maize callus. Theor Appl Genet 81:205–211

Wang Z, Taramino G, Yang D, Liu G, Tingey SV, Miao GH, Wang GL (2001) Rice ESTs with disease-resistance gene- or defense-response gene-like sequences mapped to regions containing major resistance genes or QTLs. Mol Gen Genomics 265(2):302–310

Wang D, Yu C, Zuo T, Zhang J, Weber DF, Peterson T (2015) Alternative transposition generates new chimeric genes and segmental duplications at the maize p1 locus. Genetics 201:925–935

Wang X, Cheng ZM, Zhi S, Xu F (2016) Breeding triploid plants: a review. Czech J Genet Plant Breed 52(2):41–54

Weber DF (2014) Today's use of haploids in corn plant breeding. Adv Agron 123:123–144

Wessels E, Botes WC (2014) Accelerating resistance breeding in wheat by integrating marker-assisted selection and doubled haploid technology. S Afr J Plant Soil 31:35–43

Xu Q, Yuan XP, Yu HY, Wang YP, Tang SX, Wei XH (2011) Mapping quantitative trait loci for sheath blight resistance in rice using double haploid population. Plant Breed 130(3):404–406

Yamagishi M (2002) Heterogeneous plastid genomes in anther culture-derived albino rice plants. Euphytica 123(1):67–74

Yao L, Zhang Y, Liu C, Liu Y, Wang Y, Liang D, Liu J, Sahoo G, Kelliher T (2018) OsMATL mutation induces haploid seed formation in *indica* rice. Nat Plants 4:530–533

Yi S, Wu G, Lin Y, Hu N, Liu Z (2015) Characterization of a new type of glyphosate-tolerant 5-enolpyruvyl shikimate-3-phosphate synthase from *Isoptericolava riabilis*. J Mol Catal B Enzym 111:1–8. https://doi.org/10.1016/j.molcatb.2014.11.009

Zeng YX, Xia LZ, Wen ZH, Ji ZJ, Zeng DL, Qian Q, Yang CD (2015) Mapping-resistant QTLs for rice sheath blight disease with a doubled haploid population. J Integr Agric 14(5):801–810

Zhang ZH (1989) The practicability of anther culture breeding in rice. In: Advances in plant biotechnology. CIMMYT, IRRI, Mexico, Los Baños, pp 36–37

Zhang LN, Chu QR (1984) Characteristic and chromosomal variation of somaclones and its progeny in rice (*Oryza sativa* L.). Sci Agric Sin (China) (4):14

Zheng X, Chen X, Zhang X, Lin Z, Shang J, Xu J, Zhai W, Zhu L (2004) Isolation and identification of a gene in response to rice blast disease in rice. Plant Mol Biol 54(1):99–109

Zhu DY, Pan XG (1990) Rice (*Oryza sativa* L.): guan 18: an improved variety through anther culture. In: Bajaj YPS (ed) Biotechnology in agriculture and forestry 12: rice. Springer, Berlin, pp 204–211

Zonneveld BJM, Van Iren F (2001) Genome size and pollen viability as taxonomic criteria, application to the genus *Hosta*. Plant Biol 3:176–185

Zinc-Biofortified Rice: A Sustainable Food-Based Product for Fighting Zinc Malnutrition

Mark Ian C. Calayugan, B. P. Mallikarjuna Swamy, Chau Thanh Nha, Alvin D. Palanog, Partha S. Biswas, Gwen Iris Descalsota-Empleo, Yin Myat Myat Min, and Mary Ann Inabangan-Asilo

Abstract The lack of dietary diversity among poor communities has led to nutritional consequences, particularly zinc deficiency. An adequate intake of mineral- and vitamin-rich food is necessary for achieving and maintaining good health. Zinc is one of the micronutrients considered essential to improve human health and decrease the risk of malnutrition. Biofortification of rice through breeding is a cost-effective and sustainable strategy to solve micronutrient malnutrition. The Biofortification Priority Index prepared by HarvestPlus clearly identified several countries in Asia with an immediate need for Zn biofortification. The International Rice Research Institute (IRRI) and its national partners in target countries are making efforts to develop Zn-biofortified rice varieties. The first set of high-Zn rice varieties has been released for commercial cultivation in Bangladesh, India, the Philippines, and Indonesia. Efforts have begun to mainstream grain Zn to ensure that the Zn trait becomes an integral part of future varieties. Huge scope exists to

M. I. C. Calayugan
International Rice Research Institute, Metro Manila, Philippines

University of the Philippines, Los Baños, Laguna, Philippines

B. P. M. Swamy (✉) · M. A. Inabangan-Asilo
International Rice Research Institute, Metro Manila, Philippines
e-mail: m.swamy@irri.org

C. T. Nha
Cuu Long Delta Rice Research Institute, Cần Thơ, Vietnam

A. D. Palanog
International Rice Research Institute, Metro Manila, Philippines

Philippine Rice Research Institute, Negros Occidental, Philippines

P. S. Biswas
Bangladesh Rice Research Institute, Gazipur, Bangladesh

G. I. Descalsota-Empleo
University of Southern Mindanao, North Cotabato, Philippines

Y. M. M. Min
Department of Agricultural Research, Yezin, Myanmar

© The Author(s) 2021
J. Ali, S. H. Wani (eds.), *Rice Improvement*,
https://doi.org/10.1007/978-3-030-66530-2_13

449

apply advanced genomics technologies such as genomic selection and genome editing to speed up high-Zn varietal development. An efficient rice value chain for Zn-biofortified varieties, quality control, and promotion are essential for successful adoption and consumption. The development of next-generation high-Zn rice varieties with higher grain-Zn content, stacking of multiple nutrients, along with good grain quality and acceptable agronomic traits has to be fast-tracked. Healthier rice has a large demand from all stakeholders, so we need to keep up the pace of developing nutritious rice to meet the demand and to achieve nutritional security.

Keywords Rice · Malnutrition · Biofortification · Zinc · QTL · Gene · Bayesian analysis

1 Introduction

The human body needs micronutrients for proper growth and development and to maintain good health (Maret 2017; Palanog et al. 2019). However, deficiencies in these elements and associated health risks are commonly reported in all age groups, especially in preschool children, women, and elderly people in the developing world (Caulfield et al. 2006). An estimated one-third of the global population suffers from micronutrient malnutrition, mainly because of the large dependence on cereal staples for daily nutritional needs without access to a diversified diet and supplementation (Ritchie et al. 2018). The urgent need to address micronutrient malnutrition has been widely recognized globally; hence, decreasing childhood mortality and maternal death by eradicating malnutrition is an important Sustainable Development Goal (Hanieh et al. 2020).

Among the micronutrients, zinc (Zn) is most essential for vital organs, enzymatic activity, tissue growth and development, cognitive function, immunity, etc. There is therefore a need for a regular daily supply of Zn in the required quantity to have healthy and productive populations (Prasad et al. 2014; Chasapis et al. 2020). However, an estimated two billion people suffer from Zn deficiency-related health consequences and most of them are resource-poor urban and rural dwellers (Rampa et al. 2020). The disability-adjusted life years (DALYs) due to Zn malnutrition strongly impact annual GDP growth, and hamper economic development in the developing world (Gödecke et al. 2018). Multiple interventions such as fortification of foods, micronutrient supplementation, and food diversification have been employed to mitigate Zn malnutrition; however, recurring costs and poor accessibility and awareness among the rural masses have resulted in limited success (Bouis 2017). Increasing the mineral density in the edible part of the major staple crops, which is also popularly called "biofortification," has been proven to be effective in alleviating malnutrition without much additional cost. This complementary food-based approach is the safest and cheapest way to deliver nutrients on a larger scale to the target populations (Bouis and Saltzman 2017).

Rice is among the target staple food crops for Zn biofortification in different countries of South Asia, Southeast Asia, and Africa (Siwela et al. 2020). The Biofortification Priority Index prepared by HarvestPlus clearly identified several countries in Asia with an immediate need for Zn biofortification (HarvestPlus 2020). The International Rice Research Institute (IRRI) and its national partners in target countries are making efforts to develop Zn-biofortified rice varieties. The first set of high-Zn rice varieties has been released for commercial cultivation in Bangladesh, India, the Philippines, and Indonesia (Inabangan-Asilo et al. 2019). Efforts are in place to mainstream the breeding of high-Zn rice by applying advanced breeding techniques and genomic tools to make sure Zn will be an essential component of all future varietal releases from the main breeding pipelines of IRRI (CGIAR 2018).

Over the past decade, great progress has been made in our understanding of Zn homeostasis in rice from a biofortification perspective and in the development of high-Zn rice. In this chapter, we would like to provide some insights into the recent advances in developing Zn-biofortified rice for the target countries.

2   Zn Is Critical for Human Health

Zinc plays an important role in the catalytic function of most of the enzymes needed for the structural stability and functioning of more than 3000 proteins, helps to maintain membrane stability, and protects tissues and cells from oxidative damage (Cakmak 2000; Broadley et al. 2007; Andreini et al. 2009; Maret and Li 2009).

Zinc deficiency is one of the major causes of child mortality worldwide (Black et al. 2008), which has been estimated to affect more than 178 countries (WHO 2003). Zn-deficient children are highly prone to diarrhea, respiratory ailments, poor cognitive function, and stunting (Brooks et al. 2004; Sazawal et al. 2007; Tielsch et al. 2007; Young et al. 2014). Zn deficiency during the first 1000 days for children after birth causes irreversible damage leading to less chance of survival, poor immune system and cognitive ability, and stunting (UNICEF 2013). Hence, a regular daily supply of Zn is highly essential, but this is rarely achieved by most resource-poor people. Thus, adequate Zn nutrition is essential for good health, especially for children and pregnant women for growth and development (IZiNCG 2009). The daily Zn requirement of individuals varies from 9 to 11 ppm depending on age, gender, and health conditions, but preschool children and pregnant and lactating women need more Zn (IOM 2001; Welch and Graham 2004; Iqbal et al. 2020; Alqabbani and AlBadr 2020).

3   Rice Biofortification with High Grain Zn

Rice is the single most important source of energy and nutrition for more than half of the world's population (Gross and Zhao 2014). It is a major staple crop in more than 40 countries and supplies at least 20% of the daily caloric intake of more than

3.5 billion people (FAO 2014). Asia, with 60% of the global population, consumes more than 90% of the total rice produced annually (Milovanovic and Smutka 2017). Annual per capita rice consumption exceeds 100 kg in some Asian countries (FAO 2016). However, milled rice is less nutritious; thus, most of the poor people who largely depend on rice without access to a mineral-rich diverse diet suffer from hidden hunger, including Zn deficiency.

Food-based solutions were found to be safe and effective in controlling and preventing micronutrient deficiencies, especially when multiple deficiencies occur (Torheim et al. 2010; Szymlek-Gay et al. 2009). Several studies reported that the consumption of a diverse diet and crops enriched with mineral elements provides more nutrition (Brown et al. 2002; WHO 1998). Recently, biofortification of staple crops has become a popular method for tackling malnutrition. It is the process of increasing the density of readily bioavailable mineral elements by breeding or biotechnological approaches (Garg et al. 2018) for staple food crops such as rice, which has been obtaining increased attention by breeders and policymakers in recent times. Biofortification has the lowest per capita costs vis-à-vis other interventions, and it is especially easily accessible and affordable for rural populations (Ma et al. 2008). Therefore, increasing grain-Zn content would create a significant impact on human health. One estimate suggested that an additional 8 µg/g of Zn in raw milled rice over the baseline Zn (16 µg/g) in cultivated varieties could help to reach the amount equivalent to 30% of the Estimated Average Requirement per day (HarvestPlus 2012).

4 Trait Development for High Grain Zn

4.1 High-Zn Donor Identification

Rice is endowed with abundant genetic diversity and thereby provides needed genetic variability for rice breeding programs (Rana and Bhat 2004). More than 230,000 rice accessions are maintained in global gene banks, which include landraces, cultivars, varieties, and aromatic and wild rice (Li et al. 2014). Among the different species or subgroups, wild rice, landraces, and *aus* accessions were found to be a rich source of micronutrients; they have several-fold higher nutrients than cultivated rice (Cheng et al. 2005; Banerjee et al. 2010; Descalsota-Empleo et al. 2019a, b). *Aus* accessions are genetically closer to popularly grown *indica* rice varieties, so they can be readily used by breeding programs to improve the Zn content of modern rice varieties. Some *aus* accessions such as Kaliboro, Jamir, UCP122, DZ193, and Khao ToT Long 227 have higher content of grain Zn (Norton et al. 2014; Descalsota et al. 2018). We are efficiently using *aus* germplasm in our breeding programs at IRRI and have also widely shared these donor lines with our partners for use in their breeding programs. The accessions of the 3K Genome Project, Multi-parent Advanced Generation Inter-Cross (MAGIC)-derived lines and wild rice introgression lines, were also characterized to identify valuable donors for grain Zn and used in genetic dissection studies (Bandillo et al. 2013; Swamy et al. 2018a; Descalsota et al. 2018; Zaw et al. 2019). Moreover, large scope exists for revisiting gene banks to screen for high grain Zn and other beneficial elements using advanced high-throughput phenotyping technologies. Similarly, a systematic effort to collect and characterize heirloom rice for nutritional value in partner countries will help in breeding for improved nutrition.

4.2 Association Between Yield and Zn

The development of high-yielding Zn-biofortified rice with a combination of desirable agronomic traits and tolerance of pests and diseases is a must for their successful adoption and consumption. Both yield and grain Zn are genetically complex traits and are hugely influenced by external environmental factors (Zaw et al. 2019; Descalsota-Empleo et al. 2019a). In most cases, a negative association was reported between grain-Zn content and yield, and in a few specific germplasm accessions and populations a nonsignificant negative relationship or no relationship was reported (Gregorio 2002; Norton et al. 2010; Morete et al. 2011; Anandan et al. 2011; Nha 2019). Under different soil Zn conditions and in a set of different aromatic accessions and landraces, a positive relationship between grain Zn and yield was reported (Wissuwa et al. 2008; Gangashetty et al. 2013; Sathisha 2013). Therefore, for the identification of stable high-Zn donor lines with higher or more acceptable yield, the use of appropriate breeding methods and selection strategies is needed to successfully combine yield and grain Zn.

4.3 Molecular Dissection of Grain Zn

4.3.1 QTLs and Meta-QTLs Associated with Grain Zn

Zn uptake, transport, and accumulation in the grain are governed by a complex network of quantitative trait loci (QTLs) and genes. A comprehensive review of QTLs identified for grain Zn was carried out and detailed discussion presented by Swamy et al. (2016). Several QTLs with moderate to high phenotypic variance were reported for grain Zn on all 12 chromosomes of rice. At IRRI, our group has also carried out several QTL mapping studies using biparental and multiparental populations and germplasm collections (Table 1). Swamy et al. (2018a) reported eight QTLs for grain-Zn content. All of these QTLs were distributed across the rice genome, having the lowest frequency (one QTL) on chromosomes 1, 9, and 11 and the highest frequency (seven QTLs) on chromosome 12. Chromosome 7 had the second highest number (six) of QTLs. However, the QTLs on chromosomes 7 and 12 were consistent over different backgrounds and environments. The QTLs detected on

Table 1 QTLs identified for grain-Zn content in rice

QTL	LOD/p value	PVE (%)	Additive effect (mg/kg)	Reference
$qZn_{1.1}$, $qZn_{2.1}$, $qZn_{3.1}$, $qZn_{3.2}$, $qZn_{5.1}$, $qZn_{6.1}$, $qZn_{8.1}$, $qZn_{8.2}$, $qZn_{9.1}$, $qZn_{10.1}$, $qZn_{12.1}$	2.5–12.4	3.0–36.0	0.21–6.60	Swamy et al. (2018a)
$qZn_{2.1}$, $qZn_{2.2}$, $qZn_{3.1}$, $qZn_{6.1}$, $qZn_{6.2}$, $qZn_{8.1}$, $qZn_{11.1}$, $qZn_{12.1}$, $qZn_{12.2}$	4.3–10.3	7.5–22.8	0.9–2.1	Swamy et al. (2018b)
$qZn_{1.1}$, $qZn_{2.1}$, $qZn_{4.1}$, $qZn_{6.1}$, $qZn_{6.2}$, $qZn_{7.1}$, $qZn_{12.1}$	0.001–0.0001	9.2–13.75	–	Descalsota et al. (2018)
$qZn_{1.1}$, $qZn_{6.1}$, $qZn_{12.1}$, $qZn_{12.2}$, $qZn_{12.3}$	0.0000905–0.00029	11.9–17.9	–	Descalsota-Empleo et al. (2019a)
$qZn_{2.1}$, $qZn_{3.1}$, $qZn_{5.1}$, $qZn_{5.2}$, $qZn_{7.1}$, $qZn_{8.1}$, $Zn_{9.1}$, $qZn_{11.1}$	2.77–8.99	8.6–27.7	0.81–2.06	Descalsota-Empleo et al. (2019b)
$qZn_{1.1}$, $qZn_{6.1}$, $qZn_{6.2}$	2.6–3.9	2.9–34.2	0.06–3.2	Dixit et al. (2019)
qZn_1, qZn_5, qZn_7	–	17.57–20.0	–	Zaw et al. (2019)
$qZn_{1.1}$, $qZn_{5.1}$, $qZn_{9.1}$, $qZn_{12.1}$	3.14–5.2	8.96–15.26	0.77–0.96	Calayugan et al. (2020)
$Zn_{1.1}$, $qZn_{1.2}$, $qZn_{1.3}$, $Zn_{2.1}$, $qZn_{4.1}$, $qZn_{5.1}$, $Zn_{6.1}$, $qZn_{7.1}$, $qZn_{9.1}$, $Zn_{10.1}$, $qZn_{11.1}$, $qZn_{11.2}$, $Zn_{11.3}$, $qZn_{12.1}$	3.28–15.36	12.60–46.80	2.62–4.73	Jeong et al. (2020)
$qZn_{3.1}$, $qZn_{3.1}$, $qZn_{4.2}$	4.11–9.16	9.89–24.56	0.0001–0.1	Lee et al. (2020)

PVE phenotypic variance explained

chromosome 7 contributed 5.3–35.0% of the phenotypic variance for grain-Zn content in different backgrounds, while the QTLs on chromosome 12 contributed 9–36% (Swamy et al. 2016, 2018a). In another study, Swamy et al. (2018b) detected nine QTLs responsible for Zn on chromosomes 2, 3, 6, 8, 11, and 12 through two doubled-haploid (DH) populations derived from crosses of PSBRc82 × Joryeongbyeo and PSBRc82 × IR69428. Recently, association mapping experiments using diversity panels for grain Zn led to the identification of seven QTLs on chromosomes 1, 2, 4, 6, 7, and 12 by Descalsota et al. (2018) and three QTLs on chromosomes 1, 5, and 7 by Zaw et al. (2019). All of these findings show that numerous QTLs for Zn highlight the genetic complexity of this trait.

Meta-QTL analysis provides consolidated, precise, and smaller confidence intervals for multiple QTLs reported for a trait (Goffinet and Gerber 2000; Arcade et al. 2004; Swamy et al. 2011). Jin et al. (2015) identified 22 meta-QTLs on ten different chromosomes for grain-Zn content (rMQTLs). Similarly, Raza et al. (2019) carried out meta-QTL analysis of grain-Zn QTLs reported from 24 mapping populations and three diverse germplasm sets and identified 46 MQTLs. Seven meta-QTLs ($rMQTL_{2.1}$, $rMQTL_{4.4}$, $rMQTL_{6.4}$, $rMQTL_{8.2}$, $rMQTL_{8.3}$, $rMQTL_{8.4}$, and $rMQTL_{12.4}$) were found to be common between two studies (Jin et al. 2015; Raza et al. 2019). In another study, 208 QTLs for grain Zn from 26 studies were projected on the consensus map and eventually 45 meta-QTLs were identified (Soe 2020). Overall, the confidence intervals of all the MQTLs were narrower vis-à-vis the mean values of the original QTLs. Several consistent QTLs and associated markers were identified, which are useful for efficient marker-assisted selection (MAS) programs. In addition, precise meta-QTL regions provide an opportunity to shortlist candidate genes for further functional validation.

4.3.2   Network of Metal Homeostasis Genes

Mapping of major-effect QTLs/genes for grain Zn and understanding their molecular basis can fast-track the development of Zn-biofortified rice through MAS. The genomic regions of important QTLs associated with grain Zn identified in numerous studies contained multiple hypotheticals and functionally annotated genes that function as metal chelators and ion transporters. A list of important genes associated with Zn homeostasis in rice is summarized in Swamy et al. (2016). Rice roots produce chemicals that free up mineral elements from the soil complex and promote their root uptake from the soil (Widodo et al. 2010; Nozoye et al. 2011). Several genes/gene families are involved in biosynthesis of phytosiderophores, mineral uptake, transport, and loading such as *OsDMAS*, *OsSAMS*, *OsNAS*, *OsTOM1*, and *OsNAAT* (Inoue et al. 2003, 2008; Bashir et al. 2006; Johnson et al. 2011). Zinc finger transcription factors such as *OsZIP1*, *OsZIP3*, *OsZIP4*, *OsZIP5*, and *OsZIP9* are major Zn transporters within the rice plant (Ramesh et al. 2003; Ishimaru et al. 2005; Lee et al. 2010a, b). In separate studies conducted using connected populations, ZIP family genes such as *OsZIP5* and *OsZIP9* were identified along with another 140 candidate genes (Nha 2019). In a study using DH populations, *OsZIP6* was identified as a primary candidate gene associated with grain Zn (Calayugan et al. 2020). Similarly, *OsVIT* and *OsYSL* family genes are involved in Zn transport across the tonoplast and phloem, respectively (Sasaki et al. 2011; Kakei et al. 2012; Zhang et al. 2012; Lan et al. 2013).

The well-characterized Zn metal homeostasis genes can be manipulated through genetic engineering to improve grain-Zn content in rice (Trijatmiko et al. 2016). The advanced genome editing techniques using zinc-finger nucleases (ZFNs), transcription activator-like effector nucleases (TALENs), and clustered regularly interspaced short palindromic repeats (CRISPR)/Cas systems can be used to induce modifications at specific genomic loci (Kim et al. 1996; Christian et al. 2010; Jinek et al. 2012; Chen and Gao 2013; Gao 2015). The Zn homeostasis genes can be major target sites for genome editing to improve grain-Zn content in rice.

4.3.3 Bayesian Network Analysis of Grain Yield and Zn

A Bayesian genomic prediction network (BN) provides valuable information on interactions between multiple traits and SNP markers and helps to establish relationships among them. It clearly depicts the strength and direction of associations among traits and SNP markers. In a way, it helps to validate the QTLs, genes, or trait associations identified by genome-wide association analysis (Zaw et al. 2019). In a MAGIC Plus population, a BN was used among Zn- and yield-related traits. The results clearly showed a complex relationship among traits (Fig. 1). Among the agronomic traits studied for their relationship with grain Fe and Zn, only panicle length had a direct effect on Fe and Zn content in rice (Descalsota et al. 2018). Zaw et al. (2019) conducted BN analysis in a global MAGIC population using 8110 SNP markers and 16 traits, including grain Zn. At a BN strength of more than 0.5, strong direct associations were reported among traits such as yield → zinc, zinc → filled grains, iron → zinc, and iron → grain length. Zn was associated with eight markers for each of the traits. In general, Fe and Zn content have strong positive correlations, thus providing huge opportunities to improve both minerals together. It is interesting that in both BN studies there was no direct effect of yield on Zn, indicating that combining high yield potential and high grain-Zn content is possible in order to develop successful Zn-biofortified rice varieties. We emphasize that there is a need to thoroughly dissect the influence of panicle length on grain Zn. It is commonly observed that increased yield dilutes Zn content, which results in negative correlations between these traits. There is therefore a need to make adjustments for grain-Zn mapping studies (McDonald et al. 2008).

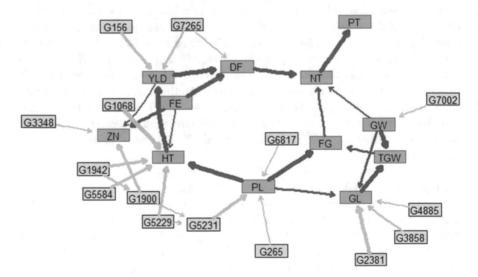

Fig. 1 Bayesian network analysis of grain Zn and agronomic traits. Note: *Zn* zinc, *Fe* iron, *HT* plant height, *DF* days to flowering, *NT* number of tillers, *PT* productive tillers, *GL* grain length, *GW* grain width, *TGW* thousand-grain weight, *YLD* grain yield

4.4   Multi-Trait Genomic Selection for Zn Biofortification

QTL mapping and GWAS methods are routinely used for molecular dissection of complex traits; however, they have limited power in detecting minor-effect loci (Bernardo 2008; Collard and Mackill 2008; Ben-Ari and Lavi 2012). In contrast, genomic selection (GS) considers genome-wide effects, including both major and minor loci, and thereby assesses the genomic estimated breeding values (GEBVs) of breeding lines (Meuwissen et al. 2001). With the recent advances in statistics, deep machine learning models are helpful in accurately estimating GEBVs and their cross-validation in training or reference sets (Montesinos-López et al. 2019). Since GS captures total genetic variance, it addresses the existing limitations of GWAS and QTL mapping to improve traits (de los Campos et al. 2009). In addition, it speeds up selection cycles, which enhances annual genetic gain and saves cost significantly (Shamshad and Sharma 2018). Therefore, great opportunity exists for employing GS-related strategies that capture both major- and minor-effect alleles to increase the genetic gain for grain yield and grain-Zn content in rice.

Selection for higher yield and other desirable agronomic traits along with high grain-Zn content is an integral part of Zn biofortification; however, both yield and Zn are genetically complex and difficult to manipulate or simultaneously improve (Garcia-Oliveira et al. 2018; Zaw et al. 2019). The rate of genetic gain for grain yield becomes stagnant at ~1% yearly. This is not sufficient to meet future demand for rice, not to mention the strong impacts of complex genetic architecture and genotype–environment interactions (Peng et al. 2000, 2004; Wassmann et al. 2009). The combined genetic gain for yield and Zn will be relatively inferior when compared with that for individual traits. Therefore, implementing multi-trait-based population improvement through genomic selection is an efficient approach.

In rice, Spindel et al. (2015) reported prediction accuracies of single-trait genomic selection (ST-GS) models for grain yield at 0.31, while Arbelaez et al. (2019) have shown predictive accuracies for grain yield at 0.36. Meanwhile, multi-trait genomic selection (MT-GS) models have illustrated higher predictive abilities than ST-GS models and the results are obvious, especially when low-heritability traits are paired with a genetically correlated secondary trait with higher heritability (Jia and Jannink 2012; Hayashi and Iwata 2013; Guo et al. 2014; Schulthess et al. 2016). Many findings have used MT-GS approaches in crop breeding, but not yet in rice. Schulthess et al. (2016) have confirmed the predictive ability of MT-GS in outperforming ST-GS pipelines for grain yield and protein content in rye. Lado et al. (2018) have verified combining two, three, and four traits in bread wheat in exploiting the benefits of MT-GS under different cross-validation scenarios. The use of correlated traits in MT-GS models gives the best prediction accuracies in a two-trait scenario. GS in maize showed higher prediction accuracy in DH populations than a GWAS panel using the same set of GBS and rAmpSeq markers, and GS outperformed MAS in predicting the performance of Zn content in maize (Guo et al. 2020). Although most of the available GS methods increased predictive ability, Zn breeders should target multiple independent phenotypes from multi-environments. Thus, multi-trait and multi-environment (MTME) models have been established to employ the information on multiple traits evaluated in multiple environments, which improves predictive ability compared to conventional, pedigree, and independent GS analysis (Montesinos-Lopez et al. 2016).

5   Development of High-Zn Rice

5.1   Phenotyping of Grain Zn

To enhance selection accuracy and to significantly improve a breeding program, reliable phenotyping is crucial. Accurate phenotyping for any trait involves a standard protocol with a set of specific standards. Usually in large biofortification breeding programs like the one at IRRI, we handle a huge number of breeding lines every season, so there is a need for quick turnover of materials with accurate phenotyping for grain Zn. Efficient high-throughput dehulling, milling, and Zn measurement protocols and equipment are needed for successful Zn biofortification of rice (Fig. 2) (Swamy et al. 2016; Guild et al. 2017). Several low-throughput qualitative, semi-quantitative, and quantitative methods are available for the estimation of grain-Zn content in rice and other cereals. Inductively coupled plasma optical emission spectrometry (ICP-OES) is used to assess nutrient density in grains (Zarcinas et al. 1987); this method is more accurate but low-throughput and input-intensive and it requires trained staff. X-ray fluorescence (XRF) is a rapid non-chemical-based

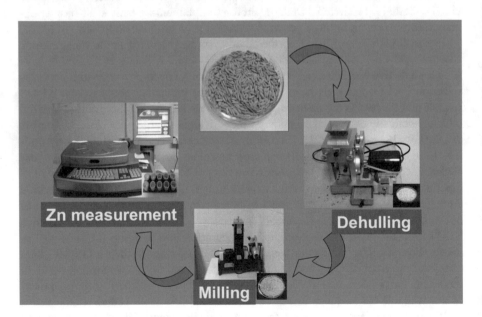

Fig. 2   Phenotyping for grain-Zn milled rice

method to measure grain-Zn content in milled rice, which has decreased cost per unit and simultaneously increased selection intensity although it still requires grain processing under a contamination-free environment of exogenous Zn sources.

5.2   Setting a Zn Target for Rice Biofortification

The development of nutritional targets for crops for biofortification breeding was established by a group of experts taking into account the food habits of the target populations, nutrient losses during food processing, and nutrient bioavailability (Hotz and McClafferty 2007). The breeding target was designed to meet the specific nutrient requirement of the target populations considering the baseline micronutrients existing in popular rice varieties and extra micronutrient content to be added to the crop of interest. Zn-biofortified rice is expected to provide >40% of the Estimated Average Requirement, which is enough to help overcome Zn-deficiency-induced health risks (Bouis and Saltzman 2017). There is a plan to release high-Zn rice varieties in three phases: the first set of varieties will have an additional Zn content of 6–8 ppm, the second wave of varieties will have 8–12 ppm, while the third wave of high-Zn rice varieties will have 12 ppm of additional grain Zn (Fig. 3).

5.3   Germplasm Enhancement and Pre-breeding for Grain Zn

Exploitable genetic variability for any trait, its systematic characterization, and efficient use are essential for a successful breeding program. Most elite modern rice varieties and their closest elite genetic pool have low grain-Zn content (Gregorio 2002). *Oryza nivara, O. rufipogon, O. longistaminata,* and *O. barthii* accessions, landraces, colored rice, and *aus* and aromatic accessions were found to have rich grain-Zn content (Swamy et al. 2016, 2018a, b; Ishikawa et al. 2017). But these accessions may not be agronomically desirable because of their poor phenotype and lower yield. Therefore, a systematic pre-breeding for grain Zn is essential to develop high-Zn rice varieties.

The advanced backcross method for genetic dissection of wild rice, and for developing high-Zn introgression lines, is an attractive approach for efficient use of wild rice accessions (Balakrishnan et al. 2020). Several wild rice-derived introgression lines with high grain Zn and yield have already been developed by several groups (Ishikawa et al. 2017; Swamy et al. 2018a). Multi-parent-derived populations to select transgressive variants with a combination of desirable traits have yielded many desirable transgressive variants for grain Zn (Gande et al. 2013; Ishikawa et al. 2017; Descalsota-Empleo et al. 2019a, b). Marker-assisted QTL deployment, QTL pyramiding, and marker-assisted recurrent selection are helpful in germplasm enhancement for grain Zn with other traits (Hill et al. 2008; Boyle et al. 2017).

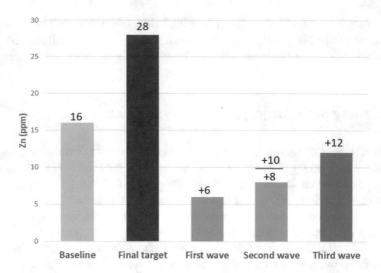

Fig. 3 Zn target set for breeding Zn-biofortified rice varieties

Several studies have characterized germplasm and advanced breeding lines for grain-Zn content (Gregorio et al. 2000; Brar et al. 2011). Garcia-Oliveira et al. (2009) identified 85 introgression lines with the highest quantities of Zn with a mean value of 27.1 ppm. Martínez et al. (2010) phenotyped grain-Zn content in 11,400 rice samples in both brown rice and milled rice and reported corresponding Zn values of 20–25 ppm and 16–17 ppm, respectively. Gande et al. (2013) identified eight transgressive lines for high Zn content (31.2–35.5 ppm). Some of these introgression lines, transgressive segregants, and breeding lines can be used as donor lines for Zn biofortification and even some can be directly tested and released as high-Zn rice varieties for commercial cultivation.

5.4   Mainstreaming of Zn Biofortification

Zinc-biofortified rice varieties have been successfully released for cultivation by farmers in some of the target countries. However, developing, releasing, and disseminating a few varieties may not create sustainable and wide-scale impact on human health. At IRRI, mainstream breeding programs are shifting from a siloed trait-based breeding approach to a modernized product development pipeline that effectively integrates the improvement of all traits necessary for market acceptance into a single variety replacement strategy. This new strategy involves using population improvement as a mechanism to drive genetic gain for complex traits, while simultaneously increasing the frequency of trait-favorable alleles. Essential to this strategy is the data-driven identification of a core set of elite lines that represent the pool of possible parental lines that can be used in the breeding process (Cobb et al. 2019). Through successive cycles of recurrent selection, mainstream breeding efforts are now able to drive genetic gain and improve the average value of all traits from a product profile in the entire elite gene pool simultaneously (Collard et al. 2017). By integrating selection for high grain-Zn content directly into the mainstream breeding effort, the mean value of grain Zn among the elite breeding lines will eventually be at or above the recommended allowance of 28 ppm in milled grains. Once this occurs, all the most recently developed new varieties released from the mainstream breeding program will have acceptable concentrations of Zn in addition to other traits more valued in the marketplace. With minimal effort, maintenance breeding for Zn can be conducted in elite breeding programs once acceptable grain-Zn content is achieved in order to ensure constant delivery of sufficient Zn to the diets of nutrition-vulnerable rice-consuming populations. Incumbent upon this strategy is the need for sufficient variation to drive genetic gain for complex traits. A three-phased approach is suggested: elite germplasm characterization, elite germplasm enhancement and selection, and mainstream breeding.

6   High-Zn Rice Testing and Release

6.1   Genotype × Environment Effects on Grain Zn

Grain-Zn content is a complex trait found to be significantly influenced by external soil and climatic factors (Chandel et al. 2010; Anuradha et al. 2012; Swamy et al. 2016; Naik et al. 2020). Meteorological factors such as temperature, relative humidity, and rainfall; soil factors such as organic matter, pH, and nutrient status; and agronomic practices such as fertilizer application, tilling, cultivation system, and irrigation (White and Broadley 2009; Joshi et al. 2010; Chandel et al. 2010) need to be taken into account. Grain-Zn content in a study conducted by Wissuwa et al. (2008) was found to be greatly influenced by the native Zn in soils, genotype, and Zn fertilizer application. In a separate study by Wang et al. (2014), water management with alternate wetting and drying (AWD) together with $ZnSO_4$ fertilization showed a positive response for higher yield coupled with higher grain-Zn content in rice. Pandian et al. (2011) conducted field experiments across three locations involving 17 genotypes of rice. The results showed that grain-Zn content varied significantly among the genotypes and locations.

Thus, G × E testing is needed to evaluate promising germplasm and the stability of mineral accumulation across generations and at multiple test sites (Gregorio 2002; Wissuwa et al. 2008; Impa et al. 2013; Naik et al. 2020). Hence, the stability of Zn-biofortified genotypes for grain-Zn content in addition to grain yield is essential for commercial release as varieties.

6.2   Zn-Biofortified Rice Varieties Released in Different Countries

Breeding efforts to biofortify rice with high grain Zn have resulted in the successful release of several high-Zn rice varieties in several countries of Asia. Five high-Zn rice varieties (BRRI dhan62, BRRI dhan64, BRRI dhan72, BRRI dhan74, and BRRI dhan84) have been released for cultivation in Bangladesh. In India, two high-Zn rice varieties (DRR Dhan45 and Chhattisgarh Zinc Rice-1) are available for farmers and consumers. Similarly, NSIC Rc 460 and Inapari Nutri Zn have been released for farmers' cultivation in the Philippines and Indonesia, respectively. All these high-Zn rice varieties have higher grain-Zn content along with desirable agronomic traits and tolerance of biotic and abiotic stresses (Swamy et al. 2016; Tsakirpaloglou et al. 2019). Several promising high-Zn lines have been successfully tested in Myanmar and Cambodia and in some African countries. We are also making efforts to develop the next wave of Zn-biofortified rice varieties with higher grain-Zn content.

6.3   High-Zn Rice Traceability and Product Control

Grain-Zn content in rice is an invisible nutritional trait and no morphological indicators differentiate Zn-biofortified rice from market rice. Maintaining the product integrity of high-Zn rice throughout the value chain is an important component of successful Zn biofortification programs. Close monitoring, supervision, and quality control are necessary with proper certification, labeling, branding, and tracing of the product (www.fao.org/tempref/codex/Meetings/CCNFSDU/ccnfsdu36/nf36_11e.pdf). The development of Zn product-specific molecular marker-based fingerprints and rapid qualitative biochemical marker kits will also help in tracing Zn-rich rice. Blockchain technology is being used in the large-scale dissemination of nutritious crops to ensure quality control and to deliver the right products to consumers (Tripoli and Schmidhuber 2018).

7   Next-Generation Multi-Nutrient Rice Varieties

Breeding for rice varieties with multiple beneficial minerals and vitamins is essential to develop them holistically for one biofortified rice product. Efforts to develop rice varieties with high Zn, high Fe, selenium, vitamin A, proteins, amino acids, etc., should be given a priority. It will also be interesting to combine high nutrient content with traits beneficial to health, such as low glycemic index, antioxidants, and resistance starch. Also, there is a need to diminish the amount of harmful elements such as arsenic and cadmium. The increase in demand for rice varieties with improved grain quality and nutrition means that a suite of rice varieties with different combinations of traits targeted to different regions should be developed.

8   Conclusions

Biofortification of rice with improved Zn content is an efficient means to tackle Zn malnutrition in predominantly rice-consuming developing countries. Some success has been achieved in understanding the molecular basis of Zn accumulation and the effects of G × E, and finally in developing and releasing Zn-biofortified rice varieties for the target countries. In all, ten high-Zn rice varieties have been released in four Asian countries. Efforts have begun to mainstream grain Zn to ensure that the Zn trait becomes an integral part of future varieties. Huge scope exists to apply advanced genomics technologies such as genomic selection and genome editing to speed up high-Zn varietal development. An efficient rice value chain for Zn-biofortified varieties, quality control, and promotion are essential for successful adoption and consumption. The development of next-generation high-Zn rice varieties with higher grain-Zn content, stacking of multiple nutrients, along with good grain quality and acceptable agronomic traits has to be fast-tracked. Healthier rice has a large demand from all stakeholders, so we need to keep up the pace of developing nutritious rice to meet the demand and to achieve nutritional security.

References

Alqabbani H, AlBadr N (2020) Zinc status (intake and level) of healthy elderly individuals in Riyadh and its relationship to physical health and cognitive impairment. Clin Nutr Exp 29:10–17. https://doi.org/10.1016/j.yclnex.2019.12.001

Anandan A, Rajiv G, Eswaran R, Prakash M (2011) Genotypic variation and relationships between quality traits and trace elements in traditional and improved rice (*Oryza sativa* L.) genotypes. J Food Sci 76(4):122–130

Andreini C, Bertini I, Rosato A (2009) Metalloproteomes: a bioinformatic approach. Acc Chem Res 42:1471–1479

Anuradha K, Agarwal S, Batchu AK, Babu AP, Swamy BPM, Longvah T, Sarla N (2012) Evaluating rice germplasm for iron and zinc concentration in brown rice and seed dimensions. J Geophys Res 4:19–25

Arbelaez JD, Dwiyanti MS, Tandayu E et al (2019) 1k-RiCA (1K-Rice Custom Amplicon) a novel genotyping amplicon-based SNP assay for genetics and breeding applications in rice. Rice 12:55. https://doi.org/10.1186/s12284-019-0311-0

Arcade A, Labourdette A, Falque M, Mangin B, Chardon F, Charcosset A, Joets J (2004) BioMercator: integrating genetic maps and QTL towards discovery of candidate genes. Bioinformatics 20(14):2324–2326

Balakrishnan D, Surapaneni M, Yadavalli VR, Addanki KR, Mesapogu S, Beerelli K, Neelamraju S (2020) Detecting CSSLs and yield QTLs with additive, epistatic and QTL × environment interaction effects from *Oryza sativa* × *O. nivara* IRGC81832 cross. Sci Rep 10(1):1–18. https://doi.org/10.1038/s41598-020-64300-0

Bandillo N, Raghava C, Muyco PA, Sevilla MAL, Lobina IT, Dilla-Ermita CJ, Tung CW, McCouch S, Thomson M, Mauleon R, Singh RK, Gregorio G, Redoña E, Leung H (2013) Multi-parent advanced generation inter-cross (MAGIC) populations in rice: progress and potential for genetics research and breeding. Rice 6:11

Banerjee S, Sharma DJ, Verulkar SB, Chandel G (2010) Use of in silico and semi quantitative RT-PCR approaches to develop nutrient rich rice (*Oryza sativa* L). Indian J Biotechnol 9:203–212

Bashir K, Inoue H, Nagasaka S, Takahashi M, Nakanishi H, Mori S, Nishizawa NK (2006) Cloning and characterization of deoxymugineic acid synthase genes from graminaceous plants. J Biol Chem 281:32395–32402

Ben-Ari G, Lavi U (2012) Marker assisted selection in plant breeding. In: Plant biotechnology and agriculture. Academic Press, Cambridge, pp 163–184

Bernardo R (2008) Molecular markers and selection for complex traits in plants: learning from the last 20 years. Crop Sci 48:1649

Black RE, Allen LH, Bhutta ZA, Caulfield LE, de Onis M, Ezzati M, Mathers C, Rivera J (2008) Maternal and child undernutrition: global and regional exposures and health consequences. Lancet 371:243–260

Bouis H (2017) An overview of the landscape and approach for biofortification in Africa. Afr J Food Agric Nutr Dev 17(2):11848–11864. https://doi.org/10.18697/ajfand.78.harvestplus01

Bouis H, Saltzman A (2017) Improving nutrition through biofortification: a review of evidence from HarvestPlus, 2003 through 2016. Glob Food Secur 12:49–58. https://doi.org/10.1016/j.gfs.2017.01.009

Boyle EA, Li YI, Pritchard JK (2017) An expanded view of complex traits: from polygenic to omnigenic. Cell 169:1177–1186. https://doi.org/10.1016/j.cell.2017.05.038

Brar B, Jain S, Singh R, Jain RK (2011) Genetic diversity for iron and zinc contents in a collection of 220 rice (Oryza sativa L.) genotypes. Indian J Genet Plant Breed 71(1):67–73

Broadley MR, White PJ, Hammond JP, Zelko I, Lux A (2007) Zinc in plants. New Phytol 173:677–702

Brooks WA, Yunus M, Santosham M et al (2004) Zinc for severe pneumonia in very young children: double-blind placebo-controlled trial. Lancet 363:1683–1688

Brown KH, Peerson JM, Kimmons JE, Hotz C (2002) Options for achieving adequate intake from home-prepared complementary foods in low income countries. In: Black RE, Fliescher Michaelson K (eds) Public health issues in infant and child nutrition. Nestle nutrition workshop series. Pediatric program, vol 48. Lippincott Williams and Wilkins; Nestec Ltd, Philadelphia, PA; Vevey, pp 239–256

Cakmak I (2000) Role of zinc in protecting plant cells from reactive oxygen species. New Phytol 146:185–205

Calayugan MIC, Formantes AK, Amparado A, Descalsota-Empleo GI, Nha CT, Inabangan-Asilo MA et al (2020) Genetic analysis of agronomic traits and grain iron and zinc concentrations in a doubled haploid population of rice (Oryza sativa L.). Sci Rep 10(1):2283. https://doi.org/10.1038/s41598-020-59184-z

Caulfield LE, Richard SA, Rivera JA et al (2006) Stunting, wasting, and micronutrient deficiency disorders. In: Jamison DT, Breman JG, Measham AR et al (eds) Disease control priorities in developing countries, 2nd edn. The International Bank for Reconstruction and Development/The World Bank, Washington, DC. Chapter 28. https://www.ncbi.nlm.nih.gov/books/NBK11761/. Co-published by Oxford University Press, New York

CGIAR (2018) 3-year system business plan companion document. CGIAR five-year biofortification strategy 2019-2023. CGIAR, Montpellier

Chandel G, Banerjee S, See S, Meena R, Sharma DJ, Verulkar SB (2010) Effects of different nitrogen fertilizer levels and native soil properties on rice grain Fe, Zn and protein contents. Rice Sci 17:213–227

Chasapis C, Ntoupa P, Spiliopoulou C, Stefanidou M (2020) Recent aspects of the effects of zinc on human health. Arch Toxicol 94(5):1443–1460. https://doi.org/10.1007/s00204-020-02702-9

Chen K, Gao C (2013) Targeted genome modification technologies and their applications in crop improvements. Plant Cell Rep 33:575–583

Cheng ZQ, Huang XQ, Zhang YZ, Qian J (2005) Diversity in the content of some nutritional components in husked seeds of three wild rice species and rice varieties in Yunnan Province of China. J Integr Plant Biol l47:1260–1270

Christian M, Cermak T, Doyle EL, Schmidt C, Zhang F, Hummel A et al (2010) Targeting DNA double-strand breaks with TAL effector nucleases. Genetics 186:757–761

Cobb JN, Juma RU, Biswas PS, Arbalaez JD, Rutkoski J, Atlin G, Hagen T, Quinn M, Ng EH (2019) Enhancing the rate of genetic gain in public-sector plant breeding programs: lessons from the breeder's equation. Theor Appl Genet 132:627. https://doi.org/10.1007/s00122-019-03317-0

Collard BC, Mackill DJ (2008) Marker-assisted selection: an approach for precision plant breeding in the twenty-first century. Philos Trans R Soc Lond Ser B Biol Sci 363:557–572

Collard BCY, Beredo JC, Lenaerts B, Mendoza R, Santelice R, Lopena V, Verdeprado H, Raghavan C, Gregorio GB, Vial L, Demomt M, Biswas PS, Iftkhauddaula KM, Rahman MA, Cobb JN, Islam MA (2017) Revisiting rice breeding methods: evaluating the use of rapid generation advance (RGA) for routine rice breeding. Plant Prod Sci 20:1. https://doi.org/10.108 0/1343943X.2017.1391705

Descalsota GIL, Swamy BPM, Zaw H, Inabangan-Asilo MA, Amparado A, Mauleon R et al (2018) Genome-wide association mapping in a rice MAGIC Plus population detects QTLs and genes useful for biofortification. Front Plant Sci 9:1–20. https://doi.org/10.3389/fpls.2018.01347

Descalsota-Empleo GI, Noraziyah AAS, Navea IP, Chung C, Dwiyanti MS, Labios RJD, Ikmal AM, Juanillas VM, Inabangan-Asilo MA, Amparado A, Reinke R, Vera Cruz CM, Chin JH, Swamy BPM (2019a) Genetic dissection of grain nutritional traits and leaf blight resistance in rice. Genes 10:30. https://doi.org/10.3390/genes10010030

Descalsota-Empleo GI, Amparado A, Inabangan-Asilo MA, Tesoro F, Stangoulis J, Reinke R, Swamy BPM (2019b) Genetic mapping of QTL for agronomic traits and grain mineral elements in rice. Crop J 7(4):560–572. https://doi.org/10.1016/j.cj.2019.03.002

Dixit S, Singh UM, Abbai R, Ram T, Singh VK, Paul A, Virk PS, Kumar A (2019) Identification of genomic region(s) responsible for high iron and zinc content in rice. Sci Rep 9:8136. https://doi.org/10.1038/s41598-019-43888-y

FAO (Food and Agriculture Organization of the United Nations) (2014) Country nutrition paper Bangladesh. In: International Conference on Nutrition, 19–21 November 2014. FAO, Rome

FAO (Food and Agriculture Organization of the United Nations) (2016) Statistical database. FAO, Rome. http://faostat3.fao.org/home/E

Gande NK, Rakhi S, Kundur PJ, Amabti R, Bekele BD, Shashidhar HE (2013) Evaluation of recombinant inbred lines of rice (*Oryza sativa* L.) for grain zinc content, yield related traits and identification of transgressant lines grown under aerobic conditions. Asian J Exp Biol Sci 4(4):567–574

Gangashetty PI, Salimath PM, Hanamaratt NG (2013) Genetic variability studies in genetically diverse non-basmati local aromatic genotypes of rice (*Oryza sativa* L.). Rice Genom Genet 4:4–8

Gao C (2015) Genome editing in crops: from bench to field. Natl Sci Rev 2:13–15

Garcia-Oliveira AL, Tan L, Fu Y, Sun C (2009) Genetic identification of quantitative trait loci for contents of mineral nutrients in rice grain. J Integr Plant Biol 51:84–92

Garcia-Oliveira AL, Chander S, Ortiz R, Menkir A, Gedil M (2018) Genetic basis and breeding perspectives of grain iron and zinc enrichment in cereals. Front Plant Sci 9:1–13. https://doi.org/10.3389/fpls.2018.00937

Garg M, Sharma N, Sharma S, Kapoor P, Kumar A, Chunduri V, Arora P (2018) Biofortified crops generated by breeding, agronomy, and transgenic approaches are improving lives of millions of people around the world. Front Nutr 5:12. https://doi.org/10.3389/fnut.2018.00012

Gödecke T, Stein AJ, Qaim M (2018) The global burden of chronic and hidden hunger: trends and determinants. Glob Food Secur 17:21–29. https://doi.org/10.1016/j.gfs.2018.03.004

Goffinet B, Gerber S (2000) Quantitative trait loci: a meta-analysis. Genetics 155(1):463–473

Gregorio GB (2002) Progress in breeding for trace minerals in staple crops. J Nutr 132:500–502

Gregorio GB, Senadhira D, Htut T, Graham RD (2000) Breeding for trace mineral density in rice. Food Nutr Bull 21:382–386

Gross BL, Zhao Z (2014) Archaeological and genetic insights into the origins of domesticated rice. Proc Natl Acad Sci U S A 111(17):6190

Guild GE, Paltridge NG, Andersson MS, Stangoulis JCR (2017) An energy-dispersive X-ray fluorescence method for analysing Fe and Zn in common bean, maize and cowpea biofortification programs. Plant Soil 419(1–2):457–466. https://doi.org/10.1007/s11104-017-3352-4

Guo G, Zhao F, Wang Y, Zhang Y, Du L, Su G (2014) Comparison of single-trait and multiple-trait genomic prediction models. BMC Genet 15:30–36

Guo R, Dhliwayo T, Mageto EK, Palacios-Rojas N, Lee M, Yu D, Ruan Y, Zhang A, San Vicente F, Olsen M, Crossa J, Prasanna BM, Zhang L, Zhang X (2020) Genomic prediction of kernel zinc concentration in multiple maize populations using genotyping-by-sequencing and repeat amplification sequencing markers. Front Plant Sci 11:1–15. https://doi.org/10.3389/fpls.2020.00534

Hanieh S, High H, Boulton J (2020) Nutrition justice: uncovering invisible pathways to malnutrition. Front Endocrinol 11:150. https://doi.org/10.3389/fendo.2020.00150

HarvestPlus (2012) International Rice Research Institute. http://www.harvestplus.org/content/zinc-rice-india

HarvestPlus (2020). https://www.harvestplus.org/knowledge-market/BPI. Accessed 20 Jun 2020

Hayashi T, Iwata H (2013) A Bayesian method and its variational approximation for prediction of genomic breeding values in multiple traits. BMC Bioinformatics 14:34

Hill WG, Michael E, Goddard ME, Visscher PM (2008) Data and theory point to mainly additive genetic variance for complex traits. PLoS Genet 4:e1000008. https://doi.org/10.1371/journal.pgen.1000008

Hotz C, McClafferty B (2007) From harvest to health: challenges for developing biofortified staple foods and determining their impact on micronutrient status. Food Nutr Bull 28(2):271–279

Impa SM, Morete MJ, Ismail AM, Schulin R, Johnson-Beebout SE (2013) Zn uptake translocation and grain Zn loading in rice (*Oryza sativa* L) genotypes selected for Zn-deficiency tolerance and high grain Zn. J Exp Bot 64:2739–2751

Inabangan-Asilo MA, Mallikarjuna Swamy BP, Amparado AF, Descalsota-Empleo GIL, Arocena EC, Reinke R (2019) Stability and G × E analysis of zinc-biofortified rice genotypes evaluated in diverse environments. Euphytica 215(3):1–17. https://doi.org/10.1007/s10681-019-2384-7

Inoue H, Higuchi K, Takahashi M, Nakanishi H, Mori S, Nishizawa NK (2003) Three rice nicotianamine synthase genes, *OsNAS1*, *OsNAS2*, and *OsNAS3* are expressed in cells involved in long-distance transport of iron and differentially regulated by iron. Plant J 36:366–381

Inoue H, Takahashi M, Kobayashi T, Suzuki M, Nakanishi H, Mori S, Nishizawa NK (2008) Identification and localisation of the rice nicotianamine aminotransferase gene *OsNAAT1* expression suggests the site of phytosiderophore synthesis in rice. Plant Mol Biol 66:193–203

IOM (Institute of Medicine) (2001) Dietary reference intakes for vitamin A, vitamin K, arsenic, boron, chromium, copper, iodine, iron, manganese, molybdenum, nickel, silicon, vanadium, and zinc. National Academy Press, Washington, DC

Iqbal S, Ali I, Rust P, Kundi M, Ekmekcioglu C (2020) Selenium, zinc, and manganese status in pregnant women and its relation to maternal and child complications. Nutrients 12(3):725. https://doi.org/10.3390/nu12030725

Ishikawa R, Iwata M, Taniko K, Monden G, Miyazaki N, Orn C et al (2017) Detection of quantitative trait loci controlling grain zinc concentration using Australian wild rice, *Oryza meridionalis*, a potential genetic resource for biofortification of rice. PLoS One 12(10):e0187224. https://doi.org/10.1371/journal.pone.0187224

Ishimaru Y, Suzuki M, Kobayashi T, Takahashi M, Nakanishi H, Mori S, Nishizawa NK (2005) OsZIP4, a novel zinc-regulated zinc transporter in rice. J Exp Bot 56:3207–3214

IZiNCG (International Zinc Nutrition Consultative Group) (2009) Systematic reviews of zinc intervention strategies. Food Nutr Bull 25:S12–S40

Jeong O-Y, Lee J-H, Jeong E-G et al (2020) Analysis of QTL responsible for grain iron and zinc content in doubled haploid lines of rice (*Oryza sativa*) derived from an intra-japonica cross. Plant Breed 139:344–355. https://doi.org/10.1111/pbr.12787

Jia Y, Jannink JL (2012) Multiple-trait genomic selection methods increase genetic value prediction accuracy. Genetics 192:1513–1522

Jin T, Chen J, Zhu L, Zhao Y, Guo J, Huang Y (2015) Comparative mapping combined with homology-based cloning of the rice genome reveals candidate genes for grain zinc and iron concentration in maize. BMC Genet 16(1):17

Jinek M, Chylinski K, Fonfara I, Hauer M, Doudna JA, Charpentier E (2012) A programmable dual-RNA-guided DNA endonuclease in adaptive bacterial immunity. Science 337:816–821

Johnson AAT, Kyriacou B, Callahan DL, Carruthers L, Stangoulis J (2011) Constitutive overexpression of the *OsNAS* gene family reveals single-gene strategies for effective iron- and zinc-biofortification of rice endosperm. PLoS One 6:e24476

Joshi AK, Crossa J, Arun B, Chand R, Trethowan R, Vargas M, Monasterio IO (2010) Genotype × environment interaction for zinc and iron concentration of wheat grain in eastern Gangetic plains of India. Field Crops Res 116:268. https://doi.org/10.1016/j.fcr.2010.01.004

Kakei Y, Ishimaru Y, Kobayashi T, Yamakawa T, Nakanshi H, Nishizawa NK (2012) OsYSL16 plays a role in the allocation of iron. Plant Mol Biol 79:583–594

Kim YG, Cha J, Chandrasegaran S (1996) Hybrid restriction enzymes: zinc finger fusions to Fok I cleavage domain. Proc Natl Acad Sci U S A 93:1156–1160

Lado B, Vázquez D, Quincke M, Silva P, Aguilar I, Gutiérrez L (2018) Resource allocation optimization with multi-trait genomic prediction for bread wheat (*Triticum aestivum* L.) baking quality. Theor Appl Genet 131:2719. https://doi.org/10.1007/s00122-018-3186-3

Lan HX, Wang ZF, Wang QH, Wang MM, Bao YM, Huang J, Zhang HS (2013) Characterization of a vacuolar zinc transporter OZT1 in rice (*Oryza sativa* L.). Mol Biol Rep 40(2):1201–1210

Lee S, Jeong HJ, Kim SA, Lee J, Guerinot ML, An G (2010a) OsZIP5 is a plasma membrane zinc transporter in rice. Plant Mol Biol 73(4–5):507–517

Lee S, Kim SA, Lee J, Guerinot ML, An G (2010b) Zinc deficiency-inducible *OsZIP8* encodes a plasma membrane-localized zinc transporter in rice. Mol Cell 29(6):551–558

Lee S-M, Kang J-W, Lee J-Y, Seo J, Shin D, Cho J-H, Jo S, Song Y-C, Park D-S, Ko J-M, Koh H-J, Lee J-H (2020) QTL analysis for Fe and Zn concentrations in rice grains using a doubled haploid population derived from a cross between rice (*Oryza sativa*) cultivar 93-11 and milyang 352. Plant Breed Biotech 8(1):69–76

Li J, Wang J, Zeigler RS (2014) The 3,000 rice genomes project: new opportunities and challenges for future rice research. Giga Sci 3:8

de los Campos G, Naya H, Gianola D, Crossa J, Legarra A, Manfredi E, Weigel K, Cotes JM (2009) Predicting quantitative traits with regression models for dense molecular markers and pedigree. Genetics 182(1):375–385

Ma G, Jin Y, Li Y, Zhai F, Kok FJ, Jacobsen E, Yang X (2008) Iron and zinc deficiencies in China: what is a feasible and cost-effective strategy? Public Health Nutr 11:632–638

Maret W (2017) Zinc in cellular regulation: the nature and significance of "zinc signals". Int J Mol Sci 18:2285

Maret W, Li Y (2009) Coordination dynamics of zinc in proteins. Chem Rev 109:4682–4707

Martínez CP, Borrero J, Taboada R, Viana JL, Neves P, Narvaez L, Puldon V, Adames A, Vargas A (2010) Rice cultivars with enhanced iron and zinc content to improve human nutrition. In: 28th International Rice Research Conference, Hanoi, Vietnam, 8–12 November 2010. OP10: Quality Grain, Health, and Nutrition

McDonald GK, Genc Y, Graham RD (2008) A simple method to evaluate genetic variation in grain zinc concentration by correcting for differences in grain yield. Plant Soil 306:49. https://doi.org/10.1007/s11104-008-9555-y

Meuwissen THE, Hayes BJ, Goddard ME (2001) Prediction of total genetic value using genome-wide dense marker maps. Genetics 157:1819–1829

Milovanovic V, Smutka L (2017) Asian countries in the global rice market. Acta Univ Agric Silvicult Mendel Brunensis 65(2):679–688. https://doi.org/10.11118/actaun201765020679

Montesinos-Lopez OA, Montesinos-Lopez A, Crossa J, Toledo FH, Perez-Hernandez O, Eskridge KM et al (2016) A genomic Bayesian multi-trait and multi-environment model. G3 6(9):2725–2744. https://doi.org/10.1534/g3.116.032359

Montesinos-López A, Montesinos-López OA, Gianola D, Crossa J, Hernández-Suárez CM (2019) Multivariate Bayesian analysis of on-farm trials with multiple-trait and multiple-environment data. Agron J 3(1):1–12. https://doi.org/10.2134/agronj2018.06.0362

Morete MJ, Impa MS, Rubianes F, Beebout SEJ (2011) Characterization of zinc uptake and transport in rice under reduced conditions in agar nutrient solution. In: 14th Philippine Society of Soil Science and Technology, Scientific Conference, 25–27 May 2011. Visayas State University, Baybay

Naik SM, Raman AK, Nagamallika M, Venkateshwarlu C, Singh SP, Kumar S, Singh SK, Ahmed HU, Das SP, Prasad K, Izhar T, Mandal NP, Singh NK, Yadav S, Reinke R, Swamy BPM, Virk P, Kumar A (2020) Genotype × environment interactions for grain iron and zinc content in rice. J Sci Food Agric 100:4150. https://doi.org/10.1002/jsfa.10454

Nha CT (2019) Dissection of QTL and genes for agronomic and biofortification traits in six connected populations of rice. Dissertation. University of the Philippines, Los Baños. 224 p

Norton GJ, Deacon CM, Xiong L, Huang S, Meharg AA, Price AH (2010) Genetic mapping of the rice ionome in leaves and grain: identification of QTLs for 17 elements including arsenic, cadmium, iron and selenium. Plant Soil 329:139–153

Norton GJ, Douglas A, Lahner B, Yakubova E, Guerinot ML et al (2014) Genome wide association mapping of grain arsenic, copper, molybdenum and zinc in rice (Oryza sativa L.) grown at four international field sites. PLoS One 9(2):e89685. https://doi.org/10.1371/journal.pone.0089685

Nozoye T, Nagasaka S, Kobayashi T, Takahashi M, Sato Y, Sato Y, Uozumi N, Nakanishi H, Nishizawa NK (2011) Phytosiderophore efflux transporters are crucial for iron acquisition in graminaceous plants. J Biol Chem 286:5446–5454

Palanog AD, Calayugan MIC, Descalsota-Empleo GI et al (2019) Zinc and iron nutrition status in the Philippines population and local soils. Front Nutr 6:81

Pandian SS, Robin S, Vinod KK, Rajeswari S, Manonmani S, Subramanian KS, Saraswathi R, Kirubhakaran APM (2011) Influence of intrinsic soil factors on genotype-by-environment interactions governing micronutrient content of milled rice grains. AJCS 5(13):1737–1744

Peng S, Laza RC, Visperas RM, Sanico AL, Cassman KG, Khush GS (2000) Grain yield of rice cultivars and lines developed in the Philippines since 1966. Crop Sci 40:307–314

Peng S, Huang J, Sheehy JE, Laza RC, Visperas RM, Zhong XH et al (2004) Rice yields decline with higher night temperature from global warming. Proc Natl Acad Sci U S A 101(27):9971–9975

Prasad R, Shivay YS, Kumar D (2014) Agronomic biofortification of cereal grains with iron and zinc. Adv Agron 125:55–91. https://doi.org/10.1016/B978-0-12-800137-0.00002-9

Ramesh SA, Shin R, Eide DJ, Schachtman P (2003) Differential metal selectivity and gene expression of two zinc transporters from rice. Plant Physiol 133:126–134

Rampa F, Lammers E, Linnemann A, Schoustra S, de Winter D (2020) African indigenous foods: opportunities for improved food and nutrition security. Food & Business Knowledge Platform. https://knowledge4food.net/african-indigenous-foods/

Rana MK, Bhat KV (2004) A comparison of AFLP and RAPD markers for genetic diversity and cultivar identification in cotton. J Plant Biochem Biotechnol 13:19–24

Raza Q, Riaz A, Sabar M, Atif RM, Bashir K (2019) Meta-analysis of grain iron and zinc associated QTLs identified hotspot chromosomal regions and positional candidate genes for breeding biofortified rice. Plant Sci 288:110214. https://doi.org/10.1016/j.plantsci.2019.110214

Ritchie H, Reay D, Higgins P (2018) Quantifying, projecting, and addressing India's hidden hunger. Front Sustain Food Syst 2:11. https://doi.org/10.3389/fsufs.2018.00011

Sasaki A, Yamaji N, Xia J, Ma JF (2011) OsYSL6 is involved in the detoxification of excess manganese in rice. Plant Physiol 157:1832–1840

Sathisha TN (2013) Genetic variation among traditional landraces of rice with specific reference to nutrition al quality. Karnataka J Agric Sci 26:474

Sazawal S, Black RE, Ramsan M, Chwaya HM, Dutta A, Dhingra U, Stoltzfus RJ, Othman MK, Kabole FM (2007) Effect of zinc supplementation on mortality in children aged 1-48 months: a community-based randomised placebo-controlled trial. Lancet 369(9565):927–934

Schulthess AW, Yu W, Miedaner T, Wilde P, Reif JC, Zhao Y (2016) Multiple-trait and selection indices genomic predictions for grain yield and protein content in rye for feeding purposes. Theor Appl Genet 129:273–287

Shamshad M, Sharma A (2018) The usage of genomic selection strategy in plant breeding. In: Next generation plant breeding. InTech, Rijeka. https://doi.org/10.5772/intechopen.76247

Siwela M, Pillay K, Govender L, Lottering S (2020) Biofortified crops for combating hidden hunger in South Africa: availability, acceptability, micronutrient retention and bioavailability. Foods 9(6):815. https://doi.org/10.3390/foods9060815

Soe YP (2020) Meta-analysis of quantitative trait loci associated with grain zinc content in rice. Dissertation. University of the Philippines, Los Baños. 180 pp

Spindel J, Begum H, Akdemir D, Virk P, Collard B, Redoña E et al (2015) Genomic selection and association mapping in rice (*Oryza sativa*): effect of trait genetic architecture, training population composition, marker number and statistical model on accuracy of rice genomic selection in elite, tropical rice breeding lines. PLoS Genet 11(2):e1004982. https://doi.org/10.1371/journal.pgen.1004982

Swamy BM, Vikram P, Dixit S, Ahmed HU, Kumar A (2011) Meta-analysis of grain yield QTL identified during agricultural drought in grasses showed consensus. BMC Genomics 12:319. https://doi.org/10.1186/1471-2164-12-319

Swamy BPM, Rahman MA, Inabangan-Asilo MA, Amprado A, Manito C, Chada-Mohanty P, Reinike R, Slamet-Loedin IH (2016) Advances in breeding for high grain Zinc in rice. Rice 9:49. https://doi.org/10.1186/s12284-016-0122-5

Swamy BPM, Kaladhar K, Anuradha K, Batchu AK, Longvah T, Sarla N (2018a) QTL analysis for grain iron and zinc concentrations in two *O. nivara* derived backcross populations. Rice Sci 25(4):197–207. https://doi.org/10.1016/j.rsci.2018.06.003

Swamy BPM, Descalsota GIL, Nha CT, Amparado A, Inabangan-Asilo MA, Manito C, Tesoro F, Reinke R (2018b) Identification of genomic regions associated with agronomic and biofortification traits in DH populations of rice. PLoS One 13(8):1–20. https://doi.org/10.1371/journal.pone.0201756

Szymlek-Gay EA, Ferguson EL, Heath AL, Gray AR, Gibson RS (2009) Food-based strategies improve iron status in toddlers: a randomized controlled trial. Am J Clin Nutr 90:1541–1551

Tielsch JM, Khatry SK, Stoltzfus RJ et al (2007) Effect of daily zinc supplementation on child mortality in southern Nepal: a community-based, cluster randomized, placebo-controlled trail. Lancet 370:1230–1239

Torheim LE, Ferguson EL, Penrose K, Arimond M (2010) Women in resource-poor settings are at risk of inadequate intakes of multiple micronutrients. J Nutr 140:2051S–2058S

Trijatmiko KR, Dueñas C, Tsakirpaloglou N, Torrizo I., Arines FM, Adeva C, Balindong J, Oliva N, Sapasap MV, Borrero J, Rey J, Francisco P, Nelson A, Nakanishi H, Lombi E, Tako E, Glahn RP, Stangoulis J, Chadha-Mohanty P, Johnson AAT, Tohme J, Barry G, Slamet-Loedin IH (2016) Biofortified indica rice attains iron and zinc nutrition dietary targets in the field. Sci Rep 6:19792. https://doi.org/10.1038/srep19792

Tripoli M, Schmidhuber J (2018) Emerging opportunities for the application of blockchain in the agri-food industry. FAO, ICTSD, Rome, Geneva. Licence: CC BY-NC-SA 3.0 IGO

Tsakirpaloglou N, Mallikarjuna Swamy BP, Acuin C, Slamet-Loedin IH (2019) Biofortified Zn and Fe rice: potential contribution for dietary mineral and human health. In: Jaiwal P, Chhillar A, Chaudhary D, Jaiwal R (eds) Nutritional quality improvement in plants. Concepts and strategies in plant sciences. Springer, Cham

UNICEF (2013) Improving child nutrition: the achievable imperative for global progress. United Nations Children's Fund, United Nations Organization, New York, NY

Wang Y, Wei Y, Dong L, Lu L, Feng Y, Zhang J et al (2014) Improved yield and Zn accumulation for rice grain by Zn fertilization and optimized water management. J Zhejiang Univ Sci B 15(4):365–374. https://doi.org/10.1631/jzus.b1300263

Wassmann R, Jagadish SVK, Sumfleth K, Pathak H, Howell G, Ismail A et al (2009) Regional vulnerability of climate change impacts on Asian rice production and scope for adaptation. Adv Agron 102:91–133

Welch RM, Graham RD (2004) Breeding for micronutrients in staple food crops from a human nutrition perspective. J Exp Bot 55:353–364

White PJ, Broadley MR (2009) Biofortification of crops with seven mineral elements often lacking in human diets – iron, zinc, copper, calcium, magnesium, selenium and iodine. Review. New Phytol 182:49–84. https://doi.org/10.1111/j.1469-8137.2008.02738.x

WHO (World Health Organization) (1998) Complementary feeding of young children in developing countries: a review of current scientific knowledge. WHO, Geneva

WHO (World Health Organization) (2003) Joint WHO/FAO Expert Consultation on diet, nutrition and the prevention of chronic diseases. WHO, Geneva

Widodo B, Broadley MR, Rose T, Frei M, Pariasca-Tanaka J, Yoshihashi T, Thomson M, Hammond JP, Aprile A, Close TJ, Ismail AM, Wissuwa MM (2010) Response to zinc deficiency of two rice lines with contrasting tolerance is determined by root growth maintenance and organic acid exudation rates, and not by zinc-transporter activity. New Phytol 186:400–414

Wissuwa M, Ismail AM, Graham RD (2008) Rice grain zinc concentrations as affected by genotype, native soil-zinc availability, and zinc fertilization. Plant Soil 306:37. https://doi.org/10.1007/s11104-007-9368-4

Young G, Mortimer E, Gopalsamy G, Alpers D, Binder H, Manary M et al (2014) Zinc deficiency in children with environmental enteropathy—development of new strategies: report from an expert workshop. Am J Clin Nutr 100(4):1198–1207. https://doi.org/10.3945/ajcn.113.075036

Zarcinas BA, Cartwright B, Spouncer LR (1987) Nitric acid digestion and multi element analysis of plant material by inductively coupled plasma spectrometry. Commun Soil Sci Plant Anal 18:131–146

Zaw H, Raghavan C, Pocsedio A, Swamy BPM, Jubay ML, Singh RK et al (2019) Exploring genetic architecture of grain yield and quality traits in a 16-way indica by japonica rice MAGIC global population. Sci Rep 9(1):1–11. https://doi.org/10.1038/s41598-019-55357-7

Zhang Y, Xu Y, Yi H, Gong J (2012) Vacuolar membrane transporters OsVIT1 and OsVIT2 modulate iron translocation between flag leaves and seeds in rice. Plant J 72:400–410

Biofortification of Rice Grains for Increased Iron Content

**Jerlie Mhay Matres, Erwin Arcillas, Maria Florida Cueto-Reaño,
Ruby Sallan-Gonzales, Kurniawan R. Trijatmiko, and Inez Slamet-Loedin**

Abstract Dietary iron (Fe) deficiency affects 14% of the world population with significant health impacts. Biofortification is the process of increasing the density of vitamins and minerals in a crop, through conventional breeding, biotechnology approaches, or agronomic practices. This process has recently been shown to successfully alleviate micronutrient deficiency for populations with limited access to diverse diets in several countries (https://www.harvestplus.org/). The Fe breeding target in the HarvestPlus program was set based on average rice consumption to fulfil 30% of the Estimated Average Requirement of Fe in women and children. In this review, we present the reported transgenic approaches to increase grain Fe. Insertion of a single or multiple genes encoding iron storage protein, metal transporter, or enzyme involved in the biosynthesis of metal chelator in the rice genome was shown to be a viable approach to significantly increase grain-Fe density. The most successful approach to reach the Fe breeding target was by overexpression of multiple genes. Despite this success, a significant effort of 8–10 years needs to be dedicated from the proof of concept to varietal release. This includes large-scale plant transformation, event selection, collection of data for premarket safety assurance, securing biosafety permits for consumption and propagation, and collection of data for variety registration.

Keywords Rice · Grain · Iron · Biofortification · Genetic engineering

J. M. Matres · E. Arcillas · M. F. Cueto-Reaño · R. Sallan-Gonzales · I. Slamet-Loedin
International Rice Research Institute, Los Baños, Laguna, Philippines

K. R. Trijatmiko (✉)
Strategic Innovation Platform, International Rice Research Institute,
Los Baños, Laguna, Philippines
e-mail: k.trijatmiko@irri.org

© The Author(s) 2021
J. Ali, S. H. Wani (eds.), *Rice Improvement*,
https://doi.org/10.1007/978-3-030-66530-2_14

1   Introduction

Nutritional deficiency is an important global health concern that affects approximately 1.86 billion people worldwide, and 61% of it is caused by dietary iron (Fe) deficiency (James et al. 2018). Iron deficiency anemia results in decreased work productivity, increased maternal mortality, increased child mortality, slowed child development, and increased susceptibility to infectious diseases (Stoltzfus 2001). Iron deficiency anemia is estimated to be responsible for 800,000 deaths/year (WHO 2002).

Although providing access to more diverse diets is the ideal solution to alleviate micronutrient deficiency, this may not be achieved in the near future in developing and less developed countries. Iron supplementation and industrial fortification have been shown to alleviate micronutrient deficiencies but require continuous significant budget allocation at the government or household level. Despite the potential to diminish iron deficiency in the population, this may not help people living in remote rural areas because of the lack of infrastructure, purchasing power, or access to markets and healthcare systems (Mayer et al. 2008). Biofortification, a process of increasing the concentration of micronutrient in the edible part of a crop through conventional plant breeding, transgenic methods, and agronomic practices, offers a feasible and cost-effective approach, complementing other efforts to reach rural populations (Bouis and Saltzman 2017).

2   Target Concentration for Fe in Polished Rice

The Estimated Average Requirement (EAR) of iron for non-pregnant, non-lactating women is 1460 mg/day, and for children 4–6 years old is 500 mg/day (WHO/FAO 2004). Current studies show that 90% of the iron remains in the grain after processing, while the nutritionist assumption is that 10% of the iron is bioavailable. With 400 g/day per capita rice consumption for an adult woman and 120 g/day for children 4–6 years old, the HarvestPlus program set 13 µg/g as the final target concentration for Fe in polished rice to achieve 30% of the EAR (Bouis et al. 2011).

3   Iron Biofortification via Conventional Plant Breeding

Initial screening of the germplasm collection at the International Rice Research Institute (IRRI) showed that the range in Fe concentration in brown rice among 1138 genotypes tested was 6.3–24.4 µg/g (Gregorio et al. 1999). However, the variation in Fe concentration in milled rice becomes narrow due to the high proportion of Fe lost during milling. A study on the iron concentration of brown and milled rice of six varieties collected from ten commercial rice mills in one province in Vietnam showed that the percentage of Fe loss due to milling ranged from 65% to 82% (Hoa and Lan 2004). Furthermore, the maximal iron concentration in milled rice was 8 µg/g over 11,337 genotypes from the International Center for Tropical Agriculture (CIAT) core collection (Martínez et al. 2010). At IRRI, the highest Fe concentration of 7.4 µg/g in polished rice was achieved by classical breeding (Virk et al. 2006, 2007).

To reach the Fe target of 13 µg/g in polished rice, transgenic approaches are potential options because of the low concentration of Fe found in the rice gene pool. Efforts have been made to study the physiology as well as the genetic basis and biochemical mechanisms involved in Fe uptake and translocation in crops and model plants. These studies have facilitated the detection of the limiting factors that could be manipulated to increase Fe concentration in rice grain.

4 Iron Uptake and Translocation

Iron is an important micronutrient required in various processes such as photosynthesis and respiration. Based on the strategy they use to uptake iron from the rhizosphere, higher plants can be categorized into three different groups (Connorton et al. 2017): (1) Strategy I plants (all dicotyledonous plants and non-graminaceous monocots) that rely on the reduction of ferric Fe(III) to ferrous Fe(II), (2) Strategy II plants (graminaceous monocots) that rely on the chelation strategy involving phytosiderophore secretion, and (3) a combination of both. *Arabidopsis* has been used as a model plant to study Strategy I plants. Some major genes responsible for iron uptake using this strategy have been identified: iron-regulated transporter 1 (*IRT1*) (Eide et al. 1996), ferric-chelate oxidase 2 (*FRO2*) (Robinson et al. 1999), and HC-ATPase (HA) genes (Kobayashi and Nishizawa 2012).

Rice has been used as a model plant to study Strategy II plants. Plants in this group that include the most important cereals in the world secrete phytosiderophore (PS) in the rhizospere. PS is a high-Fe-affinity organic molecule from the mugineic acid family (Bashir et al. 2017; Borrill et al. 2014; Kobayashi et al. 2005; Suzuki et al. 2006). Figure 1 presents the basic scheme for the genes involved in iron homeostasis in rice.

Mugineic acid is synthesized through a conserved pathway that starts from S-adenosyl-L-methionine. It is then followed by sequential reactions catalyzed by nicotianamine synthase (NAS), nicotianamine aminotransferase (NAAT), and deoxymugineic acid synthase (DMAS) enzymes, producing 20-deoxymugineic acid (DMA), a precursor of different types of mugineic acids (Bashir et al. 2006; Higuchi et al. 1999; Kobayashi and Nishizawa 2012; Takahashi et al. 1999). In rice roots, secretion of DMA is influenced by the expression of the TOM1 geneencoding efflux transporter of DMA (Nozoye et al. 2011). YELLOW STRIPE 1 (YS1) and YELLOW STRIPE 1-like (YSL1) transporters are known for their role in facilitating the uptake of Fe–MA complexes into root cells (Curie et al. 2001; Inoue et al. 2009). After reduction by ascorbate, the Fe(III)–DMA complex is likely converted

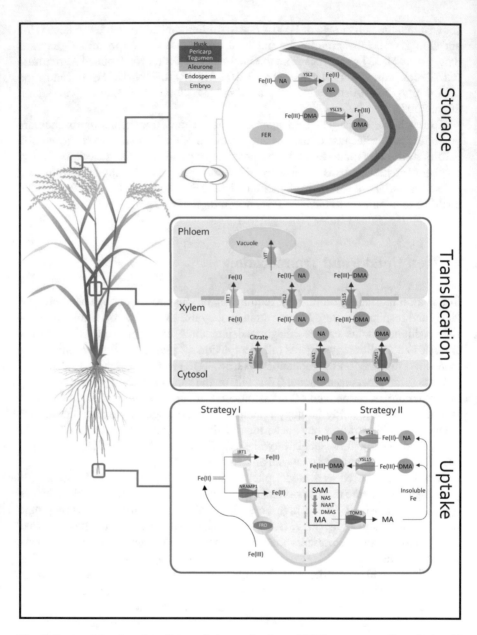

Fig. 1. Iron uptake, translocation, and storage in rice. *TOM1* transporter of mugineic acid family phytosiderophores 1, *YS1* yellow stripe 1, *YSL15* yellow stripe 1-like, *IRT1* iron-regulated transporter, *NRAMP1* natural resistance-associated macrophage protein 1, *FRO* ferric reductase oxidase, *FRDL1* ferric reductase defective-like1, *ENA* efflux transporter of nicotianamine, *VIT* vacuolar iron transporter, *FER* ferritin, *MA* mugineic acid, *DMA* 2′-deoxymugineic acid, *NA* nicotianamine, *SAM* S-adenosyl-ʟ-methionine, *NAS* nicotianamine synthase, *NAAT* nicotianamine aminotransferase, *DMAS* deoxymugineic acid synthase

to Fe(II)–NA, and then excreted to the xylem. The Fe may create complexes primarily with citrate and some with DMA for further transport (Yoneyama et al. 2015).

Aside from using Strategy II, rice may be able to uptake Fe(II) directly, as indicated by the presence of a ferrous transporter (OsIRT1) in the genome. Other metal transporters involved in Strategy I and Strategy II were also identified such as natural resistance-associated macrophage protein (NRAMP) and ZIP (zinc-regulated transporter, IRT-like protein) family (Cailliatte et al. 2010; Guerinot 2000; Lanquar et al. 2005).

Fe translocation in higher plants is a complex process involving xylem loading/unloading, phloem loading/unloading, and reabsorption (Kim and Guerinot 2007). Different chelators such as citrate, nicotianamine (NA), and MAs play an essential role in symplast metal homeostasis (Garcia-Oliveira et al. 2018). FERRIC REDUCTASE DEFECTIVE-LIKE 1 (OsFRDL1) in rice (Fig. 1) is known to encode a citrate transporter involved in the transport of Fe-citrate complex (Inoue et al. 2004; Yokosho et al. 2009).

The rice YSL family encoding influx transporter consists of 18 members (Curie et al. 2009). The OsYSL2 transporter is a carrier of Fe(II)–NA and is involved in iron transport to sink tissues (Koike et al. 2004). OsYSL15 transports Fe(III)–DMA and is involved in Fe uptake in the root and long-distance Fe transport. The Fe transporter OsYSL18 may play a specific role in fertilization, as indicated by specific expression in the pollen and pollen tubes. OsYSL18 may also be involved in Fe transport in the phloem (Kobayashi and Nishizawa 2012). As indicated by their vascular tissue expression in rice, OsIRT1 and OsTOM1 may be involved in Fe translocation within the plant as well (Ishimaru et al. 2006; Nozoye et al. 2011).

5   Iron Biofortification via Genetic Engineering

Several efforts have been conducted to increase Fe concentration in rice grains. These studies can be categorized into different approaches: (1) overexpression of geneencoding iron storage protein, (2) overexpression of geneencoding enzyme involved in the biosynthesis of metal chelator, (3) overexpression of geneencoding metal transporter, and (4) a combination of two or three approaches (Tables 1 and 2).

Fe concentration of 38.1 μg/g in brown rice was achieved by endosperm-specific expression of a soybean Fe storage protein, SoyFerH1 (Goto et al. 1999). Similar approaches using the *SoyferH1* gene driven by different promoters (OsGluB1, OsGluB4, OsGlb1, ZmUbi-1) in different backgrounds (Swarna, IR68144, BR29, IR64, M12) were reported (Slamet-Loedin et al. 2015). Stable Fe concentrations of 9.2 or 7.6 μg/g over several generations were obtained (Khalekuzzaman et al. 2006; Oliva et al. 2014). Overexpression of the *OsFer2* gene was also studied and Fe concentration of 15.9 μg/g, vis-à-vis 7 μg/g in control variety Pusa-Sugandh II, was observed (Paul et al. 2012).

Another approach in improving grain-Fe concentration in rice is by increasing the expression of genes encoding enzymes involved in the biosynthesis of metal

Table 1 Summary of transgenic approaches to develop iron-rich milled rice

	Mechanism involved	Iron ([c] in µg/g) TG	WT	Growth condition	Cultivar	References
Rice gene overexpression						
OsNAS1	Iron uptake and translocation	Up to ~40	~20	Greenhouse	Japonica cv. EYI 105	Díaz-Benito et al. (2018)
OsNAS2	Iron uptake and translocation	~10	~4	Greenhouse	Japonica cv. Dongjin	Lee et al. (2012)
OsNAS3	Iron uptake and translocation	~12	~4	Greenhouse	Japonica cv. EYI 105	Lee et al. (2009b)
OsNAS1, OsNAS2, OsNAS3	Iron uptake and translocation	Up to ~19	~4.5	Greenhouse	Japonica cv. Nipponbare	Johnson et al. (2011)
OsYSL2	Iron uptake and translocation	~7.5	~1.8	Greenhouse	Japonica cv. Tsukinohikari	Ishimaru et al. (2010)
Osfer2	Storage	Up to ~15.9	~7	Greenhouse	Indica cv. Pusa-Sugandh II	Paul et al. (2012)
Rice gene silencing/knockdown mutant						
OsVIT2	Inter-cellular/intra-cellular transport and storage	~8	~5		Japonica cv. Dongjin	Bashir et al. (2013)
OsYSL9	Inter-cellular/intra-cellular transport and storage	Up to ~2.5	1		Japonica cv. Tsukinohikari	Senoura et al. (2017)
OsDMAS1	Inter-cellular/intra-cellular transport and storage	~5	~5		Japonica cv. Dongjin	Bashir et al. (2017)
Rice gene knockout mutant						
OsVMT	Inter-cellular/intra-cellular transport and storage	9	4	Greenhouse	Japonica cv. Nipponbare	Che et al. (2019)
Overexpression of genes from different species						
HvNAS1	Iron uptake and translocation	~8.5	~4	Greenhouse	Japonica cv. Tsukinohikari	Masuda et al. (2009)

	Mechanism involved	Iron [c] in $\mu g/g$		Growth condition	Cultivar	References
		TG	WT			
HvYS1	Iron uptake and translocation	Up to ~9	~4		Japonica cv. EYI 105	Banakar et al. (2017a)
AtIRT1	Iron uptake and translocation	Up to ~4.86	~2.28	Greenhouse	Japonica cv. Taipei 309	Boonyaves et al. (2016)
SoyferH1	Storage	Up to ~37	~10	Screenhouse	Indica cv. IR68144	Vasconcelos et al. (2003)
SoyferH1	Storage	Up to ~9.2	~3.8	Greenhouse	Indica cv. BR29	Khalekuzzaman et al. (2006)
SoyferH1	Storage	Up to ~16	~6.75	Greenhouse	Indica cv. Swarna	Paul et al. (2014)
SoyferH1	Storage	Up to ~7.6	~3.3	Greenhouse	Indica cv. IR64	Oliva et al. (2014)
Multigene overexpression of genes from different species						
SoyferH2 + HvNAS1 + OsYSL2	Combined mechanisms	Up to ~4	~1	Paddy field	Japonica cv. Tsukinohikari	Masuda et al. (2012)
HvNAS1, HvNAS1 + HvNAAT, IDS3	Combined mechanisms	Up to ~7.3	~5.8	Paddy field	Japonica cv. Tsukinohikari	Suzuki et al. (2008)
SoyferH2 + HvNAS1 + OsYSL2	Combined mechanisms	~6.3 (~5.02)	~3.2 (~1.46)	Greenhouse	Tropical Japonica cv. Paw San Yin	Aung et al. (2013)
AtNAS1 + Pvfer + Afphytase	Combined mechanisms	Up to ~7	~1	Hydroponic	Japonica cv. Taipei 309	Wirth et al. (2009)
AtIRT1, Pvfer, AtNAS1	Combined mechanisms	Up to ~10.46	~2.7	Greenhouse	Japonica cv. Nipponbare	Boonyaves et al. (2017)
SoyferH1 + OsNAS2	Combined mechanisms	~15	~2.5	Paddy field	Indica cv. IR64	Trijatmiko et al. (2016)
AtNAS1 + AtFRD3 + Pvfer, AtFRD3 + Pvfer	Combined mechanisms	Up to ~11.08	~2.05	Greenhouse	Japonica cv. Nipponbare	Wu et al. (2018)

(continued)

Table 1 (continued)

	Mechanism involved	Iron ([c] in µg/g)		Growth condition	Cultivar	References
		TG	WT			
AtNAS1 + Pvfer + AtNRAMP3, Pvfer + AtNRAMP3	Combined mechanisms	Up to ~13.65	~2.72	Greenhouse	Indica cv. IR64	Wu et al. (2019)
AtNAS1 + Pvfer + ZmPSY + PaCRT1	Combined mechanisms	Up to ~6.02	~1.82	Greenhouse	Japonica cv. Nipponbare	Singh et al. (2017)
OsNAS1, HvHAATb, OsNAS1 + HvHAATb	Combined mechanisms	Up to ~18	~4	Hydroponic	Japonica cv. EYI 105	Banakar et al. (2017b)
OsNAS1, HvNAATb, OsNAS1 + HvNAATb	Combined mechanisms	Up to ~55	~20	Greenhouse	Japonica cv. EYI 105	Díaz-Benito et al. (2018)
SoyferH2 + HvNAS1 + HvNAAT-A,-B + IDS3	Combined mechanisms	Fourfold		Greenhouse	Japonica cv. Tsukinohikari	Masuda et al. (2013)

Table 2 Summary of transgenic approaches to develop iron-rich brown rice

	Mechanism involved	Iron ([c] in µg/g) TG	Iron ([c] in µg/g) WT	Growth condition	Cultivar	References
Rice gene overexpression						
OsIRT1	Iron uptake and translocation	~12	~10	Paddy field	Japonica cv. Dongjin	Lee and An (2009)
TOM1	Iron uptake and translocation	~18	~15	Hydroponic	Japonica cv. Tsukinohikari	Nozoye et al. (2011)
OsNAS1	Iron uptake and translocation	Up to ~19	~12	Paddy field	Japonica cv. Xiushui 110	Zheng et al. (2010)
OsYSL13	Iron uptake and translocation	~15	~11	Greenhouse	Japonica cv. Zhonghua 11	Zhang et al. (2018)
OsYSL15	Iron uptake and translocation	~14	~12	Paddy field	Japonica cv. Dongjin	Lee et al. (2009a)
OsbHLH	Iron uptake and translocation	~35	~18	Hydroponic	Japonica cv. Nipponbare	Kobayashi et al. (2019)
OsIRO2	Iron deficiency response	Up to ~15.5	~6	Greenhouse	Japonica cv. Tsukinohikari	Ogo et al. (2011)
TOM2	Iron deficiency response			Hydroponic	Japonica cv. Nipponbare	Nozoye et al. (2015)
Rice gene silencing/knockdown mutant						
OsVIT1	Inter-cellular/intra-cellular transport and storage	~26	~20	Paddy field	Japonica cv. Zhonghua 11	Zhang et al. (2012)
OsVIT2	Inter-cellular/intra-cellular transport and storage	~28	~20	Paddy field	Japonica cv. Dongjin	Zhang et al. (2012)
Overexpression of genes of different species						
SoyferH1	Storage	Up to ~38	~14.3	Greenhouse	Japonica cv. Kitaake	Goto et al. (1999)
SoyferH1	Storage	~18	~18	Greenhouse	Indica cv. M12	Drakakaki et al. (2000)
SoyferH1	Storage	Up to ~25	~17	Greenhouse	Japonica cv. Kitaake	Qu et al. (2005)
Multigene overexpression of genes of different species						
Pvfer + rgMT + phyA	Combined mechanisms	~22	~10	Greenhouse	Japonica cv. Taipei 309	Lucca et al. (2002)

chelator. *OsNAS1* overexpression resulted in Fe concentration of 19 µg/g in brown rice; however, the concentration decreased to only 5 µg/g after polishing (Zheng et al. 2010). Co-overexpression of *OsNAS1* and *HvNAAT* genes in japonica rice resulted in Fe concentration of 18 µg/g in the polished grain (Banakar et al. 2017b). Fe concentration of 55 µg/g in the succeeding generation was observed; however, this unusually high Fe concentration suggests either low milling degree or Fe contamination (Díaz-Benito et al. 2018). Overexpression and activation tagging of *OsNAS2*, on the other hand, resulted in 19 µg/g and 10 µg/g Fe concentration in polished rice, respectively (Johnson et al. 2011; Lee et al. 2012). Meanwhile, activation tagging of *OsNAS3* achieved 12 µg/g Fe in polished grain vis-à-vis 4 µg/g Fe in the wild type (Lee et al. 2009b).

Several studies reported increased Fe uptake and translocation by overexpression of genes encoding metal transporter, including *OsYSL2* (Ishimaru et al. 2010), *OsYSL15* (Lee et al. 2009a), and *OsYSL9* (Senoura et al. 2017). *OsYSL2* and *OsYSL15* are responsible for the uptake of Fe(II)–NA and Fe(III)–DMA, respectively, whereas *OsYSL9* is involved in the transport of both complexes. Although only a minimal Fe increase was detected in T1 brown rice of *OsYSL9* and *OsYSL15* OE lines, overexpression of *OsYSL2* resulted in a fourfold increase in Fe concentration in T1 polished rice.

Recently, two studies reported significant Fe concentration increases in rice grains by overexpressing multiple genes. Wu et al. (2019) overexpressed the *AtNAS1*, *Pvfer*, and *AtNRAMP3* genes, resulting in 13.65 µg/g Fe in polished grains under greenhouse conditions. Trijatmiko et al. (2016), on the other hand, reported an Fe concentration of 15 µg/g in polished grains under field conditions by overexpressing nicotianamine synthase (*OsNAS2*) and soybean ferritin (*SoyferH-1*) genes. This high-Fe rice event did not show a yield penalty in field trials in the Philippines and Colombia. The grain quality of the transgenic event was similar to that of the IR64 genotype background used for transformation. These two studies show the potential for further advanced development of a biofortified rice product with elevated Fe concentration.

6 Future Directions

There is little prospect of achieving the target Fe concentration to reach 30% of the EAR via conventional plant breeding because of limited genetic variation in the Fe concentration in polished grains within the global rice germplasm collection. On the contrary, recent studies show that the target concentration can be achieved via genetic engineering. Under the current regulations in different countries, it usually takes 8-10 years from proof of concept to market release of genetically modified (GM) crops (Mumm 2013). The best performing events need to be selected from large-scale transformation. These events, aside from showing stable and acceptable trait expression, should have a simple integration of transgenes and have no disruption of endogenous genes with important phenotypic manifestation. Significant efforts need to be dedicated to collecting data for premarket safety assurance of the potential product, such as detailed molecular characterization of the event, safety of newly expressed proteins, novel protein expression and dietary exposure analysis, comparative nutritional analysis, and some environmental safety data collected from multi-location and multi-season field trials. After a biosafety permit for propagation of the event has been secured, developers need to follow similar procedures as in conventional breeding of a product for variety registration.

High Fe content is a consumer trait. To facilitate adoption by farmers, this micronutrient trait needs to be combined with agronomic traits. The most prospective agronomic trait for farmer adoption is higher yield. For this purpose, the possibility to incorporate the high-Fe trait into hybrid rice needs to be explored. The high-yield trait obtained through heterosis can be combined with nutritional traits. In addition, we observed that overexpression of some genes for Fe enhancement might cause unintended effects such as a yield decrease when the plants were in a homozygous condition. However, this is not detrimental when only one allele is present in hemizygous condition, and in some cases the micronutrient concentration can be retained in the hemizygous condition. In such a situation, hybrid rice can be a solution to achieve higher micronutrient concentration using a wider gene pool.

Recent developments in the regulation of genome-edited crops in different countries have attracted many scientists to work on genome editing. In the United States, certain categories of modified plants would be exempted from the regulations if the product can also be created through conventional breeding (APHIS 2020). In Argentina, a resolution on New Breeding Techniques (NBT) was passed in 2015, which rules that, if a transgene is not used or a transgene is used but is removed in the final product, it will not be classified as a GM product (Friedrichs et al. 2019). Precise genome editing technology that produces a double-stranded break in the genome, followed by the repair of this break that leads to a mutation or deletion, may result in a product that meets the non-GM regulatory classification.

Increased Fe concentration in polished grains was observed on the T-DNA insertion mutant of *OsVIT2* (Bashir et al. 2013). The insertion of the T-DNA in the promoter region in this mutant led to the knockdown of the *OsVIT2* gene (Bashir et al. 2013). Genome editing can be used to mutate the regulatory elements of genes involved in Fe homeostasis. This type of editing could result in altered expression of the genes and consequently enhanced Fe concentration in rice grains.

Although genome editing has great potential to ease the burden of regulatory requirements, genetic engineering will still be the primary tool to achieve the target Fe concentration. Overexpression of other genes involved in Fe homeostasis needs to be explored. Special attention needs to be given to the possible yield decrease in transgenic plants. Fine-tuning the expression of the genes by choosing a moderate constitutive promoter or tissue- and/or stage-specific promoter may need to be tested to avoid any yield penalty.

References

APHIS (2020) Movement of certain genetically engineered organisms. 85 FR 29790. Federal register document: 2020-10638. APHIS, Riverdale, MD

Aung MS, Masuda H, Kobayashi T, Nakanishi H, Yamakawa T, Nishizawa NK (2013) Iron biofortification of Myanmar rice. Front Plant Sci 4:158. https://doi.org/10.3389/fpls.2013.00158

Banakar R, Alvarez Fernández Á, Abadía J, Capell T, Christou P (2017a) The expression of heterologous Fe (III) phytosiderophore transporter HvYS1 in rice increases Fe uptake, translocation and seed loading and excludes heavy metals by selective Fe transport. Plant Biotechnol J 15(4):423–432. https://doi.org/10.1111/pbi.12637

Banakar R, Alvarez Fernandez A, Díaz-Benito P, Abadia J, Capell T, Christou P (2017b) Phytosiderophores determine thresholds for iron and zinc accumulation in biofortified rice endosperm while inhibiting the accumulation of cadmium. J Exp Bot 68(17):4983–4995. https://doi.org/10.1093/jxb/erx304

Bashir K, Inoue H, Nagasaka S, Takahashi M, Nakanishi H, Mori S, Nishizawa NK (2006) Cloning and characterization of deoxymugineic acid synthase genes from graminaceous plants. J Biol Chem 281:32395. https://doi.org/10.1074/jbc.M604133200

Bashir K, Takahashi R, Akhtar S, Ishimaru Y, Nakanishi H, Nishizawa NK (2013) The knockdown of OsVIT2 and MIT affects iron localization in rice seed. Rice 6:31. https://doi.org/10.1186/1939-8433-6-31

Bashir K, Nozoye T, Nagasaka S, Rasheed S, Miyauchi N, Seki M et al (2017) Paralogs and mutants show that one DMA synthase functions in iron homeostasis in rice. J Exp Bot 68:1785. https://doi.org/10.1093/jxb/erx065

Boonyaves K, Gruissem W, Bhullar NK (2016) NOD promoter-controlled AtIRT1 expression functions synergistically with NAS and FERRITIN genes to increase iron in rice grains. Plant Mol Biol 90(3):207–215. https://doi.org/10.1007/s11103-015-0404-0

Boonyaves K, Wu TY, Gruissem W, Bhullar NK (2017) Enhanced grain iron levels in iron-regulated metal transporter, nicotianamine synthase, and ferritin gene cassette. Front Plant Sci 8:130. https://doi.org/10.3389/fpls.2017.00130

Borrill P, Connorton JM, Balk J, Miller AJ, Sanders D, Uauy C (2014) Biofortification of wheat grain with iron and zinc: integrating novel genomic resources and knowledge from model crops. Front Plant Sci 5:53. https://doi.org/10.3389/fpls.2014.00053

Bouis HE, Saltzman A (2017) Improving nutrition through biofortification: a review of evidence from HarvestPlus, 2003 through 2016. Glob Food Secur 12:49–58. https://doi.org/10.1016/J.GFS.2017.01.009

Bouis HE, Hotz C, McClafferty B, Meenakshi JV, Pfeiffer WH (2011) Biofortification: a new tool to reduce micronutrient malnutrition. Food Nutr Bull 32(1 Suppl 1):S31–S40. https://doi.org/10.1177/15648265110321S105

Cailliatte R, Schikora A, Briat JF, Mari S, Curie C (2010) High-affinity manganese uptake by the metal transporter nramp1 is essential for *Arabidopsis* growth in low manganese conditions. Plant Cell 22:904. https://doi.org/10.1105/tpc.109.073023

Che J, Yokosho K, Yamaji N, Ma JF (2019) A vacuolar phytosiderophore transporter alters iron and zinc accumulation in polished rice grains. Plant Physiol 181(1):276–288. https://doi.org/10.1104/pp.19.00598

Connorton JM, Balk J, Rodríguez-Celma J (2017) Iron homeostasis in plants: a brief overview. Metallomics 9(7):813–823. https://doi.org/10.1039/C7MT00136C

Curie C, Panaviene Z, Loulergue C, Dellaporta SL, Briat JF, Walker EL (2001) Maize yellow stripe1 encodes a membrane protein directly involved in Fe(III) uptake. Nature 409:346. https://doi.org/10.1038/35053080

Curie C, Cassin G, Couch D, Divol F, Higuchi K, Le Jean M et al (2009) Metal movement within the plant: contribution of nicotianamine and yellow stripe 1-like transporters. Ann Bot 103:1. https://doi.org/10.1093/aob/mcn207

Díaz-Benito P, Banakar R, Rodríguez-Menéndez S, Capell T, Pereiro R, Christou P, Abadia J, Fernández B, Álvarez-Fernández A (2018) Iron and zinc in the embryo and endosperm of rice (*Oryza sativa* L.) seeds in contrasting 2′-deoxymugineic acid/nicotianamine scenarios. Front Plant Sci 9:1190. https://doi.org/10.3389/fpls.2018.01190

Drakakaki G, Christou P, Stöger E (2000) Constitutive expression of soybean ferritin cDNA in transgenic wheat and rice results in increased iron levels in vegetative tissues but not in seeds. Transgenic Res 9:445. https://doi.org/10.1023/A:1026534009483

Eide D, Broderius M, Fett J, Guerinot ML (1996) A novel iron-regulated metal transporter from plants identified by functional expression in yeast. Proc Natl Acad Sci U S A 93:5624. https://doi.org/10.1073/pnas.93.11.5624

Friedrichs S, Takasu Y, Kearns P, Dagallier B, Oshima R, Schofield J, Moreddu C (2019) Meeting report of the OECD conference on "Genome Editing: Applications in Agriculture—Implications for Health, Environment and Regulation". Transgenic Res 28(3–4):419–463. https://doi.org/10.1007/s11248-019-00154-1

Garcia-Oliveira AL, Chander S, Ortiz R, Menkir A, Gedil M (2018) Genetic basis and breeding perspectives of grain iron and zinc enrichment in cereals. Front Plant Sci 9:937. https://doi.org/10.3389/fpls.2018.00937

Goto F, Yoshihara T, Shigemoto N, Toki S, Takaiwa F (1999) Iron fortification of rice seed by the soybean ferritin gene. Nat Biotechnol 17(3):282–286. https://doi.org/10.1038/7029

Gregorio G, Senadhira D, Htut T (1999) Improving iron and zinc value of rice for human nutrition. Agric Dév 23:77–81

Guerinot ML (2000) The ZIP family of metal transporters. Biochim Biophys Acta Biomembr 1465:190. https://doi.org/10.1016/S0005-2736(00)00138-3

HarvestPlus (n.d.). https://www.harvestplus.org/. Accessed 14 Jul 2020

Higuchi K, Suzuki K, Nakanishi H, Yamaguchi H, Nishizawa NK, Mori S (1999) Cloning of nicotianamine synthase genes, novel genes involved in the biosynthesis of phytosiderophores. Plant Physiol 119(2).471–479. https://doi.org/10.1104/pp.119.2.471

Hoa TTC, Lan NTP (2004) Effect of milling technology on iron content in rice grains of some leading varieties in the Mekong delta. Omonrice 12:38–44

Inoue H, Mizuno D, Nakanishi H, Mori S, Takahashi M, Nishizawa NK (2004) A rice FRD3-like (OsFRDL1) gene is expressed in the cells involved in long-distance transport. Soil Sci Plant Nutr 50:1133. https://doi.org/10.1080/00380768.2004.10408586

Inoue H, Kobayashi T, Nozoye T, Takahashi M, Kakei Y, Suzuki K et al (2009) Rice OsYSL15 is an iron-regulated iron (III)-deoxymugineic acid transporter expressed in the roots and is essential for iron uptake in early growth of the seedlings. J Biol Chem 284:3470. https://doi.org/10.1074/jbc.M806042200

Ishimaru Y, Suzuki M, Tsukamoto T, Suzuki K, Nakazono M, Kobayashi T et al (2006) Rice plants take up iron as an Fe 3+ -phytosiderophore and as Fe 2+. Plant J 45(3):335–346. https://doi.org/10.1111/j.1365-313X.2005.02624.x

Ishimaru Y, Masuda H, Bashir K, Inoue H, Tsukamoto T, Takahashi M et al (2010) Rice metal-nicotianamine transporter, OsYSL2, is required for the long-distance transport of iron and manganese. Plant J 62(3):379–390. https://doi.org/10.1111/j.1365-313X.2010.04158.x

James SL, Abate D, Abate KH, Abay SM, Abbafati C, Abbasi N et al (2018) Global, regional, and national incidence, prevalence, and years lived with disability for 354 diseases and injuries for 195 countries and territories, 1990-2017: a systematic analysis for the Global Burden of Disease Study 2017. Lancet 392:1789. https://doi.org/10.1016/S0140-6736(18)32279-7

Johnson AAT, Kyriacou B, Callahan DL, Carruthers L, Stangoulis J, Lombi E, Tester M (2011) Constitutive overexpression of the OsNAS gene family reveals single-gene strategies for effective iron- and zinc-biofortification of rice endosperm. PLoS One 6(9):e24476. https://doi.org/10.1371/journal.pone.0024476

Khalekuzzaman M, Datta K, Oliva N, Alam MF, Joarder OI, Datta SK (2006) Stable integration, expression and inheritance of the ferritin gene in transgenic elite indica rice cultivar BR29 with enhanced iron level in the endosperm. Indian J Biotechnol 5:26

Kim SA, Guerinot ML (2007) Mining iron: iron uptake and transport in plants. FEBS Lett 581:2273. https://doi.org/10.1016/j.febslet.2007.04.043

Kobayashi T, Nishizawa NK (2012) Iron uptake, translocation, and regulation in higher plants. Annu Rev Plant Biol 63:131. https://doi.org/10.1146/annurev-arplant-042811-105522

Kobayashi T, Suzuki M, Inoue H, Itai RN, Takahashi M, Nakanishi H et al (2005) Expression of iron-acquisition-related genes in iron-deficient rice is co-ordinately induced by partially conserved iron-deficiency-responsive elements. J Exp Bot 56:1305. https://doi.org/10.1093/jxb/eri131

Kobayashi T, Ozu A, Kobayashi S, An G, Jeon JS, Nishizawa NK (2019) OsbHLH058 and OsbHLH059 transcription factors positively regulate iron deficiency responses in rice. Plant Mol Biol 101:471. https://doi.org/10.1007/s11103-019-00917-8

Koike S, Inoue H, Mizuno D, Takahashi M, Nakanishi H, Mori S, Nishizawa NK (2004) OsYSL2 is a rice metal-nicotianamine transporter that is regulated by iron and expressed in the phloem. Plant J 39(3):415–424. https://doi.org/10.1111/j.1365-313X.2004.02146.x

Lanquar V, Lelièvre F, Bolte S, Hamès C, Alcon C, Neumann D et al (2005) Mobilization of vacuolar iron by AtNRAMP3 and AtNRAMP4 is essential for seed germination on low iron. EMBO J 24:4041. https://doi.org/10.1038/sj.emboj.7600864

Lee S, An G (2009) Over-expression of $OsIRT1$ leads to increased iron and zinc accumulations in rice. Plant Cell Environ 32:408. https://doi.org/10.1111/j.1365-3040.2009.01935.x

Lee S, Chiecko JC, Kim SA, Walker EL, Lee Y, Guerinot ML, An G (2009a) Disruption of $OsYSL15$ leads to iron inefficiency in rice plants. Plant Physiol 150:786. https://doi.org/10.1104/pp.109.135418

Lee S, Jeon US, Lee SJ, Kim Y-K, Persson DP, Husted S et al (2009b) Iron fortification of rice seeds through activation of the nicotianamine synthase gene. Proc Natl Acad Sci U S A 106(51):22014–22019. https://doi.org/10.1073/PNAS.0910950106

Lee S, Kim YS, Jeon US, Kim YK, Schjoerring JK, An G (2012) Activation of rice $nicotianamine$ $synthase$ 2 ($OsNAS2$) enhances iron availability for biofortification. Mol Cell 33(3):269–275. https://doi.org/10.1007/s10059-012-2231-3

Lucca P, Hurrell R, Potrykus I (2002) Fighting iron deficiency anemia with iron-rich rice. J Am Coll Nutr 21:184S–190S. https://doi.org/10.1080/07315724.2002.10719264

Martínez C, Borrero J, Taboada R, Viana J, Neves P, Narvaez L et al (2010) Rice cultivars with enhanced iron and zinc content to improve human nutrition. In: Presented at the 28th International Rice Research Conference, 8–12 November 2010. Hanoi, Vietnam

Masuda H, Usuda K, Kobayashi T, Ishimaru Y, Kakei Y, Takahashi M et al (2009) Overexpression of the barley nicotianamine synthase gene $HvNAS1$ increases iron and zinc concentrations in rice grains. Rice 2(4):155–166. https://doi.org/10.1007/s12284-009-9031-1

Masuda H, Ishimaru Y, Aung MS, Kobayashi T, Kakei Y, Takahashi M et al (2012) Iron biofortification in rice by the introduction of multiple genes involved in iron nutrition. Sci Rep 2:543. https://doi.org/10.1038/srep00543

Masuda H, Kobayashi T, Ishimaru Y, Takahashi M, Aung MS, Nakanishi H et al (2013) Iron-biofortification in rice by the introduction of three barley genes participated in mugineic acid biosynthesis with soybean ferritin gene. Front Plant Sci 4:132. https://doi.org/10.3389/fpls.2013.00132

Mayer JE, Pfeiffer WH, Beyer P (2008) Biofortified crops to alleviate micronutrient malnutrition. Curr Opin Plant Biol 11:166. https://doi.org/10.1016/j.pbi.2008.01.007

Mumm RH (2013) A look at product development with genetically modified crops: examples from maize. J Agric Food Chem 61:8254. https://doi.org/10.1021/jf400685y

Nozoye T, Nagasaka S, Kobayashi T, Takahashi M, Sato Y, Sato Y et al (2011) Phytosiderophore efflux transporters are crucial for iron acquisition in graminaceous plants. J Biol Chem 286:5446. https://doi.org/10.1074/jbc.M110.180026

Nozoye T, Nagasaka S, Kobayashi T, Sato Y, Uozumi N, Nakanishi H, Nishizawa NK (2015) The phytosiderophore efflux transporter TOM2 is involved in metal transport in rice. J Biol Chem 290(46):27688–27699. https://doi.org/10.1074/jbc.M114.635193

Ogo Y, Itai RN, Kobayashi T, Aung MS, Nakanishi H, Nishizawa NK (2011) OsIRO2 is responsible for iron utilization in rice and improves growth and yield in calcareous soil. Plant Mol Biol 75:593-605. https://doi.org/10.1007/s11103-011-9752-6

Oliva N, Chadha-Mohanty P, Poletti S, Abrigo E, Atienza G, Torrizo L et al (2014) Large-scale production and evaluation of marker-free indica rice IR64 expressing phytoferritin genes. Mol Breed 33:23. https://doi.org/10.1007/s11032-013-9931-z

Paul S, Ali N, Gayen D, Datta SK, Datta K (2012) Molecular breeding of Osfer 2 gene to increase iron nutrition in rice grain. GM Crops Food 3(4):310–316. https://doi.org/10.4161/gmcr.22104

Paul S, Ali N, Datta SK, Datta K (2014) Development of an iron-enriched high-yielding indica rice cultivar by introgression of a high-iron trait from transgenic iron-biofortified rice. Plant Foods Hum Nutr 69:203. https://doi.org/10.1007/s11130-014-0431-z

Qu LQ, Yoshihara T, Ooyama A, Goto F, Takaiwa F (2005) Iron accumulation does not parallel the high expression level of ferritin in transgenic rice seeds. Planta 222:225. https://doi.org/10.1007/s00425-005-1530-8

Robinson NJ, Procter CM, Connolly EL, Guerinot ML (1999) A ferric-chelate reductase for iron uptake from soils. Nature 397:694. https://doi.org/10.1038/17800

Senoura T, Sakashita E, Kobayashi T, Takahashi M, Aung MS, Masuda H et al (2017) The iron-chelate transporter OsYSL9 plays a role in iron distribution in developing rice grains. Plant Mol Biol 95(4–5):375–387. https://doi.org/10.1007/s11103-017-0656-y

Singh SP, Gruissem W, Bhullar NK (2017) Single genetic locus improvement of iron, zinc and β-carotene content in rice grains. Sci Rep 7(1):6883. https://doi.org/10.1038/s41598-017-07198-5

Slamet-Loedin IH, Johnson-Beebout SE, Impa S, Tsakirpaloglou N (2015) Enriching rice with Zn and Fe while minimizing Cd risk. Front Plant Sci 6:1–9. https://doi.org/10.3389/fpls.2015.00121

Stoltzfus RJ (2001) Iron-deficiency anemia: reexamining the nature and magnitude of the public health problem. J Nutr 131:69?S–700S

Suzuki M, Takahashi M, Tsukamoto T, Watanabe S, Matsuhashi S, Yazaki J et al (2006) Biosynthesis and secretion of mugineic acid family phytosiderophores in zinc-deficient barley. Plant J 48:85. https://doi.org/10.1111/j.1365-313X.2006.02853.x

Suzuki M, Morikawa KC, Nakanishi H, Takahashi M, Saigusa M, Mori S, Nishizawa NK (2008) Transgenic rice lines that include barley genes have increased tolerance to low iron availability in a calcareous paddy soil. Soil Sci Plant Nutr 54:77. https://doi.org/10.1111/j.1747-0765.2007.00205.x

Takahashi M, Yamaguchi H, Nakanishi H, Shioiri T, Nishizawa NK, Mori S (1999) Cloning two genes for nicotianamine aminotransferase, a critical enzyme in iron acquisition (strategy II) in graminaceous plants. Plant Physiol 121:947. https://doi.org/10.1104/pp.121.3.947

Trijatmiko KR, Dueñas C, Tsakirpaloglou N, Torrizo L, Arines FM, Adeva C et al (2016) Biofortified indica rice attains iron and zinc nutrition dietary targets in the field. Sci Rep 6:19792. https://doi.org/10.1038/srep19792

Vasconcelos M, Datta K, Oliva N, Khalekuzzaman M, Torrizo L, Krishnan S et al (2003) Enhanced iron and zinc accumulation in transgenic rice with the ferritin gene. Plant Sci 164:371. https://doi.org/10.1016/S0168-9452(02)00421-1

Virk P, Barry G, Das A, Lee J, Tan J (2006) Research status of micronutrient rice development in Asia. In: Proceedings of the International Symposium on Rice Biofortification: Improving Human Health Through Biofortified Rice, 15 September 2006, Suwon, Korea, pp 123–148

Virk P, Barry G, Bouis H (2007) Genetic enhancement for the nutritional quality of rice. In: Genetic enhancement for the nutritional quality of rice. Proceedings of the 26th International Rice Research Conference, 9–12 October 2006. Macmillan India Ltd, New Delhi, pp 279–285

WHO (2002) The world health report 2002, vol 16. World Health Organization, Geneva

WHO/FAO (2004) Vitamin and mineral requirements in human nutrition, 2nd edn. WHO, Geneva. https://doi.org/10.1128/AAC.03728-14

Wirth J, Poletti S, Aeschlimann B, Yakandawala N, Drosse B, Osorio S et al (2009) Rice endosperm iron biofortification by targeted and synergistic action of nicotianamine synthase and ferritin. Plant Biotechnol J 7:631. https://doi.org/10.1111/j.1467-7652.2009.00430.x

Wu TY, Gruissem W, Bhullar NK (2018) Facilitated citrate-dependent iron translocation increases rice endosperm iron and zinc concentrations. Plant Sci 270:13. https://doi.org/10.1016/j.plantsci.2018.02.002

Wu TY, Gruissem W, Bhullar NK (2019) Targeting intracellular transport combined with efficient uptake and storage significantly increases grain iron and zinc levels in rice. Plant Biotechnol J 17(1):9–20. https://doi.org/10.1111/pbi.12943

Yokosho K, Yamaji N, Ueno D, Mitani N, Jian FM (2009) OsFRDL1 is a citrate transporter required for efficient translocation of iron in rice. Plant Physiol 149:297. https://doi.org/10.1104/pp.108.128132

Yoneyama T, Ishikawa S, Fujimaki S, Yoneyama T, Ishikawa S, Fujimaki S (2015) Route and regulation of zinc, cadmium, and iron transport in rice plants (*Oryza sativa* L.) during vegetative growth and grain filling: metal transporters, metal speciation, grain Cd reduction and Zn and Fe biofortification. Int J Mol Sci 16(8):19111–19129. https://doi.org/10.3390/ijms160819111

Zhang Y, Xu YH, Yi HY, Gong JM (2012) Vacuolar membrane transporters OsVIT1 and OsVIT2 modulate iron translocation between flag leaves and seeds in rice. Plant J 72:400. https://doi.org/10.1111/j.1365-313X.2012.05088.x

Zhang C, Shinwari KI, Luo L, Zheng L (2018) *OsYSL13* is involved in iron distribution in rice. Int J Mol Sci 19:3537. https://doi.org/10.3390/ijms19113537

Zheng L, Cheng Z, Ai C, Jiang X, Bei X, Zheng Y et al (2010) Nicotianamine, a novel enhancer of rice iron bioavailability to humans. PLoS One 5(4):e10190. https://doi.org/10.1371/journal.pone.0010190

Correction to: Crop Establishment in Direct-Seeded Rice: Traits, Physiology, and Genetics

Fergie Ann Quilloy, Benedick Labaco, Carlos Casal Jr, and Shalabh Dixit

Correction to:
Chapter 6 in: J. Ali, S. H. Wani (eds.), *Rice Improvement*,
https://doi.org/10.1007/978-3-030-66530-2_6

"Owing to an error on the part of the editor and corresponding chapter author, the author sequence in the chapter opening page of chapter Crop Establishment in Direct-Seeded Rice: Traits, Physiology, and Genetics was presented wrongly. The author sequence has now been updated to be Fergie Ann Quilloy, Benedick Labaco, Carlos Casal Jr., Shalabh Dixit in the chapter opening page, table of contents, and wherever applicable throughout the book."

The updated version of this chapter can be found at
https://doi.org/10.1007/978-3-030-66530-2_6

© The Author(s) 2021
J. Ali, S. H. Wani (eds.), *Rice Improvement*,
https://doi.org/10.1007/978-3-030-66530-2_15

Index